Computational Techniques for Dental Image Analysis

K. Kamalanand
Anna University, India

B. Thayumanavan
Sathyabama University Dental College and Hospital, India

P. Mannar Jawahar
Anna University, India

A volume in the Advances in Medical Technologies and Clinical Practice (AMTCP) Book Series

IGI Global
DISSEMINATOR OF KNOWLEDGE

Published in the United States of America by
IGI Global
Medical Information Science Reference (an imprint of IGI Global)
701 E. Chocolate Avenue
Hershey PA, USA 17033
Tel: 717-533-8845
Fax: 717-533-8661
E-mail: cust@igi-global.com
Web site: http://www.igi-global.com

Library of Congress Cataloging-in-Publication Data

Names: Kamalanand, K., 1988- editor. | Thayumanavan, B., 1970- editor. | Jawahar, P. Mannar, 1953- editor.
Title: Computational techniques for dental image analysis / K. Kamalanand, B. Thayumanavan, and P. Mannar Jawahar, editors.
Description: Hershey, PA : Medical Information Science Reference, [2019] | Includes bibliographical references.
Identifiers: LCCN 2018005645| ISBN 9781522562436 (hardcover) | ISBN 9781522562443 (ebook)
Subjects: | MESH: Radiography, Dental--methods | Image Processing, Computer-Assisted--methods
Classification: LCC RK309 | NLM WN 230 | DDC 617.6/07572--dc23 LC record available at https://lccn.loc.gov/2018005645

This book is published in the IGI Global book series Advances in Medical Technologies and Clinical Practice (AMTCP) (ISSN: 2327-9354; eISSN: 2327-9370)

British Cataloguing in Publication Data
A Cataloguing in Publication record for this book is available from the British Library.

All work contributed to this book is new, previously-unpublished material. The views expressed in this book are those of the authors, but not necessarily of the publisher.

For electronic access to this publication, please contact: eresources@igi-global.com.

Advances in Medical Technologies and Clinical Practice (AMTCP) Book Series

Srikanta Patnaik
SOA University, India
Priti Das
S.C.B. Medical College, India

ISSN:2327-9354
EISSN:2327-9370

Mission

Medical technological innovation continues to provide avenues of research for faster and safer diagnosis and treatments for patients. Practitioners must stay up to date with these latest advancements to provide the best care for nursing and clinical practices.

The **Advances in Medical Technologies and Clinical Practice (AMTCP) Book Series** brings together the most recent research on the latest technology used in areas of nursing informatics, clinical technology, biomedicine, diagnostic technologies, and more. Researchers, students, and practitioners in this field will benefit from this fundamental coverage on the use of technology in clinical practices.

Coverage

- Clinical Studies
- Medical Imaging
- Clinical Data Mining
- Nutrition
- Nursing Informatics
- Biomedical Applications
- E-Health
- Clinical High-Performance Computing
- Biomechanics
- Diagnostic Technologies

IGI Global is currently accepting manuscripts for publication within this series. To submit a proposal for a volume in this series, please contact our Acquisition Editors at Acquisitions@igi-global.com or visit: http://www.igi-global.com/publish/.

The Advances in Medical Technologies and Clinical Practice (AMTCP) Book Series (ISSN 2327-9354) is published by IGI Global, 701 E. Chocolate Avenue, Hershey, PA 17033-1240, USA, www.igi-global.com. This series is composed of titles available for purchase individually; each title is edited to be contextually exclusive from any other title within the series. For pricing and ordering information please visit http://www.igi-global.com/book-series/advances-medical-technologies-clinical-practice/73682. Postmaster: Send all address changes to above address.

Titles in this Series

For a list of additional titles in this series, please visit: www.igi-global.com/book-series

Optimized Genetic Programming Applications Emerging Research and Opportunities
Bahrudin Hrnjica (University of Bihac, Bosnia and Herzegovina) and Ali Danandeh Mehr (Antalya Bilim University, Turkey)
Medical Information Science Reference • copyright 2019 • 310pp • H/C (ISBN: 9781522560050) • US $225.00 (our price)

Design and Development of Affordable Healthcare Technologies
Arindam Bit (National Institute of Technology Raipur, India)
Medical Information Science Reference • copyright 2018 • 388pp • H/C (ISBN: 9781522549697) • US $245.00 (our price)

Optimizing Health Literacy for Improved Clinical Practices
Vassilios E. Papalois (Imperial College, UK) and Maria Theodosopoulou (Imperial College, UK)
Medical Information Science Reference • copyright 2018 • 338pp • H/C (ISBN: 9781522540748) • US $235.00 (our price)

Electrocardiogram Signal Classification and Machine Learning Emerging Research and Opportunities
Sara Moein (Mount Sinai School of Medicine, USA)
Medical Information Science Reference • copyright 2018 • 196pp • H/C (ISBN: 9781522555803) • US $160.00 (our price)

Research Advancements in Pharmaceutical, Nutritional, and Industrial Enzymology
Shashi Lata Bharati (North Eastern Regional Institute of Science and Technology, India) and Pankaj Kumar Chaurasia (Babasaheb Bhimrao Ambedkar Bihar University, India)
Medical Information Science Reference • copyright 2018 • 549pp • H/C (ISBN: 9781522552376) • US $255.00 (our price)

Microbial Cultures and Enzymes in Dairy Technology
Şebnem Öztürkoğlu Budak (Ankara University, Turkey) and H. Ceren Akal (Ankara University, Turkey)
Medical Information Science Reference • copyright 2018 • 413pp • H/C (ISBN: 9781522553632) • US $265.00 (our price)

Multifunctional Nanocarriers for Contemporary Healthcare Applications
Md. Abul Barkat (K.R. Mangalam University, India) Harshita A.B. (K.R. Mangalam University, India) Sarwar Beg (Jubilant Generics, India) and Farhan J. Ahmad (Jamia Hamdard, India)

IGI Global
DISSEMINATOR OF KNOWLEDGE

701 East Chocolate Avenue, Hershey, PA 17033, USA
Tel: 717-533-8845 x100 • Fax: 717-533-8661
E-Mail: cust@igi-global.com • www.igi-global.com

Editorial Advisory Board

List of Reviewers

Table of Contents

Detailed Table of Contents

Dental radiographs suffer frequently from issues such as low contrast and non-uniform illumination. The first step, indeed a significant step, is to enhance these digital images to prepare them for successful post-processing. This pre-processing stage assists to increase the contrast between the foreground that is the teeth and bone from the background regions. In this chapter, image enhancement methods based on spatial, frequency, and spatial-frequency are implemented. The dental radiographs used are available in the public database. The performance of the enhancement methods is validated using qualitative and quantitative measures. It is observed from the results that the enhancement method aids in improving dental features such as crowns, fillings, and bridges. This enables human identification and diagnostic purpose in the way it is possible to identify a variety of diseases. It also prevents the need for a remake, saving the patient from an additional treatment.

Imaging techniques play a major role in improving the early detection and diagnostic process that helps dentists to make accurate diagnosis. One of the most useful medical images used by dentists is radiographic image, which is used for the treatment of various dental disorders. Segmentation is a fundamental step

A considerable number of heuristic procedures are widely implemented to evaluate biomedical images. This chapter proposes an evaluation procedure for digital bitewing radiography (DBR) images using the Jaya algorithm. The proposed procedure implements an image processing technique by integrating of the multi-thresholding and segmentation procedure to extract the essential tooth elements recorded with DBR. In this paper, 80 dental x-ray images are considered for the evaluation. The performance of the proposed procedure is confirmed using a relative assessment between the extracted section and its corresponding ground-truth. The results of this study confirm that, for most of the DBR cases, the proposed approach offers better values of picture likeliness measures. Hence, this technique can be considered for the automated detection of tooth elements from the DBR obtained from clinics.

Lyudmila N. Tuzova, Denti.AI, Russia
Dmitry V. Tuzoff, Steklov Institute of Mathematics in St. Petersburg, Russia
Sergey I. Nikolenko, Steklov Institute of Mathematics in St. Petersburg, Russia
Alexey S. Krasnov, Dmitry Rogachev National Research Center of Pediatric Hematology, Oncology, and Immunology, Russia

In the recent decade, deep neural networks have enjoyed rapid development in various domains, including medicine. Convolutional neural networks (CNNs), deep neural network structures commonly used for image interpretation, brought the breakthrough in computer vision and became state-of-the-art techniques for various image recognition tasks, such as image classification, object detection, and semantic segmentation. In this chapter, the authors provide an overview of deep learning algorithms and review available literature for dental image analysis with methods based on CNNs. The present study is focused on the problems of landmarks and teeth detection and classification, as these tasks comprise an essential part of dental image interpretation both in clinical dentistry and in human identification systems based on the dental biometrical information.

T. Christy Bobby, M. S. Ramaiah University of Applied Sciences, India
Shwetha V., M. S. Ramaiah University of Applied Sciences, India
Vijaya Madhavi, East Point College of Engineering and Technology, India

The stability of a dental implant is one of the most important aspects that decide the success rate of implant treatment. The stability is considerably affected by the strength of trabecular bone present in maxilla and mandible. Thus, finding of trabecular bone strength is a key component for the success of dental implants. The trabecular bone strength is usually assessed by quantity of bone in terms of bone mineral density (BMD). Recently, it has been revealed that along with quantity of bone, strength of the bone also depends on quality features commonly referred as trabecular bone microarchitecture. Since the quality of the trabecular bone is varying across the maxilla and mandible, preoperative assessment of trabecular bone microarchitecture at sub-region of maxilla and mandible are essential for stable implant treatment. Thus, in this chapter, the authors inscribe the quantitative analysis of trabecular bone quality in maxilla and mandible using CBCT images by employing contourlet transform.

Section 2
Dental Materials, Mechanics, and Instrumentation

Fluoride dental decay is the second most common disease around the world. Detection methods for early disease are very crude. Precise oral diagnosis and treatment are very strongly connected to the quality of dental imaging techniques which advances the diagnostic procedure. To study the external appearance of the teeth arches, 2D images are used. CBCT images were used to locate the bone at dental implant sites. Fiberoptic transillumination, fluorescence imaging detects caries. For qualitative and quantitative analysis of dental applications, laser-induced breakdown spectroscopy (LIBS) is used. Electron caries monitor (ECM), fiberoptic transillumination (FOTI), digital fiberoptic transillumination (DIFOTI), quantitative light-induced fluorescence (QLF) are also some of the detection methods used. Hence, in this chapter, the methodologies are analyzed and compared for easy use of the dentist.

A numerical investigation of steady state heat transfer phenomenon in human tooth is carried out in this present study. The materials generally used for repairing human tooth are ceramics, gold, structural steel, and copper. These materials are considered for the three-dimensional heat transfer analysis. The shape of the teeth is considered as perfect cubic of dimensions 5mm X 5mm X 1.5mm. The simulation results show that the teeth are subjected to temperatures of varying nature in both X and Y directions whereas Z direction is same. It is found that the temperature is maximum at the top of the teeth and minimum at bottom of the teeth (i.e., root). The contours of the heat transfer analysis for various teeth materials shows that the layers of the teeth are subjected to varying temperature. It is also found that the heat transfer characteristics depend not only on shape and boundary conditions but also on the materials used for the teeth selection.

This chapter introduces the recent trend in 3D printing (3DP) in dentistry. The advantage and disadvantages of 3DP are discussed. It elaborates on different types of 3DP techniques involved and their significance. The chapter further discuss about the biomaterial used. It also describes the complete steps involved in 3DP such as image acquisition, modeling, segmentation, and printing techniques. The merits and demerits of the different methodologies pertaining to steps involved in 3DP are illustrated. Rapid prototyping in dental implants is discussed in detail. It ends with review of a case study in implementing the technique.

The chapter gives a picture of the current data on the available anticariogenic natural products and their mechanism of action. Different phytochemicals such as phenols, flavanoids, alkaloids, terpenoids, tannins, lectins, etc. and their anticariogenic efficacy have been discussed in detail. All the data emphasise the fact that the use of natural products is emerging as an effective strategy in the prevention and treatment of dental caries. Consequently, these natural products could be incorporated in toothpastes and other oral hygiene products to promote oral health.

Foreword

It is a pleasure to write a Foreword for this reference book entitled *Computational Techniques for Dental Image Analysis*, edited by Dr. K. Kamalanand, Dr. B. Thayumanavan and Dr. P. Mannar Jawahar. Dental image analysis is an important research field since the processing of dental images is a highly challenging task and this book explores the possible techniques for processing and analysis of dental images in an efficient way.

The coordination between the contributing Authors, Reviewers, Editors and the Publisher has resulted in this excellent reference book which is highly useful for both researchers and dental practitioners. It is heartening to see that the selected peer reviewed chapters of this edited volume well covers the field of dental image processing and dental engineering. The chapters contributed to this book clearly indicates the well-planned interaction between the doctors and engineers for bringing out useful technological innovations. Further, this book covers most of the topics in dental engineering from the basic preprocessing of dental images, thresholding techniques, segmentation techniques, bone quality analysis in Maxilla and Mandible, dental instruments, 3D printing for dental implants, thermal analysis of dental materials, and phytochemicals which act against dental caries. Also, this book explains both conventional algorithms for dental image analysis and the most recent techniques such as Jaya algorithm and deep learning algorithms.

I am sure that this book will form a useful reference to biomedical engineers, researchers, dental practitioners and students to access the contemporary status on this important subject.

C. Emmanuel
Gleneagles Global Health City, India
May 2018

C. Emmanuel *was born in Trichirapalli, Tamil Nadu 1969. He is Professor and Director of the Department of Academics & Research, Gleneagles Global Health City, Chennai. Besides computational molecular medicine, his research interest includes the computer aided design and biomedical engineering. He is nationally renowned for his books, academic & scientific work, Editorial board member of scientific journals, having published many research articles also presented them in over than 150 meetings and conferences. He is the recipient of several awards and honors.*

Foreword

As a result of unstoppable developments in computer technology, it has become a critical approach to use computer-oriented techniques for achieving better problem solutions in all fields of the modern life. In this context, especially computational techniques have taken important roles to derive new solution approaches for improving already obtained results or making it possible to solve critical problems, which were not solved before. At this point, it has been also clear that both effectiveness and efficiency can be taken many steps away thanks to computer systems running many alternative forms of computational techniques. Because of that, there has been a great research interest in using such systems – techniques to solve difficult tasks in the fields requiring many practical works and efforts. The sub-fields of medical take top places in this manner, as being associated with healthcare of the humankind and also other living organisms over the world. In this context, dentistry has always been in the foreground because of its detailed and sensitive characteristics to ensure desired dental health.

Dentistry is a medical sub-field, which can be affected from technological developments rapidly and because of that, approaches, methods or techniques used for diagnosis, treatment and surgery used currently can be changed with revolutionary solution ways. It is clear that technological innovations has gained a momentum especially after the start of 21st century and dentistry has always been open to use newly developed systems and tools to make everything better for both dentists and patients. Here, it is possible to mention that effective innovations appeared within dentistry has been in the context of image analysis and equipment – material development. As it was mentioned under the first paragraph, computer systems and computational techniques employed in them have shaped and accelerated all these innovations.

This book, which is titled as Computational Techniques for Dental Image Analysis, is an important contribution to the scientific literature for enabling interested readers to have idea about what are currently done with computational methods to improve image analysis oriented works in dentistry. It is certain that accurate dental imaging is too important to have good analysis and evaluation of the encountered cases. As like many other fields employing practical tasks, there are often environmental factors and unexpected situations causing negative effects on critical tasks and dental imaging is the most widely examined issue in the dentistry. Because of that, use of computer based computational techniques is a popular research interest by causing both making effective decision making and developing effective dental equipment and materials. In the book, these problem issues are covered under 11 quality research works as classified under two sections: Dental Imaging and Analysis, and Dental Materials, Mechanics and Instrumentation, respectively. Anyone reading these chapters will have idea about the following research aspects in the context of dental image analysis:

- Image enhancement for better diagnosis, treatment and surgery phases.
- Image segmentation as an important image processing method to achieve better treatments.
- Use of heuristic and Artificial Intelligence based approaches to improve dental operations.
- Analysis that can be done over dental components.
- Evaluation of dental equipment and materials in terms of achieving better outputs in tasks.

By combining the findings – results obtained in the works done by the related authors, it is possible for readers to draw a general research environment of current computational techniques for improving dentistry and taking it steps away to build up a future with full of innovations and scientific approaches to make all medical tasks automated with more practical and effective solutions at the end.

For their efforts on providing such a valuable contribution to the scientific community, I would like to thank to the editors: Dr. Kamalanand, Dr. Thayumanavan, and Dr. Jawahar. Without their ideas and also hard works to combine different research works under a common environment, this book project would not be realized. I hope everyone interested and enrolled in that dental image analysis works will enjoy reading all valuable chapters. Let's turn the pages to open doors of innovations changing the ways of dentistry and understand how computational techniques have effective roles in better image analysis in the clinical works!

Utku Kose
Suleyman Demirel University, Turkey
June 2018

Utku Kose *received the B.S. degree in 2008 from computer education of Gazi University, Turkey as a faculty valedictorian. He received M.S. degree in 2010 from Afyon Kocatepe University, Turkey in the field of computer and D.S. / Ph. D. degree in 2017 from Selcuk University, Turkey in the field of computer engineering. Between 2009 and 2011, he has worked as a Research Assistant in Afyon Kocatepe University. Following, he has also worked as a Lecturer and Vocational School - Vice Director in Afyon Kocatepe University between 2011 and 2012 and as a Lecturer and Research Center Director in Usak University between 2012 and 2017. Currently, he is an Assistant Professor in Suleyman Demirel University, Turkey. He has more than 100 publications including articles, authored and edited books, proceedings, and reports. His research interest includes artificial intelligence, machine ethics, artificial intelligence safety, optimization, the chaos theory, distance education, e-learning, computer education, and computer science.*

Preface

The tooth is one of the most important structures of the human body since it plays an important role in the chewing, crushing and grinding of food before ingestion (Sherwood, Kell, & Ward, 2010). Hence the oral structures are significant since they help in the processing of the food in the first step of the digestion process (Pastor, 1995). Four different types of tooth are present in the oral cavity of a human being. These types include the incisors, canines, premolars and the molars (Mohamed, Christensen, & Harrison, 1983; Bhaskar, 1968). The incisors are present at the front of the oral cavity and aid in the cutting of the food into small pieces. The canines are sharp and play a vital part in tearing the food pieces (Haggard & de Boer, 2014; Cheng, Yang, Shao, Hu, & Zhou, 2009). The premolars are flat and help in crushing the food (Zachrisson, 2015). Finally, the molars function to chew and grind the food before ingesting it (Adel Kauzman, Pavone, Blanas, & Bradley, 2004). The tooth is firmly fixed inside the oral cavity by their roots implanted in the maxilla or the upper jaw and the mandible, which is the lower jaw (Zahradnicek, Horacek, & Tucker, 2012). Each tooth has four main parts namely the enamel, dentin, cementum and the pulp (Bastone, Freer, & McNamara, 2000). The enamel is a tough and mineralized tissue and is the outermost structure of the tooth. Cementum is a special material which shelters the root of the tooth and dentin is a porous material surrounding the dental pulp. The dental pulp is composed of soft connective tissue containing blood vessels and nerves. Hence the dental structures are highly complex and are multifunctional in nature (Schour & Massler, 1940; Rubin, Krishnamurthy, Capilouto, & Yi, 1983).

Even though the dental structures are highly important and useful for proper processing of the food, these structures are prone to several problems and pathologies. The most common tooth pathologies include dental plaque and dental caries (Selwitz, Ismail, & Pitts, 2007). Every now and then, large amounts of bacteria collect on the tooth and develop as a biofilm known as the plaque. If the plaque is not removed, severe complications may manifest. Dental caries is one of the most common health problems of today's world. Dental caries, commonly known as cavities is a dental infection which damages the tooth and causes severe pain (Featherstone, 1999). Hence, it is an important to identify and diagnose the diseased tooth effectively. In general, the diagnostic techniques need to be non-invasive and must cause minimal discomfort without any side effects (Syed & Loesche, 1972).

Dental image processing is an important field of research since it is very difficult to segment dental structures from dental images (Kopelman & Taub, 2005). The algorithms required for processing and analyzing dental images need to be highly efficient and cost effective (Rajinikanth, Satapathy, Fernandes, & Nachiappan, 2017; Raja, Rajinikanth, & Latha, 2014; Rajinikanth & Couceiro, 2015a, 2015b; Satapathy, Raja, Rajinikanth, Ashour, & Dey, 2016; Rajinikanth, Raja, & Latha, 2014; Manic, Priya, & Rajinikanth, 2016; Ambikapathy & Krishnamurthy, 2018). In recent years, the advances in image processing techniques have given rise to efficient diagnostic solutions (Manickavasagam, Sutha, & Kamalanand,

2014; Vaishnavi, Jeevananthan, Begum, & Kamalanand, 2014). Further, in some cases, computational algorithms, have reduced the burden of resource constrained environments and have increased the diagnostic efficiency (Kamalanand & Ramakrishnan, 2015; Kamalanand & Jawahar, 2013a, 2013b). Also, soft computing approaches such as artificial neural networks and swarm intelligence techniques have proved to be highly efficient in finding solutions to problems in the field of biomedical analysis such as biomedical diagnostics (Kamalanand & Jawahar, 2014a, 2014b, 2012; Kamalanand & Mannar Jawahar, 2015) and therapy planning (Kamalanand & Mannar Jawahar, 2016).

This edited book entitled *Computational Techniques for Dental Image Analysis*, deals with the state of art approaches in dental image analysis and dental engineering and is sectioned into two sections. The first section entitled "Dental Imaging and Analysis" collects seven chapters on the algorithms and procedures involved in the processing of dental images. The second section entitled "Dental Materials, Mechanics, and Instrumentation" collects four chapters on the dental imaging instruments, numerical analysis of dental materials, 3D printing for dental implants and phytochemicals for dental caries treatment.

This book consists of 11 excellent chapters in the field of dental engineering. The first chapter entitled "Digital Image Enhancement Techniques for Dental Radiographs: A Support to Clinicians", authored by E. Priya, deals with the preprocessing of dental images, which is the first step in dental image analysis. In common, the dental radiographs suffer from problems such as low contrast and non-uniform illumination. This chapter discusses the image enhancement methods based on spatial, frequency and spatial-frequency, for preparing the dental images for further post processing.

In second chapter entitled "Thresholding Techniques for Dental Radiographic Images: A Comparative Study", authored by Sukanya & Krishnamurthy, explores another preprocessing step known as thresholding for analysis of dental images. This chapter presents and explains various thresholding techniques such as Otsu's method for global thresholding and Niblack's, Bersen's, Sauvola's techniques for local thresholding, applied to dental images.

Image segmentation is a highly important procedure in the analysis of medical images. The third chapter entitled "Image Segmentation Using Contour Models: Dental X-Ray Image Segmentation and Analysis", authored by Kavitha et al., presents the case studies of multiple region segmentation from dental X-rays using contour models.

Chapter 4 entitled "Dental Image Segmentation Using Clustering Techniques and Level Set Methods" authored by Prabha Sathees describes the clustering techniques and level set methods with different edge maps for segmentation of dental images. This chapter proposes the segmentation techniques for the root canal treatment. Also, the integration of coherence enhanced diffusion filtering in basic level set segmentation methodology for improving the segmentation performance of dental images, is discussed in detail.

In the fifth chapter entitled "Jaya Algorithm-Assisted Evaluation of Tooth Elements Using Digital Bitewing Radiography Images", the authors Suresh Manic have proposed proposes an evaluation procedure for Digital Bitewing Radiography (DBR) images using a recent heuristic algorithm namely the Jaya algorithm. The proposed technique can be considered for the automated detection of tooth elements from Digital Bitewing Radiography (DBR) images.

In Chapter 6 entitled "Teeth and Landmarks Detection and Classification Based on Deep Neural Networks", the authors Lyudmila N Tuzova et al., focus on the problems of landmarks and teeth detection and classification since they encompass an indispensable part of dental image understanding applied to clinical dentistry and also human identification systems constructed on the dental biometric information. The authors discuss on the deep learning algorithms such as the convolutional neural networks for the analysis of dental images.

The dental implant's stability is the most important aspect in the success of the implant treatment. In this regard, the strength of trabecular bone present in maxilla and mandible affect the stability of the dental implants. In the seventh chapter, entitled "Dental Cone Beam Computed Tomography for Trabecular Bone Quality Analysis in Maxilla and Mandible", the authors, Christy Bobby et al., quantitatively analyze the quality of the trabecular bone in maxilla and mandible using dental Cone Beam Computed Tomography (CBCT) images using contourlet transform.

In Chapter 8 entitled "Comparison and Analysis of Dental Imaging Techniques", the author D. Najumnissa, has presented a comparative analysis on the different detection techniques for dental analysis namely, dental radiography, Cone Beam Computed Tomography (CBCT), Fiberoptic Transillumination (FOTI), Digital Imaging Fiber-Optic Transillumination (DIFOTI), Electrical Caries Monitor (ECM), Laser Induced Breakdown Spectroscopy (LIBS) and Quantitative Light-Induced Fluorescence (QLF) along with their usage, advantages and disadvantages.

The study of heat transfer properties in the human teeth is highly important for the design of dental implants since the dental structures are subjected to varying thermal loads in day to day life. The ninth chapter entitled "Thermal Analysis of Artificial Dental Materials Using Numerical Simulation" authored by Thanigaiarasu and Narendra Kumar, presents a numerical investigation of steady state heat transfer phenomenon in human tooth. The authors have considered artificial tooth materials such as ceramics, gold, structured steel and copper for the three-dimensional heat transfer analysis.

In the next chapter entitled "Recent Trends in 3D Printing of Dental Models: Rapid Prototyping in Dental Implants", authored by Kayalvizhi Mohan, presents the recent advances in #D printing for dental implants. This chapter also discusses the advantages and disadvantages of 3D printing techniques for dentistry, along with the different types of 3D printing techniques. The chapter further presents the steps involved in 3D printing technology for dental applications.

In the final chapter, entitled, "Phytochemicals: Their Therapeutic Potential Against Dental Caries", authored by Karthikeyan Ramalingam and Bennett T Amaechi, a detailed presentation on the current data on the available anticariogenic natural products and their mechanism of action, is provided. The authors have discussed the anticariogenic efficacy of different phytochemicals such as phenols, flavanoids, alkaloids, terpenoids, tannins, lectins etc. This chapter emphasizes that the natural products might be utilized in toothpastes and other oral health products for the promotion of oral hygiene.

The topics covered in this book are highly diverse and cover most of the topics in dental imaging and technology. It has been a pleasure to serve as editors for this reference book and we hope that it serves as a useful reference material researchers, academicians and students working in the field of dental engineering and technology.

K. Kamalanand
Anna University, India

B. Thayumanavan
Sathyabama University Dental College and Hospital, India

P. Mannar Jawahar
Anna University, India

REFERENCES

Adel Kauzman, B. D. S., Pavone, M., Blanas, N., & Bradley, G. (2004). Pigmented lesions of the oral cavity: Review, differential diagnosis, and case presentations. *Journal - Canadian Dental Association, 70*(10), 682–683. PMID:15530266

Ambikapathy, B., & Krishnamurthy, K. (2018). Analysis of electromyograms recorded using invasive and noninvasive electrodes: A study based on entropy and Lyapunov exponents estimated using artificial neural networks. *Journal of Ambient Intelligence and Humanized Computing*, 1–9.

Bastone, E. B., Freer, T. J., & McNamara, J. R. (2000). Epidemiology of dental trauma: A review of the literature. *Australian Dental Journal, 45*(1), 2–9. doi:10.1111/j.1834-7819.2000.tb00234.x PMID:10846265

Bhaskar, S. N. (1968). Oral pathology in the dental office: Survey of 20, 575 biopsy specimens. *The Journal of the American Dental Association, 76*(4), 761–766. doi:10.14219/jada.archive.1968.0119 PMID:5237768

Cheng, R., Yang, H., Shao, M. Y., Hu, T., & Zhou, X. D. (2009). Dental erosion and severe tooth decay related to soft drinks: A case report and literature review. *Journal of Zhejiang University. Science. B., 10*(5), 395–399. doi:10.1631/jzus.B0820245 PMID:19434767

Featherstone, J. D. (1999). Prevention and reversal of dental caries: Role of low level fluoride. *Community Dentistry and Oral Epidemiology, 27*(1), 31–40. doi:10.1111/j.1600-0528.1999.tb01989.x PMID:10086924

Haggard, P., & de Boer, L. (2014). Oral somatosensory awareness. *Neuroscience and Biobehavioral Reviews, 47*, 469–484. doi:10.1016/j.neubiorev.2014.09.015 PMID:25284337

Kamalanand, K., & Jawahar, P. M. (2012). Coupled jumping frogs/particle swarm optimization for estimating the parameters of three dimensional HIV model. *BMC Infectious Diseases, 12*(1Suppl 1), 82. doi:10.1186/1471-2334-12-S1-P82 PMID:22471518

Kamalanand, K., & Jawahar, P. M. (2013a). Particle swarm optimization based estimation of HIV-1 viral load in resource limited settings. *African Journal of Microbiological Research, 7*(20), 2297–2304. doi:10.5897/AJMR12.1924

Kamalanand, K., & Jawahar, P. M. (2013b). Resource limited estimation of HIV-1 viral load from CD4 cell count using coupled bacterial foraging/jumping frogs algorithm. *The World Allergy Organization Journal, 6*(S1), 189. doi:10.1186/1939-4551-6-S1-P189

Kamalanand, K., & Jawahar, P. M. (2014a). A graphical user interface for resource limited estimation of HIV-1 viral load using swarm intelligence techniques. *Journal of Bioinformatics and Intelligent Control, 3*(2), 110–116. doi:10.1166/jbic.2014.1073

Kamalanand, K., & Jawahar, P. M. (2014b). Hybrid BFPSO algorithm based estimation of optimal drug dosage for antiretroviral therapy in HIV-1 infected patients. *BMC Infectious Diseases, 14*(S3), E14. doi:10.1186/1471-2334-14-S3-E14

Kamalanand, K., & Mannar Jawahar, P. (2015). Comparison of Swarm Intelligence Techniques for Estimation of HIV-1 Viral Load. *IETE Technical Review*, *32*(3), 188–195. doi:10.1080/02564602.2014.1000981

Kamalanand, K., & Mannar Jawahar, P. (2016). Comparison of particle swarm and bacterial foraging optimization algorithms for therapy planning in HIV/AIDS patients. *International Journal of Biomathematics*, *9*(02), 1650024. doi:10.1142/S1793524516500248

Kamalanand, K., & Ramakrishnan, S. (2015). Effect of gadolinium concentration on segmentation of vasculature in cardiopulmonary magnetic resonance angiograms. *Journal of Medical Imaging and Health Informatics*, *5*(1), 147–151. doi:10.1166/jmihi.2015.1370

Kopelman, A., & Taub, E. (2005). *U.S. Patent No. 6,845,175*. Washington, DC: U.S. Patent and Trademark Office.

Manic, K. S., Priya, R. K., & Rajinikanth, V. (2016). Image multithresholding based on Kapur/Tsallis entropy and firefly algorithm. *Indian Journal of Science and Technology*, *9*(12). doi:10.17485/ijst/2016/v9i12/89949

Manickavasagam, K., Sutha, S., & Kamalanand, K. (2014). Development of systems for classification of different plasmodium species in thin blood smear microscopic images. *Journal of Advanced Microscopy Research*, *9*(2), 86–92. doi:10.1166/jamr.2014.1194

Mohamed, S. E., Christensen, L. V., & Harrison, J. D. (1983). Tooth contact patterns and contractile activity of the elevator jaw muscles during mastication of two different types of food. *Journal of Oral Rehabilitation*, *10*(1), 87–95. doi:10.1111/j.1365-2842.1983.tb00103.x PMID:6572242

Pastor, M. (1995). *The digestive system*. Academic Press.

Raja, N., Rajinikanth, V., & Latha, K. (2014a). Otsu based optimal multilevel image thresholding using firefly algorithm. *Modelling and Simulation in Engineering*, *2014*, 37.

Rajinikanth, V., & Couceiro, M. S. (2015a). RGB histogram based color image segmentation using firefly algorithm. *Procedia Computer Science*, *46*, 1449–1457. doi:10.1016/j.procs.2015.02.064

Rajinikanth, V., & Couceiro, M. S. (2015b). Optimal multilevel image threshold selection using a novel objective function. In *Information Systems Design and Intelligent Applications* (pp. 177–186). New Delhi: Springer. doi:10.1007/978-81-322-2247-7_19

Rajinikanth, V., Raja, N. S. M., & Latha, K. (2014). Optimal multilevel image thresholding: An analysis with PSO and BFO algorithms. *Australian Journal of Basic and Applied Sciences*, *8*(9), 443–454.

Rajinikanth, V., Satapathy, S. C., Fernandes, S. L., & Nachiappan, S. (2017). Entropy based segmentation of tumor from brain MR images–a study with teaching learning based optimization. *Pattern Recognition Letters*, *94*, 87–95. doi:10.1016/j.patrec.2017.05.028

Rubin, C., Krishnamurthy, N., Capilouto, E., & Yi, H. (1983). Stress analysis of the human tooth using a three-dimensional finite element model. *Journal of Dental Research*, *62*(2), 82–86. doi:10.1177/00220345830620021701 PMID:6571871

Satapathy, S. C., Raja, N. S. M., Rajinikanth, V., Ashour, A. S., & Dey, N. (2016). Multi-level image thresholding using Otsu and chaotic bat algorithm. *Neural Computing & Applications*, 1–23.

Schour, I., & Massler, M. (1940). Studies in tooth development: The growth pattern of human teeth part II. *The Journal of the American Dental Association*, *27*(12), 1918–1931. doi:10.14219/jada.archive.1940.0367

Sherwood, L., Kell, R., & Ward, C. (2010). *Human physiology: From cells to systems*. Cengage Learning.

Syed, S. A., & Loesche, W. J. (1972). Survival of human dental plaque flora in various transport media. *Applied Microbiology*, *24*(4), 638–644. PMID:4628799

Vaishnavi, G. K., Jeevananthan, K., Begum, S. R., & Kamalanand, K. (2014). Geometrical analysis of schistosome egg images using distance regularized level set method for automated species identification. *Journal of Bioinformatics and Intelligent Control*, *3*(2), 147–152. doi:10.1166/jbic.2014.1080

Zachrisson, B. U. (2015). Esthetics in tooth display and smile design. In Esthetics and Biomechanics in Orthodontics (2nd ed.; pp. 54-73). Academic Press.

Zahradnicek, O., Horacek, I., & Tucker, A. S. (2012). Tooth development in a model reptile: Functional and null generation teeth in the gecko Paroedura picta. *Journal of Anatomy*, *221*(3), 195–208. doi:10.1111/j.1469-7580.2012.01531.x PMID:22780101

Acknowledgment

Firstly, as the Editors of this book, we thank IGI Global Publishers for providing us the opportunity to edit this work. The Editors thank the contributing authors for their excellent chapters constituting this book. We thank the Editorial Review Board members of this book for spending their valuable time to review and enhance each chapter. Also, we thank the Editorial Advisory Members for their valuable suggestions. We thank Ms. Jan Travers, Director of Intellectual Property and Contracts, IGI Global and Ms. Mariah, Assistant Development Editor of IGI Global for the valuable support extended by them to bring out this book. Finally, we thank Dr. C. Emmanuel, Director, Academics and Research, Global Hospitals and Health City, Chennai, India and Dr. Utku Köse, Suleyman Demirel University, Turkey, for writing Forewords for this book.

K. Kamalanand
Anna University, India

B. Thayumanavan
Sathyabama University Dental College and Hospital, India

P. Mannar Jawahar
Anna University, India

Section 1
Dental Imaging and Analysis

Chapter 1
Digital Image Enhancement Techniques for Dental Radiographs:
A Support to Clinicians

E. Priya
Sri Sairam Engineering College, India

ABSTRACT

Dental radiographs suffer frequently from issues such as low contrast and non-uniform illumination. The first step, indeed a significant step, is to enhance these digital images to prepare them for successful post-processing. This pre-processing stage assists to increase the contrast between the foreground that is the teeth and bone from the background regions. In this chapter, image enhancement methods based on spatial, frequency, and spatial-frequency are implemented. The dental radiographs used are available in the public database. The performance of the enhancement methods is validated using qualitative and quantitative measures. It is observed from the results that the enhancement method aids in improving dental features such as crowns, fillings, and bridges. This enables human identification and diagnostic purpose in the way it is possible to identify a variety of diseases. It also prevents the need for a remake, saving the patient from an additional treatment.

INTRODUCTION

Human identification is a crucial and prime issue these days as traditional authentication systems may confirm tentative identity. Biometrics is one such technology that is used in identification for uniquely recognizing individuals based on the inherent, physical or behavioural qualities, such as fingerprint, iris, hand vein and voice. Dental biometrics has leading edge when compared to other modalities, which has more complex features and might lead to high error rate (Vijayakumari, Ulaganathan, & Banumathi, 2011; Lin, Lai, & Huang, 2010; Lin, Lai, & Huang, 2012; Rehman, Akram, Faraz, & Riaz, 2015).

DOI: 10.4018/978-1-5225-6243-6.ch001

Dental biometrics is the primary biometric technique to identify individuals on the basis of their dental characteristics as dental features of human beings are naturally unique. They can be used to authenticate humans precisely or more or less to the highest promising similarity (Rehman, Akram, Faraz, & Riaz, 2015). Thus dental biometrics gets priority over other approaches. Dental biometrics utilizes the evidence revealed by dental radiographs for human identification. This evidence includes tooth contours, relative positions of neighboring teeth, and shapes of the dental features such as crowns, fillings and bridges (Jain, & Chen, 2004).

The conventional biometric characteristics may not be reliable or promising in authentication of humans deceased during natural disasters. In such situations dental features is considered a useful tool with highest accuracy rate (Rehman, Akram, Faraz, & Riaz, 2015). In addition, forensic identification based on dental features relies on the morphology aid in identifying the victims (Zhou, 2010).

Dental radiographs endure frequently from issues such as low contrast and non-uniform illumination that set difficulty to automatically extract the boundaries of teeth for further processing (Rad, Rahim, Rehman, & Saba, 2016; Jain, Chen, & Minut, 2003). Since dental radiographs often suffer from poor quality and uneven exposure, the accuracy of the identification degrades. This complicate the prime stage in further processing such as the task of segmentation, which is the most challenging step (Zhou, 2010). Also for these poor quality images the next vital stage is shape extraction which is an intricate hitch for dental radiographs, where some tooth contours are indiscernible (Jain, & Chen, 2004). Fortunately, the radiographs not only give us the information about the shape of the teeth, but also other information such as the artificial prosthesis of the teeth, the striae patterns and trabecular patterns, etc (Jain, Chen, & Minut, 2003).

The objective of this work is to analyze different image enhancement techniques and to find the most suitable method for dental image radiographs. Image pre-processing is the essential pace in image processing and computer vision, as it makes the processed image more appropriate to exhibit for further analysis. This analysis includes primitive operations such as denoising, contrast enhancement, image smoothing and sharpening, and advanced operations such as image segmentation and others (Adatrao, & Mittal, 2016).

Pre-processing is the first and foremost step in dental image processing and has its importance as the whole processing is based on it. Images of bad quality make difficulties at every stage of feature extraction and matching (Rehman, Akram, Faraz, & Riaz, 2015). This pre-processing stage assists to increase the contrast between the foreground that is the teeth and bony from the background regions. The image enhancement step is performed before segmentation to improve the quality of the dental X-ray images and to separate the teeth in each jaw. This step is crucial as the accuracy of the extracted features depends on the results of image enhancement (Rad, Rahim, Rehman, & Saba, 2016; Zhou, 2010).

The application of dental radiographs could assist clinician or the specialist for detection of any abnormalities in root and bone structure. It helps to develop an automated dental identification system for identifying missing and unidentified people based on dental characteristics and providing automated search and matching capabilities for dental radiographs. Dental features aid in postmortem biometric identification and they represent a suitable repository for unique reliable identifying feature in forensic science (Rad, Rahim, Rehman, & Saba, 2016; Zhou, 2010).

Apart from identification systems developed, the dental radiographs are widely used for diagnostic purpose in the way to identify diversity of diseases. It also prevents the need for a remake, saving the patient from an additional dose of treatment. Digital imaging from dental radiographs makes the diagnosis and treatment process reliable with more accuracy. A content-based image archiving and retrieval system for assisting in human identification using dental radiographs has also gained popularity (Gormez, & Yilmaz, 2009; Rad, Rahim, Rehman, & Saba, 2016; Zhou, 2010).

In this work an effort has been made to improve the visual quality of the digital dental radiographs so that the tooth contours are discernible which could be used in wide variety of applications such as individual identification in mass disasters, forensic science etc.

BACKGROUND

The pre-processing is described as a type of processing performed on raw data, which could be a signal or image to prepare it for the next processing step. Thus, pre-processing is the primary procedure which converts the data into a form that would be more easily and effectively processed. Hence, pre-processing is a vital stage prior to any other process as it controls the suitability and feasibility of the outputs for the succeeding stages (Alginahi, 2010). The techniques thus tailored for pre-processing enhances the selected features extracted from the data thereby eliminate the irrelevant or redundant ones (Hall, Kruger, Dwyer, Hall, Mclaren, & Lodwick, 1971; Wenzel, 1988).

Image pre-processing facilities thus restore images that are degraded by factors such as motion or focus blur, interference patterns or improper exposure. Pre-processing facilitates to observe more diagnostic information from the radiograph without having to expose the patient to further examinations. In oral radiology digital image processing and is subsequent stages has mostly been used in relation with the computer-assisted diagnostic aid for the interpretation and study of periodontal processes (Wenzel, 1988).

The prime advantage of digital radiography is its ability to process the image data so that the information content of the image is more accessible to the human perception. It is considered the most essential one, as automation in digital imaging techniques plays a vital role in characterization and analysis of digital radiographic images (Gormez, & Yilmaz, 2009). Wenzel (1988) proved that diagnostic accuracy improved while utilizing enhanced intraoral radiographs of varying density when compared with originally impaired density.

Recently many work has been reported on dental image analysis in the framework of image processing techniques such as denoising, image enhancement, segmentation, shape based analysis, feature extraction, machine learning and pattern recognition techniques.

Chen & Jain (2005) presented an automatic method for matching dental radiographs and used area-based metrics to establish the identity of the deceased subjects along with anisotropic diffusion as pre-processing technique to enhance the images. The dental radiographs are segmented by mixture of Gaussians for human identification. They suggested using few other features such as the shape of mandibular (lower jaw) canals and maxillary (upper jaw) sinus for subjects with missing teeth.

Literature reveals that a semi-automatic contour extraction method for shape extraction and pattern matching. The hitch in their approach is that their algorithm is not practically feasible if the image is too blurred and even slight angle deviation in the ante-mortem and post-mortem images are not addressed in their work (Banumathi, Vijayakumari, Geetha, Shanmugavadivu, & Raju, 2007).

An algorithm to classify and enumerate the teeth in bitewing images was developed (Lin, Lai, & Huang, 2012). Bayesian classification scheme is used by them to classify the molar tooth and assigned an absolute number to each tooth according to the common numbering system prescribed in the field of dentistry.

Three steps namely radiograph segmentation, pixel classification and contour matching was proposed for automated dental identification system (Banumathi, Vijayakumari, Geetha, Shanmugavadivu, & Raju, 2007). The authors automated the analysis of identification of persons based on the ante and post mortem reports. In their work anisotropic-based Gaussian filter was implemented for the removal of noise.

Filters such as Butterworth and homomorphic have been implemented to improve both contrast and intensity illumination unevenness (Vijayakumari, Ulaganathan, & Banumathi, (2011). The results of their work found application in person identification using dental patterns.

A simple tooth extraction algorithm and a matching technique for person identification were proposed (Pushparaj, Gurunathan, & Arumugam, 2013). They used a Butterworth high pass filter of order two followed by homomorphic filter for removal of uneven illumination in the dental radiographs. Their proposed work helped in forensic dentistry to identify missing persons in certain situations such as critical mass disaster. They reported that their work is not suitable for occluded dental pattern images.

In general other than dental radiographs literature reveals different techniques for noise reduction. Lee (1980) developed computational techniques involving contrast enhancement and noise filtering on images based on their local mean and variance for real time digital image processing applications.

Hemanth & Anita (2012) suggested suitable methodologies for image pre-processing and feature extraction for Magnetic Resonance (MR) brain image analysis. Morphology based techniques are adopted to eliminate extra-cranial (skull) tissues which normally interfere with brain tissues leading to performance reduction of the system.

Ramasubramanian & Selvaperumal (2016) compared five diverse pre-processing methods of color retinal images. They found that even though morphological top hat procedure offers a good result it failed to remove the noise completely. The green channel output is processed by a combination of median and Contrast Limited Adaptive Histogram Equalization (CLAHE) and reported this method as an optimum technique for pre-processing of retinal images.

The sputum smear digital tuberculosis raw images suffer from non-uniform illumination, and hence these images are pre-processed (Priya, Srinivasan, & Ramakrishnan, 2014). The authors suggested that surface fitting method suited well for these images and the performance is analyzed using multifractal analysis. The results demonstrate that this method aid in further identification and classification of sputum smear images.

The multi-scale stabilization noise removal technique was proposed using wavelets, ridgelets and curvelets for removing Poisson noise from simulated, Barbara and fluorescent tubulins images (Zhang, Fadili, & Starck, 2008). It is observed that their algorithm compete with state-of-the-art approaches, with low computational burden.

Fuzzy connectedness based intensity non-uniformity correction has been implemented by Zhou, & Bai, (2007). A framework that comprises of atlas registration, fuzzy connectedness segmentation and parametric bias field correction are used in this approach. Results reveal that the proposed technique can be used only if the intensity variations between the images are of a limited range. In a similar way yet another work contributed a hybrid pre-processing technique for cardiac MR images. The analytical result shows that the proposed methods give better delineation of both the ventricles of cardiac MR images thereby their contribution improved the clinical relevance in diagnosing ventricles (Nageswararao, Srinivasan, Babu peter, & Priya, 2016).

The literature review discussed so far is factual and promising that highlights, reveals and makes the reader to understand the immense application of image enhancement, a division of image pre-processing technique.

IMAGE ENHANCEMENT TECHNIQUES FOR DENTAL RADIOGRAPHS

Dataset

The digital dental radiographic images used in this work are obtained from the publically available database (Vahab – LabArchives). The digital X-rays have a higher degree of accuracy over traditional X-ray films. Digital imaging makes the diagnosis and treatment process reliable with more accuracy (Rad, Rahim, Rehman, & Saba, 2016). In general there are three types of dental images according to the manner the dental features are captured or viewed. Most commonly used dental radiographs are panoramic, periapical and bitewing view.

The panoramic view has complete upper and lower jaws, but it does not convey the fine details of teeth information and gives broader overview of entire dentition. It not only shows teeth but also sinus, upper and lower jaw bone. The periapical on the other hand shows entire view of the specific teeth including crowns and roots which in turn includes bones, teeth, air and fillings. This view especially has only few teeth. The third is the bitewing view which presents the crowns and parts of the roots of molar and pre-molar teeth that are more distinctive than other teeth. Usually bitewing dental radiographs are taken during routine check-ups (Dighe, & Shriram, 2012; Jain, Chen, & Minut, 2003).

The periapical dental radiograph shows the complete tooth; from crown to beyond the end root where actually the tooth is anchored in the jaw. In addition to it, these radiograph exhibits all dimension of tooth and consist of all teeth either upper or lower jaw. This type of dental radiograph imaging is common and inexpensive, which is normally used for dental disease diagnosis and abnormalities detection. The database used in this chapter consists of one hundred and twenty diverse periapical radiograph images from top to bottom jaw along with ground truth (Rad, Rahim, Rehman, & Saba, 2016; Vahab – LabArchives).

Typical Representation of Dental Radiograph

Typical representative periapical digital dental radiograph is shown in Figure 1. The radiographs are of Red Green Blue (RGB) color space images with a dimension of 748 × 512 pixels. The last twelve pixel rows of the image contain information about the images and it could be removed in the processing stages. It is noted that each image in the database contains at least one to four complete teeth either from maxillary or mandibular.

A tooth has two main parts as seen from the digital dental radiograph, the crown which is above the gum line and the root which sits in the bone below the gum line. Other than this the radiograph has three distinctive regions namely background (the air), teeth, and bones and in some images the dental features like fillings are also visible (Zhou, 2010; Jain, & Chen, 2004).

Figure 2 (a) represents three of the dental radiographs from the database. In this chapter for the sake of explanation of different image enhancement technique these three images are chosen, but otherwise the enhancement techniques are implemented for the whole database consisting of one hundred and twenty images.

Figure 1. Parts of periapical dental radiograph

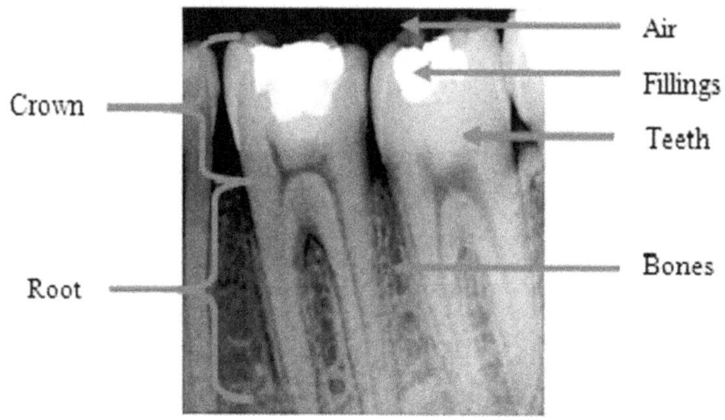

Figure 2a. Periapical dental radiographic images

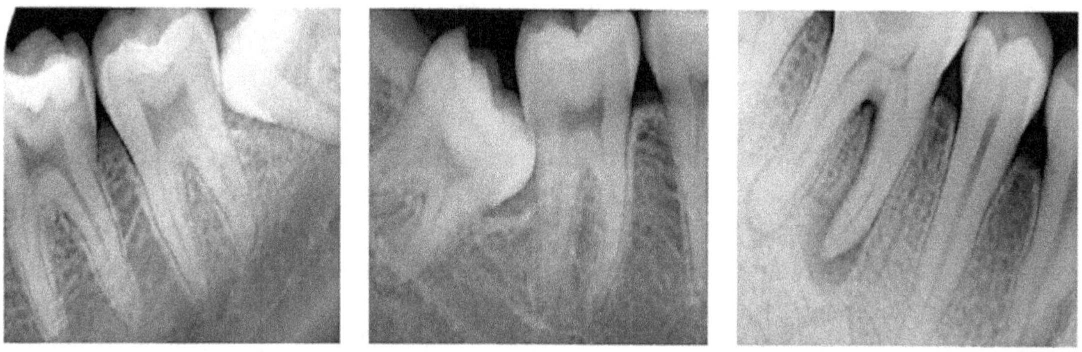

Figure 2b. Corresponding 2-D histogram

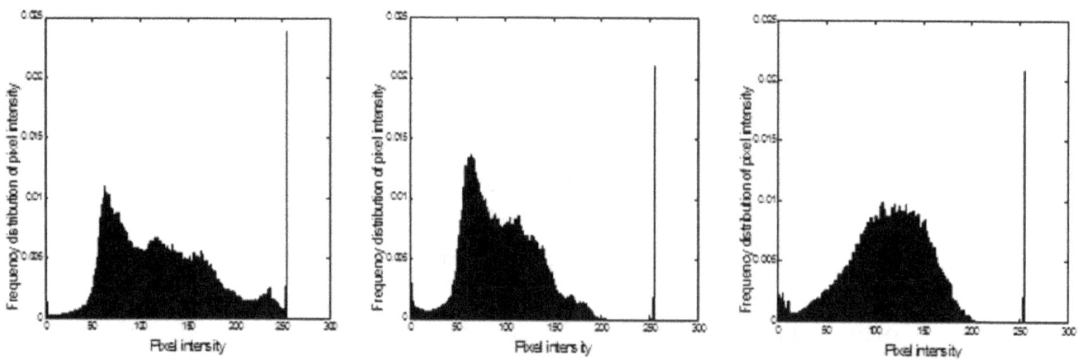

The histogram which is a plot of number of pixel values versus gray levels is shown in Figure 2 (b) for the dental radiographs presented in Figure 2 (a). The brightest pixel refers to the fillings in the first image and the same is inferred from the histogram. The first and second histograms are left skewed compared to the third, as the crown portion of the images is visible which is only partially visible in the third.

Figure 2c. A line in the images

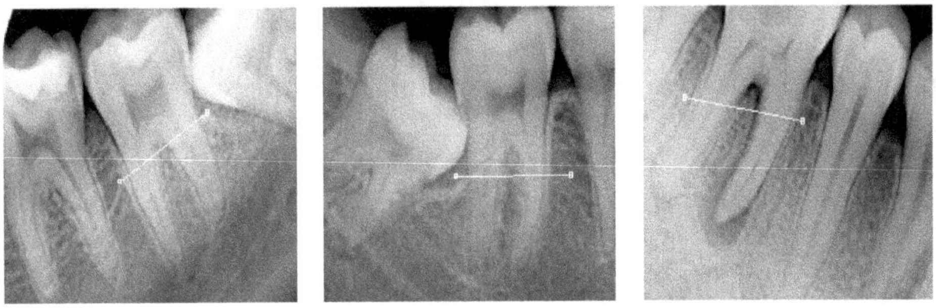

Figure 2d. Respective intensity profile along the line

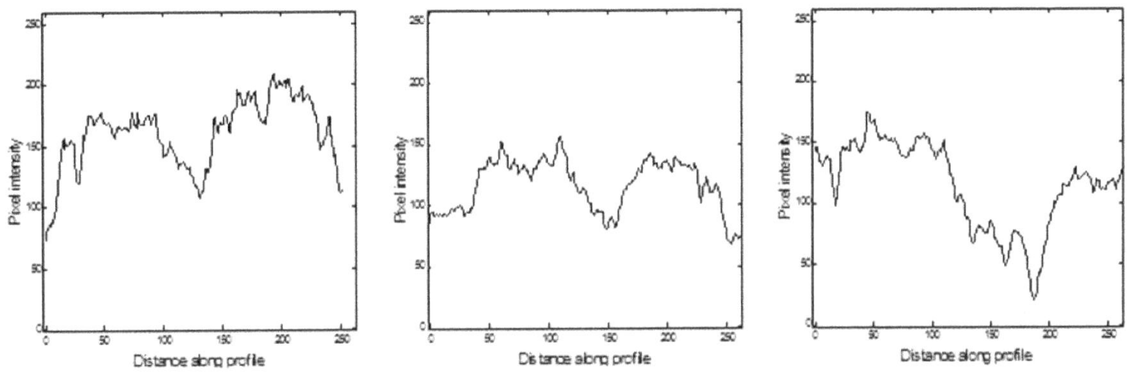

A line is drawn in the images as shown in Figure 2 (c) to examine the intensity profiles along them and the 2-D profiles are presented in Figure 2 (d). The profile shows two distinct peaks which represent the root portion of the image whereas the valley symbolizes the bone portion of the radiograph.

The following topic discusses briefly about different image enhancement techniques implemented in this chapter so that it makes the dental radiographs ready for the next processing stage.

Image Enhancement Techniques

The image enhancement techniques aim to improve the interpretability or insight of information in images for human observers and make the raw image more suitable for any specific application. Image enhancement techniques can be broadly classified into three categories. They are spatial domain methods which operate directly on pixels and frequency domain methods which operate on the transformed domain of an image. The third operates on the spatial-frequency domain. In general qualitatively by human perception and quantitatively by applying statistical and error measures the performance of the enhancement techniques can be judged to find the most appropriate (Das, 2015).

The teeth regions in the dental radiographs have the highest intensity and the bone regions have high intensity that may be even similar to that of the teeth whereas the background has a distinctively low intensity. Threshold based techniques can be used to separate the background from the image. But these methods fail to demarcate teeth from bones because their intensities are sometimes similar, especially in cases of irregular exposure. The first step to prepare the image for successful segmentation is to improve the image contrast by enhancing the teeth from the background regions (Zhou, 2010).

Image Enhancement Techniques Used in this Chapter

Different image pre-processing methods that includes spatial filtering, filtering in frequency and spatial-frequency domain are compared based on their ability to remove noise. These methods are aimed to enhance the dental radiographic images both in terms of qualitative and quantitative measures.

Median Filter

Median filter reduces noise rather than image blurring. Here the gray level of each pixel is replaced by the median of the gray level in the neighborhood of that pixel. This filtering operation is particularly effective when the noise pattern consists of strong spike like components and the characteristic to be preserved is edge sharpness (Banumathi, Vijayakumari, Geetha, Shanmugavadivu, & Raju, 2007).

The sequence is as follows:

1. Choose $m \times m$ window template, where m is odd. Slide the window template and roam for a pixel position that coincides with the template centre and store the gray value of each pixel corresponding to $w_{(x,y)}$ in the template.
2. Arrange the gray values of all pixels from small to large order to generate monotonic data sequence $\left\{w_{(x,y)}\right\}$ and find the middle which represents the $median\left\{w_{(x,y)}\right\}$.
3. The value of $median\left\{w_{(x,y)}\right\}$ is assigned to the middle of the centre pixel corresponding to the window template (Zeng, Liu, Fan, & Tang, 2012).

Figure 3a. Pre-processed output of periapical dental radiographic images using median filter

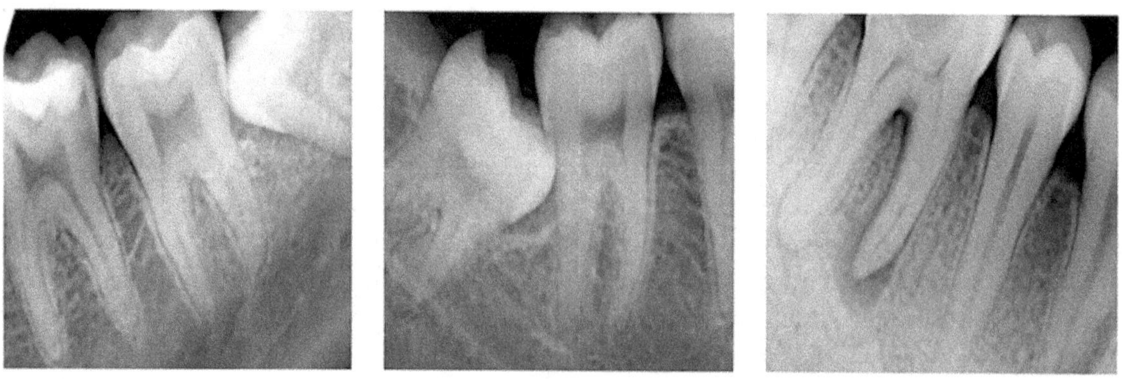

Figure 3b. Corresponding 2-D histogram

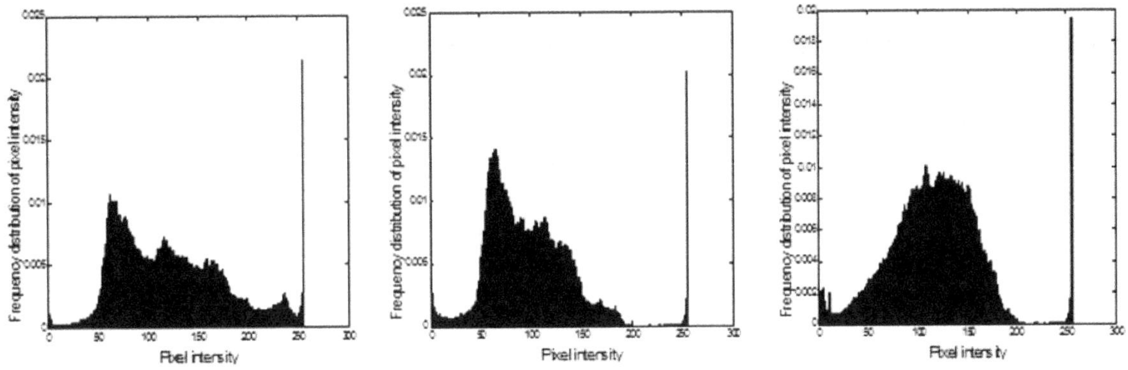

Figure 3c. A line in the images

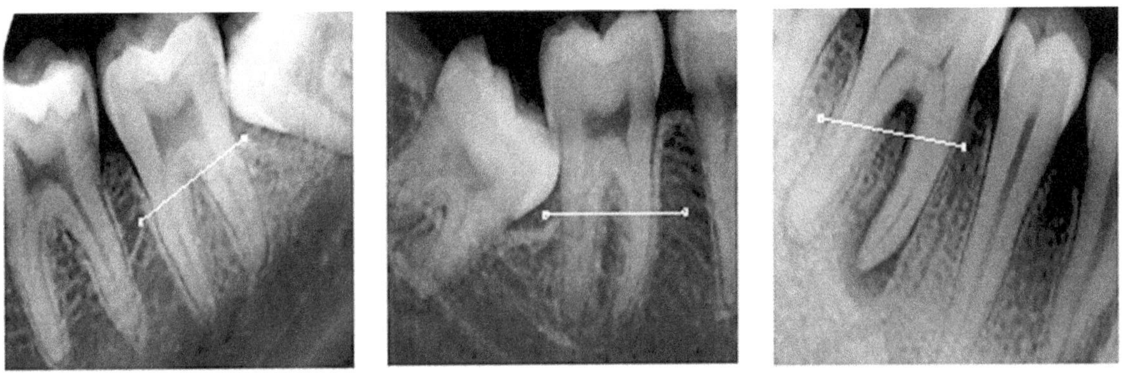

Figure 3d. Respective intensity profile along the line

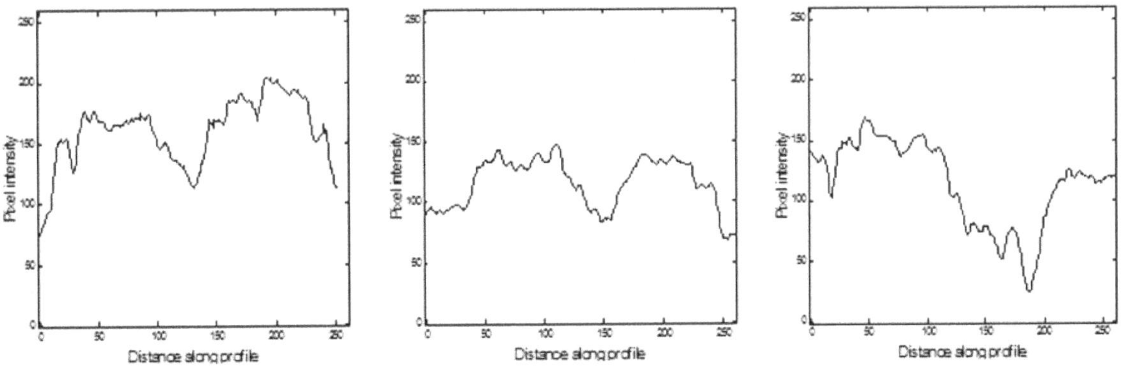

Qualitative Inference of Median Filter

The 2-D histogram plot of median filter method observed from Figure 3 (b) points out distinct peaks of foreground and background. Almost the value of valley is zero in many cases as this again distinguishes the foreground from background.

The 2-D intensity plot in Figure 3 (d) shows distinct plot between foreground and background. This is due to the deepness between two bumps in the intensity profile as observed from the third plot. The smoothness between the boundaries is well observed from the 2-D intensity profile in Figure 3 (d) compared with Figure 2 (d).

Top-Hat and Bottom-Hat

Mathematical morphology methods use mathematical principles and relationships between categories based on the structural properties of objects to extract the components of an image, which are useful in describing the shape of zones. Morphologic operators can either be defined for gray scale images or for binary images. Two morphological filters such as top-hat and bottom-hat filter extracts low intensity foreground objects (or dark ones) on a dark (or light) but slowly changing background (Hassanpour, Samadiani, & Salehi, 2015; Deserno, 2010; Zhou, 2010).

The top-hat and bottom-hat filter are expressed in terms of morphological closing and opening operators respectively as follows:

$$Tophat(X) = X_{TH} = X - \left(X \circ Y\right)$$

$$Bottomhat(X) = X_{BH} = \left(X \bullet Y\right) - X \tag{1}$$

where X and Y corresponds to the gray level input image and the structuring element. Also $\circ$ refers to the opening and $\bullet$ the closing operators respectively. The opening operator removes weak boundaries between objects and small details while the closing operator removes small holes and fills the disjoint boundaries (Hassanpour, Samadiani, & Salehi, 2015).

The top-hat filter returns the difference between the result of morphological opening operation and the original image X. The bottom-hat filter is performed by subtracting the original image from the result of gray scale closing operation. The top-hat filter is applied to enhance the bright regions that correspond to the teeth. Similarly bottom-hat filter is applied to produce large pixel values to enhance the dark bones and air areas. The enhanced image I_E is obtained by adding the original image with the result of top-hat filter and subtracting the result of the bottom-hat filter which is expressed as follows (Zhou, 2010):

$$I_E = X + X_{TH} - X_{BH} \tag{2}$$

A disk shaped structuring element is used to enhance the image so as to adjust the non-uniform illumination in these images.

Figure 4a. Pre-processed output of periapical dental radiographic images using top-hat and bottom-hat morphological filter

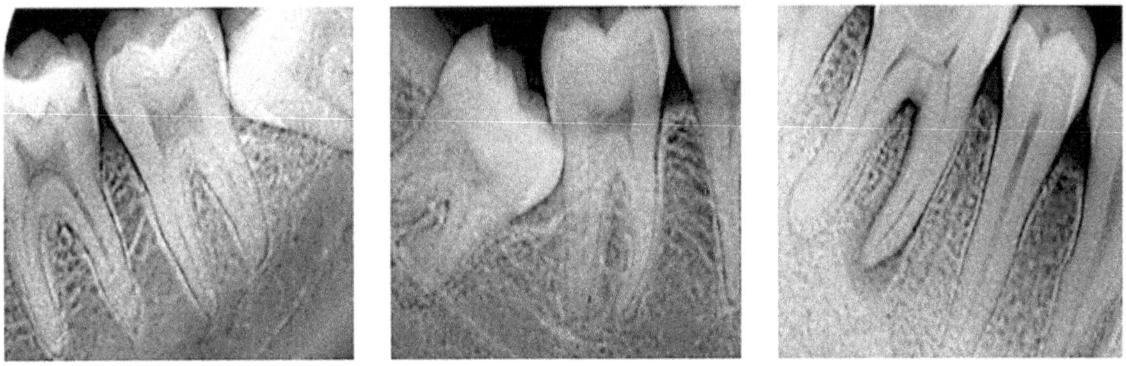

Figure 4b. Corresponding 2-D histogram

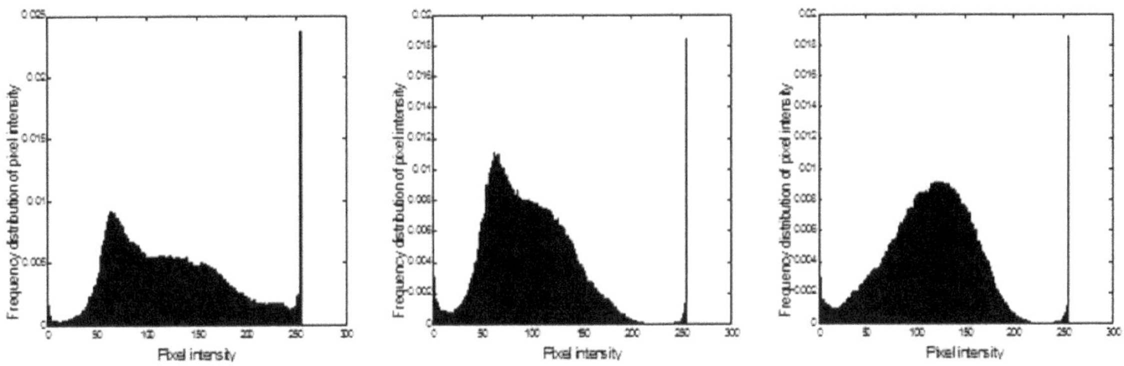

Figure 4c. A line in the images

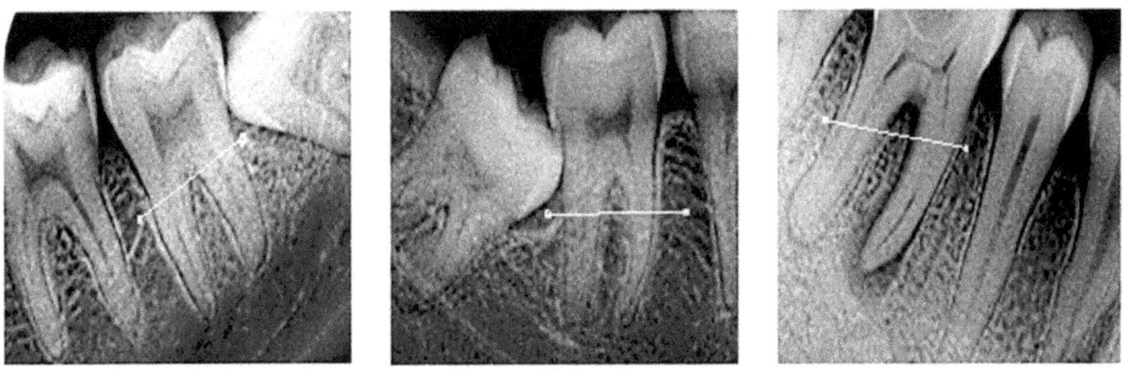

Figure 4d. Respective intensity profile along the line

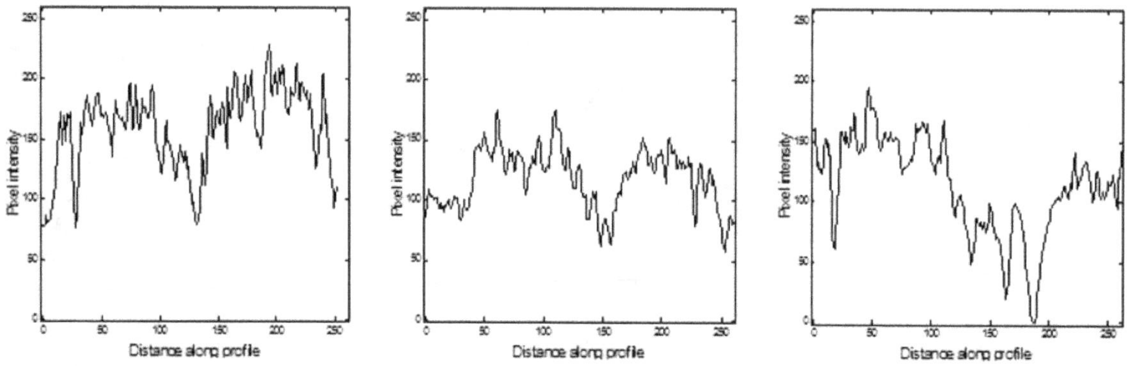

Qualitative Inference of Top-Hat and Bottom-Hat

Though the contrast is improved with the aid of the morphological top and bottom-hat filter, it is observed from Figure 4 (c) that the dental features such as fillings are not much prominent.

The noisy intensity profile in Figure 4 (d) is because of the dual process performed by these filters. The reason behind this is, the top-hat enhances bright foreground teeth regions that correspond to the teeth and the bottom-hat filter provides large pixel values and enhances the dark bones and air areas of the dental radiographs.

Contrast Limited Adaptive Histogram Equalization

CLAHE is derived through improvisations from the conventional or earlier methods such as histogram equalization and adaptive histogram equalization. It proves to be an efficient method compared to these methods as it utmost avoids the redundant oversaturation of bright pixels in the image (Bhat, & Patil, 2014).

In the earlier methods the enhancement function is applied all over the neighbourhood pixels and thereby the transformation function is derived. This is different from CLAHE because of its contrast limiting and the redistribution in gray level image, enhancing the contrast (Yadav, Maheshwari, & Agarwal, 2014). The CLAHE algorithm is discussed as follows:

1. Compute the dynamic range (number of bins used in histogram transform function), the clip limit N_{Clip}, and distribution type after combining all input values such as the pixel values in row and column direction separately.
2. Divide the image into contextual regions and compute gray level mapping and clipped histogram. Then in contextual region, number of pixels is equally divided in each gray level and the average number of pixels in the gray level is described as follows:

$$N_{avg} = \frac{N_{CR-X_P} * N_{CR-Y_P}}{N_g} \tag{3}$$

where N_{avg} represents average number of pixels, N_g refers to number of gray levels in contextual regions, N_{CR-X_p} and N_{CR-Y_p} indicates number of pixels in X and Y direction of contextual regions respectively.

3. Compute the actual clip limit as follows:

$$N_{CL} = N_{Clip} * N_{avg} \tag{4}$$

4. Interpolate gray level mapping in order to get the enhanced image.

Qualitative Inference of CLAHE

The histogram plot in Figure 5 (b) shows the contrast being increased in the dental radiographs, as this is visible from the two distinct vertical lines corresponding to the foreground and background. This is also evident from the wide range of pixel intensities occupied by the CLAHE processed images.

Figure 5a. Pre-processed output of periapical dental radiographic images using CLAHE

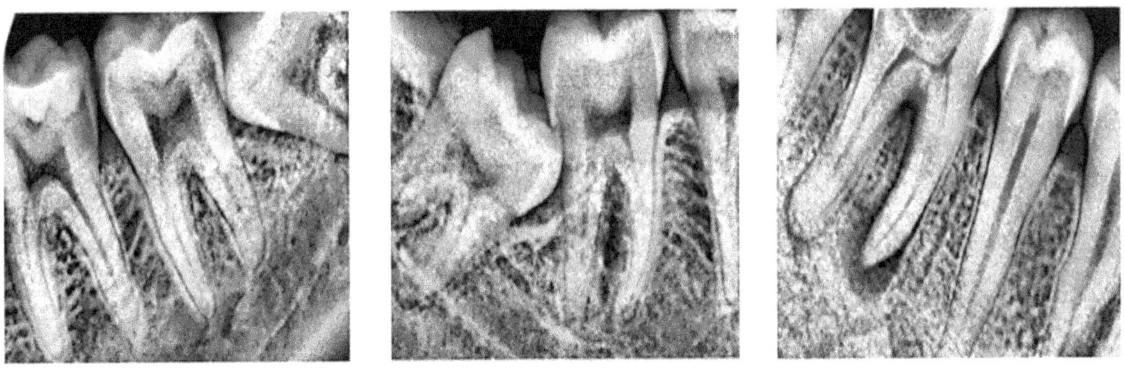

Figure 5b. Corresponding 2-D histogram

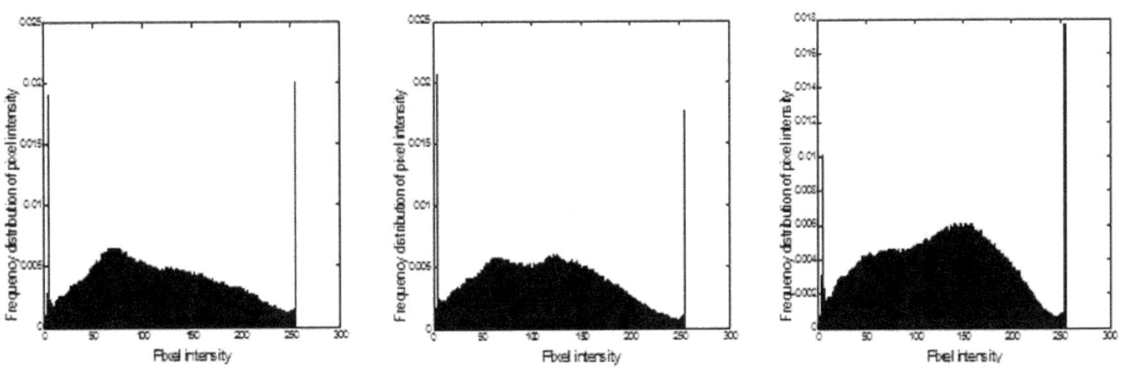

Figure 5c. A line in the images

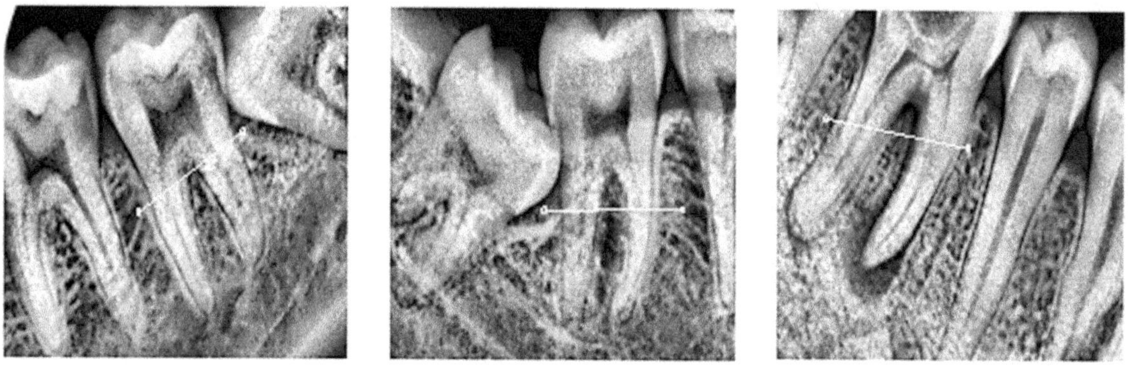

Figure 5d. Respective intensity profile along the line

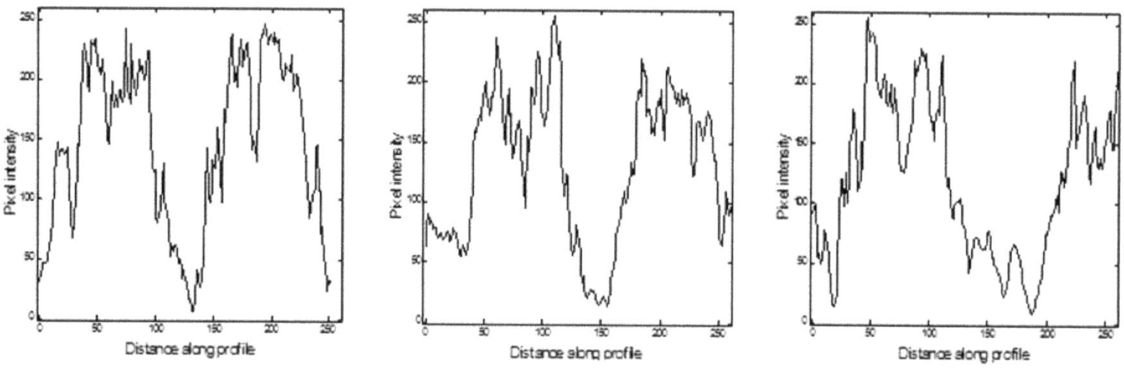

Figure 5 (d), the intensity profile is not smooth as compared with Figure 3 (d), yet the contrast in these radiographs have improved a lot, as it is observed from the large bumps corresponding to the tooth root structures.

Cubic Spline Interpolation

Cubic spline interpolation method is quite familiar and widely used for its minimal error and smoother curve development. Spline interpolation is a form of interpolation where the interpolant is a unique type of piecewise polynomial called a spline which is popular in the mathematical field of numerical analysis. Spline interpolation is often chosen over polynomial interpolation since the interpolation error can be made minimal even when using lower degree polynomials for the spline. Spline interpolation evades the problem of Runge's phenomenon where in wavering occur between points while interpolating using higher degree polynomials. Let r be the input pixel intensity and s be the output pixel intensity, then the piecewise cubic polynomial for the curve is given as:

$$s = ar^3 + br^2 + cr + d \tag{5}$$

where a, b, c and d are the piecewise polynomial constants. These will differ for different piecewise portions of curve (Panda, 2016).

Qualitative Inference of Cubic Spline Interpolation

Similar to median filter shown in Figure 3 (b), zero valley is observed in the histogram plot as presented in Figure 6 (b). Because of the interpolation technique, the bone portion of radiograph gets merged with the background, thus enhances the visibility of the crown portion.

The 2-D intensity plot in Figure 6 (d) shows distinct demarcation between foreground and background. The deepness between two bumps in the intensity profile as observed from the third plot might be due to the interpolation method.

Anisotropic Diffusion Filter

The extension of Gaussian filters is the anisotropic diffusion which is otherwise known as bilateral filtering or Partial Differential Equation (PDE)-based anisotropic diffusion that preserves smoothing. Anisotropic filters are preferred compared to Gaussian filters as Gaussian blurs the image edges while smoothing image noise. Gaussian filters are isotropic in the sense that all surrounding pixels affect the

Figure 6a. Pre-processed output of periapical dental radiographic images using cubic spline interpolation

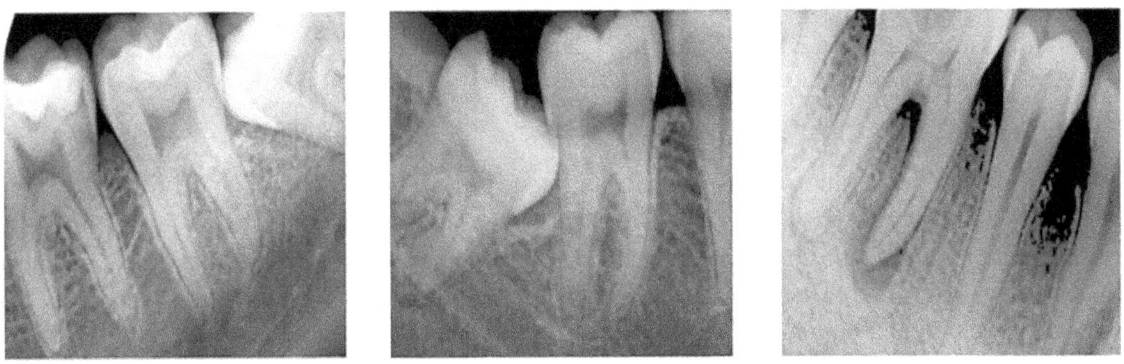

Figure 6b. Corresponding 2-D histogram

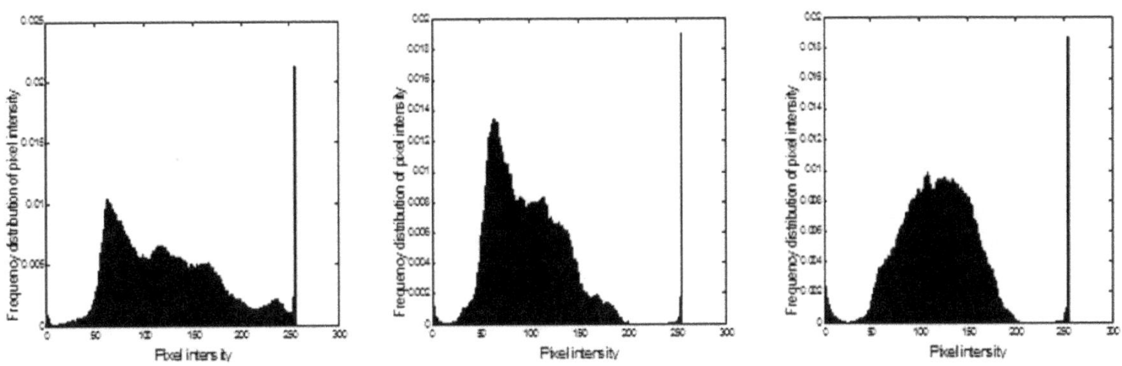

Figure 6c. A line in the images

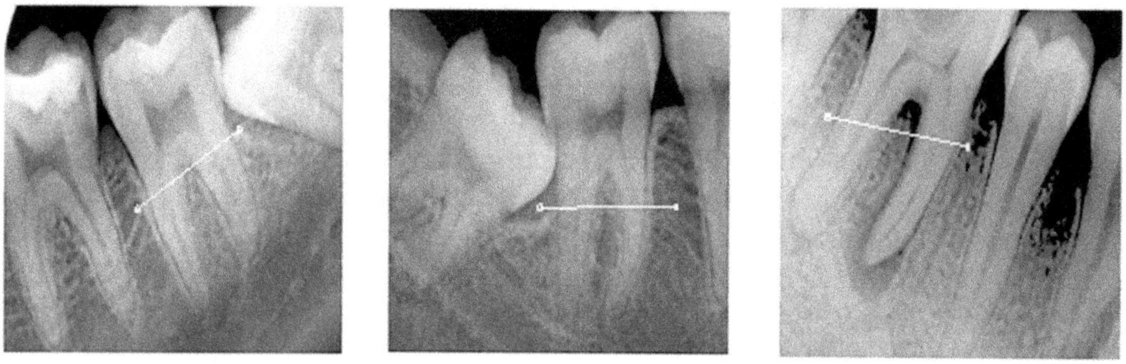

Figure 6d. Respective intensity profile along the line

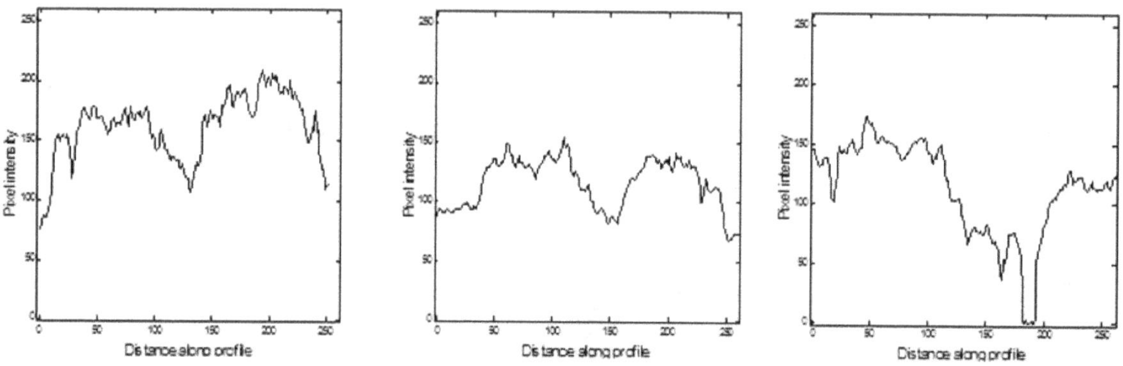

centre pixel in a similar fashion regardless of their intensity variation. This results in noise reduction as well as edge blurring (Banumathi, Vijayakumari, Geetha, Shanmugavadivu, & Raju, 2007).

Anisotropy in diffusion means that the smoothing induced by the PDE can be favoured in some directions and prevented in others. This is specified by local eigenvectors and eigenvalues of the diffusion field and is location and direction dependent. Anisotropic non-linear diffusion is an influential image processing technique, which simultaneously removes noise and enhances sharp features in two or three dimensional images. They also reflect the local orientation of image features (Mirebeau, Fehrenbach, Risser, & Tobji, 2015).

Perona and Malik proposed a non-linear diffusion method for avoiding the blurring and localization problems of linear diffusion filtering. They applied an inhomogeneous process that reduces the diffusivity at those locations which have a larger likelihood to be edges measured by $\left|\nabla u\right|^2$ (Weickert, 1998). The Perona Malik filter is based on the equation:

$$\partial_t u = div\left(g\left(\left|\nabla u\right|^2\right)\nabla u\right) \tag{6}$$

and it uses diffusivities such as

$$g\left(s^2\right) = \frac{1}{1 + s^2/\lambda^2}\left(\lambda > 0\right) \tag{7}$$

Considering the anisotropic diffusion equation

$$I_t = div\left(c\left(x, y, t\right)\nabla I\right) = c\left(x, y, t\right)\Delta I + \nabla c \cdot \nabla I \tag{8}$$

where div the divergence operator, ∇ and Δ represents gradient and Laplacian operators respectively with respect to the space variables. This reduces to the isotropic heat diffusion equation $I_t = c\Delta I$ if $c\left(x, y, t\right)$ is a constant. If at the time (scale) t the location of the region boundaries appropriate for that scale are known then the smoothing within the region is preferred compared to smoothing across the boundaries. This could be achieved by setting the conduction coefficient to be 1 in the interior of each region and 0 at the boundaries. The blurring would then take place separately in each region with no interaction between regions and hence the region boundaries would remain sharp (Perona, & Malik, 1990).

Qualitative Inference of Anisotropic Diffusion Filter

The histogram plot presented in Figure 7(b) shows the difference in the foreground and background intensities when compared with Figure 2 (b). It is in such a way that the background intensities are smoothed while the boundaries between them and foreground is preserved, as this is the advantage of anisotropic diffusion.

The smoothness between the boundaries is well observed from the 2-D intensity profile in Figure 7 (d) compared with Figure 2 (d). This is analogous to median filter as observed from the intensity profile in Figure 3 (d).

Figure 7a. Pre-processed output of periapical dental radiographic images using anisotropic diffusion filter

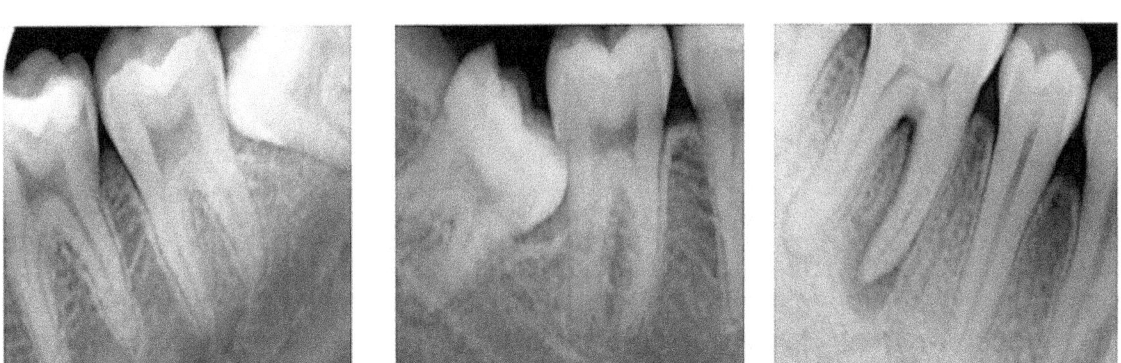

Figure 7b. Corresponding 2-D histogram

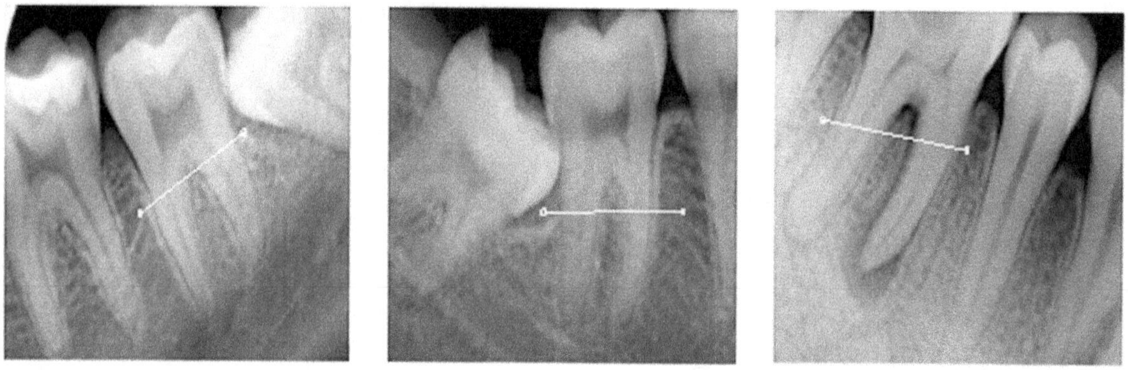

Figure 7c. A line in the images

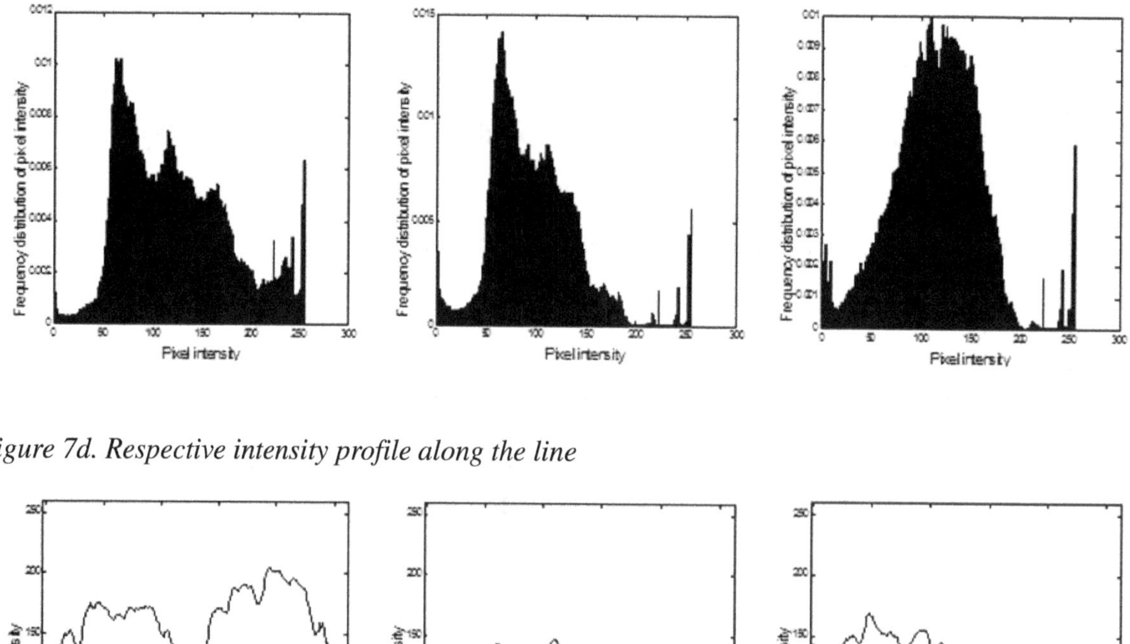

Figure 7d. Respective intensity profile along the line

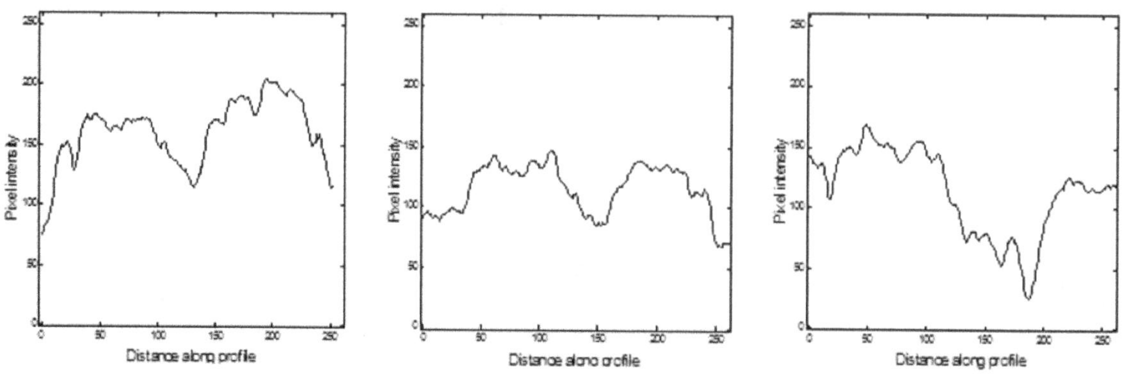

Homomorphic Filter

Frequency domain filters provides better contrast and illumination evenness in the image. Homomorphic filters are widely used in image processing for compensating the effect of non-uniform illumination in an image (Vijayakumari, Ulaganathan, & Banumathi, 2011; Pushparaj, Gurunathan, & Arumugam,

2013). To express the process of homomorphic filtering first the basic image model is chosen, which is expressed as

$$f(x,y) = f_i(x,y) \cdot f_r(x,y) \quad (9)$$

where $f_i(x,y)$ refers to the lighting and $f_r(x,y)$ the reflection function. Since multiplication cannot be directly transformed into frequency domain logarithmic is taken first followed by transformation in Fourier domain and is represented as

$$F[\ln f(x,y)] = F[\ln f_i(x,y) + \ln f_r(x,y)] \quad (10)$$

The above equation in frequency domain is referred as

$$F(U,V) = F_I(U,V) + F_R(U,V) \quad (11)$$

The first term infers to the slowly varying low frequency component and the second infers to the frequent changes corresponding to high frequency.

If the image has inhomogeneous illumination, the details of dark regions are difficult to discriminate. Thus to eliminate this heterogeneity the gray scale range of illumination function can be compressed which weakens its composition in frequency domain. This enhances the composition of reflection function so that the image details of dark regions can be enhanced. The following equation expresses it.

$$G_{\ln}(U,V) = F_{\ln}(U,V) \cdot H(U,V)$$

$$= F_{i,\ln}(U,V) H(U,V) + F_{r,\ln}(U,V) H(U,V) \quad (12)$$

The homomorphic filtered image is derived by taking inverse Fourier and antilog of the above equation (Wu, Merchant, & Castleman, 2010),

$$g(x,y) = \exp\left\{F^{-1}\left[G_{\ln}(U,V)\right]\right\} \quad (13)$$

Qualitative Inference of Homomorphic Filter

Results presented in Figure 8 (b) shows that the homomorphic filter enhances the distribution of reflection function and hence image details of dark regions are also enhanced.

The same is reflected in the intensity profile as Figure 8 (d) presents an approximate plot. Because of this the root and bone structure of the radiograph is not distinguishable. Also the intensity profile seems to be blunt instead of peaks.

Figure 8a. Pre-processed output of periapical dental radiographic images using homomorphic filter

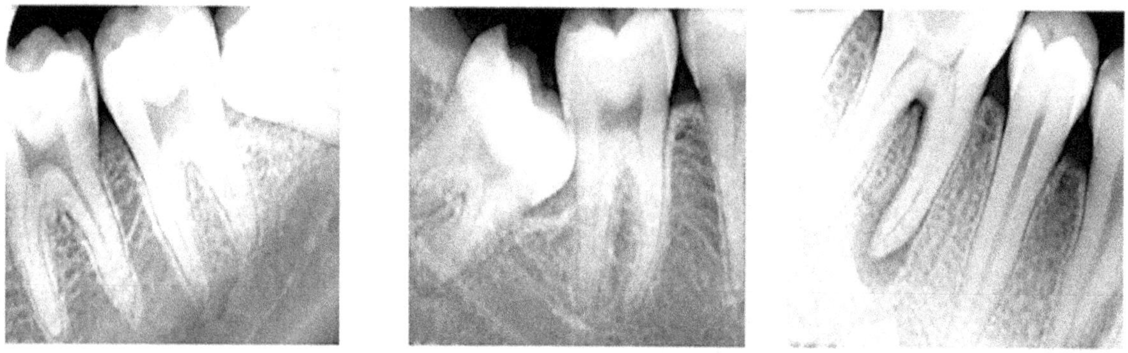

Figure 8b. Corresponding 2-D histogram

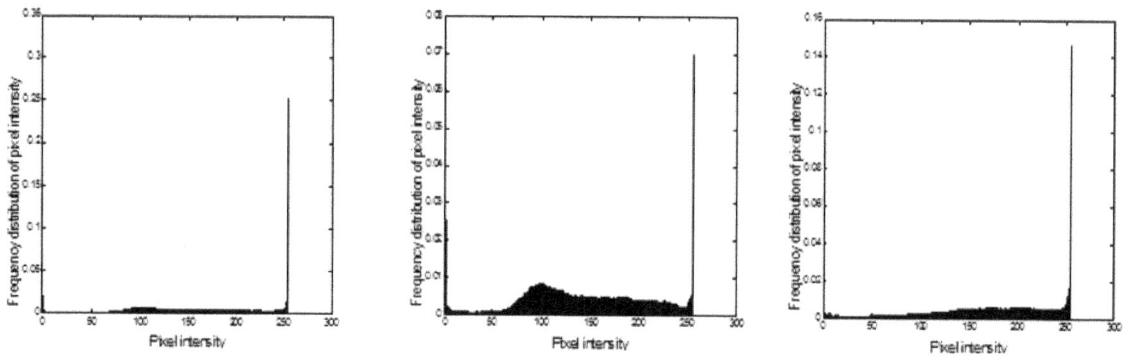

Figure 8c. A line in the images

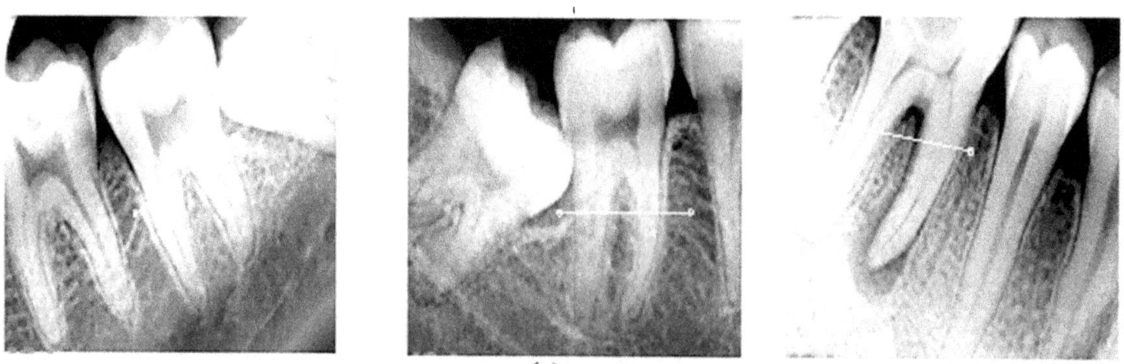

Wavelet and SVD

The growth of multi-resolution wavelet transform methods has made an impact on several areas of signal and image processing (Zeng, Dong, Chi, & Xu, 2004).

Figure 8d. Respective intensity profile along the line

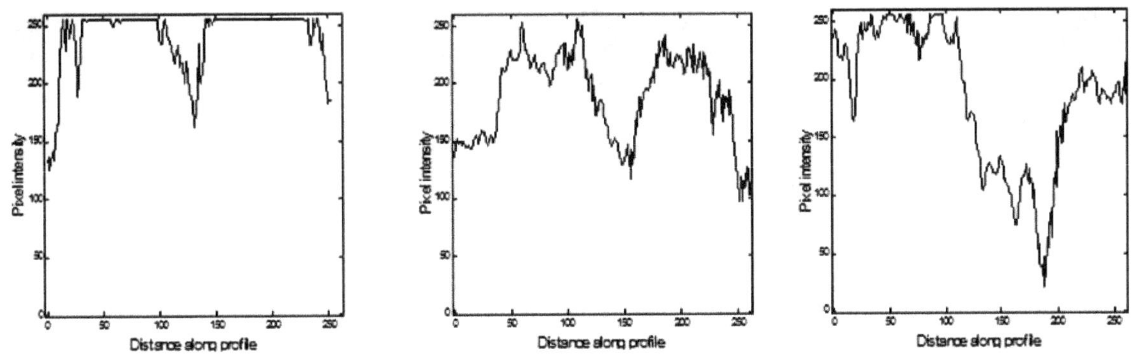

The 2-D Discrete Wavelet Transform (DWT) of an image is performed by performing 1-D DWT first along the rows and then along the columns. This procedure results in four sub-band images namely: low-low (LL), low-high (LH), high-low (HL) and high-high (HH). The frequency components of these sub-bands include the full frequency spectrum of the original image. Figure 9 shows the block diagram representation of the 2-D DWT.

Among the four sub-band images Singular Value Decomposition (SVD) is applied to the LL sub-band image. Then by computing SVD of $m \times n$ real or complex matrix M results in the factorization of the form

$$A = U * S * V^T \tag{14}$$

where U is an $m \times m$ real or complex unitary matrix, $\sum$ is an $m \times n$ rectangular diagonal matrix with non-negative real numbers on the diagonal and V^* (conjugate transpose of V) is a $n \times n$ real or complex unitary matrix. The transformation factor obtained after this is the new enhanced LL sub-band. Finally the output is combined by resizing the new SVD enhanced LL sub-band to the size of interpolated estimated other three sub-bands by applying IDWT (Inverse DWT) technique and generate the final resolution of the enhanced image (Sharma, & Khunteta, 2016).

Figure 9. Block diagram representation of 2-D DWT

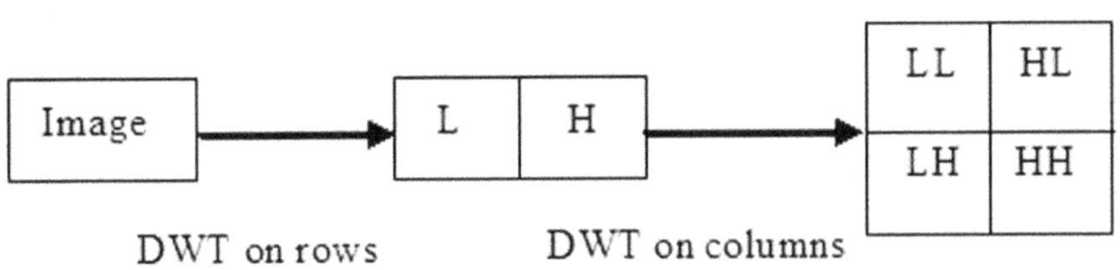

Figure 10a. Pre-processed output of periapical dental radiographic images using wavelet denoising

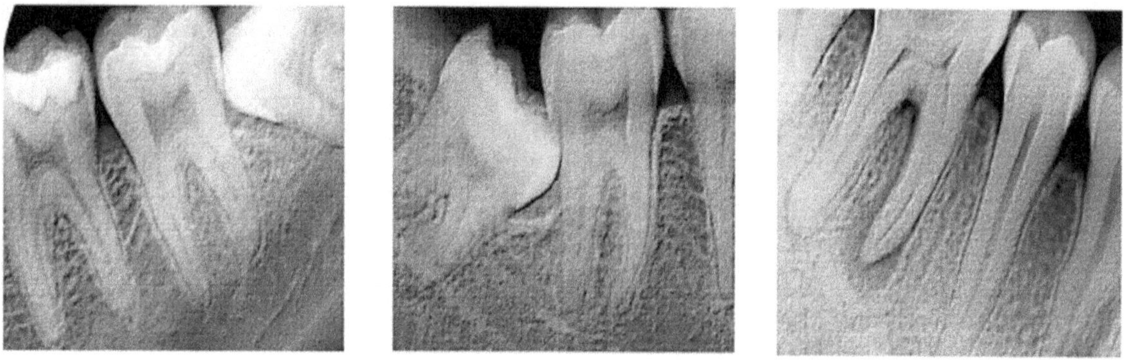

Figure 10b. Corresponding 2-D histogram

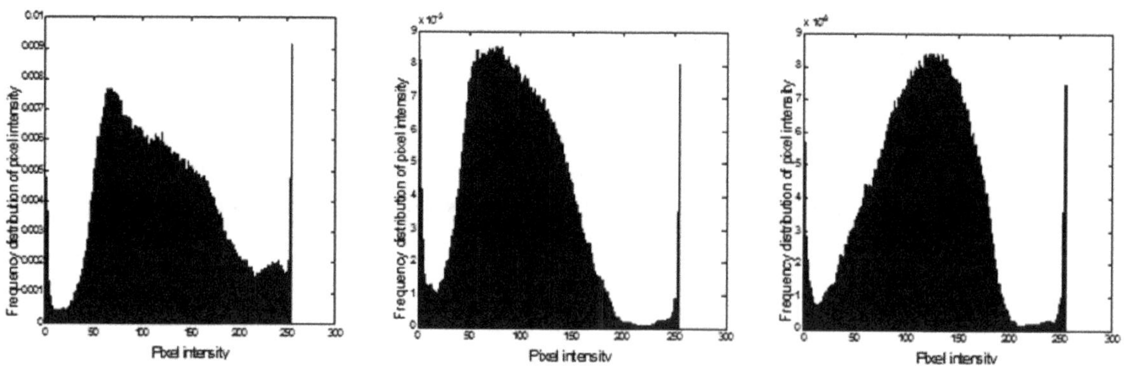

Figure 10c. A line in the images

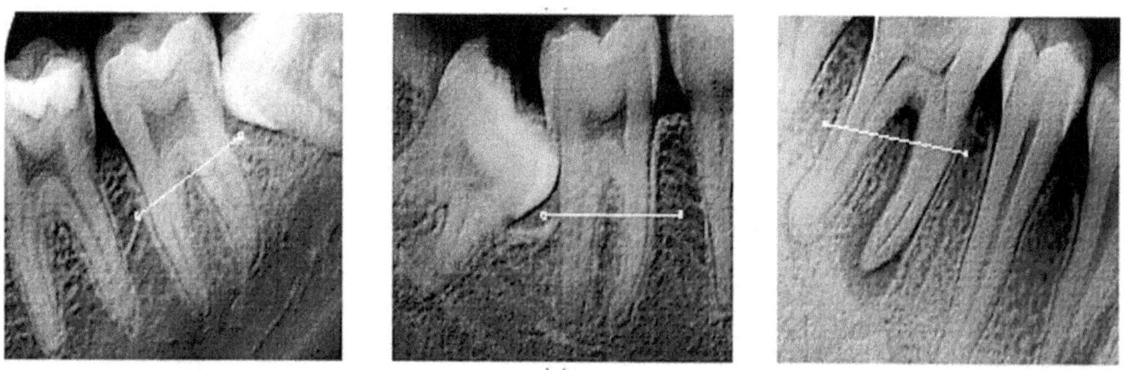

Qualitative Inference of Wavelet and SVD

It is observed from Figure 10 (b) that since the valley depth of image histogram is not dominant much difference is not noticed between the crown and the root of the radiographs. Similarly the intensity profile presented in Figure 10 (d) shows not much difference between the foreground and background. Hence the qualitative result of wavelet and SVD combination is not much appealing for the dental radiographs.

Figure 10d. Respective intensity profile along the line

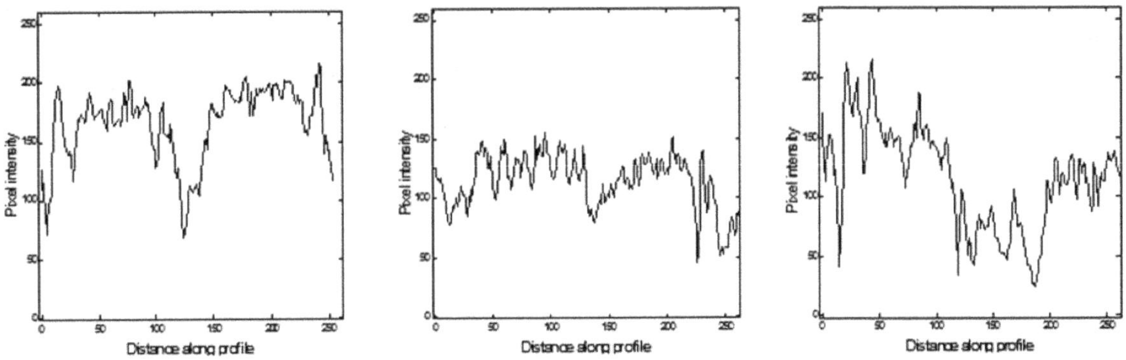

Histogram Based Statistical Measures

The histogram shows the frequency distribution of pixel values (e.g., gray scales) and it is simple to define pixel transforms using a histogram. The shape of gray-level histogram gives an idea about overall appearance of an image. The contrast of an image could be improved through the stretching of gray scales (Deserno, 2010). In this chapter the histogram based features are extracted from the pre-processed images, as histogram-based analysis is useful for shape analysis of objects.

The shape of an image histogram provides many clues. The skewness measures shape of histogram. Positively skewed histogram indicates light image whereas negatively skewed histogram indicates bright image and the mean measures the average intensity. Kurtosis measures degree of flatness or peakedness of a distribution, in other words it gives the overall representation of frequencies in the middle of the range compared to normal distribution (Dighe, & Shriram, 2012).

The energy feature measures the uniformity of intensity level distribution. If energy value is high, then the distribution is of small number of intensity levels. The entropy measures the randomness of the distribution of the coefficients values over the intensity levels. If the entropy is high, then the distribution is among more intensity levels in the image and this histogram-based statistical feature is the inverse of energy (Seetha, Devi, & Sunitha, 2012).

Inference From Histogram-Based Statistical Measures

The normalized histogram based statistical features before image enhancement is presented in Table 1. The value of mean and variance is observed to be low as the brightness of the dental radiographs is low.

Results demonstrate that the average skewness is around 0.59 and thus for all images in the dataset, dental radiograph parts such as crown, root and air are visible. Not much sharpness around the edge is observed for the raw images and hence the kurtosis is low.

The energy feature measures the uniformity of intensity level distribution. Since the number of intensity levels distribution are lower as observed from Figure 2 (b) the statistical measure entropy is less. Table 2 shows the average and standard deviation of normalized histogram-based statistical features for different image enhancement techniques.

Table 1. Normalized histogram based statistical features before image enhancement

Average ± SD Value Before Image Enhancement	Histogram Based Statistical Features					
	Mean	Variance	Skewness	Kurtosis	Energy	Entropy
	0.48 ±b 0.17	0.35 ± 0.21	0.59 ± 0.18	0.39 ± 0.23	0.14 ± 0.12	0.55 ± 0.24

*SD-standard deviation

Table 2. Normalized histogram based statistical features for different image enhancement techniques

Image Enhancement Methods		Histogram Based Statistical Features					
		Mean	Variance	Skewness	Kurtosis	Energy	Entropy
Average ± SD values	Median filter	0.46 ± 0.17	0.35 ± 0.21	0.58 ± 0.17	0.21 ± 0.23	0.14 ± 0.12	0.58 ± 0.23
	Top-hat & bottom-hat filter	0.53 ± 0.16	0.37 ± 0.20	0.59 ± 0.16	0.26 ± 0.16	0.12 ± 0.11	0.60 ± 0.21
	CLAHE	0.55 ± 0.17	0.48 ± 0.19	0.53 ± 0.18	0.38 ± 0.18	0.05 ± 0.08	0.83 ± 0.10
	Cubic spline interpolation	0.60 ± 0.23	0.45 ± 0.22	0.44 ± 0.15	0.23 ± 0.19	0.21 ± 0.16	0.55 ± 0.23
	Anisotropic diffusion filter	0.54 ± 0.15	0.36 ± 0.21	0.53 ± 0.15	0.33 ± 0.21	0.08 ± 0.08	0.68 ± 0.16
	Homomorphic filter	0.57 ± 0.16	0.36 ± 0.20	0.60 ± 0.15	0.23 ± 0.15	0.20 ± 0.19	0.67 ± 0.19
	Wavelet & SVD	0.55 ± 0.16	0.34 ± 0.20	0.62 ± 0.16	0.34 ± 0.19	0.06 ± 0.05	0.69 ± 0.16

*SD-standard deviation

It is observed from the table that the mean is high for cubic spline interpolation, thus enabling the brightness of these images to increase and thereby the features of the dental radiographs are highlighted. Since the contrast of CLAHE processed images is high the histogram based variance feature for CLAHE is found to be high among the other methods.

The methods other than cubic spline interpolation exhibit positive skewness hence it exhibits uniform background intensity values. The lower skewness for cubic spline is because of the difference in intensity values between foreground and background. Sharp change in gray level is observed between various dental parts in CLAHE enhanced images and thus kurtosis is found to be high but in cubic spline sharp changes around edges are not observed.

The number of intensity levels is found to be quiet low for cubic spline hence the histogram based energy feature is low. Also, the distribution of intensity levels are more and thus the entropy is observed to be low. In general, cubic spline performs better than other methods in terms of histogram based statistical measures.

Image Quality and Error Measures

Apart from the qualitative measures, few calculable quantitative measures for comparing the performance of the image enhancement methods are aimed at by using image quality and error measures. The error measures attempted in this chapter includes, Mean Square Error (MSE), Peak Signal to Noise Ratio (PSNR), Average Difference (AD), Maximum Difference (MD), Laplacian Mean Square Error (LMSE) and Normalized Absolute Error (NAE). The quality factors include, Structural Content (SC), Normalized Cross Correlation (NCC) and Structural SIMilarity (SSIM).

The mathematical formulations of these measures are defined in this section. MSE is obtained from the average square intensity of the original image I_R and the resultant enhanced image I_E and can be written as,

$$MSE = \frac{1}{XY}\sum_{y=1}^{Y}\sum_{x=1}^{X} e^2\left(x,y\right) \tag{15}$$

where $e\left(x,y\right)$ is the error difference between original raw (I_R) and enhanced image (I_E).

PSNR value is obtained by computing the difference between the original and enhanced image in which the SNR of all its pixel value is the maximum value of the most appropriate.

$$PSNR = 10\log\frac{s^2}{MSE} \tag{16}$$

where, $s = 255$ for 8-bit image.

$$AD = \frac{1}{XY}\sum_{y=1}^{Y}\sum_{x=1}^{X} e\left(x,y\right) \tag{17}$$

$$MD = \max\left\{\left|e\left(x,y\right)\right|\right\} \tag{18}$$

$$LMSE = \frac{\sum_{y=1}^{Y-1}\sum_{x=2}^{X-1}\left[O\left\{e\left(x,y\right)\right\}\right]^2}{\sum_{y=1}^{Y-1}\sum_{x=2}^{X-1} O\left\{I_E\right\}^2} \tag{19}$$

where $O\left\{e\left(x,y\right)\right\}$ is the sum of 4-neighborhood pixel values. $O\left\{I_E\right\}$ can be I_R or $I_R^{1/3}$ or $H\left\{\left(U^2+V^2\right)^{1/2}\right\}F\left(U,V\right)$ in cosine transform domain.

$$NAE = \frac{\sum_{y=1}^{Y}\sum_{x=1}^{X}\left|O\left\{e\left(x,y\right)\right\}\right|}{\sum_{y=1}^{Y}\sum_{x=1}^{X}\left|O\left\{I_E\right\}\right|} \tag{20}$$

$$SC = \frac{\sum_{y=1}^{Y}\sum_{x=1}^{X}\left[I_R\right]^2}{\sum_{y=1}^{Y}\sum_{x=1}^{X}\left[I_E\right]^2} \tag{21}$$

$$NCC = \frac{\sum_{y=1}^{Y}\sum_{x=1}^{X} I_R \cdot I_E}{\sum_{y=1}^{Y}\sum_{x=1}^{X}\left[I_R\right]^2} \tag{22}$$

SSIM is obtained by dividing the original image and the enhanced image in the block size of 8x8.

$$SSIM = \frac{\left(2\mu_x\mu_y + c_1\right)\left(2\sigma_{xy} + c_2\right)}{\left(\mu_x^2 + \mu_y^2 + c_1\right)\left(\sigma_x^2 + \sigma_y^2 + c_2\right)} \tag{23}$$

where μ_x and μ_y refers to mean along x and y direction, σ_x and σ_y the standard deviation along x and y direction and σ_{xy} is the covariance (Wang, 2004; Nevriyanto, & Purnamasari, 2017; Eskicioglu, & Fisher, 1995).

Inference From Image Quality and Error Measures

The image quality measures and error measures are quantified with the help of box plot presented in Figure 11. The x-axis refers to the measures and the y-axis refers to the normalized average value computed for the all the images in the dataset.

The MSE value which represents the error difference is less for cubic spline compared to other methods as inferred from the box plot. Also the peak signal to noise ratio is high which clearly explains that the cubic spline interpolation method suppresses the noise present in the images. These measures prove that the false positive in identifying the dental features could be reduced by the cubic spline.

It is observed from the plot that the error measures such as LMSE and NAE are absolutely low for cubic spline. Though the Laplacian MSE is less for homomorphic filter and wavelet methods the NAE measure is observed to be high for these methods as the dark regions of the images are enhanced. This observation demonstrates the invisibility of the root structure of the dental radiographs using the homomorphic filter and wavelet methods.

The box plot demonstrates that except for cubic spline the AD and MD seems to be high. This describes the similarity between the cubic spline interpolation method with that of the reference ground truth images. In a same way the NCC is observed to be high as this measures the correlation with the ground truth images. In median, anisotropic and homomorphic filter methods the average value of MD

Figure 11a. Normalized values of image quality measures for median filter

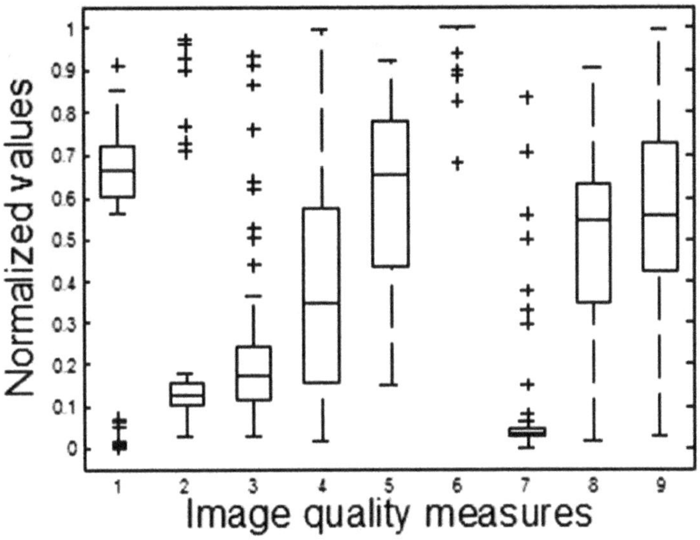

Figure 11b. Normalized values of image quality measures for top-hat and bottom-hat filter

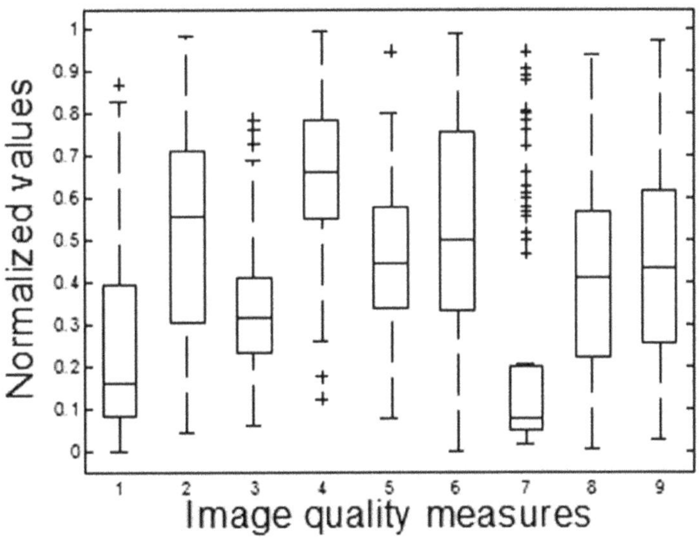

is not well defined as most of them fall as outliers, this proves that there is less uniformity among the enhanced images by these methods.

SC measures the percentage of similarity as this is a measure of correlation. The SC is observed to be high for top-hat and bottom-hat filter than other methods including cubic spline. The SSIM average value of cubic spline is close to 1, hence the image sharpness is expected to be high.

Figure 11c. Normalized values of image quality measures for CLAHE

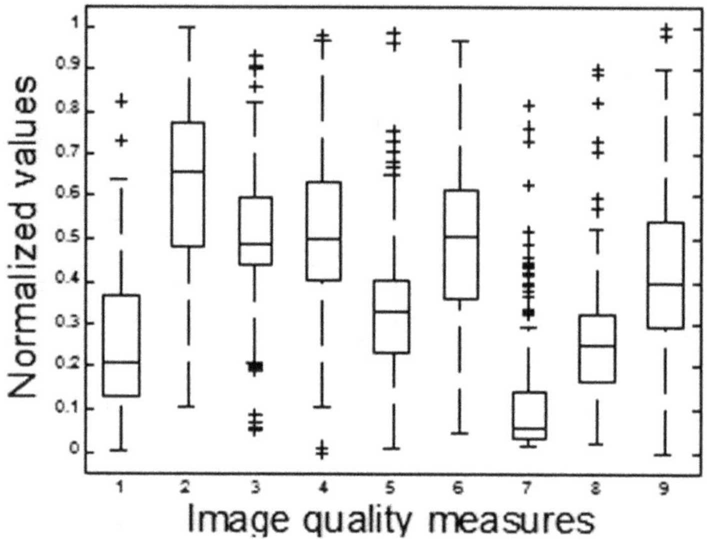

Figure 11d. Normalized values of image quality measures for cubic spline interpolation

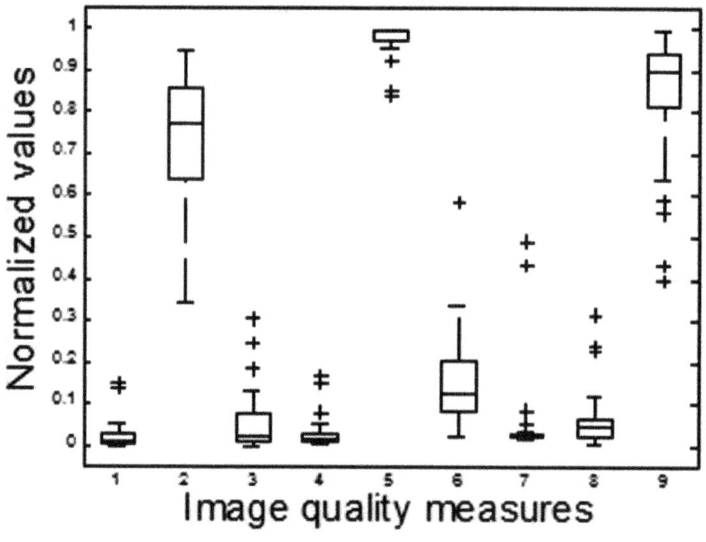

FUTURE RESEARCH DIRECTIONS

The purpose of image enhancement filters is to improve image quality by removing noise. Noise is an unrelated component of an image that hinder identification and elucidation of the data of interest. Noise can either be a low frequency component that represents the gradual intensity changes or high frequency noise which could be the speckle noise. Filters that sharpen an image remove low frequency noise and intended to enhance the detail in an image. Filters that smoothen an image remove high frequency noise and propose to reduce the amplitude of small detail in an image (Zhang, Fadili, & Starck, 2008).

Figure 11e. Normalized values of image quality measures for anisotropic diffusion

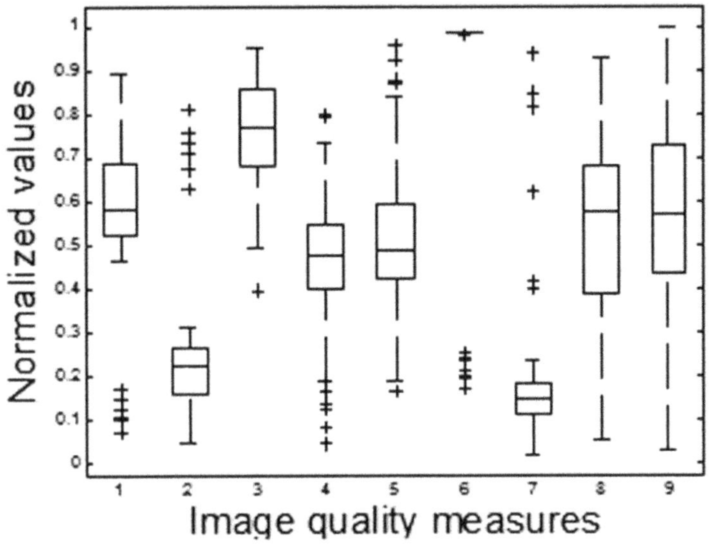

Figure 11f. Normalized values of image quality measures for homomorphic filter

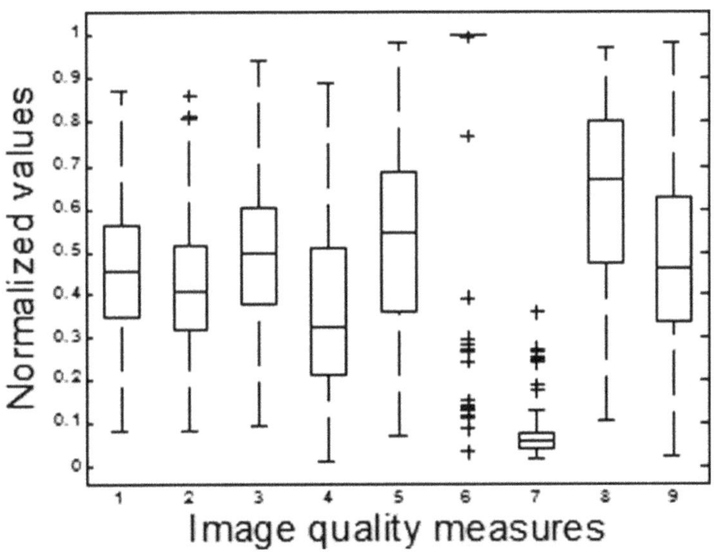

It is expected that if the distribution of noise is studied thoroughly then the filter that is suitable to remove or reduce noise could be implemented. This information could improve the identification of dental features in the radiographs. Also image restoration algorithm to handle poor quality radiographs could be attempted after image enhancement.

Figure 11g. Normalized values of image quality measures for wavelet and SVD (x axis label from 1 to 9 corresponds to MSE, PSNR, AD, SC, NCC, MD, LMSE, NAE and SSIM)

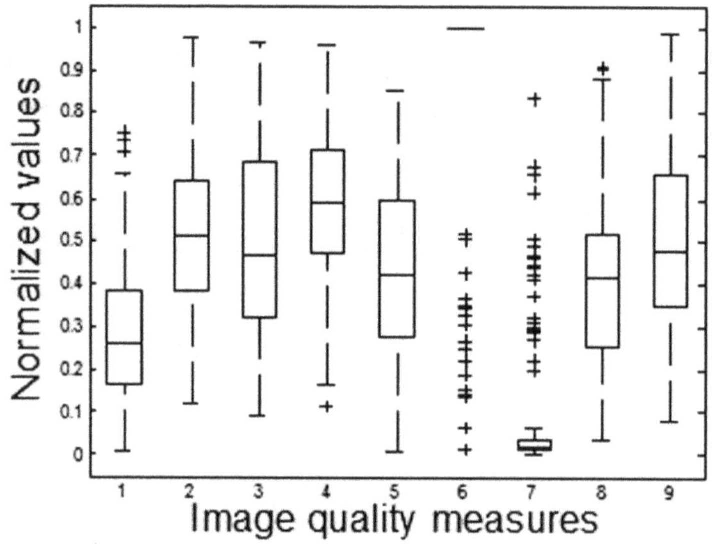

CONCLUSION

Diagnostic accuracy depends on true and false positive decisions. If a diagnostic method is capable of increasing the number of true positive diagnoses, it may also increase the number of false positives (Wenzel, 1988). In this chapter with the help of appropriate image enhancement scheme the true positives and false negatives can be increased thereby the identification and characterization of dental radiographs could be improved by the human observers.

Different image enhancement schemes few from spatial, frequency and spatial-frequency domain is attempted. The spatial domain filters attempted include median, morphological method based on top-hat and bottom-hat, CLAHE, cubic spline interpolation and anisotropic diffusion. The homomorphic filter is a frequency based approach. The combination of wavelet and SVD is a spatial-frequency domain filter.

Both qualitative and quantitative measures are implemented to reveal the best method among them. Results demonstrate that cubic spline image enhancement method is suited best for the dataset used in this chapter. Inferences from box plot reveal that cubic spline interpolation out performs other methods. This is also true since the major advantages of the metrics used in quantitative measures are its simplicity and mathematical compliance.

In addition to the human identification from dental radiographs and clinical inferences, because of the recent advancements in image processing procedures such as the image enhancement techniques the radiation dose or the frequent exposure to radiation may also be reduced substantially.

REFERENCES

Adatrao, S., & Mittal, M. (2016, October). An analysis of different image preprocessing techniques for determining the centroids of circular marks using Hough transform. In *Frontiers of Signal Processing (ICFSP), International Conference on* (pp. 110-115). IEEE. 10.1109/ICFSP.2016.7802966

Alginahi, Y. (2010). Preprocessing techniques in character recognition. In *Character Recognition*. In-Tech. doi:10.5772/9776

Banumathi, A., Vijayakumari, B., Geetha, A., Shanmugavadivu, N., & Raju, S. (2007). Performance analysis of various techniques applied in human identification using dental X-rays. *Journal of Medical Systems*, *31*(3), 210–218. doi:10.100710916-007-9057-0 PMID:17622024

Bhat, M., & Patil, T. (2014, May). Adaptive clip limit for contrast limited adaptive histogram equalization (CLAHE) of medical images using least mean square algorithm. In *Advanced Communication Control and Computing Technologies (ICACCCT), 2014 International Conference on* (pp. 1259-1263). IEEE.

Chen, H., & Jain, A. K. (2005, January). Dental biometrics: Alignment and matching of dental radiographs. In Application of Computer Vision, 2005. WACV/MOTIONS'05 Volume 1. Seventh IEEE Workshops on (Vol. 1, pp. 316-321). IEEE.

Chen, H., & Jain, A. K. (2005, January). Dental biometrics: Alignment and matching of dental radiographs. *IEEE Transactions on Pattern Analysis and Machine Intelligence*, *27*(8), 1319–1326. doi:10.1109/TPAMI.2005.157 PMID:16119269

Das, A. (2015). *Guide to signals and patterns in image processing: foundations, methods and applications*. Springer. doi:10.1007/978-3-319-14172-5

Deserno, T. M. (2010). Fundamentals of biomedical image processing. In *Biomedical Image Processing* (pp. 1–51). Berlin: Springer. doi:10.1007/978-3-642-15816-2_1

Dighe, S. C., & Shriram, R. (2012, November). Dental biometrics for human identification based on dental work and image properties in Periapical radiographs. In TENCON 2012-2012 IEEE Region 10 Conference (pp. 1-6). IEEE. doi:10.1109/TENCON.2012.6412216

Digital Dental Periapical X-Ray Database for Caries Screening. (n.d.). Vahab – LabArchives, Your Electronic Lab Notebook, Biomed Central Edition.

Eskicioglu, A. M., & Fisher, P. S. (1995). Image quality measures and their performance. *IEEE Transactions on Communications*, *43*(12), 2959–2965. doi:10.1109/26.477498

Gormez, O., & Yilmaz, H. H. (2009). Image post-processing in dental practice. *European Journal of Dentistry*, *3*(4), 343. PMID:19826609

Hall, E. L., Kruger, R. P., Dwyer, S. J., Hall, D. L., Mclaren, R. W., & Lodwick, G. S. (1971). A survey of preprocessing and feature extraction techniques for radiographic images. *IEEE Transactions on Computers*, *100*(9), 1032–1044. doi:10.1109/T-C.1971.223399

Hassanpour, H., Samadiani, N., & Salehi, S. M. (2015). Using morphological transforms to enhance the contrast of medical images. *The Egyptian Journal of Radiology and Nuclear Medicine*, *46*(2), 481–489. doi:10.1016/j.ejrnm.2015.01.004

Hemanth, D. J., & Anitha, J. (2012). Image Pre-processing and Feature Extraction Techniques for Magnetic Resonance Brain Image Analysis. In *Computer Applications for Communication, Networking, and Digital Contents* (pp. 349–356). Berlin: Springer. doi:10.1007/978-3-642-35594-3_47

Jain, A. K., & Chen, H. (2004). Matching of dental X-ray images for human identification. *Pattern Recognition*, *37*(7), 1519–1532. doi:10.1016/j.patcog.2003.12.016

Jain, A. K., Chen, H., & Minut, S. (2003, June). Dental biometrics: human identification using dental radiographs. In *International Conference on Audio-and Video-Based Biometric Person Authentication* (pp. 429-437). Springer. 10.1007/3-540-44887-X_51

Lee, J. S. (1980). Digital image enhancement and noise filtering by use of local statistics. *IEEE Transactions on Pattern Analysis and Machine Intelligence*, *PAMI-2*(2), 165–168. doi:10.1109/TPAMI.1980.4766994 PMID:21868887

Lin, P. L., Lai, Y. H., & Huang, P. W. (2010). An effective classification and numbering system for dental bitewing radiographs using teeth region and contour information. *Pattern Recognition*, *43*(4), 1380–1392. doi:10.1016/j.patcog.2009.10.005

Lin, P. L., Lai, Y. H., & Huang, P. W. (2012). Dental biometrics: Human identification based on teeth and dental works in bitewing radiographs. *Pattern Recognition*, *45*(3), 934–946. doi:10.1016/j.patcog.2011.08.027

Mirebeau, J. M., Fehrenbach, J., Risser, L., & Tobji, S. (2015). *Anisotropic diffusion in ITK*. arXiv preprint arXiv:1503.00992

Nageswararao, A. V., Srinivasan, S., Babu, S., & Priya, E. (2016, January). Multifractal Analysis Based Segmentation of Short Axis CMRI Images. In *2nd International Conference on Biomedical Systems, Signals and Images*. Department of Applied Mechanics, IIT Madras.

Nevriyanto, A., & Purnamasari, D. (2017, August). Image enhancement using the image sharpening, contrast enhancement, and Standard Median Filter (Noise Removal) with pixel-based and human visual system-based measurements. In *Electrical Engineering and Computer Science (ICECOS), 2017 International Conference on* (pp. 114-119). IEEE.

Panda, S. P. (2016, March). Image contrast enhancement in spatial domain using fuzzy logic based interpolation method. In *Electrical, Electronics and Computer Science (SCEECS), 2016 IEEE Students' Conference on* (pp. 1-4). IEEE. 10.1109/SCEECS.2016.7509315

Perona, P., & Malik, J. (1990). Scale-space and edge detection using anisotropic diffusion. *IEEE Transactions on Pattern Analysis and Machine Intelligence*, *12*(7), 629–639. doi:10.1109/34.56205

Priya, E., Srinivasan, S., & Ramakrishnan, S. (2014). Retrospective Non-Uniform Illumination Correction Techniques in Images of Tuberculosis. *Microscopy and Microanalysis*, *20*(5), 1382–1391. doi:10.1017/S1431927614012896 PMID:25115957

Pushparaj, V., Gurunathan, U., & Arumugam, B. (2013). An effective dental shape extraction algorithm using contour information and matching by mahalanobis distance. *Journal of Digital Imaging*, *26*(2), 259–268. doi:10.100710278-012-9492-4 PMID:22695751

Rad, A. E., Rahim, M. S. M., Rehman, A., & Saba, T. (2016). Digital dental X-ray database for caries screening. *3D Research*, *7*(2), 18.

Ramasubramanian, B., & Selvaperumal, S. (2016, April). A comprehensive review on various preprocessing methods in detecting diabetic retinopathy. In *Communication and Signal Processing (ICCSP), 2016 International Conference on* (pp. 0642-0646). IEEE. 10.1109/ICCSP.2016.7754220

Rehman, F., Akram, M. U., Faraz, K., & Riaz, N. (2015, April). Human identification using dental biometric analysis. In *Digital Information and Communication Technology and its Applications (DICTAP), 2015 Fifth International Conference on* (pp. 96-100). IEEE. 10.1109/DICTAP.2015.7113178

Seetha, D. M., Devi, G. M., & Sunitha, D. K. (2012, February). Performance Assessment of Feature Based Image Retrieval using Neural Networks. In Issue ICTM 2011 (Vol. 2). Academic Press.

Sharma, A., & Khunteta, A. (2016, September). Satellite Image Enhancement using Discrete Wavelet Transform, Singular Value Decomposition and its Noise Performance Analysis. In *Micro-Electronics and Telecommunication Engineering (ICMETE), 2016 International Conference on* (pp. 594-599). IEEE.

Vijayakumari, B., Ulaganathan, G., & Banumathi, A. (2011). An effective shape extraction algorithm for dental radiographs using contour information. *International Journal of Computer Science and Telecommunications*, *2*(4), 311–316.

Wang, Z. (2004). Image quality assessment from error measurement to structural similarity. *IEEE Transactions on Image Processing, 13*(4), 600-612.

Weickert, J. (1998). *Anisotropic diffusion in image processing* (Vol. 1). Stuttgart, Germany: Teubner.

Wenzel, A. (1988). Effect of image enhancement for detectability of bone lesions in digitized intraoral radiographs. *European Journal of Oral Sciences*, *96*(2), 149–160. doi:10.1111/j.1600-0722.1988.tb01422.x PMID:3162600

Wu, Q., Merchant, F., & Castleman, K. (2010). *Microscope image processing*. Academic press.

Yadav, G., Maheshwari, S., & Agarwal, A. (2014, September). Contrast limited adaptive histogram equalization based enhancement for real time video system. In *Advances in Computing, Communications and Informatics (ICACCI), 2014 International Conference on* (pp. 2392-2397). IEEE. 10.1109/ICACCI.2014.6968381

Zeng, H., Liu, Y. Z., Fan, Y. M., & Tang, X. (2012). An improved algorithm for impulse noise by median filter. *AASRI Procedia*, *1*, 68–73. doi:10.1016/j.aasri.2012.06.014

Zeng, P., Dong, H., Chi, J., & Xu, X. (2004, August). An approach for wavelet based image enhancement. In *Robotics and Biomimetics, 2004. ROBIO 2004. IEEE International Conference on* (pp. 574-577). IEEE.

Zhang, B., Fadili, J. M., & Starck, J. L. (2008). Wavelets, ridgelets, and curvelets for Poisson noise removal. *IEEE Transactions on Image Processing*, *17*(7), 1093–1108. doi:10.1109/TIP.2008.924386 PMID:18586618

Zhou, J. (2010). *Biometric recognition systems employing novel shape-based features* (Doctoral dissertation). University of Miami.

Zhou, Y., & Bai, J. (2007). Atlas-based fuzzy connectedness segmentation and intensity nonuniformity correction applied to brain MRI. *IEEE Transactions on Biomedical Engineering*, *54*(1), 122–129. doi:10.1109/TBME.2006.884645 PMID:17260863

ADDITIONAL READING

Abaza, A., Ross, A., & Ammar, H. (2009, November). Retrieving dental radiographs for post-mortem identification. In Image Processing (ICIP), 2009 16th IEEE International Conference on (pp. 2537-2540). IEEE. 10.1109/ICIP.2009.5414011

Akhoondali, H., Zoroofi, R. A., & Shirani, G. (2009). Fully automatic extraction of panoramic dental images from CT-scan volumetric data of the head. *Journal of Applied Sciences (Faisalabad)*, *9*(11), 2106–2114. doi:10.3923/jas.2009.2106.2114

Al-sherif, N., Guo, G., & Ammar, H. H. (2012, December). A new approach to teeth segmentation. In Multimedia (ISM), 2012 IEEE International Symposium on (pp. 145-148). IEEE. 10.1109/ISM.2012.35

Bowen, R. L. (1991). U.S. Patent No. 5,057,018. Washington, DC: U.S. Patent and Trademark Office.

Brooks, R. A., & Di Chiro, G. (1976). Beam hardening in x-ray reconstructive tomography. *Physics in Medicine and Biology*, *21*(3), 390–398. doi:10.1088/0031-9155/21/3/004 PMID:778862

Choorat, P., Chiracharit, W., & Chamnongthai, K. (2012, January). A single tooth segmentation using structural orientations and statistical textures. In *Biomedical Engineering International Conference (BMEiCON)*, 2011 (pp. 294-297). IEEE. 10.1109/BMEiCon.2012.6172074

Clarke, L. P., Qian, W., & Li, L. (1998). U.S. Patent No. 5,799,100. Washington, DC: U.S. Patent and Trademark Office.

Dighe, S., & Shriram, R. (2012). Preprocessing, segmentation and matching of dental radiographs used in dental biometrics. *International Journal of Science and Applied Information Technology*, *1*(2).

Dighe, S., & Shriram, R. (2012). Preprocessing, segmentation and matching of dental radiographs used in dental biometrics. *International Journal of Science and Applied Information Technology*, *1*(2).

Dobbins, J. T. III, & Godfrey, D. J. (2003). Digital x-ray tomosynthesis: Current state of the art and clinical potential. *Physics in Medicine and Biology*, *48*(19), R65–R106. doi:10.1088/0031-9155/48/19/R01 PMID:14579853

Dobbins, J. T. III, & Godfrey, D. J. (2003). Digital x-ray tomosynthesis: Current state of the art and clinical potential. *Physics in Medicine and Biology*, *48*(19), R65–R106. doi:10.1088/0031-9155/48/19/R01 PMID:14579853

Fonseka, M., Jayasinghe, R., Abeysekara, W. P. M. M., & Wettasinghe, K. A. (2013). Evaluation of the radiographic quality of roots filling, performed by undergraduates in the Faculty of Dental Sciences, University of Peradeniya, Sri Lanka. *International Journal of Research in Medical Sciences*, *1*(3), 12–16.

Göbel, P. M., Belbachir, A. N., & Truppe, M. (2005). Background removal in dental panoramic x-ray images by the a-trous multiresolution transform. In Circuit Theory and Design, 2005. Proceedings of the 2005 European Conference on (Vol. 1, pp. I-331). IEEE. 10.1109/ECCTD.2005.1522977

Goebel, P. M., Belbachir, A. N., & Truppe, M. (2005, July). Noise estimation in panoramic X-ray images: An application analysis approach. In Statistical Signal Processing, 2005 IEEE/SP 13th Workshop on (pp. 996-1001). IEEE.

Goebel, P. M., Belbachir, N. A., & Truppe, M. (2005, August). Blind background subtraction in dental panoramic x-ray images: an application approach. In *Joint Pattern Recognition Symposium* (pp. 434-441). Springer, Berlin, Heidelberg. 10.1007/11550518_54

Hart, D., Hillier, M. C., & Wall, B. F. (2009). National reference doses for common radiographic, fluoroscopic and dental X-ray examinations in the UK. *The British Journal of Radiology*, *82*(973), 1–12. doi:10.1259/bjr/12568539 PMID:18852213

Hart, D., & Wall, B. F. (2002). *Radiation exposure of the UK population from medical and dental X-ray examinations* (p. W4). Chilton, UK: NRPB.

Indraswari, R., Arifin, A. Z., Navastara, D. A., & Jawas, N. (2015, September). Teeth segmentation on dental panoramic radiographs using decimation-free directional filter bank thresholding and multistage adaptive thresholding. In Information & Communication Technology and Systems (ICTS), 2015 International Conference on (pp. 49-54). IEEE. 10.1109/ICTS.2015.7379870

Ji, Z., Dasgupta, D., Yang, Z., & Teng, H. (2006, July). Analysis of dental images using artificial immune systems. In Evolutionary Computation, 2006. CEC 2006. IEEE Congress on (pp. 528-535). IEEE.

Jongsma, F. H., Lambrechts, P., & Vanherle, G. (1983, July). Contouring of tooth imprints by means of a fluorescence technique adapted to a spatially filtered moire illumination. In *Biostereometrics' 82* (Vol. 361, pp. 76–81). International Society for Optics and Photonics. doi:10.1117/12.966001

Kelbauskas, E., Andriukaitiene, L., & Nedzelskiene, I. (2009). Quality of root canal filling performed by undergraduate students of odontology at Kaunas University of Medicine in Lithuania. *Stomatologija*, *11*(3), 92–96. PMID:19996675

Kolehmainen, V., Siltanen, S., Järvenpää, S., Kaipio, J. P., Koistinen, P., Lassas, M., ... Somersalo, E. (2003). Statistical inversion for medical x-ray tomography with few radiographs: II. Application to dental radiology. *Physics in Medicine and Biology*, *48*(10), 1465–1490. doi:10.1088/0031-9155/48/10/315 PMID:12812458

Lai, Y. H., & Lin, P. L. (2008, October). Effective segmentation for dental X-ray images using texture-based fuzzy inference system. In *International Conference on Advanced Concepts for Intelligent Vision Systems* (pp. 936-947). Springer, Berlin, Heidelberg. 10.1007/978-3-540-88458-3_85

Lakhani, K., Minocha, B., & Gugnani, N. (2016). Analyzing edge detection techniques for feature extraction in dental radiographs. *Perspectives on Science*, *8*, 395–398. doi:10.1016/j.pisc.2016.04.087

Li, H., Sun, G., Sun, H., & Liu, W. (2012, October). Watershed algorithm based on morphology for dental X-ray images segmentation. In Signal Processing (ICSP), 2012 IEEE 11th International Conference on (Vol. 2, pp. 877-880). IEEE.

Li, S., Fevens, T., Krzyżak, A., & Li, S. (2006). An automatic variational level set segmentation framework for computer aided dental X-rays analysis in clinical environments. *Computerized Medical Imaging and Graphics*, *30*(2), 65–74. doi:10.1016/j.compmedimag.2005.10.007 PMID:16500077

Lin, P. L., & Lai, Y. H. (2009, May). An effective classification system for dental bitewing radiographs using entire tooth. In Intelligent Systems, 2009. GCIS'09. WRI Global Congress on(Vol. 4, pp. 369-373). IEEE. 10.1109/GCIS.2009.390

Molteni, R. (1993). Direct digital dental x-ray imaging with Visualix/VIXA. *Oral Surgery, Oral Medicine, and Oral Pathology*, *76*(2), 235–243. doi:10.1016/0030-4220(93)90211-L PMID:8361738

Møystad, A., Svanaes, D. B., Risnes, S., Larheim, T. A., & Gröndahl, H. G. (1996). Detection of approximal caries with a storage phosphor system. A comparison of enhanced digital images with dental X-ray film. *Dento Maxillo Facial Radiology*, *25*(4), 202–206. doi:10.1259/dmfr.25.4.9084274 PMID:9084274

Mozzo, P., Procacci, C., Tacconi, A., Martini, P. T., & Andreis, I. B. (1998). A new volumetric CT machine for dental imaging based on the cone-beam technique: Preliminary results. *European Radiology*, *8*(9), 1558–1564. doi:10.1007003300050586 PMID:9866761

Nassar, D. E. M., & Ammar, H. H. (2003, May). A prototype automated dental identification system (ADIS). In *Proceedings of the 2003 annual national conference on Digital government research* (pp. 1-4). Digital Government Society of North America.

Niinimäki, K., Siltanen, S., & Kolehmainen, V. (2007). Bayesian multiresolution method for local tomography in dental x-ray imaging. *Physics in Medicine and Biology*, *52*(22), 6663–6678. doi:10.1088/0031-9155/52/22/008 PMID:17975290

Nomir, O., & Abdel-Mottaleb, M. (2005). A system for human identification from X-ray dental radiographs. *Pattern Recognition*, *38*(8), 1295–1305. doi:10.1016/j.patcog.2004.12.010

Nomir, O., & Abdel-Mottaleb, M. (2007). Human identification from dental X-ray images based on the shape and appearance of the teeth. *IEEE Transactions on Information Forensics and Security*, *2*(2), 188–197. doi:10.1109/TIFS.2007.897245

Nomir, O., & Abdel-Mottaleb, M. (2007). Human identification from dental X-ray images based on the shape and appearance of the teeth. *IEEE Transactions on Information Forensics and Security*, *2*(2), 188–197. doi:10.1109/TIFS.2007.897245

Nomir, O., & Abdel-Mottaleb, M. (2008). Fusion of matching algorithms for human identification using dental X-ray radiographs. *IEEE Transactions on Information Forensics and Security*, *3*(2), 223–233. doi:10.1109/TIFS.2008.919343

Nurtanio, I., Astuti, E. R., Ketut Eddy Purnama, I., Hariadi, M., & Purnomo, M. H. (2013). Classifying cyst and tumor lesion using support vector machine based on dental panoramic images texture features. *IAENG International Journal of Computer Science*, *40*(1), 29–37.

Pattanachai, N., Covavisaruch, N., & Sinthanayothin, C. (2012, May). Tooth recognition in dental radiographs via Hu's moment invariants. In Electrical Engineering/Electronics, Computer, Telecommunications and Information Technology (ECTI-CON), 2012 9th International Conference on (pp. 1-4). IEEE. 10.1109/ECTICon.2012.6254347

Pushparaj, V., Gurunathan, U., & Arumugam, B. (2012, December). Human forensic identification with dental radiographs using similarity and distance metrics. In India Conference (INDICON), 2012 Annual IEEE (pp. 329-334). IEEE. 10.1109/INDCON.2012.6420638

Pushparaj, V., Gurunathan, U., Arumugam, B., Baskaran, A., & Valliappan, A. (2013, July). An effective numbering and classification system for dental panoramic radiographs. In Computing, Communications and Networking Technologies (ICCCNT), 2013 Fourth International Conference on (pp. 1-8). IEEE. 10.1109/ICCCNT.2013.6726480

Rad, A. E., Mohd Rahim, M. S., Rehman, A., Altameem, A., & Saba, T. (2013). Evaluation of current dental radiographs segmentation approaches in computer-aided applications. *IETE Technical Review*, *30*(3), 210–222. doi:10.4103/0256-4602.113498

Rad, A. E., Rahim, M. S. M., Kumoi, R., & Norouzi, A. (2012). Dental x-ray image segmentation and multiple feature extraction. In 2nd world conference on innovation and computer sciences (pp. 10-14).

Rad, A. E., Rahim, M. S. M., & Norouzi, A. (2014). Level set and morphological operation techniques in application of dental image segmentation. World Academy of Science, Engineering and Technology, International Journal of Medical, Health, Biomedical. *Bioengineering and Pharmaceutical Engineering*, *8*(4), 182–185.

Raju, J., & Modi, C. K. (2011, June). A proposed feature extraction technique for dental x-ray images based on multiple features. In Communication Systems and Network Technologies (CSNT), 2011 International Conference on (pp. 545-549). IEEE. 10.1109/CSNT.2011.116

Said, E., Fahmy, G. F., Nassar, D., & Ammar, H. (2004, August). Dental x-ray image segmentation. In *Biometric Technology for Human Identification* (Vol. 5404, pp. 409–418). International Society for Optics and Photonics. doi:10.1117/12.541658

Said, E. H., Nassar, D. E. M., Fahmy, G., & Ammar, H. H. (2006). Teeth segmentation in digitized dental X-ray films using mathematical morphology. *IEEE Transactions on Information Forensics and Security*, *1*(2), 178–189. doi:10.1109/TIFS.2006.873606

Said, E. H., Nassar, D. E. M., Fahmy, G., & Ammar, H. H. (2006). Teeth segmentation in digitized dental X-ray films using mathematical morphology. *IEEE Transactions on Information Forensics and Security*, *1*(2), 178–189. doi:10.1109/TIFS.2006.873606

Syriopoulos, K., Velders, X. L., Sanderink, G. C., & van Der Stelt, P. F. (2001). Sensitometric and clinical evaluation of a new F-speed dental X-ray film. *Dento Maxillo Facial Radiology*, *30*(1), 40–44. doi:10.1038j.dmfr.4600575 PMID:11175272

Tohnak, S., Mehnert, A. J. H., Mahoney, M., & Crozier, S. (2007). Synthesizing dental radiographs for human identification. *Journal of Dental Research*, *86*(11), 1057–1062. doi:10.1177/154405910708601107 PMID:17959896

Versteeg, K. H., Sanderink, G. C., Velders, X. L., van Ginkel, F. C., & van der Stelt, P. F. (1997). In vivo study of approximal caries depth on storage phosphor plate images compared with dental x-ray film. *Oral Surgery, Oral Medicine, Oral Pathology, Oral Radiology, and Endodontics*, *84*(2), 210–213. doi:10.1016/S1079-2104(97)90071-8 PMID:9269024

Wang, C. W., Huang, C. T., Lee, J. H., Li, C. H., Chang, S. W., Siao, M. J., ... Fischer, P. (2016). A benchmark for comparison of dental radiography analysis algorithms. *Medical Image Analysis*, *31*, 63–76. doi:10.1016/j.media.2016.02.004 PMID:26974042

Yousefi, B., Hakim, H., Motahir, N., Yousefi, P., & Hosseini, M. M. (2012). Visibility enhancement of digital dental X-ray for RCT application using bayesian classifier and two times wavelet image fusion. *The Journal of American Science*, *8*(1).

Zhou, J., & Abdel-Mottaleb, M. (2005). A content-based system for human identification based on bitewing dental X-ray images. *Pattern Recognition*, *38*(11), 2132–2142. doi:10.1016/j.patcog.2005.01.011

KEY TERMS AND DEFINITIONS

Denoising: Denoising is a signal processing method that extract signal from a mixture of signal and noise thus preserving the useful information.

Dental Radiographs: Dental radiographs are x-rays of the teeth, bones, and soft tissues. It helps to find issues related to teeth, mouth, and jaw which cannot be seen during a visual examination.

Frequency Domain: Frequency domain refers to the analysis of mathematical functions on 1-D or 2-D signals with respect to frequency rather than time or spatial domain.

Histogram: An image histogram is a graphical representation that plots the number of pixels for each intensity value. It is also called as the intensity distribution of the image.

Image Enhancement: Image enhancement improves the quality of images for human perception by removal of noise, reduction of blurring, increase in contrast, and provides more detail information.

Image Pre-Processing: Pre-processing is a common name assigned for the basic operations with images at the lowest level of abstraction which aims at the improvement of image data that suppresses unwanted distortions or enhances some of the image features important for further processing.

Image Quality Measures: It is a characteristic of an image that measures the perceived image degradation when compared to an ideal or perfect image.

Non-Uniform Illumination: It is the low frequency intensity variations in the image caused by the sources of noise induced by the sensor being used to record the image.

Similarity Measures: It quantifies the similarity between two images and shows how self-similar an image is with the query or processed image.

Spatial Domain: Spatial resolution refers to the number of pixels required in construction of a digital image. The images having higher spatial resolution comprise with a greater number of pixels than those of lower spatial resolution.

Chapter 2
Thresholding Techniques for Dental Radiographic Images: A Comparative Study

Arockia Sukanya
Madras Institute of Technology, India

Kamalanand Krishnamurthy
Madras Institute of Technology, India

ABSTRACT

Imaging techniques play a major role in improving the early detection and diagnostic process that helps dentists to make accurate diagnosis. One of the most useful medical images used by dentists is radiographic image, which is used for the treatment of various dental disorders. Segmentation is a fundamental step as it involves separation of an image into regions corresponding to the objects. A simple and natural way to segment such regions is through thresholding. In this chapter, various thresholding techniques such as Otsu's method for global thresholding and Niblack's, Bersen's, and Sauvola's techniques for local thresholding are extensively explained with the help of dental radiographic images.

INTRODUCTION

Tooth is the hardest substances in the human body. Tooth plays an important role in chewing as well as in speech. There are various regions present in the teeth. These include enamel, dentin, pulp, cementum and ligaments. The white and firm part of the teeth is called enamel, which is made up of a chemical called calcium phosphate, a rock-hard mineral. A hard tissue, which is next layer to the enamel, is dentin that contains microscopic tubes (Rad, Rahim & Norouzi, 2013). A soft structure of the teeth where nerves and blood vessels pass through is called pulp. Cementum is a tissue that connects the gums, jaw bones and root of the teeth strongly. Structures surrounding the teeth, condition of the teeth and dental disorders like lesions which is difficult to identify medically can be easily detected using dental imaging.

DOI: 10.4018/978-1-5225-6243-6.ch002

Dental disorders are considered to be a major problem in recent days. Some of the common dental problems faced by humans are tooth decay, oral cancers, tooth erosion etc. There are various techniques to diagnose dental problems and the most commonly used technique is dental radiography (Amer & Aqel, 2015; Rad, Rahim & Norouzi, 2013; Said, Fahmy, Nassar & Ammar, 2004). Dental X- ray segmentation is important and necessary in medical diagnosis, which gives information to the doctors regarding dental diseases. Treatments such as dental crown, restoration and root canal are most commonly used diagnostic techniques. There are various modalities for diagnosing these disorders. One such modality is diagnosing it with the help of dental radiographs. Dental radiography is one of the most useful dental exams performed by dentists as it allows the identification of structures which are not revealed on the surfaces of the teeth.

A dental radiography is a technique where the production of radiographs of the teeth and adjacent structures are obtained due to exposure to x-rays (Rad, Rahim & Norouzi, 2013). A dental image provides information during dental procedures such as root canal therapy, placement of dental implants (Ribeiro, Dias, de Best, da Silva, Neves & Street, 2014). Dental imaging allows instant and easy transmission of images and electronic storage. They are used to detect wounds, disorders and surroundings of teeth and other parts of the mouth which is not identified medically. Segmentation of dental x-rays does not work well by using traditional techniques (Lai & Lin, 2008). In past decades, continuous evolution in radiological image lead to the development of image processing for medical images, which became a popular tool for researchers in medical fields, includes Dentistry. In the field of computer vision, digital images acts as a medium for carrying information (Subramanyam, Prasad & Anuradha, 2014). In the field of dentistry, segmentation is used to identify various structures such as jaws, wounds, abscess, bones etc (Subramanyam, Prasad & Anuradha, 2014).

Conversion from binary image to scanned grayscale image, while removing the background and preserving the target is an important step in image segmentation (Graham Leedham, Takru, Tan & Mian, 2003). It segments an image into separate regions or objects. Image understanding and content analysis is a critical step in segmentation of an image (Senthilkumaran & Vaithegi, 2016). The process of dividing an image with respect to some criterion, into group of pixels which are similar is called segmetation. Each segment should not intersect with each other, and adjacent segments must be heterogeneous. Thus segmentation deals with splitting an image into meaningful regions (Jayaraman, 2009). The first step in pre-processing of images is done by image segmentation where the objects of interest are extracted for further analysis (Sukhdeep Kaur, Manjit Sandhu & Jaipreet Kaur, 2016). Edge-based and region based approaches are two types on which segmentation is classified (Senthilkumaran & Vaithegi, 2016). Discontinuity and similarity are two properties on which the image segmentation is categorized Segmentation based on similarity is called Region method and on discontinuities is called as boundary method (Subramanyam, Prasad & Anuradha, 2014). In recent years, entropy based approaches have also gained significant interest for solving problems in biomedical image and signal processing (Rajinikanth, Raja, & Kamalanand, 2017; Alagumariappan, Krishnamurthy, Kandiah, & Ponnuswamy, 2017; Kamalanand, & Ramakrishnan, 2015; Alagumariappan, Rajagopal, & Krishnamurthy, 2016; Ambikapathy, & Krishnamurthy, 2018).

CLASSIFICATION OF IMAGE SEGMENTATION

Segmentation of image is classified broadly into two types:

- Local segmentation method;
- Global segmentation method.

Segmentation of a whole image into number of small regions or windows is called local segmentation (Firdousi & Parveen, 2014; Chaubey, 2016; Senthilkumaran & Vaithegi, 2016; Subramanyam, Prasad & Anuradha, 2014). A complete image is considered for the segmentation called global segmentation. As the segmentation is done for whole image, a large number of pixels are being considered.

Group of pixels which has similar property in an image is known as region. In a particular object or region a pixel is being assigned. Boundaries between regions are located directly in boundary based approach. Edges are identified and boundaries are formed in edge based detection technique (Subramanyam, Prasad & Anuradha, 2014).

Among these segmentation techniques available, a region based approach called thresholding technique is considered to be one of the simplest and extensively used techniques.

BACKGROUND

Biomedical image segmentation plays a major role in recent years. There are various ways to segment an image. One among them is using thresholding techniques. The simplest form of segmentation of image is thresholding. It is broadly classified into two categories, namely Global and Local thresholding. Number of thresholding methods has been developed in recent years based on both types. In global thresholding only one threshold value is applied to entire image while distinct threshold applied to distinct region is called local threshold. The thresholding is applied to the image which is determined by the pixel's neighborhood. Recent research on thresholding techniques has delivered better results in segmentation of images. Wang *et al.* (2016) have compared various dental radiography algorithms. The authors have concluded that automatic segmentation and analysis of dental radiographs is a challenging task. Rad *et al.* (2012) worked on implementation different segmentation techniques on dental x-rays and some of the similarity measures have also been found. Lai *et al.* (2008) developed a method, where segmentation of teeth from dental x-rays has been done using fuzzy entropy based technique.

Harandi *et al.* (2011) worked on segmenting the upper and lower jaws from x- rays using morphological techniques and active contours, and found that obtained results are performing well in the process of segmentation. Tuan, (2016) worked on a method called semi supervised fuzzy entropy technique which has been implemented to find the edges of the teeth from x- rays. Subramanyam *et al.* (2014) worked on segmentation of medical image using thresholding techniques, where the authors have performed

Figure 1. Classification of Image Segmentation methods

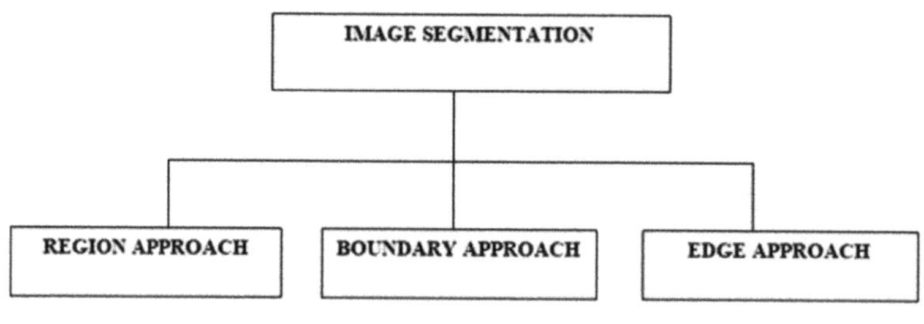

a detailed comparison of Sauvola and Niblack local thresholding algorithm on medical images. These techniques aim at removal of the noises from the background of the images. Comparing these two techniques Niblacks method performs well in the noise removal. The performance of these methods has been compared with standard segmentation parameters.

Chaubey, (2016) presented an approach to segment mammogram images and to extract texture features present in that image by comparing both local and global thresholding methods. The author concluded that there are two factors namely correctness and stability which is used for the segmentation of larger objects from image and the obtained results performed better segmentation. Graham Leedham *et al.* (2003) worked on comparing some of the thresholding methods, where various thresholding techniques are used to split the target and background from an image and concluded that this method outperforms the others where the illumination is not constant for the images. Firdousi *et al.* (2014) worked on comparison of various local thresholding techniques. The aim is to remove background noises form the image documents. The authors have compared various local thresholding algorithms and also presented a new binarization algorithm.

Otsu, (1979) proposed a technique where to maximise the seperability of resultant classes in gray level, a best threshold value is selected. The proposed method is simple and straightforward and it is extended for the use of multilevel thresholding. Tao *et al.* (2003) worked on multilevel segmentation of image using fuzzy entropy and genetic algorithm. Here, three regions from an image is considered and a fuzzy function is assigned for individual regions and based on these function an optimal fuzzy function has been identified and it is used for further analysis. The proposed algorithm produce better results and performance is also better. Similarly the authors have worked on object segmentation using fuzzy entropy and ant colony optimization algorithm. This method performs better segmentation among the different entropy-based thresholding techniques available. Garg, (2013) proposed a method where a comparison on existing various thresholding methods and their measurements are done. The author concluded that global thresholding method is better for calculating threshold values of a gray scale images but it gives poor results for color images due to illumination while using local thresholding it is eliminated.

Subramanyam *et al.* (2014) proposed different segmentation techniques available for extraction of dental image. These methods are implemented on dental radiographs. Kaur *et al.* (2016) using MATLAB analysis on various image segmentation techniques was done. The segmentation is carried out using MATLAB software and operations were performed on each pixel and the objects were analyzed in the best manner. Manic *et al.* (2016) proposed a work using Kapur/Tsallis entropy and firefly algorithm for multi thresholding of an image where the authors have proposed a method for maximizing the entropy values and found that it provides better convergence values and performance values are also better. Singh *et al.* (2012) worked on thresholding techniques in image binarization using on local contrast and mean method. Here different local thresholding methods have been compared so that it could be used for both color and grayscale images based on the intensities and the results obtained were perfect.

These were some of the work done by various researchers on different types of thresholding techniques. Each method gives different ways to perform segmentation and to extract different region of interests. These thresholding techniques were implemented on dental radiographic images and found that the various regions present in the images were segmented and variations from the background and object are also obtained.

IMAGE SEGMENTATION BASED ON THRESHOLDING

The segment, which has pixels with similar intensity, is obtained by thresholding technique. In this technique, the boundaries of an image are obtained by separating solid objects resting on contrasting background. Different types of segmentation procedures are available to separate the object from its background. Among those thresholding is simplest and easiest way (Singh, Roy, Singh, Sinam, & Singh, 2012). In this technique, the region which is to be segmented should be of homogeneous intensity and a back-ground should be of different intensity level. Hence, this type of image can be segmented into two distinct parts by simple thresholding.

The thresholds are given by,

$$T = T\left[x, y, p\left(x, y\right), f\left(x, y\right)\right] \tag{1}$$

where, T represents the threshold value. x,y are the coordinates of the threshold value. P(x,y), f(x,y) are the gray level points. An image's threshold g(x,y) is defined as

$$y = \begin{cases} 1 iff\left(x, y\right) > 1 \\ 0 iff\left(x, y\right) \leq 0 \end{cases} \tag{2}$$

The thresholding of grayscale images is divided into two categories namely local thresholding and global thresholding. Aimed at the whole image, only one threshold value is calculated (Garg, 2013). Each pixel is allotted to image foreground or background. Simple thresholding schemes compare a single global threshold with each pixels gray level. The value of Local threshold T depends on both p(x,y) and f(x,y) (Amer & Aqel, 2015).

VARIOUS TYPES OF THRESHOLDING TECHNIQUES

See Figure 2.

Figure 2. Classification of Thresholding Procedures

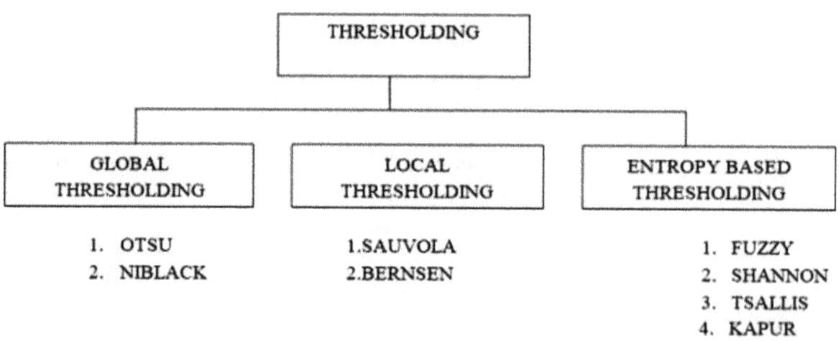

GLOBAL THRESHOLDING

Global thresholding method also called single thresholding is used where the distribution of intensities is discrete between the objects of background and foreground. A single threshold is used to differentiate the objects apart, where the difference between image foreground and background is distinct. Here, threshold depends purely on the gray level value of the image. Thus, the value of threshold T depends purely depends on the grey level value and pixel of the image (Firdousi & Parveen, 2014; Amer & Aqel, 2015). It is simple and most widely used of all possible segmentation methods. The threshold value Θ for global thresholding is chosen and the following condition is applied:

$$f\left(m,n\right)=\begin{cases}1 & iff\left(m,n\right)\geq\theta \\ 0 & else\end{cases} \tag{3}$$

The above equation is the complete description of a binarization algorithm; it does not indicate on selection of the threshold parameter Θ. The value of Θ has to be selected in an optimal manner. It is simplest and most widely used technique. The disadvantage of this method is that it is difficult to choose the threshold, when pixels of the segment overlap within the same intensity value (Firdousi & Parveen, 2014; Amer & Aqel, 2015). This overlap occurs due to noise, which is reduced by minimum-error technique. It evaluates the essential parameters and chooses the thresholds to minimalize the classification errors. Variable thresholding can be used if the overlap occurs due to variation in illumination across image (Firdousi & Parveen, 2014; Amer & Aqel 2015).

ADAPTIVE THRESHOLDING

Global thresholding works well for those images which has two distinct regions. Many cases this may not be suitable for all conditions where object's contrast vary. In such case threshold chosen at one region might not work for other regions. Hence to overcome this condition, adaptive thresholding is used wher the threshold is chosen as a changing function of position in the image (Firdousi & Parveen, 2014; Amer & Aqel 2015; Garg, 2013; Izzaty & Hisham, 2015).

HISTOGRAM BASED THRESHOLD SELECTION

When an object is resting on a contrasting back-ground, a bimodal histogram is formed. Two peaks are obtained where each peak corresponds to comparatively large number of pixels inside and outside the object (Annadurai, 2007). The difference between the peaks corresponds to the few points around the boundary of the object (Amer & Aqel, 2015). This difference is generally used to identify the threshold gray level (Firdousi & Parveen, 2014; Amer & Aqel, 2015).

The histogram is the derivative of the area function for an object whose boundary is defined by thresholding:

$$H(D) = -\frac{d}{dD}A\left(D\right) \tag{4}$$

Where D represents the gray level, A (D) represents the area of the object obtained by thresholding at gray level D, and H(D) represents the histogram. If the threshold increases from D to D+ Δ D., a slight decrease in area is found if D is near the dip of the histogram. The threshold is placed at the dip of the histogram to minimize errors (Annadurai, 2007).

Apart from these methods which are commonly used some of the other techniques namely Otsu's method, Iterative and Multistage method for global thresholding and Niblack's technique, Bersen's technique, Sauvola's technique for local thresholding, and some entropy based techniques such as Tsallis, Fuzzy, Kapurs etc is extensively explained with the help of dental radiographic images.

OTSU'S METHOD

Otsu method, entropy-based thresholding, etc are some of the commonly used global thresholding methods. Among these, popular thresholding technique is Otsu's algorithm.

In 1979, Otsu's image thresholding was coined. The best values are obtained by maximizing the between-class variance function. The computation is fast as it functions directly on the gray level histogram (Otsu, 1979). Some of the application of this method is in the binarization of document pattern identification etc. This technique is used for pre-processing of image where further quantification and feature analysis of images done (Li, Huang, Ding, Gatenby, Metaxas & Gore, 2011). Here, the threshold value is chosen such that intra-class variances is minimized for the segmentation of an image. Two different peaks from the histogram is obtained where, peaks correspond to the target and background of the image (Li, Huang, Ding, Gatenby, Metaxas & Gore, 2011). The threshold is obtained by minimalizing the intra class variance. The class which has the maximum variance is used in determining the thresholds (Otsu, 1979; Bhandari, Kumar & Singh, 2015). Otsu's method creates suboptimal results. By satisfying some conditions, the optimization procedure is utilized to identify possible and best threshold in the histogram (Otsu, 1979; Bhandari, Kumar & Singh, 2015).

Let L be the intensity value in the range {0,1,2, ...L-1}, for a given RGB image. The probability distribution function P_i^C is defined as:

$$P_i^C = \frac{h_i}{N}\sum\nolimits_{i=0}^{L-1} p_i\mathrm{C} = 1 \tag{5}$$

where, i= specific intensity ranges from $\{0 \leq i \leq L\text{-}1\}$,C = {R,G,B}N = total number of pixels in the image, andh_i^C=number of pixels for the particular intensity level I in component C.

The total mean of each element of the image is computed as:

$$\mu_T^C \sum\nolimits_{i=0}^{L-1} ip\ P_i^C = 1$$

The *m* - level thresholding displays *m-1*threshold levels $t_j^C$where, the value of *j*= 1,2,…,*m-1*, and the action is performed as:

$$F^C \text{ x,y)=} \begin{cases} 0 \ , f^C(x,y) \le t_1^C \\ \frac{1}{2}\left(t_1^C + t_2^C\right) t_1^C f^C(x,y) \le t_2^C \\ . \\ . \\ . \\ \frac{1}{2}\left(t_{m-2}^{\ \ C} + t_{m-1}^{\ \ C}\right) t_{m-2}^{\ \ C} f^C(x,y) \le t_{m-1}^{\ \ C} \\ L-1 \ , f^C(x,y) > t_{m-1}^{\ \ C} \end{cases} \tag{6}$$

where, *y* and *x* are the height (H) and width (W), in pixels, of the image of size W × H represented by *f C (x, y)* with *L* intensity levels for every element.

The possible outcomes W_j^C of classes D_i^C………….D_m^C are denoted by:

$$w_j^{\ C} = \begin{cases} \sum_{i=0}^{t_j^C} p_i^{\ C} j = 1 \\ \sum_{i=0}^{L-1} t_{j-1}^{\ \ C} + p_i^{\ C} j = m \\ \sum_{i=}^{t^c_j} t_{j-1}^{\ \ C} + p_i^{\ C} 1 < j < m \end{cases} \tag{7}$$

For each class the mean μ_j^C can then be calculated as:

$$\mu_j^{\ C} = \begin{cases} \sum_{i=0}^{t_j^C} \frac{p_i^{\ C}}{w_j^{\ C} q} j = 1 \\ \sum_{i=t_j^C-1}^{t_j^C} +1 \frac{p_i^{\ C}}{w_j^{\ C} q} 1 < j < m \\ \sum_{i=t^c_j-1}^{L-1} +1 \frac{p_i^{\ C}}{w_j^{\ C} q} j = m \end{cases} \tag{8}$$

In Otsu's technique each component's between class variance of can be given as:

$$\sigma_{\mathrm{B}}^{c^2} = \sum_{j=1}^{m} w_j^C \left(\mu_j^C - \mu_T^C\right)^2 \tag{9}$$

where, w_j^C is the possibility of occurrence, and μ_j^C is the mean value

Lastly, the m – level thresholding is an optimization problem for finding t_j^C, which maximizes the cost function of every image component C, and is given by:

$$_{\phi}C = \max_{1<t_i^C<\ldots L-1} \sigma_B^{c^2}(t_j^C) \tag{10}$$

Hence, the threshold is calculated

1. According to the threshold, the pixels are separated into two clusters.
2. Mean is calculated for individual cluster.
3. Squared mean difference is obtained.
4. Multiplication of number of pixels with one cluster to another.

The optimal threshold value is one that maximizes the between-class variance (or, on the contrary, minimizes the within-class variance).

NIBLACK'S METHOD

Niblack's procedure computes threshold using local standard deviation and means (Gayathri, Menon & Viswa, 2014; Metaxas & Gore, 2011; Senthilkumaran & Kirubakaran, 2014). The threshold is obtained by the equation:

$$T(x,y) = m(x,y) + k \circ s(x,y) \tag{11}$$

where, m(x,y) and s(x,y) are the average of a local area and standard deviation values, respectively. To preserve local particulars and to suppress noise size of the neighborhood should be small. The value of k is used to adjust how much of the total print object boundary is taken as a portion of the assumed object (Naidu, Kumar, & Chiranjeevi, 2017). An enhanced version of the Niblack's algorithm:

$$T(x,y) = m(x,y) \circ \left[1 + k \circ \left(1 - \frac{s(x,y)}{R}\right)\right] \tag{12}$$

where, k and R are empirical constants. In order to preserve the local details of an image, the size of the neighborhood should be of small in size, and at the same time it should be large enough to suppress noise. To determine how much of the total print object boundary is taken as a part of the given object the value of k is adjusted. The improved Niblack's technique utilizes parameters k and R to reduce its sensitivity to noise (Senthilkumaran & Vaithegi, 2016).

SAUVOLA'S TECHNIQUE

In this technique, the threshold T(x,y) is computed by considering standard deviation $\delta\left(x,y\right)$ and mean m(x,y) of the pixel intensities in a window w x w centered around the pixel at (x,y)and expressed as:

$$T\left(x,y\right)=m\left(x,y\right)\left[1+k\frac{\delta\left(x,y\right)}{R}-1\right]\tag{13}$$

This technique is improved version of Niblack's technique to add more local deviation to the mean value. The value R eliminates the effect of local deviation. Where R is the maximum value of standard deviation and k is a constant which takes positive values (Naidu, Kumar & Chiranjeevi, 2017). In this technique, dark region of the background is removed when the threshold T(x,y) goes below the mean value .

BERNSEN'S TECHNIQUE

This is a local binarization technique proposed by Bernsen, where the threshold value is determined using local contrast value (Chaubey, 2016). For each pixel (x,y) local threshold is computed by the expression

$$\text{T(x,y)}=\frac{\text{Imax}+\text{Imin}}{2}\tag{14}$$

Wherein a w x w window centered at (x,y) respectively I_{min} and I_{max} are the minimum and maximum gray level value, based on the local contrast value the threshold is assigned and hence it can be given as (Garg, 2013),

$$\text{T(x,y)=GT}\tag{15}$$

Here, local maximum and minimum pixel values are related. It is not appropriate integral image. The computational time is dependent on local window size, which works like other local techniques.

ENTROPY BASED THRESHOLDING

Segmentation process is assessed based on decrease the homogeneity across the regions and increase the homogeneity of pixels within each segmented region. Entropy is a measure of spontaneous dispersal of energy that is amount of energy spread during a process (Bhandari, Kumar & Singh, 2015). The term entropy comes in second law of thermodynamics. The efficiency of information transferred is measured using entropy (Tao, Tian & Liu, 2003). Different types of entropy-based techniques are explained along with the mathematical equations.

Fuzzy Entropy

It is an entropy function, where the sharpness of argument fuzzy set is enhanced when the fuzzy sets become smaller. It was introduced by Luca and Termini (De Luca & Termini, 1972; Bhandari, Kumar & Singh, 2015). Without the statistical impression, fuzzy set plays an important role in many processes (Tao, Tian & Liu, 2003; Prasad, Krishna & Reddy, 2013).

Let D = {(i,j) =0,1,....M-1; j=0,1,....N-1} and G={0,1,....l-1], where M,N and l are three positive integers. Let I(x,y) be the gray level value of the image at the pixel (x,y) and D_k={(x,y):I(x,y)=k,(x,y) $\in D$., k=0,1,....l-1.

Let T be the threshold, that segments an image target and other parts. Domain D of the considered image is divided into two parts E_d and E_b. E_d is pixels with low gray levels and E_b is pixels with high gray levels. π.={E_d,E_b} is an unidentified probabilistic partition of D, whose probability distribution is defined as

$$p_d=P(E_d),\ p_b=P(E_b) \tag{16}$$

The membership function has three parameters a, b and c based on these values the threshold depends. For each k=0,1,....255

$$D_{kd}=\{(x,y):I(x,y)\leq T,(x,y)\in ._k\} \tag{17}$$

$$D_{kb}=\{(x,y):I(x,y)>,(x,y)\in ._k\} \tag{18}$$

The probability equations are:

$$p_{kd}=P(D_{kd})=p_k{*}P_{dlk} \tag{19}$$

$$p_{kb}=P(D_{kb})=p_k{*}P_{blk} \tag{20}$$

Shannon's Entropy

Claude Shannon introduced an entropy constructed on the Boltzmann's H theorem and is named as Shannon entropy (Gayathri, Menon & Viswa, 2014). Rate of compression is high in this method. Speed of transmission is also high. Reliant on the number of repetitive information or message, number of bits required is compressed (Manic, Priya & Rajinikanth, 2016; Tao, Tian & Liu, 2003; Prasad, Krishna & Reddy, 2013; Gayathri, Menon & Viswa, 2014; Metaxas & Gore, 2011; Senthilkumaran & Kirubakaran, 2014; Bhandari, Kumar & Singh, 2015). Let X is discrete random variable with elements {X_1, X_2 ...,X_n}, then probability mass function P(X) is expressed by:

$$H\left(X\right) = E\left[I\left(X\right)\right] = E\left[-\ln\left(P\left(X\right)\right)\right]. \quad (21)$$

where, E is the anticipated value operator, I is the information content and I(X) is a random variable (Tao, Tian & Liu, J, 2003; Prasad, Krishna & Reddy, 2013; Gayathri, Menon & Viswa, 2014; Metaxas & Gore, 2011; Senthilkumaran & Kirubakaran, 2014; Bhandari, Kumar & Singh, 2015).

$$H\left(X\right) = \sum_{i-1}^{n} P(x)I(x) = -\sum_{i=1}^{n} P(x)logP(x_i) \quad (22)$$

where, *b* base of the algorithm and it is equal to 2. If P (x_i) = 0, then

$$\lim_{p\to 0-1} plog\left(p\right) = 0 \quad (23)$$

The above equation is discrete for X and some of them are appropriate for continuous function by replacing it by summation to integration.

Tsallis Entropy

A new thresholding method called Tsallis entropy-based function was proposed. Tsallis entropy function is adopted as the objective function during the multi-level thresholding process. Tsallis is an entropy based new thresholding method is proposed (Prasad, Krishna & Reddy, 2013; Martınez, Nicolás, Pennini & Plastino, 2000; Zhang & Wu 2011). When applied to noisy images this method produces good result (Manic, Priya & Rajinikanth, 2016). This method uses the objective property and global of the histogram (Prasad, Krishna, & Reddy, 2013; Martınez, Nicolás, Pennini & Plastino, A. 2000; Zhang& Wu 2011). Let L represents gray levels in a given image and these gray level ranges from {0, 1, 2, …… (L-1)}. Let h(i) represents number of pixels corresponding to the gray-level i and N represents total number of pixels in the image given by:

$$N = \sum_{i=0}^{L-1} h(i) \quad (24)$$

and

$$P_i = \frac{h(i)}{N} \quad (25)$$

where P_i is an estimation of the probability of occurrence of intensity level *i* in the image.

The multilevel thresholding based on Tsallis objective function is given by:

$$f(t) = \arg\max[S_q^A(t) + S_q^B(t) + \ldots (1-q).S_q^A(t).S_q^B(t)] \quad (26)$$

where:

$$S_q^A(t) = \frac{1-\sum_{i=0}^{t_1-1}\left(\frac{P_i}{P^A}\right)^q}{q-1} \quad ; P^A = \sum_{i=0}^{t_1-1} P_i \, ; \tag{27}$$

$$S_q^B(t) = \frac{1-\sum_{i=t_1}^{t_2-1}\left(\frac{P_i}{P^B}\right)^q}{q-1} \quad ; P^B = \sum_{i=t_1}^{t_2-1} P_i \, ; \tag{28}$$

$$S_q^M(t) = \frac{1-\sum_{i=t_M}^{L-1}\left(\frac{P_i}{P^M}\right)^q}{q-1} \quad ; P^M = \sum_{i=t_M}^{L-1} P_i \, ; \tag{29}$$

In the above equations, t_1, t_2,.....,t_M represent threshold levels. Further, $t_1 t_2 < ... < t_{M..}$. Tsallis objective function is obtained by o maximizing equation (25).

Kapur's Entropy

In 1985, Kapur's entropy was found in order to segment the gray scale image by maximizing the entropy of histogram (Bhandari, Kumar & Singh, 2015). Let, Th = [th_1, th_2, ..., th_{k-1}] is a vector of the image thresholds. Then, the Kapur's entropy is given by:

$$J_{\max} = f_{kapur}(Th) = \sum_{j=1}^{k} H_j^c \tag{30}$$

Generally, each entropy is computed independently. Hence, for multi-level thresholding problem, it is given by:

$$H_1^c = \sum_{j=1}^{th_1} \frac{Ph_j^c}{\omega_0^c} \ln\left(\frac{Ph_j^c}{\omega_0^c}\right) \tag{31}$$

$$H_2^c = \sum_{j=th_1 1}^{th_2} \frac{Ph_j^c}{\omega_1^c} \ln\left(\frac{Ph_j^c}{\omega_1^c}\right) \tag{32}$$

$$H_k^c = \sum_{j=th2+1}^{L} \frac{Ph_j^c}{\omega_{k-1}^c} \ln\left(\frac{Ph_j^c}{\omega_{k-1}^c}\right) \tag{33}$$

Where Ph_j^C is the probability distribution of the intensity levels, C is unity (1) for gray level images and $w_0^c, w_0^c, \ldots w_{k-1}^c$ probability occurrence for k levels. The FA based exploration randomly adjusts the values of threshold until J_{max} is reached (Bhandari, Kumar & Singh, 2015).

Figure 3 shows a sample dental radiograph which is obtained from the data set. Using this radiographic image some of the recent thresholding techniques have been implemented and is shown in the figure. A gray scale image is segmented into many distinct regions which is called multilevel thresholding. Here, for a single image several thresholds are chosen which is applied for segmentation of an image. Segments are formed by grouping the pixels which has same intensity. Multilevel thresholding works well for objects with complex and color background, where results produced by bi-level thresholding are not satisfactory (Arora, Acharya, Verma & Panigrahi, 2008). Here, to these sample dental radiographs various entropy-based thresholding methods have been implemented. Multilevel value ranges from 2 to 5 have been chosen for performing segmentation. The results are shown in the figure where comparisons of different techniques with various thresholding levels have been implemented.

The above figure shows the application of entropy techniques for thresholding of dental radiographs. Results demonstrate that entropy based techniques are performing well in the segmentation of an image. Multi-level thresholding have been implemented in the dental radiographs. Some of the well-known thresholding techniques such as fuzzy entropy, Shannon's entropy and kapur's entropy have been implemented on the dental radiographs.

ADVANTAGES AND LIMITATIONS OF THRESHOLDING TECHNIQUES

The threshold is obtained based on histogram of the image. It does not need any former information about the image and is the simplest of all the methods. It is inexpensive in terms of computation. Further, it is simple and fast in terms of implementation. Most of the real time application uses this technique. Thresholding does not provide any relationship between pixel values. It mainly depends on the intensity of an image which is a major drawback of this method. Spatial information of an image is been neglected, thus it cannot guaranty the segmented regions are contiguous. Sensitivity to noise is high in this technique.

Figure 3. A sample Dental Radiograph (https://grand-challenge.org/)

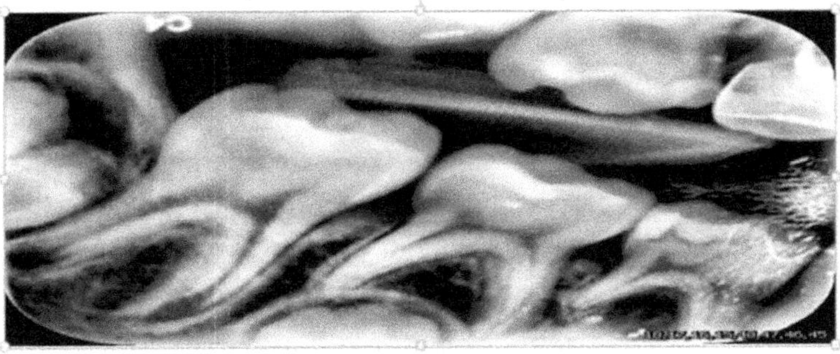

Figure 4. Multilevel thresholding of radiographs for entropy-based thresholding methods

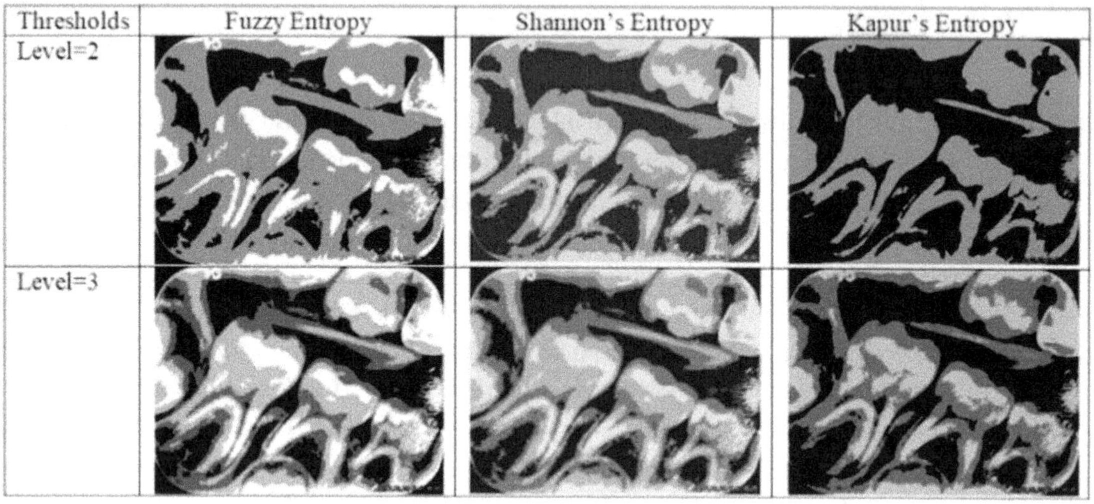

Table 1. Performance measures of Fuzzy, Kapur and Shanon Entropy based thresholding techniques

Technique	Th	MSE	RMSE	PSNR	NCC	AD	SSIM	MD	NAE
FUZZY	2	1450.6	38.0872	16.5152	0.9738	7.3596	0.9767	67	0.2767
	3	911.6259	30.1931	18.5326	1.0546	1.4936	0.8641	63	0.2058
KAPUR	2	4774.9	69.1006	11.3412	0.5789	59.5329	2.4453	128	0.4930
	3	2378.6	48.7709	14.3676	0.7234	41.8177	1.7487	91	0.3463
SHANON	2	595.0808	24.3943	20.3850	0.9674	0.0493	1.0352	48	0.1738
	3	328.2652	18.1181	22.9686	0.9816	0.0067	1.0199	35	0.1284

Improper selection of threshold will lead to over or under segmentation as selecting a threshold value is a crucial process. Thus proper care should be taken in the selection of thresholds.

FUTURE RESEARCH DIRECTIONS

In the field of biomedical, thresholding techniques are used for applications such as diagnosis and for further analysis. There are various methodologies available on thresholding techniques based on both segmentation methods namely, local and global thresholding (Chaubey, 2016; Graham Leedham, Takru, Tan & Mian, 2003; Otsu,1979; Subramanyam, Prasad, & Anuradha, 2014; Firdousi & Parveen, 2014; Senthilkumaran & Vaithegi, 2016; Prasad, Krishna & Reddy, 2013) . Each technique has its own pros and cons. Among this entropy based thresholding methods are emerging in recent times and the results obtained are also perfect for further processing. The results obtained can be further enhanced by using some of the optimization techniques such as PSO, Firefly algorithm, Bat algorithm and BFO etc.

CONCLUSION

Extraction of information from an image without the assistance of humans refers to image processing. There are different steps available in processing of an image. Among the methods available, an important step considered for processing of image is called segmentation. It divides the image into various parts and analysis is made on theses separated regions. In the field of medicine, segmentation acts a major role, as segmenting a medical image is a tedious process. Dental radiograph segmentation becomes difficult as identifying the affected regions and segmenting is a time consuming process. Many researchers have found number of ways to perform segmentation. Segmentation of image is classified into different types (Amer & Aqel, 2015; Chaubey, 2016). Among those, thresholding technique is a preliminary method in image segmentation process (Firdousi & Parveen, 2014; Garg, 2013). In this chapter different thresholding techniques based on both local and global methods have been discussed along with the mathematical equations (Pal & Pal, 1993; Sahoo, Soltani & Wong, 1988). A detailed comparative study has been made for both local and global thresholding methods. All the mentioned methods have its pros and cons. Some of the most significant and recently used entropy based thresholding techniques are explained extensively with the help of sample dental radiographs. As a result of comparison, it is found that entropy based thresholding techniques provide better results. For the results obtained, image quality measures have been found. It provides better performance and the computation time taken to perform segmentation is also fast. Results demonstrate that this type of thresholding techniques are simple and give more information regarding the sample data and it is useful for further segmentation process. The further process can be enhanced by performing some of the well-known optimization techniques on the dental radiographs.

REFERENCES

Abdel-Mottaleb, J. Z. M. (2004). Automatic Human identification based on Dental X-ray images. *SPIE Conference on Defense and Security-Biometric Technology for human Identification.*

Alagumariappan, P., Krishnamurthy, K., Kandiah, S., & Ponnuswamy, M. J. (2017). Effect of electrode contact area on the information content of the recorded electrogastrograms: An analysis based on Rényi entropy and Teager-Kaiser Energy. *Polish Journal of Medical Physics and Engineering*, *23*(2), 37–42. doi:10.1515/pjmpe-2017-0007

Alagumariappan, P., Rajagopal, A., & Krishnamurthy, K. (2016). Complexity Analysis on Normal and Abnormal Electrogastrograms Using Tsallis Entropy. In *3rd International Electronic and Flipped Conference on Entropy and Its Applications*. Multidisciplinary Digital Publishing Institute. 10.3390/ecea-3-A003

Ambikapathy, B., & Krishnamurthy, K. (2018). Analysis of electromyograms recorded using invasive and noninvasive electrodes: A study based on entropy and Lyapunov exponents estimated using artificial neural networks. *Journal of Ambient Intelligence and Humanized Computing*, 1–9.

Amer, Y. Y., & Aqel, M. J. (2015). An Efficient Segmentation Algorithm for Panoramic Dental Images. *Procedia Computer Science*, *65*, 718–725. doi:10.1016/j.procs.2015.09.016

Anjna, E., & Kaur, E. R. (2017). Review of Image Segmentation Technique. *International Journal (Toronto, Ont.)*, *8*(4).

Annadurai, S. (2007). *Fundamentals of digital image processing*. Pearson Education India.

Bhandari, A. K., Kumar, A., & Singh, G. K. (2015). Modified artificial bee colony based computationally efficient multilevel thresholding for satellite image segmentation using Kapur's, Otsu and Tsallis functions. *Expert Systems with Applications*, *42*(3), 1573–1601. doi:10.1016/j.eswa.2014.09.049

Bhandari, A. K., Kumar, A., & Singh, G. K. (2015). Tsallis entropy based multilevel thresholding for colored satellite image segmentation using evolutionary algorithms. *Expert Systems with Applications*, *42*(22), 8707–8730. doi:10.1016/j.eswa.2015.07.025

Chaubey, A. K. (2016). Comparison of The Local and Global Thresholding Methods in Image Segmentation. *World Journal of Research and Review, 2*(1), 1-4.

Chauhan, A. S., Silakari, S., & Dixit, M. (2014, April). Image segmentation methods: A survey approach. In *Communication Systems and Network Technologies (CSNT), 2014 Fourth International Conference on* (pp. 929-933). IEEE. 10.1109/CSNT.2014.191

Chi, Z., Yan, H., & Pham, T. (1996). *Fuzzy algorithms: with applications to image processing and pattern recognition* (Vol. 10). World Scientific. doi:10.1142/9789812830111

De Albuquerque, M. P., Esquef, I. A., & Mello, A. G. (2004). Image thresholding using Tsallis entropy. *Pattern Recognition Letters*, *25*(9), 1059–1065. doi:10.1016/j.patrec.2004.03.003

De Luca, A., & Termini, S. (1972). A definition of a nonprobabilistic entropy in the setting of fuzzy sets theory. *Information and Control*, *20*(4), 301–312. doi:10.1016/S0019-9958(72)90199-4

Firdousi, R., & Parveen, S. (2014). Local Thresholding Techniques in Image Binarization. *International Journal Of Engineering And Computer Science*, *3*(3), 4062–4065.

Garg, N. (2013). Binarization Techniques used for grey scale images. *International Journal of Computers and Applications*, *71*(1).

Gayathri, V., Menon, H. P., & Viswa, A. (2014). *Challenges in Edge Extraction of Dental X-Ray Images Using Image Processing Algorithms–A Review*. Academic Press.

Gonzalez, R. C., & Woods, R. E. (2002). *Digital image processing*. Academic Press.

Gonzalez, Woods, & Eddins. (2010). Digital Image Processing Using MATLAB (2nd ed.). Tata McGraw Hill Education Private Limited.

Graham Leedham, C. Y., Takru, K., Tan, J. H. N., & Mian, L. (2003, August). Comparison of some thresholding algorithms for text/background segmentation in difficult document images. In *Proceedings of the seventh international conference on document analysis and recognition* (*Vol. 2*, pp. 859-864). Academic Press. 10.1109/ICDAR.2003.1227784

Harandi, A. A., Pourghassem, H., & Mahmoodian, H. (2011, December). Upper and lower jaw segmentation in dental X-ray image using modified active contour. In *Intelligent Computation and Bio-Medical Instrumentation (ICBMI), 2011 International Conference on* (pp. 124-127). IEEE. 10.1109/ICBMI.2011.88

Holambe, S., & Kumbhar, P. (2016). Comparison between Otsu's Image Thresholding Technique and Iterative Triclass. *International Journal of Computer Trends and Technology, 33*(2), 80–82. doi:10.14445/22312803/IJCTT-V33P117

Izzaty, N., & Hisham, N. (2015). *Image segmentation using adaptive thresholding*. Academic Press.

Jain, A. K. (1989). *Fundamentals of digital image processing*. Prentice-Hall, Inc.

Jain, A. K., & Chen, H. (2004). Matching of dental X-ray images for human identification. *Pattern Recognition, 37*(7), 1519–1532. doi:10.1016/j.patcog.2003.12.016

Jiang, X., Zhang, R., & Nie, S. (2012). Image segmentation based on level set method. *Physics Procedia, 33*, 840–845. doi:10.1016/j.phpro.2012.05.143

Jumb, V., Sohani, M., & Shrivas, A. (2014). Color Image Segmentation Using K-Means Clustering and Otsu's Adaptive Thresholding. *International Journal of Innovative Technology and Exploring Engineering, 3*(9), 72–76.

Kamalanand, K., & Ramakrishnan, S. (2015). Effect of gadolinium concentration on segmentation of vasculature in cardiopulmonary magnetic resonance angiograms. *Journal of Medical Imaging and Health Informatics, 5*(1), 147–151. doi:10.1166/jmihi.2015.1370

Kapur, J. N., Sahoo, P. K., & Wong, A. K. (1985). A new method for gray-level picture thresholding using the entropy of the histogram. *Computer Vision Graphics and Image Processing, 29*(3), 273–285. doi:10.1016/0734-189X(85)90125-2

Kaur, N., & Kaur, R. (2011). A review on various methods of image thresholding. *International Journal on Computer Science and Engineering, 3*(10), 3441.

Khan, M. W. (2014). A survey: Image segmentation techniques. *International Journal of Future Computer and Communication, 3*(2), 89–93. doi:10.7763/IJFCC.2014.V3.274

Khurshid, K., Siddiqi, I., Faure, C., & Vincent, N. (2009). Comparison of Niblack inspired binarization methods for ancient documents. *DRR, 7247*, 1–10.

Kom, G., Tiedeu, A., & Kom, M. (2007). Automated detection of masses in mammograms by local adaptive thresholding. *Computers in Biology and Medicine, 37*(1), 37–48. doi:10.1016/j.compbiomed.2005.12.004 PMID:16487954

Lai, Y. H., & Lin, P. L. (2008, October). Effective segmentation for dental X-ray images using texture-based fuzzy inference system. In *International Conference on Advanced Concepts for Intelligent Vision Systems* (pp. 936-947). Springer. 10.1007/978-3-540-88458-3_85

Li, C., Huang, R., Ding, Z., Gatenby, J. C., Metaxas, D. N., & Gore, J. C. (2011). A level set method for image segmentation in the presence of intensity in homogeneities with application to MRI. *IEEE Transactions on Image Processing, 20*(7), 2007–2016. doi:10.1109/TIP.2011.2146190 PMID:21518662

Li, X., Abaza, A., Nassar, D. E., & Ammar, H. (2006). Fast and accurate segmentation of dental x-ray records. *Lecture Notes in Computer Science, 3832*, 688–696. doi:10.1007/11608288_92

Manic, K. S., Priya, R. K., & Rajinikanth, V. (2016). Image multithresholding based on kapur/tsallis entropy and firefly algorithm. *Indian Journal of Science and Technology*, *9*(12). doi:10.17485/ijst/2016/v9i12/89949

Martınez, S., Nicolás, F., Pennini, F., & Plastino, A. (2000). Tsallis' entropy maximization procedure revisited. *Physica A*, *286*(3), 489–502. doi:10.1016/S0378-4371(00)00359-9

Naidu, M. S. R., Kumar, P. R., & Chiranjeevi, K. (2017). *Shannon and fuzzy entropy based evolutionary image thresholding for image segmentation*. Alexandria Engineering Journal.

Ntirogiannis, K., Gatos, B., & Pratikakis, I. (2014). A combined approach for the binarization of handwritten document images. *Pattern Recognition Letters*, *35*, 3–15. doi:10.1016/j.patrec.2012.09.026

Otsu, N. (1979). A threshold selection method from gray-level histograms. *IEEE Transactions on Systems, Man, and Cybernetics*, *9*(1), 62–66. doi:10.1109/TSMC.1979.4310076

Pal, N. R., & Pal, S. K. (1993). A review on image segmentation techniques. *Pattern Recognition*, *26*(9), 1277–1294. doi:10.1016/0031-3203(93)90135-J

Pandey, V., & Gupta, V. (2014). MRI image segmentation using shannon and non shannon entropy measures. *International Journal of Application or Innovation in Engineering & Management*, *3*(7), 41–46.

Prasad, M. S., Krishna, V. R., & Reddy, L. S. S. (2013). Investigations on Entropy Based Threshold Methods. *Asian Journal of Computer Science & Information Technology, 1*(5).

Rad, A. E., Rahim, M. S. M., & Norouzi, A. (2013). Digital dental x-ray image segmentation and feature extraction. *Indonesian Journal of Electrical Engineering and Computer Science*, *11*(6), 3109–3114.

Raja, N., Rajinikanth, V., & Latha, K. (2014). Otsu based optimal multilevel image thresholding using firefly algorithm. *Modelling and Simulation in Engineering*, *2014*, 37.

Raja, N. S. M., Kavitha, G., & Ramakrishnan, S. (2012). Analysis of Vasculature in Human Retinal Images Using Particle Swarm Optimization Based Tsallis Multi-level Thresholding and Similarity Measures. *SEMCCO*, *7677*, 380–387.

Raja, N. S. M., Sukanya, S. A., & Nikita, Y. (2015). Improved PSO based multi-level thresholding for cancer infected breast thermal images using Otsu. *Procedia Computer Science*, *48*, 524–529. doi:10.1016/j.procs.2015.04.130

Rajinikanth, V., Aashiha, J. P., & Atchaya, A. (2014). Gray-level histogram based multilevel threshold selection with bat algorithm. *International Journal of Computers and Applications*, *93*(16).

Rajinikanth, V., & Couceiro, M. S. (2015). RGB histogram based color image segmentation using firefly algorithm. *Procedia Computer Science*, *46*, 1449–1457. doi:10.1016/j.procs.2015.02.064

Rajinikanth, V., Raja, N. S. M., & Kamalanand, K. (2017). Firefly Algorithm Assisted Segmentation of Tumor from Brain MRI using Tsallis Function and Markov Random Field. *Journal of Control Engineering and Applied Informatics*, *19*(3), 97–106.

Rajinikanth, V., Raja, N. S. M., & Latha, K. (2014). Optimal multilevel image thresholding: An analysis with PSO and BFO algorithms. *Australian Journal of Basic and Applied Sciences*, *8*(9), 443–454.

Rajinikanth, V., Raja, N. S. M., & Satapathy, S. C. (2016). Robust color image multi-thresholding using between-class variance and cuckoo search algorithm. In *Information systems design and intelligent applications* (pp. 379–386). New Delhi: Springer. doi:10.1007/978-81-322-2755-7_40

Ribeiro, M. R., Dias, M. A., de Best, R., da Silva, E. A., Neves, C. D. T. T., & Street, R. S. (2014). Enhancement and segmentation of dental structures in digitized panoramic radiography images. *International Journal of Applied Mathematics*, *27*(4), 387–406. doi:10.12732/ijam.v27i4.6

Sahoo, P. K., Soltani, S. A. K. C., & Wong, A. K. (1988). A survey of thresholding techniques. *Computer Vision Graphics and Image Processing*, *41*(2), 233–260. doi:10.1016/0734-189X(88)90022-9

Said, E., Fahmy, G. F., Nassar, D., & Ammar, H. (2004, April). Dental x-ray image segmentation. *Proceedings of the Society for Photo-Instrumentation Engineers*, *5404*, 409–417. doi:10.1117/12.541658

Sarkar, S., & Das, S. (2013). Multilevel image thresholding based on 2D histogram and maximum Tsallis entropy—a differential evolution approach. *IEEE Transactions on Image Processing*, *22*(12), 4788–4797. doi:10.1109/TIP.2013.2277832 PMID:23955760

Sathya, P. D., & Kayalvizhi, R. (2010). Optimum multilevel image thresholding based on tsallis entropy method with bacterial foraging algorithm. *International Journal of Computational Science*, *7*(5), 336–343.

Sauvola, J., & Pietikäinen, M. (2000). Adaptive document image binarization. *Pattern Recognition*, *33*(2), 225–236. doi:10.1016/S0031-3203(99)00055-2

Senthilkumaran, N., & Kirubakaran, C. (2014). Efficient implementation of Niblack thresholding for MRI brain image segmentation. *International Journal of Computer Science and Information Technologies*, *5*, 2173–2176.

Senthilkumaran, N., & Vaithegi, S. (2016). Image segmentation by using thresholding techniques for medical images. Computer Science & Engineering. *International Journal (Toronto, Ont.)*, *6*(1).

Sezgin, M. (2004). Survey over image thresholding techniques and quantitative performance evaluation. *Journal of Electronic Imaging*, *13*(1), 146–168. doi:10.1117/1.1631315

Shafarenko, L., Petrou, H., & Kittler, J. (1998). Histogram-based segmentation in a perceptually uniform color space. *IEEE Transactions on Image Processing*, *7*(9), 1354–1358. doi:10.1109/83.709666 PMID:18276345

Shah, S., Abaza, A., Ross, A., & Ammar, H. (2006, September). Automatic tooth segmentation using active contour without edges. In *Biometric consortium conference, 2006 biometrics symposium: Special session on research at the* (pp. 1–6). IEEE. doi:10.1109/BCC.2006.4341636

Singh, O. I., Sinam, T., James, O., & Singh, T. R. (2012). Local Contrast and Mean Thresholding in Image Binarization. *International Journal of Computers and Applications*, *51*(6).

Singh, T. R., Roy, S., Singh, O. I., Sinam, T., & Singh, K. (2012). *A new local adaptive thresholding technique in binarization.* arXiv preprint arXiv:1201.5227

Sparavigna, A. C. (2015). Tsallis entropy in bi-level and multi-level image thresholding. *International Journal of Sciences, 4*(1), 40–49. doi:10.18483/ijSci.613

Stathis, P., Kavallieratou, E., & Papamarkos, N. (2008). An Evaluation Technique for Binarization Algorithms. *J. UCS, 14*(18), 3011–3030.

Subramanyam, R. B., Prasad, K. P., & Anuradha, B. (2014). Different Image Segmentation Techniques for Dental Image Extraction. *Int. Journal of Engineering Research and Applications, 4*, 9622.

Tao, W. B., Tian, J. W., & Liu, J. (2003). Image segmentation by three-level thresholding based on maximum fuzzy entropy and genetic algorithm. *Pattern Recognition Letters, 24*(16), 3069–3078. doi:10.1016/S0167-8655(03)00166-1

Tuan, T. M. (2016). A cooperative semi-supervised fuzzy clustering framework for dental X-ray image segmentation. *Expert Systems with Applications, 46*, 380–393. doi:10.1016/j.eswa.2015.11.001

Tuan, T. M., Son, L. H., & Dung, L. B. (2016). Dynamic semi-supervised fuzzy clustering for dental X-ray image segmentation: An analysis on the additional function. *Journal of Computer Science and Cybernetics, 31*(4), 323.

Wang, C. W., Huang, C. T., Lee, J. H., Li, C. H., Chang, S. W., Siao, M. J., ... Fischer, P. (2016). A benchmark for comparison of dental radiography analysis algorithms. *Medical Image Analysis, 31*, 63–76. doi:10.1016/j.media.2016.02.004 PMID:26974042

Zhang, Y., & Wu, L. (2011). Optimal multi-level thresholding based on maximum Tsallis entropy via an artificial bee colony approach. *Entropy (Basel, Switzerland), 13*(4), 841–859. doi:10.3390/e13040841

Zhou, J., & Abdel-Mottaleb, M. (2005). A content-based system for human identification based on bitewing dental X-ray images. *Pattern Recognition, 38*(11), 2132–2142. doi:10.1016/j.patcog.2005.01.011

KEY TERMS AND DEFINITIONS

Binarization: It is the process of converting an image pixel into binary image.

Caries: Cavity formation in the teeth caused by bacteria that attach to teeth, form acids, and cause tooth decay.

Crown: A crown is a type of dental restoration which completely caps or encircles a tooth or dental implant.

Dentin: A layer underlying the enamel. Dentin is made of living cells, which secrete a hard mineral substance.

Enamel: The hardest, white outer part of the tooth. Enamel is mostly made of calcium phosphate, a rock-hard mineral.

Entropy: It is the quantity which is used to describe the amount of information which must be coded for compression algorithm.

Histogram: It is a graph showing the number of pixels in an image at each different intensity value found in the image.

Multilevel Thresholding: A process that segments a gray level into several distinct regions.

Pulp: The softer, living inner structure of teeth. Blood vessels and nerves run through the pulp of the teeth.

Radiography: It is an imaging technique using x-rays to view the internal structure of the tooth.

Restoration: A dental restoration or dental filling is a treatment to restore the morphology of missing tooth structure resulting from caries or replacement of structure supported by dental implants.

Root Canal: It is a treatment procedure where the pulp is replaced with some inert material.

Segmentation: It is a process of partitioning a digital image into multiple segments to simplify the representation of image into meaningful and easier for analysis.

Thresholds: It is a process of converting a gray level image into binary image.

Chapter 3
Image Segmentation Using Contour Models:
Dental X-Ray Image Segmentation and Analysis

Kavitha G.
Anna University, India

Muthulakshmi M.
Anna University, India

Latha M.
Anna University, India

ABSTRACT

Image segmentation is an important task in image processing, which is widely used in medical applications such as abnormality detection and after treatment progress monitoring. Conventionally, texture, region, and edge information are used for segmentation. Recently, the majority of image segmentation uses contour-based models. The problem of efficient segmentation in medical images is of great importance in disease diagnosis. Medical images suffer from weak boundaries, and placement of initial contour is a major issue. Level method is an effective method for segmentation of image as it has ability to tackle complex geometries. It helps to detect the precise location of the target region and help to prevent the boundary leakage problem. This chapter presents an overview of the advanced region and edge-based level set segmentation algorithms and their application in the dental x-ray images. Computer-aided diagnosis from x-ray images are of interest to clinicians in detection and accurate decision making. Case studies of multiple region segmentation from dental x-rays are presented.

DOI: 10.4018/978-1-5225-6243-6.ch003

INTRODUCTION

Computer aided diagnosis system based on X-Ray images is significant to make accurate decision of diseases and therapeutic interventions (Son, 2018). X-ray technique is cheaper, widely used by radiologist, easily accessible electronically and has lower radiations compared to CT scan. A dental X-ray image consists of teeth area, dental structural area with bone, soft tissues and background area (Son & Tuan, 2016). They play a significant role in detection of periodontitis, root fracture, bone loss, oral cancer and carries that is difficult to be seen by a visual dental examination (Jain & Chen, 2004). Periodontitis results from the progression of gingivitis, which involves the inflammation and infection of the ligaments and bones that support the teeth. The jaw bones such as mandible and maxilla suffer from odontogenic lesions (Rajendran, 2009). Caries is caused by specific type of bacteria that destroys tooth enamel and layer underneath it. Prolonged bone loss and root decay may lead to tooth loss or erosion of the jaw bone. If bone loss and root decay are detected at an early stage, a suitable remedy can be provided by appropriate dental procedures such as root canal treatment.

One of the most interesting applications of dental X-rays is forensic identification and age estimation (Babshet, 2010; Jain, 1999; Zhou & Mottaleb, 2005). It is used in human identification system (Chen & Jain, 2004; Fahmy, 2004). In forensic dentistry, post-mortem and ante-mortem dental records are compared to confirm the identity of an individual (Pretty & Sweet, 2001). Dental images can be classified into bitewing, periapical, and panoramic views. Conventionally the teeth and the bones were segmented in the bitewing view. Further, each tooth is separated into crown and root, and the contours of the teeth are stored in the database to identify victims. The extraction of root is challenging than crown as it overlaps with the jaws due to lower difference in tissue density (Cavalcanti, 1999; Goaz & White, 1982). Each tooth can be classified into different types such as molar, premolar, canine and incisor (Lin, 2010). Wisdom tooth is an extra tooth and commonly affects other teeth while eruption (Nelson & Ash, 2010). Dental development based methods are applicable to age estimation for individuals under 21 years of age (Gustafson, 1950; Kvaal, 1995; Shahin, 2013).

There are certain challenges in the manual inspection of dental X-rays by dentists such as: expensive, requires specialized training, inter-observer variability and non-quantitative measurement (Jain, 2003). Segmentation is the initial task in the field of medical image processing, as the segmented regions are essential for diagnosis of disease pathology (Zhou, 2017). Dental X-ray image segmentation is absolutely necessary to analyze these images in order to obtain valuable information regarding the medical diagnosis and other recognition system (Said, 2004). The segmentation of dental X-rays is challenging due to noise, low contrast, sample artifacts, complicated topology and weaker boundaries between teeth. In addition, the segmentation of these images are difficult due to partial volume effect, intensity inhomogeneity and noise (Chen, 2012).

Appropriate enhancement techniques can be applied to dental radiographs that suffer from low contrast and uneven exposure to reduce the complexity involved in the segmentation task. Suitable morphological operations can be performed to enhance the contrast of the image that makes the teeth regions brighter and suppresses the intensity in the bone and the background regions (Gonzalez & Woods, 2008). The dental images were enhanced using morphological processing and wavelet transformation was used for segmentation (Patanachai, 2010). Edge extraction is essential for root canal treatment (Gayathri & Menon, 2014). Intensity inhomogeneity occurs due to imperfections in image acquisition (Dhawan, 2011). Hence bias correction can be carried out separately or interleaved with segmentation. Thus, there is a need for accurate and efficient algorithm to perform segmentation of dental X-rays to allow earlier interventions. In addition tooth isolation and shape refinement techniques are used to improve the segmentation outcome (Lin, 2013).

The features extracted from the segmented dental image provide information about different teeth's parts such as enamel, cementum, gum and root canal. Moments based and statistical features extracted from dental X-rays along with Bayesian classifier were used for segmentation (Lira, 2014). The shape and texture features can be extracted from these images for further classification (Duda, 2001). Classification of tooth into different types also plays a major role in forensic and dental biometrics. This can be carried out with various supervised and un-supervised classification algorithms (Dougherty, 2009). Effective image processing methods are required to develop computerized dental X-ray image analysis system (Lpez-Lpez, 2012; Nakamoto, 2008; Wriedt, 2012).

BACKGROUND

X-ray dental image consists of three main parts such as teeth area, dental structural area and background (Scott, 1977). Several methods have been attempted in literature to segment the region of interest from X-ray images. Image based methods partition the image with respect to pixel intensities (Ali, 2017; Son & Tuan, 2017). It is difficult to process blank holes in missing teeth by conventional thresholding methods (Kondo, 2004). They are less accurate due to overlapping intensities. Adaptive thresholding is used to automate segmentation in human identification system (Nomir & Abdel-Mottaleb, 2005). Fuzzy inference system is used in various literatures for dental diagnosis (Lin & Lai, 2008). Here a given input data set is mapped to an output data set using fuzzy logic (Lai & Lin, 2008). However, experts' experience is required to form fuzzy rules and knowledge, to avoid duplication of rules and as well as to provide accurate diagnosis (Grabisch, 2013). These issues can be solved by using a combination of segmentation, classification and decision making algorithms that reduces the vagueness in determination of rules (Pretty & Maupomé, 2004). Barboza et al. used graph based segmentation algorithm for dental recognition (2012).

The accuracy of segmentation can be further improved by region based methods where a seed point is selected and surrounding pixels with similar morphology are combined (Alazab, 2009; Mahoor & Abdel-Mottaleb, 2005). Region based techniques work well only when the predefined curves were provided (Said, 2006). However, the segmentation results are more sensitive to initial seed selection and noises. Gradient of an image is used as a constraint in edge based methods (Lakhani, 2016). Narrow regions and weaker boundaries are difficult to be segmented by these methods.

Recently ACM is preferred over other medical segmentation methods (Nurtanio, 2011; Rad, 2013; Shah, 2006). Li et al. employed pathological level set modeling to segment the dental X-ray images into three distinct regions (2005). Initial contour trained by support vector machine is provided as input to coupled level set (Li, 2006; Li, 2007). The thresholding along with coupled level set is used to detect lesion region in dental radiographs (Lin, 2012). The basic level set was utilized to segment tooth region from the extracted individual tooth (Rad, 2014). The outcome of fuzzy C-means algorithm is used as initial contour for level set to detect carries (Chaabene, 2017). Local Gaussian distribution fitting energy is added with level set to overcome intensity inhomogeneity problem in dental X-rays (Pandey, 2017).

Machine learning methods provide better segmentation results with less user interaction, but require more time and large training data set (Miki, 2017). Bayesian network was attempted to diagnose toothache (Chattopadhyay, 2012; Chattopadhyay, 2016). SVM was used to predict the osteoporosis from panoramic view dental images (Kavitha, 2012). Biologically inspired artificial intelligence technique have been used for dental radiograph segmentation (Keshtkar & Gueaieb, 2006).

CONTOUR BASED METHODS

In the ACM, an initial curve move towards the object boundaries via minimization of energy (Kass, 1988). The ACM based image segmentation techniques are classified as edge and region based methods. The edge based segmentation method is based on gradient of the image. Gradient based methods detect edges from the maximum and minimum of first derivative of an image. Laplacian method of edge detection searches for zero crossings in the second derivative of the image to find edges. Region based active contour method looks for uniformity of intensity, texture or color within a sub region. It uses statistical information of image intensity within each subset instead of searching geometrical boundaries. Level set method of active contour was proposed by Osher & Sethian to avoid the split and merge problem in the previous method (1988). This level set method drives an initial curve based on gradient or region information to the image boundaries. It is sensitive to placement of initial contour. A good initialization avoids local solutions and also reduces the iterative times the curve evolution occurs.

LEVEL SET METHOD

The variational level set approach can be adopted to perform segmentation and bias correction simultaneously. The major categories include region based, edge based and hybrid level set.

Region Based Level Set Methods

Images with intensity inhomogeneity can be described as (Li, 2008):

$$I = bJ + n \quad (1)$$

where J is the original image intensity, I is the measured image intensity, b is the bias field and n is the noise. The following assumptions are made for the original image J and bias field b:

1. The bias field b varies slowly in the entire image domain Ω.
2. The original intensity J in each tissue takes a unique value $J(x) \approx$ cased on the physical property of the tissue.

The image domain Ω has to be separated into N disjoint regions, Ω_i, i = 1, 2, …, N based on the measured image I and evolution of the zero level contour φ. The image domain is partitioned by minimization of the following energy function:

$$E(\varphi,b,c)=\int\left(\sum_{i=1}^{N}\int K(x-y)\left|I(y)-b(x)c_i\right|^2 M_i(\varphi)dy\right)dx \tag{2}$$

where K(x-y) is the Gaussian kernel function of radius ρ, c_i is the intensity mean in the ith region. $M_i(\varphi)$ is the regional characteristic function that satisfies the conditions $\cup_{i=1}^{N}$. M1 and $M_iM_j = 0$ (i≠j). For N=4, $M_i(\varphi)$ is given as $M_1(\varphi) = H(\varphi_1)H(\varphi_2)$, $M_2(\varphi) = H(\varphi_1)(1-H(\varphi_2))$, $M_3(\varphi) = (1-H(\varphi_1))H(\varphi_2)$ and $M_4(\varphi) = (1-H(\varphi_1))(1-H(\varphi_2))$. Here H(φ) is the Heaviside function with contant ε = 1.

The level set evolution can be made more appropriate by the addition of length L(φ) and distance regularization term P(φ). The length regularization term is used to avoid small and isolated regions in the final segmentation. The next term regularizes the zero level contour by penalizing its deviation from a signed distance function. L(φ) and P(φ) are defined as:

$$L(\varphi)=\int\left|\nabla H(\varphi)\right|dx \tag{3}$$

$$P(\varphi)=\int\frac{1}{2}\left(\left|\nabla\varphi\right|-1\right)^2 dx \tag{4}$$

Therefore, the entire energy function is given as

$$E(\varphi,b,c)=\varepsilon(\varphi,b,c)+\upsilon L(\varphi)+\mu P(\varphi) \tag{5}$$

where μ, υ are positive constants. The image segmentation can be performed by minimization of the above energy function. The minimization of E(φ,b,c) for fixed c and b can be obtained by using standard gradient descent method.

$$\frac{\partial\varphi}{\partial t}=-\frac{\partial E}{\partial\varphi} \tag{6}$$

where $\frac{\partial E}{\partial\varphi}$ the Gateaux derivative of the energy E. The Gateaux derivative is computed and the corresponding gradient flow equation is expressed as:

$$\frac{\partial\varphi_k}{\partial t}=-\sum_{i=1}^{N}\frac{\partial M_i(\varphi)}{\partial\varphi_k}e_i+\upsilon\delta(\varphi_k)div\left(\frac{\nabla\varphi_k}{\left|\nabla\varphi_k\right|}\right)+\mu div\left(d_p\left|\nabla\varphi_k\right|\nabla\varphi_k\right) \tag{7}$$

For fixed φ and b, the optimal c_i that minimizes the energy function. $E(\varphi,b,c)$,s given by:

$$c_i = \frac{\int \left(b * K\right) I M_i\left(\varphi\right) dy}{\int \left(b^2 * K\right) M_i\left(\varphi\right) dy} \tag{8}$$

Similarly for fixed φ and c, the optimal b that minimizes the energy function . $E\left(\varphi, b, c\right)$,s specified by:

$$b = \frac{\left(IJ^{(1)}\right) * K}{J^{(2)} * K} \tag{9}$$

where $J^{(1)} = \sum_{i=1}^{N} c_i u_i$ d $J^{(2)} = \sum_{i=1}^{N} c_i^2 u_i$. .The performance of this local k-means (LKM) model is affected by initialization of contours, weaker edges and time taken for convergence of contour to object boundary (Mitiche & Ayed, 2011). Inorder to overcome these problems, different variation of multiphase level set methods are discussed to partition the image domain.

The above discussed method uses the local image information as the constraint. Zhang et al. utilized the local image information for the construction of LIF model, where the constraint is the difference between the fitting image and the original image (2010). The energy function of the LIF model is given by:

$$E^{LIF}\left(\varphi\right) = \frac{1}{2} \int \left(I\left(x\right) - I^{LFI}\left(x\right)\right)^2 dx \tag{10}$$

where I is the original image and I^{LFI} is the local fitted image, that can be formulated as:

$$I^{LFI} = m_1 H\left(\varphi\right) + m_2\left(1 - H\left(\varphi\right)\right) \tag{11}$$

Here φ is the level set function, H(φ) is the Heaviside function, m_1 and m_2 are the local average intensities inside and outside the curve. This model can segment images with intensity inhomogeneity. However, the bias field information is not included in this model. Therefore, image fitting model based on bias field can be used for image segmentation (Li, 2011). The placement of initial contour is a challenging task in level set method. FCM clustering has been used to extract the initial contour in autistic MR brain images (JacFredo, 2015b). Further, the fuzzy membership function can be incorporated into the energy fitting term to improve the accurate detection of regions (Bhadauria, 2013). Here the intensity information in local region is used to guide the evolution of contour. The fuzzy based energy descriptor model is given by:

$$E\left(\varphi, b, c\right) = \int \sum_{i=1}^{N} e_i\left(x\right) M_i\left(\varphi\left(x\right)\right) dx \tag{12}$$

where e_i is the fuzzy based intensity descriptor based on the bias field b and mean of the cluster region c, that is given by:

$$e_i = \left| I - bc_i \right|^2 \tag{13}$$

Here the level set function φ is defined by fuzzy membership function output u_{ik} and cluster center v_i. This fuzzy based distance regularized multiphase level set has been used for segmentation in diagnosis of brain disorder (JacFredo, 2014a).

The fuzzy GM model based energy descriptor is utilized in Eqn. 13 for segmentation of subcortical regions in autistic brain image to minimize the computation time (JacFredo, 2014b). Here the fuzzy GM model based intensity descriptor with σ as variance is given by (Ji, 2012):

$$e_i = K\left(log\left(\sqrt{2\pi\sigma}\right) + \frac{\left| I - bc_i \right|^2}{2\sigma^2} \right) \tag{14}$$

In conventional level set methods, level set function develops irregularities when it evolves. Re-initialization of degraded level set function is required in order to overcome this. However re-initialization affects the numerical accuracy of level set evolution. Level set function without re-initialization can be obtained by addition of reaction diffusion term to the level set function that is given by (Zhang, 2013):

$$\frac{\partial\varphi}{\partial t} = \varepsilon\Delta\varphi + \frac{1}{\varepsilon}\left(\lambda\delta\left(\varphi\right) div\left(g\frac{\nabla\varphi}{\left|\nabla\varphi\right|} \right) + \upsilon g\delta\left(\varphi\right) \right) \tag{15}$$

where ε, λ and υ are positive constants and g. a stopping function with a Gaussian kernel G_σ, of standard deviation σ ($g = \frac{1}{1+\left|\nabla G_\sigma * I\right|^p}$., ≥ 1). This method has been attempted in the segmentation of corpus callosum from MR brain images for autism and Alzheimer's subjects (JacFredo, 2015a; Anandh, 2016a). Anandh et al. used reaction diffusion level set to differentiate normal and neuro degenerative subjects (2016b).

Estimation of bias field and tissue segmentation can be simultaneously achieved by an energy minimization process that optimizes the estimates of two multiplicative components of an MR image namely bias field b and true image J. The true image describes the physical property of the tissues and the bias field specifies the intensity inhomogeneity and spatial properties. The energy function of this multiplicative intrinsic component optimization (MICO) method with fuzzifier q ≥ 1 is represented as (Li, 2014):

$$E_q\left(u, a, w\right) = \int \sum_{i=1}^{N} \left| I\left(x\right) - w^T G\left(x\right) a_i \right|^2 u_i^q dx \tag{16}$$

Here constant a and fuzzy membership function u_i represent $J(x)$ i.e. . $J(x) = \sum_{i=1}^{N} a_i u_i$ .n addition bias field is represented as $b(x) = w^T G(x)$. ,here G(x) and w^T denotes smooth basis function and weighting coefficient respectively.

The optimization of bias field and true image can be done by minimizing the energy function F with respect to u_i, a_i, and w. The optimization of u_i, a_i, and w are carried out as:

$$a_i = \frac{\int I(x) b(x) u_i^q(x)\, dx}{\int b^2(x) u_i^q(x)\, dx}, i = 1, \ldots, N \tag{17}$$

$$w = \left(\int G(x) G^T(x) \left(\sum_{i=1}^{N} a_i^2 u_i^q(x) \right) dx \right)^{-1} \int G(x) I(x) \left(\sum_{i=1}^{N} a_i u_i^q(x) \right) dx \tag{18}$$

$$u_i = \frac{\left(\delta_i(x)\right)^{\frac{1}{1-q}}}{\sum_{j=1}^{N} \left(\delta_j(x)\right)^{\frac{1}{1-q}}} \quad i=1,\ldots,N \tag{19}$$

$$\delta_i(x) = \left| I(x) - w^T G(x) a_i \right|^2 \tag{20}$$

MICO has been applied in ventricle segmentation from schizophrenic MR brain images (Latha & Kavitha, 2017).

MICO is sensitive to noise as it is based on pixel intensities. Hence to deal with intensity inhomogeneity and noise, a region based level set method with neighborhood clustering is used for segmentation and bias correction. Here, the bias field is estimated from a linear combination of smooth orthogonal basis functions. Local clustering criterion is used to cluster the homogeneous regions (Feng, 2017). The energy of this LINC level set function is given by:

$$E(\varphi, b, c) = \int \left(\sum_{i=1}^{N} \lambda_i \int K(x-y) \left| I(y) - w^T G(y) a_i \right|^2 dy \right) M_i(\varphi(x))\, dx \tag{21}$$

where I, K and λ are the image, Gaussian kernel and multipliers respectively. For a fixed φ and bias field, the optimal w and a that minimizes the energy function are evaluated. The desirable amount of contextual information in an image is controlled by inclusion of regularization parameter in the fuzzy membership function (Elazab, 2015).

Edge Based Level Set Methods

The need for re-initialization can be eliminated by level set evolution with distance regularization term (Li, 2010). The energy function with regularization term is given by:

$$E(\varphi) = \mu R_p(\varphi) + E_{ext}(\varphi) \tag{22}$$

where the distance regularization term is generally defined by $R_p(\varphi) = \int p(|\nabla\varphi|)\,dx$ th a potential function $p : [0,\infty] \rightarrow \Re$ sh that for all s. The level set evolution equation with distance regularization term is given by:

$$\frac{\partial\varphi}{\partial t} = \mu div\left(d_p\left(|\nabla\varphi|\nabla\varphi\right)\right) + \lambda\delta_\varepsilon(\varphi)\,div\left(g\frac{\nabla\varphi}{|\nabla\varphi|}\right) + \alpha g\delta_\varepsilon(\varphi) \tag{23}$$

where the first term on the right hand side is the Gateaux derivative of $Rp(\varphi)$ and μ, λ and α are positive constants. The g. g an edge indicator function with a Gaussian kernel G_σ of standard deviation σ ($g = \dfrac{1}{1 + |\nabla G_\sigma * I|^2}$). The DRLSE is used to segment ventricle region for analysis of chronic brain disease (Kayalvizhi, 2013).

The potential function of regularization term is generally considered as $p(|\nabla\varphi|) = \frac{1}{2}(|\nabla\varphi| - 1)^2$ this is used to penalize the level set function for deviation from the signed distance function. Though signed distance property of level set is preserved here, it produces undesirable effect at locations far away from zero level set. In order to overcome this side effect penality term with double well property is used in DRLSE-M.

$$p(|\nabla\varphi|) = \begin{cases} \frac{1}{2}(|\nabla\varphi| - 1)^2\ ; & |\nabla\varphi| \geq 1 \\ \frac{1}{(2\pi)^2}\left(1 - \cos\left(2\pi|\nabla\varphi|\right)\right); & |\nabla\varphi| \leq 1 \end{cases} \tag{24}$$

The modified DRLSE is used for segmentation of ventricle to study progression of brain disorder (Kayalvizhi, 2015). The faster evolving speed is achieved by consideration of two competing components during the evolution in forward and backward diffusion rate. The DRLSE-E (Wang, 2014) is thus given by:

$$p(|\nabla\varphi|) = \begin{cases} \frac{1}{2}(|\nabla\varphi|^2 - 1) - \ln|\nabla\varphi|; & |\nabla\varphi| \geq 1 \\ \frac{1}{2}(|\nabla\varphi|(|\nabla\varphi| - 1)^2; & |\nabla\varphi| \leq 1 \end{cases} \tag{25}$$

Enhanced DRLSE has faster evolving speed than DRLSE and DRLSE-M.

Hybrid Level Set Methods

The inappropriate initialization of zero level set reduces the speed of curve evolution and it gets stuck in local minima. Thus if the initial contour is set closer to the boundary of the object, less number of iterations are required for segmentation. The edge information can be added to the region based level set to obtain hybrid level set. Edge information can be obtained using various edge operators (Ding, 2017). Optimized LoG operator smoothes the homogeneous regions and enhances the edge of the object. The weighted optimized LoG term is added to the energy function of Eqn. 6 to get optimized LoG driven local region level set. The energy function of local region level set driven by optimized boundary is given as:

$$E^{LROB}\left(\varphi,b,c\right)=\varepsilon\left(\varphi,b,c\right)+\omega L_{\Delta}\left(\varphi\right)+\upsilon L\left(\varphi\right)+\mu P\left(\varphi\right) \tag{26}$$

where ω is a constant and $L_{\Delta}\left(\varphi\right)$ the optimized LoG term that is obtained by solving the following iterative equation.

$$\frac{\partial L}{\partial t}=g\left(\left|\nabla I\right|\right)\times L-\left(1-g\left(\left|\nabla I\right|\right)\right)\times\left(L-\beta\times\Delta\left(G_{\sigma}*I\right)\right) \tag{27}$$

Here L is the original LoG, $\boldsymbol{g}\left(\left|\nabla \boldsymbol{I}\right|\right)$ a edge indicator and β is a positive constant that enhances the object edges.

RESULTS AND DISCUSSION ON DENTAL X-RAYS

In this section level set methods such as local k-means model (LKM), optimized boundary driven level set (OBDLS) and MICO are used for dental X-ray segmentation. The performances of these methods have been compared visually and analysed using similarity measures such as Dice coefficient and Jaccard index.

The dental X-ray images are obtained from the online dataset provided by Rad et al. (2016) (http://dx.doi.org/10.6070/H47H1GJ4). The database contains 120 periodical dental X-ray images that are useful for dental disease detection and diagnosis. The images are in periapical view that covers from top to bottom jaw. The age group of subjects is from 25 to 35 years. The carries region in each image is specified by experts. The image dimension is 748 x 512 and they are stored in "jpeg" format. The pathological conditions considered here include three types of dental diseases such as enamel, dentinal and pulpal caries. In enamel caries microbial dental plaque is formed whereas dentinal and pulpal caries occur in dentinal tubules and root surfaces respectively.

Sample dental images are shown in Figure 1. The database has normal, carries, missing tooth and teeth with metal fillings. The typical images under these categories are shown in Figure 1 (a) – (d) respectively. The groundtruth was obtained from standard tools. As a preliminary step in dental diagnosis, segmentation of dental X-rays has been shown here. The segmentation of teeth will help in detailed

Figure 1a. Sample dental images: Normal

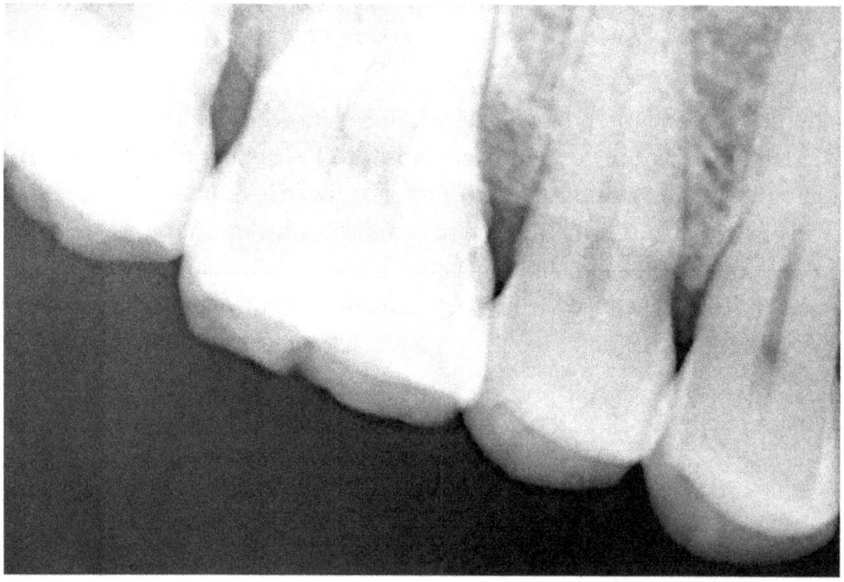

Figure 1b. Carries

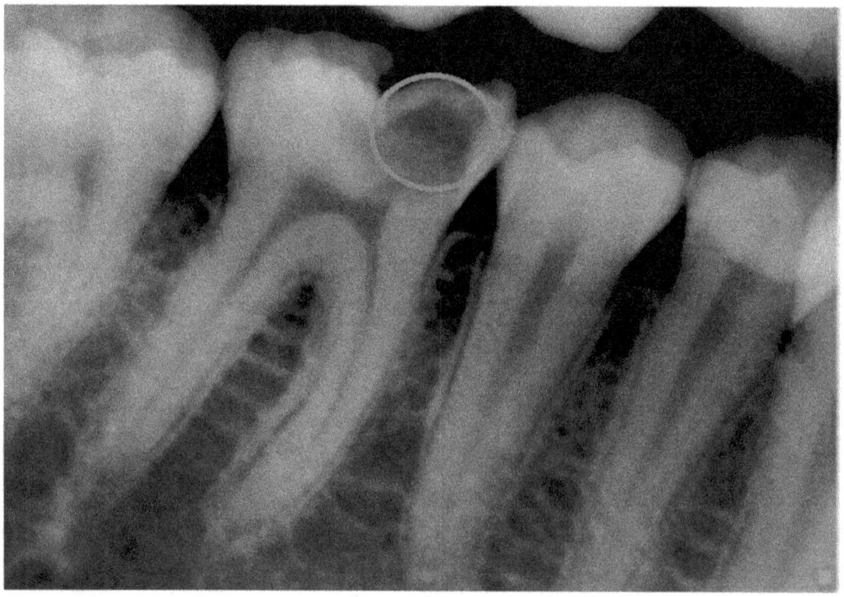

analysis of different dental abnormalities. Further individual tooth can be delineated from the segmented region by placement of ROI.

The segmentation results for a sample dental X-ray image using LKM and OBDLS method is shown in Figure 2. The local k-means model makes use of local region scalable fitting energy. The parameters such as σ, μ, υ and time step influence the evolution of contour. The scale parameter σ defines the size

Figure 1c. Missing tooth

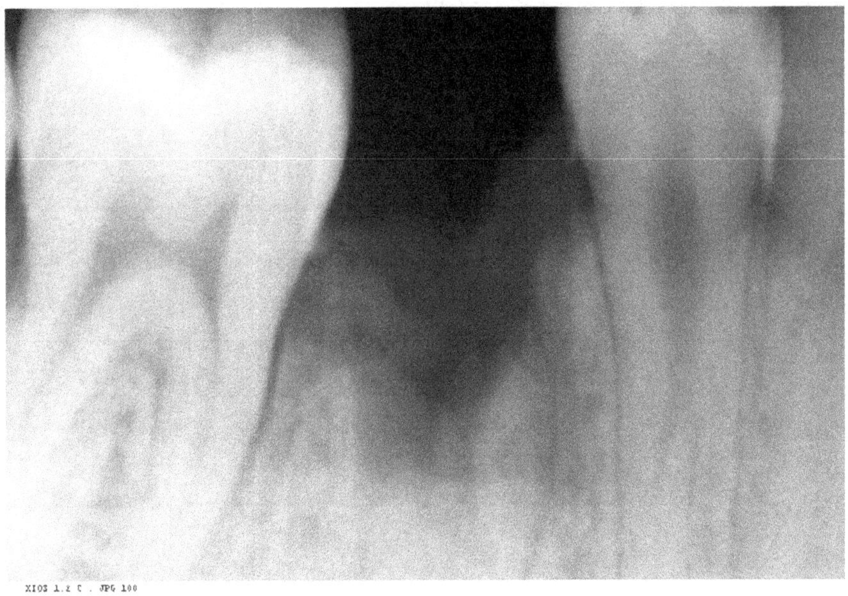

Figure 1d. Teeth with metal fillings

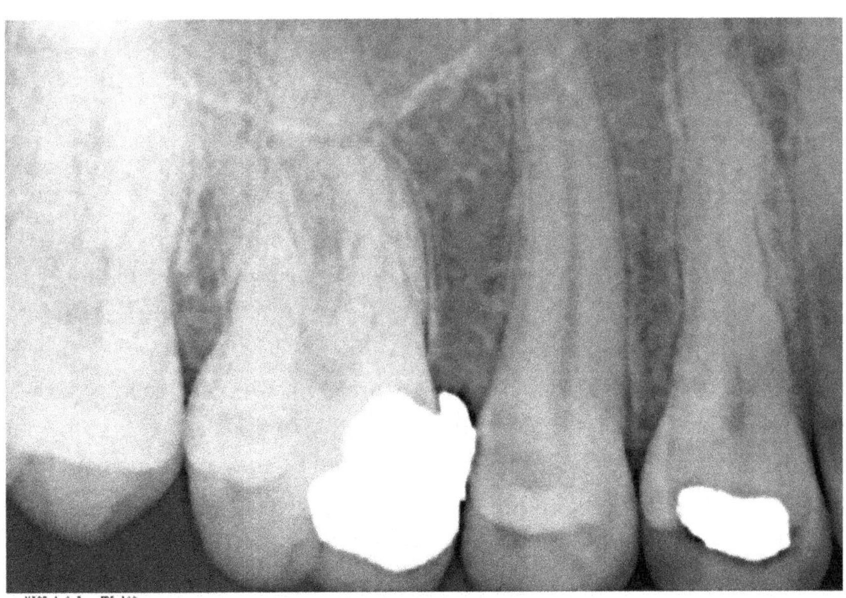

of the neighborhood. The coefficient of distance regularization term is represented by μ. The υ controls the arc length term. The parameters of the LKM model are chosen as follows: $\sigma = 10$, $\mu = 1$, $\upsilon = 6.05$ and time step = 0.1.

The optimized boundary driven level set (OBDLS) method makes use of the optimized LoG to obtain the edge term. The addition of edge term makes easier the evolution of contour towards the target bound-

Figure 2. Segmentation of a typical periapical x-ray image using LKM and OBDLS methods: (e) & (h) - Contour at 10 iterations, (f) & (i) - Contour at 50 iterations and (g) & (j) - Segmentation results

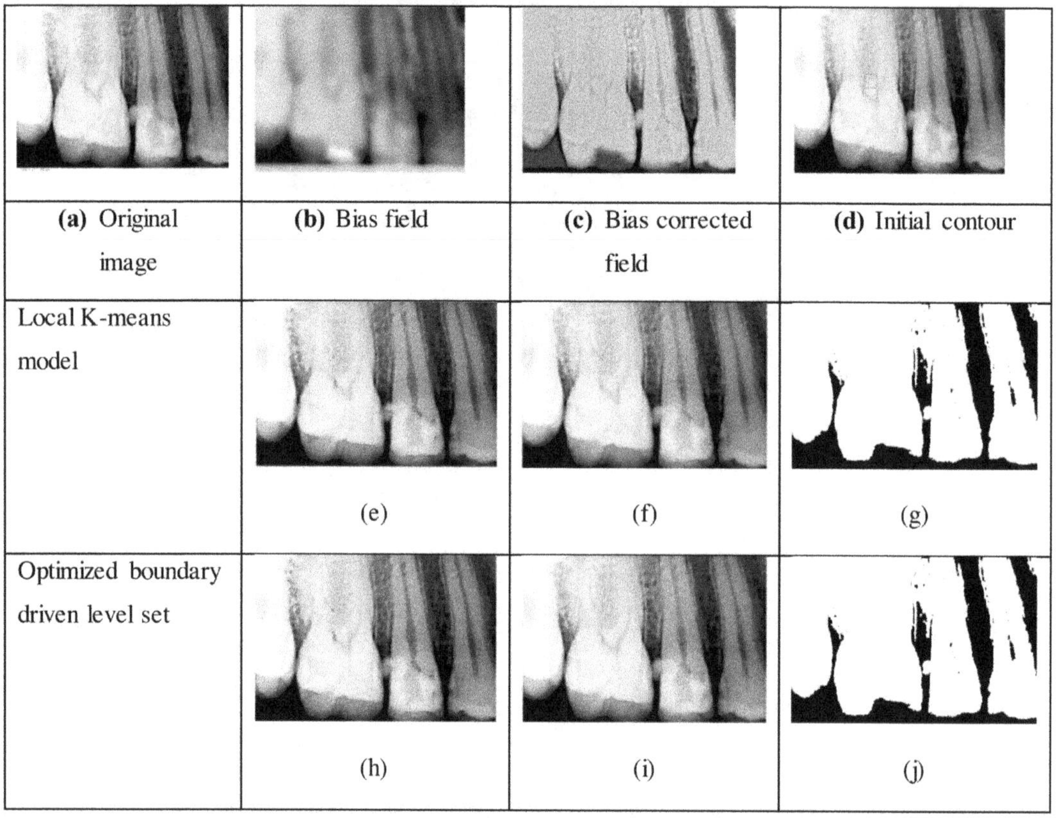

ary. The evolution of contour is controlled by parameters such as σ, μ, υ, ω and time step. The weight of the optimized LoG term in the energy function is ω. The parameters of the level set are chosen as follows: $\sigma = 10$, $\mu = 1$, $\upsilon = 0.001*255^2$, $\omega = 5$ and time step = 0.1. This makes sure that the evolution of the contours is attracted towards the edges.

The sample raw dental X-ray image is given in Figure 2(a). The bias field is shown in Figure 2(b) and the bias corrected image is represented in Figure 2 (c). The bias correction mechanism is similar in both LKM and OBDLS method. The level set is initialized by manual placement of a square contour within the tooth region. This initial contour placement is depicted in Figure 2(d). The evolution of contour at 10 and 50 iterations is shown in Figure 2(e) and 2(f) respectively for LKM method. Similarly for OBDLS algorithm it is shown in Figure 2(h) and 2(i) respectively. The final segmented output is shown for LKM and OBDLS method in Figure 2(g) and (j) respectively. Overall both the LKM and OBDLS provides final contour close to the manual contour. However, later method produces contour closer to the object boundary.

MICO uses energy minimization for bias field estimation and segmentation. It is employed for the segmentation of dental X-rays. The results are shown in Figure 3. The original image, bias field and the bias corrected image are shown in Figure 3 (a) - (c) respectively. The segmented region of teeth is shown in Figure 3 (d). It provides results similar to OBDLS method as the energy minimization framework

Figure 3a. Segmentation of a typical periapical X-ray image using MICO: Original image

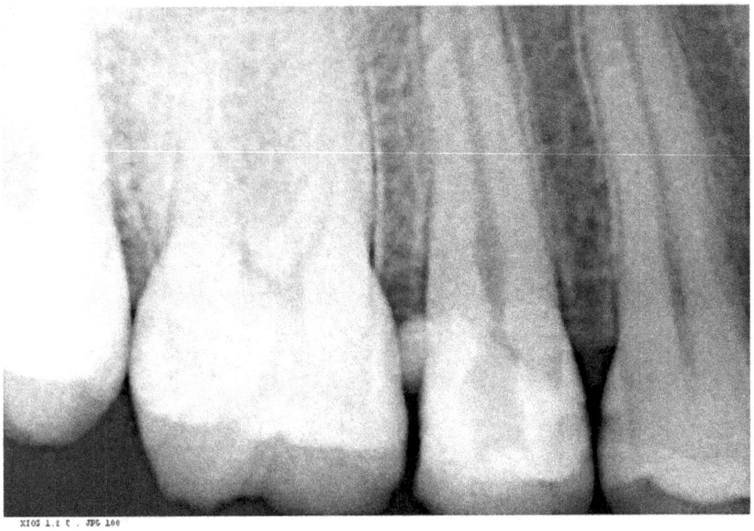

Figure 3b. Bias field

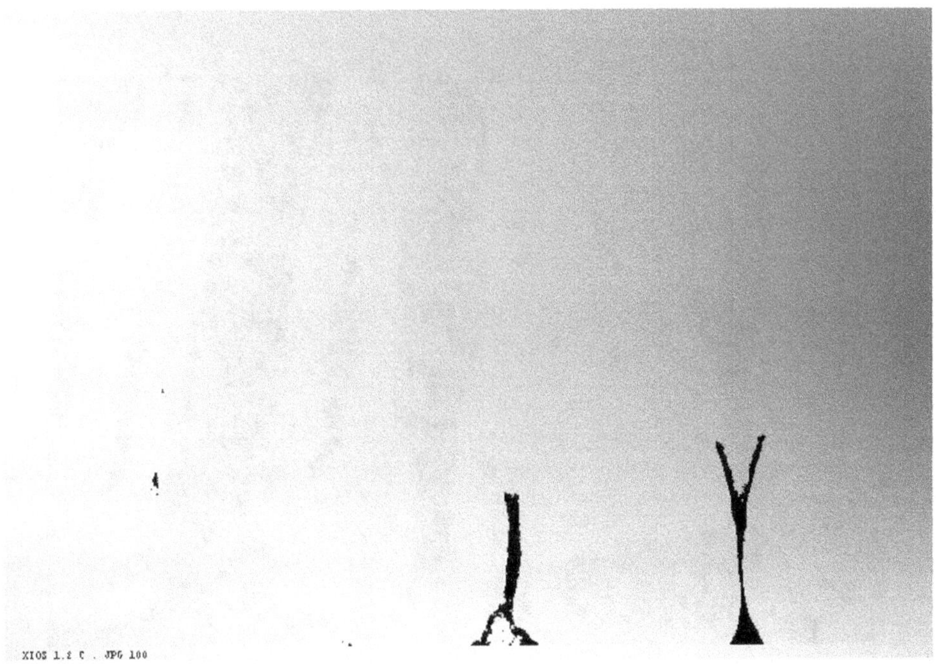

is able to better capture the object boundaries. It is obvious that the recent level set based methods can provide appropriate segmentation of dental images.

The performance of these algorithms is quantitatively validated using Dice coefficient and Jaccard index. The average values of these similarity measures obtained for LKM, OBDLS, and MICO methods is shown in Figure 4. The average Dice coefficient is obtained as 0.897, 0.907 and 0.902 for LKM,

Figure 3c. Bias corrected image

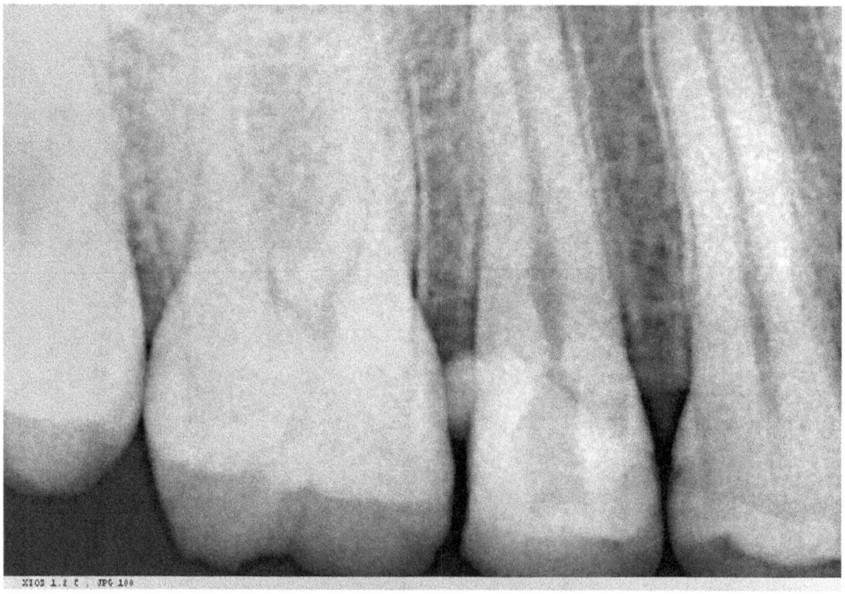

Figure 3d. Segmented results

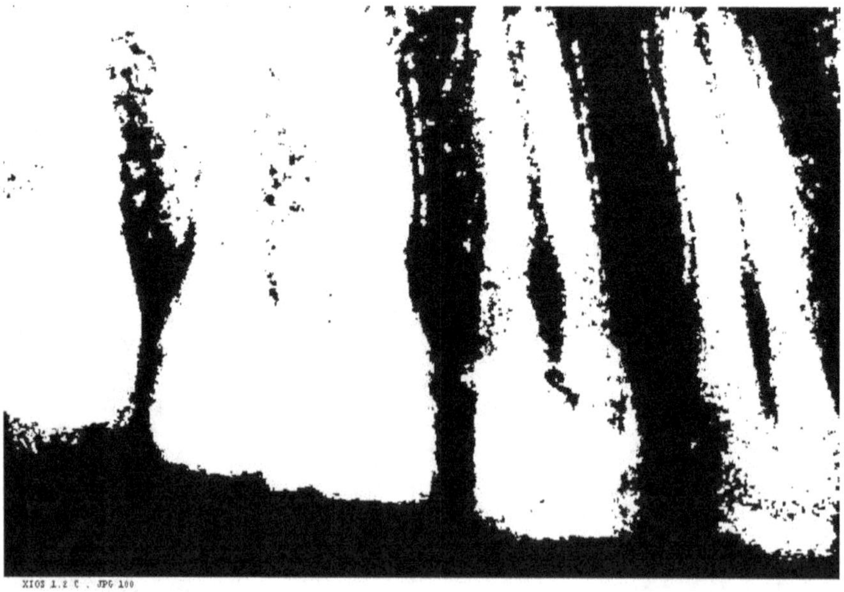

OBDLS, and MICO methods respectively. Similarly the mean Jaccard index is evaluated as 0.818, 0.833 and 0.823 respectively. The OBDLS and MICO methods perform better than LKM. It is difficult to obtain appropriate boundary of the teeth for root canal procedure due to intensity homogeneity and its geometry. Both the proposed segmentation algorithms MICO and OBDLS methods incorporate bias correction that overcomes intensity inhomogeneity problem. The segmentation performance of OBDLS and MICO is similar. Hence based on the requirement either region based MICO or edge based OBDLS can be used for segmentation of dental images. However, the placement of initial contour is challenging due to variation in teeth dimension from subject to subject. This is taken care by OBDLS as the edge based energy drives the level set close to the boundary.

The local K-means model and optimized boundary driven level set segmentation have been applied to dental X-ray images. The performance is analysed visually and quantitatively. The algorithm parameters are kept almost the same on different images. The fine tuning of these parameters might lead to better results with large dataset. Further measures can be taken to reduce the influence of the parameters. The performance of the two methods is assessed by comparing them visually to each other and with similarity measures. The manual curve however can be drawn by an expert.

In conclusion it can be said that different variants of level set can be used for dental image segmentation. Appropriate segmentation methods improve the quality of treatment and save cost. The processing time of the algorithm should be considerably small to make it usable practically. Hence, the processing time of the segmentation algorithm also plays a significant role.

FUTURE RESEARCH DIRECTIONS

The future research directions in the field of dental X-ray image analysis involve several issues: choice of seed points, memory management, computation time, initialization, training, less user interaction and

Figure 4. Performance comparison of different methods: LKM – Local K-means model, OBDLS – Optimized boundary driven level set, MICO – Multiplicative intrinsic component optimization

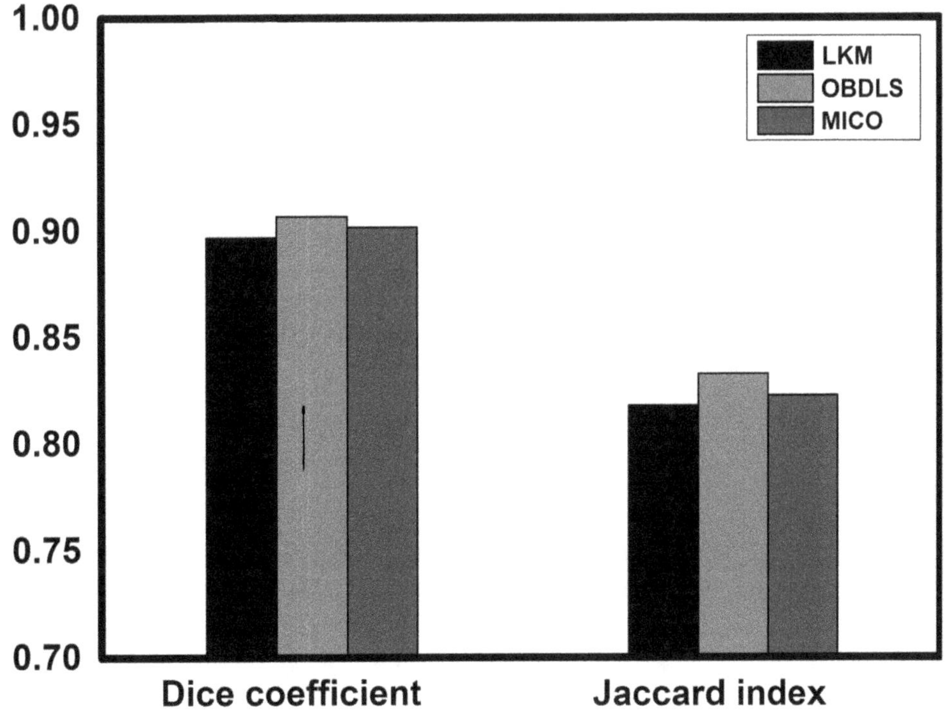

computational complexity. The structure of dental images differs a lot. These images are sensitive to noise such as Gaussian, salt and pepper noise. The segmentation algorithms should be developed that are robust to noise and data presentation.

Various tools can be used for segmentation. Even if many segmentation algorithms are explained here, they also outline a natural guide for future research. Specialized datamining methods can be investigated for dental X-ray image segmentation.

The dental segmentation results can be use in various real world applications such as dental diagnosis system, automated dental identification system, forensic, teeth numbering, dental age estimation and analysis of dental plaque. Accurate segmentation plays a major role in all these systems. It aids effective treatment strategy and reliable person identification. It can assist dental clinicians in their professional work to a great extent. This is an initial step in development of a hybrid approach for dental diagnosis problem combining segmentation, classification and decision making methods.

CONCLUSION

The aim of this chapter was to analyse the performance of edge and region based level set algorithm for segmentation of dental X-rays. Here, region based MICO and edge based OBDLS algorithms are studied for dental image segmentation. These algorithms reduce the impact of intensity inhomogeneity and initialization of contour for level set. From the segmented image, individual tooth can be extracted for further processing.

REFERENCES

Alazab, M., Islam, M., & Venkatraman, S. (2009). Towards Automatic Image Segmentation Using Optimised Region Growing Technique. In AI 2009: Advances in Artificial Intelligence. AI 2009. Springer. doi:10.1007/978-3-642-10439-8_14

Ali, M., Son, L. H., Khan, M., & Tung, N. T. (2017). Segmentation of Dental X-ray Images in Medical Imaging using Neutrosophic Orthogonal Matrices. *Expert Systems with Applications*. doi:10.1016/j.eswa.2017.09.027

Anandh, K. R., Sujatha, C. M., & Ramakrishnan, S. (2016). Laplace Beltrami eigen value based classification of normal and Alzheimer MR images using parametric and non-parametric classifiers. *Expert Systems with Applications*, *59*(C), 208–216.

Anandh, K. R., Sujatha, C. M., & Ramakrishnan, S. (2016). A Method to Differentiate Mild Cognitive Impairment and Alzheimer in MR Images using Eigen Value Descriptors. *Journal of Medical Systems*, *40*(25). PMID:26547845

Babshet, M., Acharya, A. B., & Naikmasur, V. G. (2010). Age estimation in Indians from pulp/tooth area ratio of mandibular canines. *Forensic Sci. Int.*, *197*(1-3), 125e1–e4.

Barboza, E., Marana, A. & Oliveira, D. (2012). A Multibiometric Approach in a Semi-Automatic Dental Recognition Using DIFT Technique and Dental Shape Features. *SIBGRAPI 2012 - Workshop of Theses and Dissertations*, 13-18.

Bhadauria, H. S., Singh, A., & Dewal, M. L. (2013). An integrated method for haemorrhage segmentation from brain CT imaging. *Computers & Electrical Engineering*, *39*(5), 1527–1536. doi:10.1016/j.compeleceng.2013.04.010

Cavalcanti, M., Ruprecht, A., Johnson, W., Thomas, S., Jakobsen, J., & Paulo, S. (1999). Oral and maxillofacial radiology. *Oral Surgery, Oral Medicine, and Oral Pathology*, *88*, 353–357. doi:10.1016/S1079-2104(99)70042-9

Chaabene, M., Ali, R. B., Ejbali, R., & Zaied, M. (2017). Hybrid approach for detection of dental caries based on the methods FCM and level sets. *SPIE 10341, Ninth International Conference on Machine Vision*. doi: 10.1117/12.2268437

Chattopadhyay, C., Kim, D. W., Gombos, D. S., Oba, J., Qin, Y., Williams, M. D., & Patel, S. P. (2016). Uveal melanoma: From diagnosis to treatment and the science in between. *Cancer*, *122*(15), 2299–2312. doi:10.1002/cncr.29727 PMID:26991400

Chattopadhyay, S., Davis, R. M., Menezes, D. D., Singh, G., Acharya, R. U., & Tamura, T. (2012). Application of Bayesian classifier for the diagnosis of dental pain. *Journal of Medical Systems*, *36*(3), 1425–1439. doi:10.100710916-010-9604-y PMID:20945154

Chen, H., & Jain, A. K. (2004). Tooth contour extraction for matching dental radiographs. Proc. of International Conference on Pattern Recognition, 3, 522-525.

Chen, Y., Zhang, J., & Yang, J. (2012). An anisotropic images segmentation and bias correction method. *Magnetic Resonance Imaging*, *30*(1), 85–95. doi:10.1016/j.mri.2011.09.003 PMID:22055751

Dhawan, A. P. (2011). *Medical Image Analysis* (2nd ed.). Wiley-IEEE Press. doi:10.1002/9780470918548

Ding, K., Xiao, L., & Weng, G. (2017). Active contours driven by region-scalable fitting and opti-mized Laplacian of Gaussian energy for image segmentation. *Signal Processing*, *134*, 224–233. doi:10.1016/j.sigpro.2016.12.021

Dougherty, G. (2009). *Digital Image Processing for Medical Applications*. Cambridge University Press.

Duda, R. O., Hart, P. E., & Stork, D. G. (2001). *Pattern Classification*. New York-John Wiley & Sons.

Elazab, A., Wang, C., Jia, F., Wu, J., Li, G., & Hu, Q. (2015). Segmentation of Brain Tissues from Magnetic Resonance Images Using Adaptively regularized Kernel-Based Fuzzy *C*-Means Clustering. *Computational and Mathematical Methods in Medicine, 485495,* 1–12. doi:10.1155/2015/485495 PMID:26793269

Fahmy, G., Nassar, D., Haj-Said, E., Chen, H., Nomir, O., & Zhou, J. (2004). Towards an automated dental identification system (ADIS). *International conference on biometric authentication*, *3*, 522–525. 10.1007/978-3-540-25948-0_107

Feng, C., Zhao, D., & Huang, M. (2017). Image segmentation and bias correction using local inhomogeneous iNtensity clustering (LINC): A region-based level set method. *Neurocomputing*, *219*(C), 107–129. doi:10.1016/j.neucom.2016.09.008

Fredo, A. R. J., Kavitha, G., & Ramakrishnan, S. (2014). Segmentation and morphometric analysis of sub-cortical regions in autistic MR brain images using fuzzy Gaussian model based distance regularized multi-phase level set. *International Journal of Biomedical Engineering and Technology*, *15*(3), 211–223. doi:10.1504/IJBET.2014.064647

Fredo, A. R. J., Kavitha, G., & Ramakrishnan, S. (2015). Segmentation and analysis of corpus callosum in autistic MR brain images using reaction diffusion level sets. *Journal of Medical Imaging and Health Informatics*, *5*(4), 737–741. doi:10.1166/jmihi.2015.1442

Gayathri, V., & Menon, H. P. (2014). Challenges in edge extraction of dental x-ray images using image processing algorithms - a review. *Int. J. Comput. Sci. Inf. Technol.*, *5*, 5355–5358.

Goaz, P. W., & White, S. C. (1982). *Oral Radiology-Principles and Interpretation*. St. Louis, MO: The C.V. Mosby Company.

Gonzalez, R. C., & Woods, R. E. (2006). *Digital Image Processing* (3rd ed.). Prentice-Hall Inc.

Grabisch, M., Nguyen, H. T., & Walker, E. A. (2013). Fundamentals of Uncertainty Calculi with Applications to Fuzzy Inference. Springer Science & Business Media.

Gustafson, G., & Malmö, D. O. (1950). Age determination on teeth. *The Journal of the American Dental Association*, *41*(1), 45–54. doi:10.14219/jada.archive.1950.0132 PMID:15428197

Huh, J., Nam, H., Kim, J., Park, J., Shin, S., & Lee, R. (2015). Studies of automatic dental cavity detection system as an auxiliary tool for diagnosis of dental caries in digital x-ray image. *Progress in Medical Physics*, *25*(1), 52–58. doi:10.14316/pmp.2015.26.1.52

Jac Fredo, A. R., Kavitha, G., & Ramakrishnan, S. (2014). Segmentation and analysis of brain subcortical regions using regularized multi-phase level set in autistic MR Images. *International Journal of Imaging Systems and Technology*, *24*(3), 256–262. doi:10.1002/ima.22101

Jac Fredo, A. R., Kavitha, G., & Ramakrishnan, S. (2015). Automated segmentation and analysis of corpus callosum in autistic MR images using fuzzy c-means based level set. *Journal of Medical and Biological Engineering*, *35*(3), 331–337. doi:10.100740846-015-0047-2

Jain, A., Bolle, R., & Pankanti, S. (1999). *Biometrics-Personal Identi'cation in Networked Society*. Dordrecht: Kluwer Academic Publishers.

Jain, A. K., & Chen, H. (2004). Matching of dental X-ray images for human identification. *Pattern Recognition*, *37*(7), 1519–1532. doi:10.1016/j.patcog.2003.12.016

Jain, A. K., Chen, H., & Minut, S. (2003). Dental biometrics: human identi'cation using dental radiographs. *Proc. Fourth International Conference on AVBPA*, 429-437.

Ji, Z., Xia, Y., Sun, Q., Xia, D., & Feng, D. D. (2012). Local Gaussian distribution fitting based FCM algorithm for brain MR image segmentation. In Intelligent Science and Intelligent Data Engineering, 7202, 318-325. doi:10.1007/978-3-642-31919-8_41

Kass, M., Witkin, A., & Terzopoulos, D. (1988). Snakes: Active contour models. *International Journal of Computer Vision, 1*(4), 321–331. doi:10.1007/BF00133570

Kavitha, M. S., Asano, A., Taguchi, A., Kurita, T., & Sanada, M. (2012). Diagnosis of osteoporosis from dental panoramic radiographs using the support vector machine method in a computer-aided system. *BMC Medical Imaging, 12*(1), 2–11. doi:10.1186/1471-2342-12-1 PMID:22248480

Kayalvizhi, M., & Kavitha, G., Sujatha, C. M., & Ramakrishnan, S. (2015). Study of Alzheimer's Disease progression in MR brain images based on segmentation and analysis of ventricles using Modified DRLSE Method and Minkowski Functionals. *Biomedical Sciences Instrumentation, 51*, 332–340. PMID:25996736

Kayalvizhi, M., Kavitha, G., & Sujatha, C. M. (2013). Analysis of ventricle regions in Alzheimer's brain MR images using level set based methods. *International Journal of Biomedical Engineering and Technology, 12*(3), 300–319. doi:10.1504/IJBET.2013.057266

Keshtkar, F., & Gueaieb, W. (2006). Segmentation of dental radiographs using a swarm intelligence approach. *IEEE Canadian Conference on Electrical and Computer Engineering*, 328-331. 10.1109/CCECE.2006.277656

Kondo, T., Ong, S. H., & Foong, K. W. (2004). Tooth segmentation of dental study models using range images. *IEEE Transactions on Medical Imaging, 23*(3), 350–362. doi:10.1109/TMI.2004.824235 PMID:15027528

Kvaal, S. I., Kolltveit, K. M., Thomsen, I. O., & Solheim, T. (1995). Age estimation of adults from dental radiographs. *Forensic Science International, 74*(3), 175–185. doi:10.1016/0379-0738(95)01760-G PMID:7557754

Lai, Y. H., & Lin, P. L. (2008). *Effective Segmentation for Dental X-ray Images Using Texture-based Fuzzy Inference System. In Advanced Concepts for IntelligentVision Systems* (pp. 936–947). Berlin: Springer.

Lakhani, K., Minocha, B., & Gugnani, N. (2016). Analyzing edge detection techniques for feature extraction in dental radiographs. *Perspectives on Science, 8*, 395–398. doi:10.1016/j.pisc.2016.04.087

Latha, M., & Kavitha, G. (2017). Segmentation and analysis of ventricles in Schizophrenic MR brain images using optimal region based energy minimization framework. *IEEE Proc. 4th International Conference on Signal Processing, Communications and Networking (ICSCN -2017)*. 10.1109/ICSCN.2017.8085735

Li, C., Gore, J. C., & Davatzikos, C. (2014). Multiplicative intrinsic component optimization (MICO) for MRI bias field estimation and tissue segmentation. *Magnetic Resonance Imaging, 32*(7), 913–923. doi:10.1016/j.mri.2014.03.010 PMID:24928302

Li, C., Huang, R., Ding, Z., Gatenby, C., Metaxas, D., & Gore, J. (2008). A variational level set approach to segmentation and bias correction of images with intensity inhomogeneity. *Medical image computing and computer-assisted intervention, 11*(Pt 2), 1083-1091.

Li, C., Huang, R., Ding, Z., Gatenby, J. C., Metaxas, D. N., & Gore, J. C. (2011). A level set method for image segmentation in the presence of intensity inhomogeneities with application to MRI. *IEEE Transactions on Image Processing*, *20*(7), 2007–2016. doi:10.1109/TIP.2011.2146190 PMID:21518662

Li, C., Xu, C., Gui, C., & Fox, M. D. (2010). Distance Regularized Level Set Evolution and Its Application to Image Segmentation. *IEEE Transactions on Image Processing*, *19*(12), 3243–3254. doi:10.1109/TIP.2010.2069690 PMID:20801742

Li, S., Fevens, T., Krzyzak, A., Jin, C., & Li, S. (2007). Semi-automatic computer aided lesion detection in dental X-rays using variational level set. *Pattern Recognition*, *40*(10), 2861–2873. doi:10.1016/j.patcog.2007.01.012

Li, S., Fevens, T., Krzyz˙ak, A., & Li, S. (2005). Level set segmentation for computer aided dental X-ray analysis. SPIE conference on medical imaging, 5747, 580-589. doi:10.1117/12.595537

Li, S., Fevens, T., Krzyzak, A., & Li, S. (2006). An automatic variational level set segmentation framework for computer aided dental X-rays analysis in clinical environments. *Computerized Medical Imaging and Graphics*, *30*(2), 65–74. doi:10.1016/j.compmedimag.2005.10.007 PMID:16500077

Lin, P. L., Huang, P. W., Cho, Y. S., & Kuo, C. H. (2013). An automatic and effective tooth isolation method for dental radiographs. *Opto-Electronics Review*, *21*(1), 126–136. doi:10.247811772-012-0051-9

Lin, P. L., Huang, P. Y., & Huang, P. W. (2012). An automatic lesiondetection method for dental X-ray images by segmentationusing variational level set. *Proc. 2012 International Conference on Machine Learning and Cybernetics*, 1821-1826.

Lin, P. L., & Lai, Y. H. (2008). Effective segmentation for dental X-rayimages using texture-based fuzzy inference system. *Advanced Concepts for Intelligent Visions System*, *5259*, 936–947.

Lin, P. L., Lai, Y. H., & Huang, P. W. (2010). An effective classification and numbering system for dental bitewing radiographs using teeth region and contour information. *Pattern Recognition*, *43*(4), 1380–1392. doi:10.1016/j.patcog.2009.10.005

Lira, P., Giraldi, G., Neves, L. & Feijoo, R. (2014). Dental R-Ray Image Segmentation Using Texture Recognition. *Latin America Transactions*, *12*(4).

Lpez-Lpez, J. (2012). Computer-aided system for morphometric mandibular index computation (using dental panoramic radiographs). *Med. Oral Patol. Oral*, *17*, e624–e632. doi:10.4317/medoral.17637 PMID:22322489

Mahoor, M. H., & Abdel-Mottaleb, M. (2005). Classification and numbering of teeth in dental bitewing images. *Pattern Recognition*, *38*(4), 577–586. doi:10.1016/j.patcog.2004.08.012

Miki, Y., Muramatsu, C., Hayashi, T., Zhou, X., Hara, T., Katsumata, A., & Fujita, H. (2017). Classification of teeth in cone-beam CT using deep convolutional neural network. *Computers in Biology and Medicine*, *80*, 24–29. doi:10.1016/j.compbiomed.2016.11.003 PMID:27889430

Mitiche, A., & Ayed, I. B. (2011). *Variational and Level Set Methods in Image Segmentation. In Springer Topics in Signal Processing*. Springer-Verlag Berlin Heidelberg. doi:10.1007/978-3-642-15352-5

Nakamoto, T., Taguchi, A., Ohtsuka, M., Suei, Y., Fujita, M., Tsuda, M., ... Tanimoto, K. (2008). A computer-aided diagnosis system to screen for osteoporosis using dental panoramic radiographs. *Dento Maxillo Facial Radiology*, *37*(5), 274–281. doi:10.1259/dmfr/68621207 PMID:18606749

Nelson, S. J., & Ash, M. M. (2010). *Wheeler's Dental Anatomy, Physiology, and Occlusion* (9th ed.). Saunders Elsevier.

Nomir, O., & Abdel-Mottaleb, M. (2005). A system for human identification from X-ray dental radiographs. *Pattern Recognition*, *38*(8), 1295–1305. doi:10.1016/j.patcog.2004.12.010

Nurtanio, I., Purnama, I. K. E., Hariadi, M., & Purnomo, M. H. (2011). Cyst and tumor lesion segmentation on dental panoramic images using active contour models. *IPTEK J. Technol. Sci.*, *22*(3).

Osher, S., & Sethian, J. A. (1988). Fronts Propagating with Curvature Dependent speed: Algorithms Based on Hamilton-Jacobi Formulation. *Journal of Computational Physics*, *79*(1), 12–49. doi:10.1016/0021-9991(88)90002-2

Pandey, P., Bhan, A., Dutta, M. K., & Travieso, C. M. (2017). Automatic Image Processing Based Dental Image Analysis Using Automatic Gaussian Fitting Energy and Level Sets. *International Conference and Workshop on Bioinspired Intelligence*. 10.1109/IWOBI.2017.7985529

Patanachai, N., Covavisaruch, N., & Sinthanayothin, C. (2010). Wavelet transformation for dental X-ray radiographs segmentation technique. *8th International Conference on ICT and Knowledge Engineering*, 103-106.

Pretty, I. A., & Sweet, D. (2001). A look at forensic dentistry-Part 1: The role of teeth in the determination of human identity. *British Dental Journal*, *190*(7), 359–366. PMID:11338039

Pretty, J. A., & Maupomé, G. (2004). A closer look at diagnosis in clinical dental practice:part 1. Reliability, validity, specificity and sensitivity of diagnostic procedures. *Journal - Canadian Dental Association*, *70*(4), 251–256. PMID:15120020

Rad, A. E., Mohd Rahim, M. S., Rehman, A., Altameem, A., & Saba, T. (2013). Evaluation of current dental radiographs segmentation approaches in computer-aided applications. *IETE Technical Review*, *30*(3), 210–222. doi:10.4103/0256-4602.113498

Rad, A. E., Rahim, M. S. M., & Norouzi, A. (2014). Level set and morphological operation techniques in application of dental image segmentation. *Int. Sch. Sci. Res.Innov.*, *8*(4), 177–180.

Rad, A. E., Rahinm, M. S. M., Rehman, A., & Saba, T. (2016). Digital Dental X-ray Database for Caries Screening. *3D Res*, *7*, 18.

Rajendran, R. (2009). Shafer's textbook of oral pathology (6th ed.). Academic Press.

Said, E., Fahmy, G. F., Nassar, D., & Ammar, H. (2004). Dental x-ray image segmentation. Proc. International society for optics and photonics defense and security, 409-417.

Said, E. H., Nassar, D. E. M., Fahmy, G., & Ammar, H. H. (2006). Teeth segmentation in digitized dental X-ray films using mathematical morphology. *IEEE Transactions on Information Forensics and Security*, *1*(2), 178–189. doi:10.1109/TIFS.2006.873606

Sethian, J. A. (1999). *Level Set Methods and Fast Marching Methods*. Cambridge, UK: Cambridge University Press.

Shah, S., Abaza, A., Ross, A., & Ammar, H. (2006). Automatic tooth segmentation using active contour without edges. *IEEE Biometrics Symposium*. 10.1109/BCC.2006.4341636

Shahin, K. A., Chatra, L., & Shenai, P. (2013). Dental and craniofacial imaging in forensics. *JOFRI*, *1*(2), 56–62.

Son, L. H., & Tuan, T. M. (2016). A cooperative semi-supervised fuzzy clustering framework for dental X-ray image segmentation. *Expert Systems with Applications*, *46*, 380–393. doi:10.1016/j.eswa.2015.11.001

Son, L. H., & Tuan, T. M. (2017). Dental segmentation from X-ray images using semi-supervised fuzzy clustering with spatial constraints. *Engineering Applications of Artificial Intelligence*, *59*, 186–195. doi:10.1016/j.engappai.2017.01.003

Son, L. H., Tuan, T. M., Fujita, H., Dey, N., Ashour, A. S., Ngoc, V. T. N., ... Chu, D. T. (2018). Dental diagnosis from X-Ray images: An expert system based on fuzzy computing. *Biomedical Signal Processing and Control*, *39*, 64–73. doi:10.1016/j.bspc.2017.07.005

Wang, X., Shan, J., Niu, Y., Tan, L., & Zhang, S. X. (2014). Enhanced distance regularization for re-initialization free level set evolution with application to image segmentation. *Neurocomputing*, *141*, 223–235. doi:10.1016/j.neucom.2014.03.011

Wriedt, S., Jaklin, J., Al-Nawas, B., & Wehrbein, H. (2012). Impacted upper canines: Examination and treatment proposal based on 3d versus 2d diagnosis. *Journal of Orofacial Orthopedics*, *73*(1), 28–40. doi:10.100700056-011-0058-8 PMID:22246048

Zhang, K., Zhang, L., Song, H., & Zhang, D. (2013). Reinitialization-free level set evolution via reaction diffusion. *IEEE Transactions on Image Processing*, *22*(1), 258–271. doi:10.1109/TIP.2012.2214046 PMID:22910114

Zhang, K., Zhang, L., & Zhang, S. (2010). A variational multiphase level set approach to simultaneous segmentation and bias correction. *Proc. 17th IEEE International Conference on Image Processing (ICIP)*, 4105-4108. 10.1109/ICIP.2010.5651554

Zhou, J., & Mottaleb, M. A. (2005). A content-based system for human identification based on bitewing dental X-ray images. *Pattern Recognition*, *38*(11), 2132–2142. doi:10.1016/j.patcog.2005.01.011

Zhou, S., Wang, J., Zhang, M., Cai, Q., & Gong, Y. (2017). Correntropy-based level set method for medical image segmentation and bias correction. *Neurocomputing*, *234*, 216–229. doi:10.1016/j.neucom.2017.01.013

ADDITIONAL READING

Amer, Y. Y., & Aqel, M. J. (2015). An Efficient *Segmentation* Algorithm for Panoramic *Dental* Images. *Procedia Computer Science*, *65*, 718–725. doi:10.1016/j.procs.2015.09.016

Cunha, P., Guevara, M. A., Messias, A., Rocha, S., Reis, R., & Nicolau, P. M. G. (2013). A method for segmentation of dental implants and crestal bone. *International Journal of Computer Assisted Radiology and Surgery*, *8*(5), 711–721. doi:10.100711548-012-0802-6 PMID:23212460

Gumus, E. (2016). Segmentation and root localization for analysis of dental radiographs. *Signal, Image and Video Processing*, *10*(6), 1073–1079. doi:10.100711760-016-0861-1

Lin, P. L., Huang, P. Y., Huang, P. W., Hsu, H. C., & Chen, C. C. (2014). Teeth segmentation of dental periapical radiographs based on local singularity analysis. *Computer Methods and Programs in Biomedicine*, *113*(2), 433–445. doi:10.1016/j.cmpb.2013.10.015 PMID:24252317

Marroquin, T. Y., Karkhanis, S., Kvaal, S. I., Vasudavan, S., Kruger, E., & Tennant, M. (2017). Age estimation in adults by dental imaging assessment systematic review. *Forensic Science International*, *275*, 203–211. doi:10.1016/j.forsciint.2017.03.007 PMID:28410514

Mathworks (2015). Dental Image Segmentation. Available at: http://www.mathworks.com/matlabcentral/fileexchange/51714-dental-image-segmentation

Oliveira, J., & Proença, H. (2011). Caries Detection in Panoramic Dental X-ray Images. In J. Tavares & R. Jorge (Eds.), *Computational Vision and Medical Image Processing. Computational Methods in Applied Sciences* (Vol. 19, pp. 175–190). Dordrecht: Springer. doi:10.1007/978-94-007-0011-6_10

Scott, J. H. (1977). *Introduction to dental anatomy*. Edinburgh: Churchill Livingstone.

Shaheen, E., Khalil, W., Ezeldeen, M., Casteele, E. V. D., Sun, Y., Politis, C., & Jacobs, R. (2017). Accuracy of segmentation of tooth structures using 3 different CBCT machines. *Oral Surgery, Oral Medicine, Oral Pathology and Oral Radiology*, *123*(1), 123–128. doi:10.1016/j.oooo.2016.09.005 PMID:27938942

Tuan, T. M., Ngan, T. T., & Son, L. H. (2016). A novel semi-supervised fuzzy clustering method based on interactive fuzzy satisficing for dental x-ray image segmentation. *Applied Intelligence*, *45*(2), 402–428. doi:10.100710489-016-0763-5

Wang, C. W., Huang, C. T., Lee, J. H., Li, C. H., Chang, S. W., Siao, M. J., ... Lindner, C. (2016). A benchmark for comparison of dental radiography analysis algorithms. *Medical Image Analysis*, *31*, 63–76. doi:10.1016/j.media.2016.02.004 PMID:26974042

Zhang, Z., Ong, S. H., Zhong, X., & Foong, K. W. C. (2016). Efficient 3D dental identification via signed feature histogram and learning keypoint detection. *Pattern Recognition*, *60*, 189–204. doi:10.1016/j.patcog.2016.05.007

Chapter 4
Dental Image Segmentation Using Clustering Techniques and Level Set Methods

Prabha Sathees
Hindustan Institute of Technology and Science, India

ABSTRACT

Segmentation is necessary for dental images for finding the parts of the teeth, surrounding tissues, and bones. The human identification system in dental methodology is a tedious and time-consuming process. The automatic identification system is the best solution for dental diagnosis and dental treatment systems. Choosing an appropriate region of interest with high accuracy and success rate is a challenging one. This can be attained with the help of proper segmentation methodologies. The segmentation techniques proposed for the root canal treatment are analyzed and compared. Clustering techniques and level set methods with different edge maps are implemented for the proper analysis of segmentation in dental images. Finally, the integration of coherence-enhanced diffusion filtering in basic level set segmentation methodology seems to be effective in improving the segmentation performance of dental images.

INTRODUCTION

Identification of decayed tooth is the vital part in the analysis of dental images. Root canal treatment is necessary due to the inflammation or infection in the pulp. Issues in the root canal may also arise due to gum disease which leads to the need for root canal treatment (Tiwari & Yardi, 2006). The observation of dental diagnosis and treatment has been made easy from the information of dental X-ray image. These radiographs are very useful for medical specialists of analyzing dental images from hidden dental structures (Jiayin & Zhicheng, 2010; Joao & Hugo, 2011; Shuo, Thomas, Adam, 2006; Lai & Lin, 2006; Lai, Lin, Huang, 2010; Hui & Oksam, 2010). The natural cavity that present within the center of the tooth is nothing but root canal. The soft tissue within the root canal is pulp. The development of root of the tooth depends on blood vessels and nerves present in the pulp (Cremers, 2003). The normal tooth is classified as three regions that is crown, neck and root part. The top part of the tooth is crown and the

DOI: 10.4018/978-1-5225-6243-6.ch004

bottom end is root part. The pulp expands from top to bottom end of the tooth thereby connecting the tip of surrounding tissues.

Root Canal Treatment (RCT) in endodontics is suggested for severely infected tooth pulp. If any one of the teeth are broken, decayed, or having loose filling, bacteria can infiltrate the pulp, causing it to become infected. To ensure the efficacy of any future dental treatment, infected tooth pulp must be treated as soon as possible. Root canal treatment involves the removal of the infected or inflamed dental nerve followed by the cleaning, shaping, disinfection and obturation (filling) of the space that the nerve occupied within the pulp of the tooth (Johnson & Guttmann, 2007). This procedure is usually carried out when a tooth has an abscess or when the nerve at the centre of the tooth has become inflamed due to decay, fracture or a crack. Often the only alternative to this treatment is having the tooth extracted. After clearing the area of bacteria, the tooth is treated with a special filling material that seals out potential pathogens that may cause re-infection. Placement of crown at atop in treated tooth for reinstate the outer structure as well as functionality back to the tooth (Michen, Pearson, Rahbaran, Gulabilava, 2003). If there is damage in middle and bottom part of the tooth, then root canal treatment is recommended. The size and shape of the root canals vary from person to person. This is the reason for the precise recognition of root canal as a key step for further analysis.

The decayed teeth extraction method is the regular process for the cleaning of roots in order to avert the damage transferred to neighborhood teeth. The step involved in the treatment procedure of the tooth contains the removal ofthe damaged area, appareling and cleansing and the final stage contains the filling and closing it (Shafer, 2006). For this process, radiographs help in determining the prognosis by comparison with post operative and follow up action. Various image processing techniques are employed for accurate detection of location and shape of the root canal. The primary identification in root canal treatment is to asset the aspect of impairment area in tooth (Ahmed, Taib, Khalid, Ahmad, Taib, 2011). The need for segmentation of image is an important and a tough assignment for medical practitioners since areas of gum and pulp drop down into similar level of intensities (Jain & Chauhan, 2017).

Segmentation is an important step in processing of a variety of medical images (Manickavasagam, Sutha, & Kamalanand, 2014; Kamalanand, & Ramakrishnan, 2015; Vaishnavi, Jeevananthan, Begum, & Kamalanand, 2014; Rajinikanth, Fernandes, Bhushan, & Sunder, 2018; Rajinikanth, Satapathy, Dey, & Vijayarajan, 2018) .The dental x-ray image segmentation would be complex owing to the pattern and intensity distinction within the single dental x-ray images and also across different dental x-ray images. Dental image analysis systems are analyzed by many researches however it has to be considered to find the suitable technique (Lai & Lin, 2008). A segmentation method is implemented for the segmentation of tooth using integral projection and Bayes rule where a semi-automatic approach is used for the confinement of tooth (Anil & Hong, 2004). In the reported work of segmentation, three steps have been carried out, first one is image enhancement, second one is region of interest localization and final step results from the Snake method after performing morphological operations (Jindan & Mohamed, 2005). In order to segment dental x-ray, a method depends on thresholding has been proposed using the adaptation technique (Omaima & Mohamed, 2005). The Swarm-intelligence is combined with cellular-automata model approach has been introduced to segment dental images (Keshtkar & Gueaieb, 2006). Mathematical morphology approach is introduced in the crisis of teeth segmentation, thereby improvement in segmentation is achieved by a set of morphology refinement activityand connected components are used for investigation to achieve the ambition of interested area (Eyad, DiaaEldin, Gamal & Hany, 2006). In order to identify the region of lesions, semi-automatic lesion detection structure is generated in dental images from the coupling of two level set functions and then the derivation of initial contour obtained from support vector machine (Li, Fevens, Krzyzak, Jin, & Li, 2007). The focus in the dental image is on the automatic detection of the root canal and its proper segmentation for effectual endodontic intervention. The accurate identification of root canal shows the way to less chair time, decreased clinical burden and cost.

FCM CLUSTERING AND LEVEL SET SEGMENTATION

The clustering methods are emerging as a valuable tool in the pattern recognition techniques for the identification of similar and dissimilar patterns in the image. FCM methods are widely used to segment the dissimilar regions in images. The weight parameter estimation and bias field correction are simultaneously achieved during the segmentation (Szilagyi, Szilagyi, & Benyo 2007). K-means algorithms are fused with fuzzy set theory in order to segments an image by the implementation of fuzzy c-means algorithm (Siyal & Lin, 2005). The training data is not required in clustering techniques however initialization is necessary from user and also it is user dependent. Also in the region growing methods, intensity inhomogeneity and noise is pretentious one in clustering technique.

The segmentation of ROI leads to the extraction of desired region from the surrounding tissues. Low contrast images are inherently amorphous and lack clear edges (Borchartt, Conci, Lima, Resmini, & Sanchez., 2012; Zhou & Aggarwal, 2004). The identification of desired area from similar intensities surrounding boundaries is complex one (Machado, Giraldi, Novotny, Marques, & Conci, 2013). Hence, methods that aid in automatic segmentation from low contrast images is an open challenge (Borchartt, Conci, Lima, Resmini, & Sanchez., 2012).

Level Set Methods (LSM) is adequate for approaching complicated topological modification in curves as active action (Li, Xu, Gui, & Fox, 2010). Snake method with gradient vector flow and LSM are generally used to extract organs from different imaging modalities and delineate tumors with good accuracy. The merits of deformable models are insensitive to noise, false edges and small intensity inhomogeneity (Laura, Rubio, Carbayo, Pascau, Tellado, Ramón, Descoand, & Santos, 2009; Zhuang, Valentino, & Toga, 2006; Jayadevappaet, Srinivas Kumar, & Murty, 2011)

Active contour methods have been proposed to extract the smooth shapes using partial differential equations (Caselles, Catte, Coll, & Dibos, 1993). Subsequently, Geodesic active contours have been introduced for image segmentation. This technique established as the contour emerging based on geometrical measures of the image with respect to time (Caselles, Ron, & Sapiro, 1997). LSM has been applied for segmentation of stomach from CT images (Ravikanth, Sethian, & Vemuri, 1995).

Variational level set method is introduced in order to force the curve to move towards the boundary, thereby avoiding the re-initialization procedure (Li, Xu, Guiand, Fox, 2005). A modified LSM is introduced to avoid redundant curve evolution. The curve is made dynamic by incorporating the geometric information into the regularization term (Kuo, Maryellen, Ingrid. Boone, Lindfors, Yang, & Edwards, 2014).

Then, the new regularization namely Distance Regularized Level Set Method (DRLSM) was introduced for avoiding periodic re-initialization using double well potential function and an energy due to external term has been added for driving the required function to reaching desired boundary (Li, Xu, Gui, & Fox, 2010). In some particular cases, when the curve is flat, then initial curve moves far away the desired boundary. To address this limitation, an improved DRLSM is proposed in which a new penalty term is introduced (Wu, Wu, & Huang, 2012).

Perona and Malik have proposed the non-linear anisotropic diffusion based filtering process for detection of edges and representing the images using scale space. Smoothing operation has been performed by considering the local gradient information. By doing so, the intra-regions in the image undergo smoothing process and consequently, the edges of inter-regions are preserved. This method has been widely used in the areas such as edge detection, smoothing, enhancement and image segmentation (Suganthi & Ramakrishnan, 2014; Prabha, Sujatha, & Ramakrishnan, 2015).

Weickert (1999a) has proposed coherence enhancing diffusion filtering for enhancing coherent structures that contains edges at different orientations. Coherence diffusion along with the integration of diffusion tensors and structure tensors significantly helps to achieve orientation smoothing (Prabha & Sujatha, 2014). This process also results in high coherence and good scale space representation. This non-linear coherence enhancing diffusion filter concurrently performs noise reduction as well as preservation of edges. Despite the existence of various non-linear diffusion processes, this method has been reported to the preserve the sharpness of edges and retains the fine scale details of image (Weickert. 1999b).

SEGMENTATION

The significant region of root canal treatemnet is segmented from the dental images with the use of FCM and level set methods. FCM method and level set method based on edge technique are utilized for the performance of segmentation.

FCM Clustering

The number of clusters c is partitioned as $\left\{z_k\right\}_{k=1}^{G}$ for forming the objective function of FCM (Bhadauria, Singh, & Dewal, 2013) is given by

$$O = \sum_{i=1}^{c}\sum_{k=1}^{G} M_{ik}^{m}\left\|z_k - c_i\right\|^2$$

where z_k is the kth pixel of the gray value, G number of grey levels, c_i is the ith cluster centers, M_{ik} is the membership function of fuzzy with k^{th} pixel corresponding to i^{th} cluster which is strained to $\sum_{i=1}^{c} M_{ik} = 1\forall k;$ $M_{ik} \in [0,1]$, m is the criterion to control the results segmentation using fuzzy property. The membership function of fuzzy M_{ik} is given by:

$$M_{ik} = \frac{\left\|z_k - c_i\right\|^{-2/m-1}}{\sum_{i=1}^{c}\left\|z_k - c_i\right\|^{-2/m-1}}$$

The cluster centres c_i are iteratively calculated as:

$$c_i = \frac{\sum_{k=1}^{G} M_{ik}^{m} z_k}{\sum_{k=1}^{G} M_{ik}^{m}}$$

In general, the cluster centres with less number of iterations are initially choose as random one to diminish the objective function. The iteration number has been carried out for 50 and the cluster centre is eight. In this work, the root canals are extracted using FCM method.

Level Set Segmentation

The segmentation of ROI aims to extract the root canal from other parts of the image. LSM is proposed and adopted for root canal extraction in dental images. It is a numerical technique proposed by (Osher and Sethian, 1988) that is used to track complex shapes and interfaces. It is mainly adapted in the fields of processing the images, computer vision, fluid dynamics and computational geometry. It was first applied in medical imaging to segment stomach from CT images (Ravikanth, Sethian, & Vemuri, 1995).

Zero level set function is represented by closed contour as a higher order is known as level set function (LSF). The LSM manipulates this closed contour implicitly through LSF. Signed distance functions (SDF) are implicit functions which possess the property $\left|\nabla_{,}\right| = 1$. where θ .is the zero level set curve. In traditional level set formulation, certain regions of LSF can develop shocks, due to very sharp and flat gradients. This results in highly inaccurate computation. In order to reduce these errors, periodic reinitialization of the LSF is recommended. In order to make LSF as a stable one, it is important to implement reinitialization procedure. The application and sensitivity plays a major role in the reinitialization of LSF based on their accuracy (Prabha, Suganthi, & Sujatha, 2015). The addition of extra diffusion in reinitialization procedure leads to the sharp corner discovery as difficult one. The next one drawback is the execution of contour is extreme from the actual boundary, it may take more iteration to reinitialize the LSF to SDF due to small gradient. It might also be the case that the reinitialization may not occur at all and the LSF may not move at all.

To overcome the complex, expensive and undesirable results of periodic reinitialization, finite difference strategy is implemented for the replacement of reinitialization procedure in level set method (Li, Xu, Guiand, & Fox, 2005). This method was adopted and applied on dental images in order to delineate root canal treatment.

Level Set Method Without Re-Initialization

Dental images are subjected to level set method without re-initialization. In order to avoid reinitialization problem, the diffusion equation in forward and backward direction are organized with penalty term in LSE equation thereby controlling the signed distance property thereby costly reinitialization is avoided. The evolution equation in level set is represented by

$$\frac{\partial \theta}{\partial t} + A\left|\nabla \theta\right| = 0$$

where, $\theta = \theta\left(x, y, t\right)$ is the level set function (LSF) and A is the speed term. The penalty term (Li, Xu, Guiand, & Fox, 2005; Caselles, Ron, & Sapiro, 1997; Malladi, Sethian, & Vemuri, 1995) is evolved to control the signed distance property (SDF) $\left|\nabla \varphi\right| = 1$ which is given as

$$\varepsilon\left(\theta\right) = \mu m\left(\theta\right) + E_{ext}\left(\theta\right)$$

where $\mu > 0$ maintains SDF (Evans, 1998), $E_{ext}\left(\theta\right)$ is the external energy functional and the penalty term $m\left(\theta\right)$ is described below

$$m\left(\theta\right) = \int_{\Omega} \frac{1}{2}\left(\left|\nabla \theta\right| - 1\right)^2 dxdy$$

The variation of calculus is used to attain the steady state solution which is represented as

$$\frac{\partial \varphi}{\partial t} = -\frac{\partial \varepsilon}{\partial \theta}$$

The aim of zeroth level set curve is reaching to the desired boundary by the influence of external energy functional (Li, Xu, Guiand, & Fox, 2005) and is given by

$$E_{ext}\left(\theta\right) = \lambda_{LS} H_g\left(\theta\right) + eB_g\left(\theta\right)$$

where, λ_{LS} 0 and e are constants, the terms $H_g\left(\theta\right)$ means surface integral and $B_g\left(\theta\right)$ means area integral respectively.

The energy function is minimized by the gradient flow (Li, Xu, Guiand, & Fox, 2005) which is given by

$$\frac{\partial \theta}{\partial t} = \mu\left[\Delta\theta - div\left(\frac{\nabla\theta}{\left|\nabla\theta\right|}\right)\right] + \lambda\delta\left(\theta\right) div\left(g\frac{\nabla\theta}{\left|\nabla\theta\right|}\right) + \nu g\delta\left(\theta\right)$$

where g is the edge indicator function which is defined as

$$g = \frac{1}{1 + \left|\nabla G_\sigma * I\right|^2}$$

where the image I is convolved with Gaussian filer G_σ which gives the evolution equation of LSF. The zero level set contour is reaching the object boundary by the influence of energy functional which is determined by the gradient flow. In the above equation, the positive constants μ and λ play less significant roles in the movement of initial contour whereas the constant ν .decides the direction of curve movement, either inward or outward. The effect of the diffusion rate $\left(1-\frac{1}{\left|\nabla\varphi\right|}\right)$ in the first term and g the edge indicator, plays a major role in moving the initial contour towards accurate boundary. In ordinary level set method, the intensity gradient of Gaussian smoothed image is defined as an edge indicator and when the diffusion rate is positive, i.e., $\left|\nabla\varphi\right| > 1$ the gradient is reduced, otherwise the gradient is increased.

Dental images doesn't have clear boundary and the image is a combination of similar intensity level. Therefore edge indicator has been described by intensity method is greater than zero which makes the contour leads to leakage problems. As the accuracy of segmentation depends on the edge indicator and diffusion rate of basic level set method, in this thesis, attempts are made to improve the edge indicator such that initial contour can evolve towards the desired and true boundary.

Edge Enhancement Methods

In the conventional level set methods, edge indicator function g is achieved by the convolution of raw image performed with Gaussian kernel. The diffusion algorithms are implemented in partial differential equations for the removal of noise by processing the raw image. The initial condition of an image $I\left(x,y,0\right)$ having linear isotropic structure in diffusion equation is described as

$$\frac{\partial I\left(x,y,t\right)}{\partial t} = div\left(\nabla I\right)$$

where $I\left(x,y\right)$ is the raw image with x and y co-ordinates, t specifies the time bound and ∇I is the gradient of image. The filtering consequence of an image is described by $I\left(x,y,t\right) = I_0\left(x,y\right) * G\left(x,y;t_g\right)$ where $I_0\left(x,y\right)$ is the raw image and $G\left(x,y;t_g\right)$ is the Gaussian kernel of variance t_g. For high significance of t_g, more number of images is determined at coarser level. Hence, the Gaussian filtering results as smearing in the image (Perona, Shiota, & Malik, 1994). Unfortunately, this results in over smoothing of images leading to loss of edge information. To overcome this drawback, different edge enhancement methods are attempted (Srinivasan & Swaminathan, 2014; Prabha, Anandh,, Sujatha, & Ramakrishnan, 2014).

Anisotropic Diffusion Filtering

In order to achieve the regions of an image as smooth operation and also to avoid the edge blurring strategy, the non-linear anisotropic framework in diffusion equation is implemented for the replacement of drawbacks in linear isotropic structure (Perona & Malik, 1990) that is shown below

$$\frac{\partial I\left(x,y,t\right)}{\partial t} = div\left[f\left(\| \nabla I \|\right)\nabla I\right)]$$

where $\| \nabla I \|$ is the magnitude of gradient and $g\left(\| \nabla I \|\right)$ is the stopping function for edges and is given by

$$f\left(\| \nabla I \|\right) = \exp\left(-\left(\frac{\| \nabla I \|}{k_T}\right)^2\right) k_T$$

represents the threshold of strength of edges. The stopping criterion for the edges in the diffusion process are allowed as g (x)$\rightarrow 0$ then $x \rightarrow \infty$. The implementation of non-linear anisotropic framework in diffusion process resulted as filtered image which is represented as I_t. The edge map 'g' is described as

$$g = \frac{1}{1+ | \nabla I_t |^2}$$

Coherence Enhancing Diffusion Filtering

Diffusion filters with coherence enhancing features are useful in enlarge flow like organization and extra edges are obtained at diverse directions (Weickert, 1999a). The creamy detail of the gradient of Gaussian processed image $u\left(x,t\right)$ those tensor flow is adopted for the description of framework is given by

$$J_0\left(\nabla u_\sigma\right) := \nabla u_\sigma \otimes \nabla u_\sigma := \nabla u_\sigma \nabla u_\sigma{}^T$$

In the above equation, ∇u_σ is the version of smoothed part of gradient image $u\left(x,t\right)$. the Gaussian K_ρ are component wise convolved with diverse directions which is finally averaged as

$$J_\rho : \left(\nabla u_\sigma\right) := K_\rho * \nabla u_\sigma \otimes \nabla u_\sigma \left(\rho \geq 0\right)$$

The coherence framework information is obtained by the eigen values of J_ρ which describes the definite measure of anisotropy. It become large for robustly altered eigen values and the value is nearly zero for isotropic formation. The tensor analysis framework $J_\rho : \left(\nabla u_\sigma\right)$ is ingrained into non linear diffusion filter for describing flow like organization in an image (Weickert, 1999b).

RESULTS AND DISCUSSION

Anatomical structures are manually traced by medical experts which are time consuming process. In order to facilitate these issues, segmentation of ROI is done by advanced image processing algorithms. Sirona HE- LIODENT DS SIDEXIS machine is used to acquire 120 images from 120 patients and stored that image in TIFF format. In order to conduct the study of dental images, Ethical approval (IRB Number 1-102-05-017) was procured by ethics committee of the Tri- Service General Hospital in Taipei, Taiwan (Wang, Huang, Lee, Li, Chang, Siao, & Fischer, 2016). Generally, the dental image processing such as segmentation is bitewing radiograph is a complex processsince the data discrepancy is high and teeth are sometimes labeled as background that makes the task of learning as complicated one.

Segmentation Using FCM Method

The bitewing Radiograph dental image is shown in Figure.1 (a). The FCM method is implemented for the above mentioned dental image. Initially, the FCM with all possible number of cluster centers are analyzed. Finally, FCM with eight and nine cluster center is implemented as shown in Figure.1 (b-c). Ground truth images have been formed by medical experts is shown in Figure. 1(d).

The ground truth region are selected by medical experts is seven however FCM method are considering background also as one cluster so this is the reason for choosing eight clusters in this method. Here, the aim is to extract the root canal treatment, however even choosing nine clusters also it is difficult for as to extract the root canal treatment. The FCM method fails to segment the root canal treatment since the intensity level of root canal is similar to the intensities of crown, restoration and enamel. So there is a need for a method that can extract the Region of Interest (ROI) based on the energy associated within the region. Therefore we are choosing LSM that has the energy associated with the contour, image and edges.

Figure 1. (a) Dental bitewing Radiograph Image (b) FCM with eight clusters (c) FCM with nine clusters and (d) Ground truth for dental image

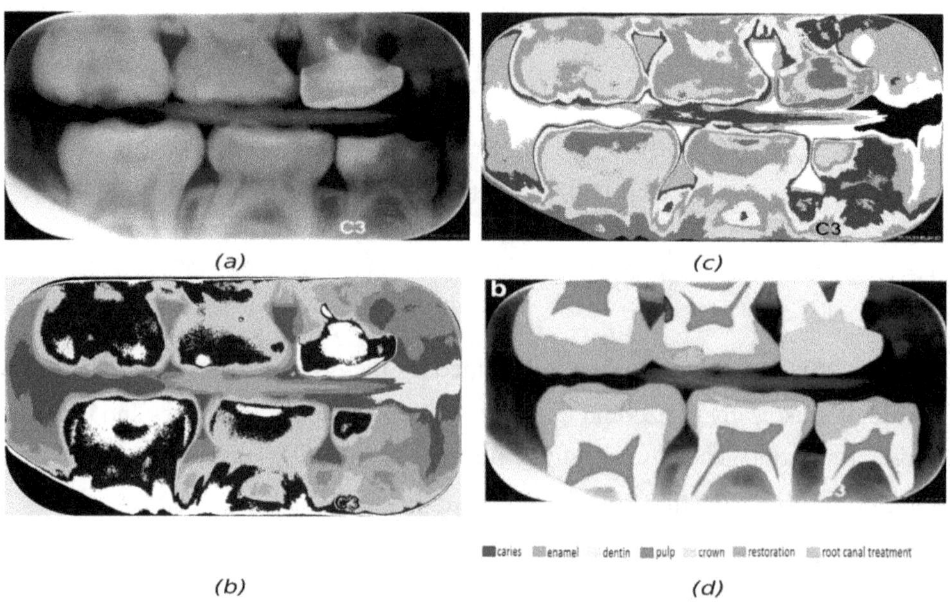

**For a more accurate representation see the electronic version.*

SEGMENTATION USING LEVEL SETS

Segmentation Using Basic Level Set Method Without Re-Initialization

The level set function evolves over several iterations according to the edge map. The image gradient in the edge map acts as edge stopping function for the level set to stop its evolution. The evolved level set contour mask is finally created at the end of iterations. This represents the segmented region of interest. This section describes the segmentation results obtained using BLSM. The results of the segmentation algorithm are shown in Figure. 2. The generated edge map is shown in Figure. 2 (c) which represents the intensity gradient of the Gaussian filtered image. A final binary mask evolved after last iteration is superimposed with input image in order to generate the segmented outputas shown in Figure. 2 (d). Ground truth images have been formed by medical experts for root canal treatment is shown in Figure. 2 (b).

In BLSM, the intensity gradient of raw image is used to progress the level set contour initially towards the borders. It results in spurious edges which act as hurdle for the free flow of required function to reach desired boundary. The boundary of the root canal results in thick edges and hence, the level set function tends to settle at the false boundary which is observed even when the number of iteration is increased. Thus, mostly under or over segmentation is observed using Gaussian edge map. Image is observed with discontinuous edges near dentin of neighboring tissues and contour leakage is observed.

This section discusses about the optimal parameters that are selected for evolving the level set equation according to evolution equation. The various parameters that need to be initialized in the evolution equation are the coefficients of regularization, area and line integral terms and the Dirac delta function.

The Dirac delta function helps restrict the zero level set into a local neighborhood. The width of this function is defined by a positive constant ε. If it is too small, there are higher chances of the energy functional getting stuck at a local minima. On the contrary, for large value of ε, the final contour may not be accurate. Therefore, the parameter ε is set to 1.5 throughout the experiment.

Figure 2. (a) Dental bitewing Radiograph Image (b) Gaussian edge map (c) Ground truth for root canal treatment and (d) Segmented output for Gaussian edge map

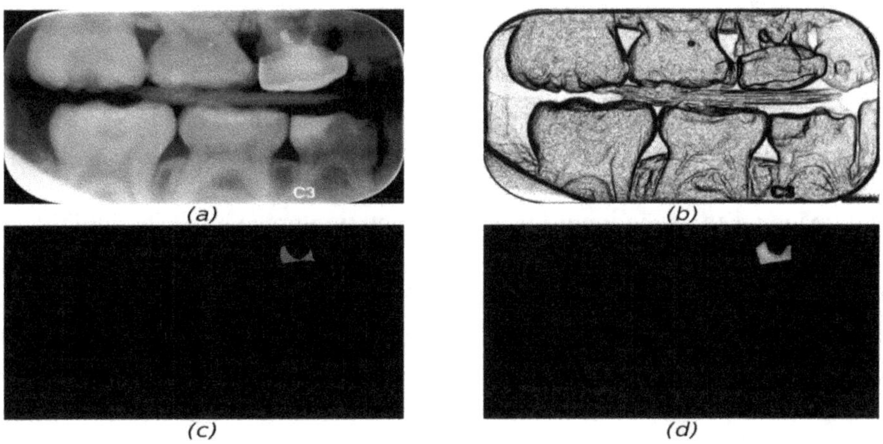

The coefficient of regularizing term is a positive constant μ and it satisfies the Courant-Friedrichs-Lewy's (CFL) law given by $\mu \nabla t < 0.25$. In this work, the time step is set to 5 trading off between number of iterations and accurate segmentation. Therefore, following the CFL law, μ is set to 0.04. The movement of initial level set contour either to expand or shrink is defined by the constant α which is a real number. If the initial contour is placed inside the ROI, the coefficient α in the weighted area term constant has to be positive, so that the zero level contour can shrink in the level set evolution; and it should take negative constant to expand the contour when placed outside the ROI. In this work, the initial contour is defined outside ROI and therefore a positive value of α is initialized. Small value of α restricts the movement of the initial contour while a large value of α leads to the leakage of contour at the edges. The most accurate segmentation is obtained when $\alpha = 0.4$ and is maintained constant for the entire experimentation.

The coefficient of line integral λ helps in detecting objects of different sizes. Small objects can be detected when λ is small and conversely, to detect large objects, the value of λ has to be larger. In order to root canal from pulp, it is necessary to detect small changes and therefore a smaller value of λ is preferred. Low value of λ, led to the leakage of contour beyond the desired edges, while higher values restricted the movement of the contour. The value of 5 for λ gave better results and hence, the coefficient of line integral is set as 5.0. In BLSM framework, the image is initially smoothed by Gaussian kernel. The filtered image entirely results in blurring for increased value of sigma due to the performance of same smoothing operation in all regions.

There is significant improvement in filtering the noise as sigma is increased. At the same time, the regions near root canal treatment of neighboring tissues such as dentin are smoothed to greater extent resulting in blurred edges. Therefore, sigma value is manually chosen to be 0.8 as optimal one according to the best visual smoothed image and the optimal chosen edge map is interlinked with LSM. Dental images are subjected to BLSM based segmentation procedure. The typical segmented results are displayed in Figure. 2 (d). As the intensity edge map results in thick and spurious edges, the level set contour stuck to the false edge boundaries. The movement of level set function is restricted due to false edge detail and the absence of clear edges, the level set performed either over or under segmentation.

Segmentation Using Anisotropic Diffusion Filter Based Level Set Method Without Re-Initialization

This section describes the segmentation results obtained using ADLSM. The algorithm evolves the initialized contour over several iterations. At the end of the iteration a mask of the final level contour is created and multiplied with the original image, to get the segmented output.

The procedure of segmentation on a representative image is shown in Figure. 3. Figure. 3 (a) represents the dental image taken as the input for the segmentation procedure. The initial level set contour on the input image. This initial level set function is manipulated according to the level set equation to obtain the evolving contour at different time instants. Figure. 3 (c) represents the edge map that is used to evolve the initial level set contour. The edge map is obtained as the inverse gradient of anisotropic diffusion filtered image.

The initialized contour evolves over a few iterations and sticks to the edges of the root canal. Finally, a mask is generated using this contour and multiplied with the original image to obtain the final segmented root canal treatment as shown in Figure. 3 (d). The results show that the segmentation algorithm could able to delineate the regions of interest properly than the integration of Gaussian edge map in LSM.

The choice of filter parameters also plays a significant role and hence, the filtered images are analyzed with the corresponding edge maps obtained. The parameters that influence in ADF, are K (edge strength threshold) and T (number of iterations). The edge threshold is varied for diffusion results with specified number of iterations (T=20). The diffusion result shows considerable alteration yet with small variations in K value that occurs owing to the range of the gradient operator which takes value between 0 and 255. The value of K is found to be optimal for K = 8 depend on the best visual diffusion results. The diffusion with small values of edge threshold results in under-smoothed image. Similarly, over- smoothed image is observed in both edges and details of the root canal for large values of K.

An inadequate diffusion process with less number of iterations results in an unfinished removal of the background and noise component at edges. Good diffusion results can be occurred in terms of well-preserved edges maintaining the details of the root canal at T = 20. The filtered image lost most of the fine details within the pulp region due to over diffusion effect for larger values of T.

The edges are preserved, smoothed and sharpened by ADF operation. The optimal parameters are selected such that it smoothes the inner-region with lower gradient and the diffusion effect is stopped at the edges with higher gradient. The efficacy of ADF on dental images is further validated by extracting the root canal treatment using level sets. The representative segmented results using ADLSM is shown in Figure. 3 (d). For the selected optimum parameters, the edge map shows the true edges present even near dentin. Thus, the algorithm could effectively move the contour towards the true edges and resulted with more accurate segmentation results than Gaussian edge map results.

Figure 3. (a) Dental bitewing Radiograph Image (b) Gaussian edge map (c) Ground truth for root canal treatment and (d) Segmented output for Gaussian edge map

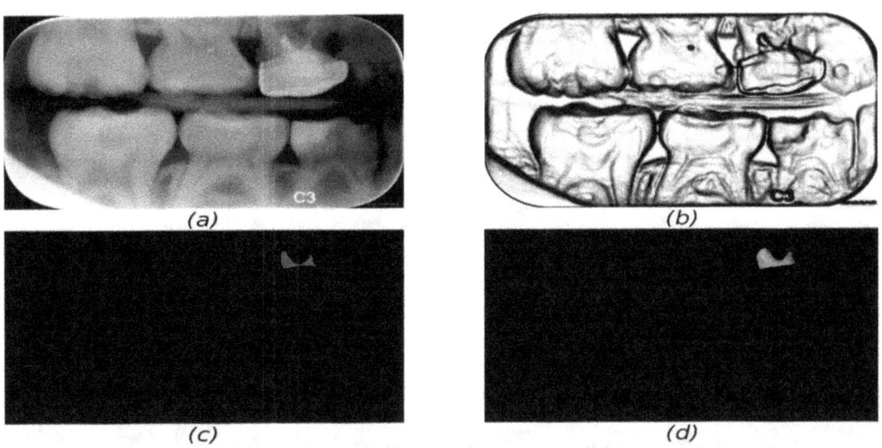

Segmentation Using Coherence Diffusion Filter Based Level Set Method Without Re-Initialization

This section describes the segmentation results obtained using coherence enhancing diffusion filtering. The procedure of segmentation on a representative image is shown in Figure. 4. Figure. 4 (a) represents the dental image taken as the input for the segmentation procedure. The initial level set contour on the input image. This initial level set function is manipulated according to the level set equation to obtain the evolving contour at different time instants. Figure. 4 (c) represents the edge map that is used to evolve the initial level set contour. The edge map is obtained as the inverse gradient of coherence enhancing diffusion filtered image. The initialized contour evolves over a few iterations and sticks to the edges of the root canal. Finally, a mask is generated using this contour and Multiplied with the original image to obtain the final segmented root canal treatment as shown in Figure. 4 (d). It is found that, the edge map is observed to be distinctive and more inaccessible structures from the close by structures. The results show that the segmentation algorithm could able to delineate the regions of interest properly than the integration of Gaussian and anisotropic diffusion edge map in LSM. This might be due to the ability of the basic level set in segmenting the root canal regions by using image gradient obtained from coherence enhanced diffusion filtering. The precise segmentation of root canal treatment is facilitated by the integration of orientation based edge map in the level set framework. By doing so, it is possible for the zero level set to correctly settle down on the edges of the root canal without any contour leakage.

Figure. 4 shows the results of edge map generation using coherence enhancing diffusion filtering. It can be observed that, the boundaries of root canal are found to be thin and continuous. In addition to the contrast and edge enhancement, it could also avoid undesired blurring resulted by the traditional Gaussian type filtering. Edge stopping function is influenced by the parameter sigma value which is chosen as 10. The amount of diffusion depends upon the time parameter which is defined here as 3. Standard derivation parameter assigned for gradient calculation is set as 1. Good diffusion results are observed in terms of well-preserved edges for these selected optimum parameters. Coherence enhancing diffusion filtering is able to enhance the contrast of edges from its background in 100 iterations.

Figure 4. (a) Dental bitewing Radiograph Image (b) Gaussian edge map (c) Ground truth for root canal treatment and (d) Segmented output for Gaussian edge map

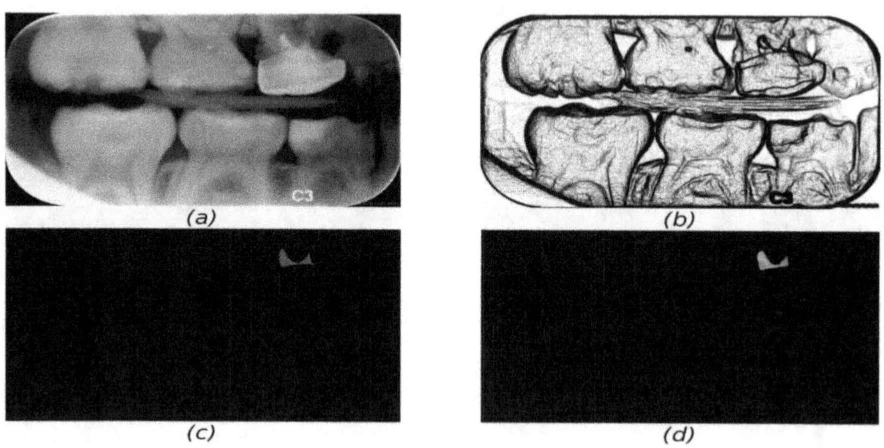

These edge maps are found to be continuous, sharp, distinct and of high contrast. The orientation information makes this method is insensitive to the magnitude of boundaries in the images. Hence, this method could provide the enhanced edge information that includes both the strong and weak edges. This would permit the precise segmentation of anatomical structures since ddirection smoothing in Gaussian method is replaced by orientation smoothing in coherence diffusion.

Discussion on Different Edge Maps

The edge map generated represents the intensity gradient of the Gaussian filtered image and is adopted as stopping boundary for the level set function to settle in desired boundaries. Unfortunately, the edge map obtained using Gaussian smoothing results in the blurring of edges. This is because, the Gaussian kernel performs uniform smoothing and results in loss of edge information. Consequently, there are chances for the level set function to have leakage of contour and inaccurately segment the dental structures leading to erroneous quantification of shape changes.

To overcome the drawback of uniform smoothing, non-linear diffusion processes such as anisotropic diffusion filtering and coherence enhancing diffusion are attempted. The results of edge map using different techniques and their corresponding segmentation results of root canals are shown in Figure. 2, Figure. 3 and Figure. 4 respectively. The choice of parameters in the generation of edge maps influence the performance of the segmentation.

The input image is convolved with the Gaussian filter to smooth the noise and intensity variations. The edge map is obtained by taking the gradient of the smoothened image. Gaussian edge map images are shown in Figure. 2. The edge map parameters are observed as standard deviation $\sigma_G = 1.5$ and contrast parameter $k_T = 65$. For the higher values of sigma, over smoothing of edges takes place owing to blurring of boundaries. The obtained edge map using Gaussian filter seems to be blurring and discontinuous in nature. The discontinuity of edge map seems to be more predominant in the boundary of root canal. This is seriously undesired as this might result in leakage of contours. Also, the presence of false edges might cause difficulties for the free flow of LSF towards the edges resulting in either over or under segmentation.

Non-linear anisotropic diffusion process is used to perform region specific smoothing on the image. The main parameter involved in the diffusion process is k_T which represents the edge strength threshold. To identify the optimum parameter values for these images, different choices of k_T are attempted during the experiment. It is observed that every increment of k_T has resulted in both the region-specific smoothing and increased the contrast of the image gradient. Based on this observation, the choice of k_T has been finalized as 8 throughout the segmentation process. Figure 3 show the results of edge map obtained using anisotropic diffusion filtering. Compared to the edge map generated using Gaussian filtering, this edge map seems to be continuous and sharp. This might be due to the region-specific smoothing performed by this filter. The sharpness of the image has increased slightly compared to anisotropic diffusion filtering. For further examination, the edge maps are observed to be stronger, sharper and more distinct with good edge continuity.

Figure. 4 shows the results of edge map generation using coherence enhancing diffusion filtering. It can be observed that, the boundaries of root canal are erect to be thick and continuous. In order to further enhance the contrast and edges, it could also avoid displeasing blurring resulted by the conventional type of Gaussian filtering. The optimal parameter value of Gaussian K_{ρ}, and diffusion time parameter which could influence the amount of diffusion are fixed as 10 and 3 respectively. The parameter standard derivation used for the pre and post smoothing computation is fixed as 1 and chosen iteration is 100 for enhancing the contrast of edges from its background. It is also found that, the coherence enhancing framework integrated with the diffusion process could perform faster diffusion on the images compared to linear diffusion model.

The segmented root canal using different filtering techniques are shown in Figure. 2 to Figure. 4 respectively. As observed, the root canal extractions are found to be enlarged in the images. The edge maps of dental images using different filtering techniques such as Gaussian, anisotropic diffusion and coherence enhanced diffusion are shown in Figure. 2 (b), Figure. 3 (b) and Figure. 4 (b) respectively. The corresponding segmented results are shown in Figure. 2 (d), Figure. 3 (d) and Figure. 4 (d). Gaussian edge map obtained with standard deviation σ_{G} of 1.5 and contrast parameter k_T of 65 appear to be blurred and discontinuous for all the subjects. The discontinuity of edge map using Gaussian filtering seems to be more predominant in the root canal treatment. The edge map obtained using anisotropic diffusion seems to be continuous and sharp for edge strength threshold value of 8 in dental subjects.

In order to enhance the edge contrast, coherence enhancing diffusion filtering is modified further for sigma of 10 and time parameter of 3. It can be observed that among all the edge maps, coherence enhanced diffusion filtering is found to be continuous, sharper and of high contrast. The edge stopping process in the level set evolution is decided by the edge map obtained using different approaches. Zero level set function undergoes number of evolutions and stops at the boundaries of region of interest. The segmented root canal using different filtering techniques are shown in Figure. 2 (d) to Figure. 4(d) respectively.

Conversely, inconsistent performance is observed in all measures of other techniques. For example, Gaussian smoothing is found to perform poor in all the cases due to its uniform smoothing. Anisotropic diffusion-based filtering could perform better than Gaussian kernel in all the subjects. It is also observed that, the discontinuity near edges are found to be less for edge maps obtained using coherence enhancing diffusion filtering. The thick edges produce by this filter and the presence of closely associated boundaries of other regions in the images might have caused challenges to the level set in performing accurate segmentation of root canal treatment.

Among all the techniques, the results obtained using coherence enhancing diffusion filtering edge map is found to have high magnitude values for all images. Similarly, it is also noted that, the Gaussian kernel resulted in low performance compared to the results of anisotropic based non-linear diffusion filters. Rather, the performance in accurate segmentation of root canal treatment by coherence enhancing diffusion filtering is found to be high than anisotropic diffusion filter for dental subjects. The segmentation of performance of Gaussian and anisotropic diffusion filtering is low, this might be due to the homogeneous pattern of edges with the nearby neighboring structures. Also, the edges are found to be thick and this might have resulted in the inaccurate segmentation of root canal treatment.

CONCLUSION

Segmentation techniques have gained a widespread importance in the recent years, for the development of systems for medical diagnosis and analysis (Manickavasagam, Sutha, & Kamalanand, 2014; Rajinikanth, Satapathy, Fernandes, & Nachiappan, 2017; Rajinikanth, Raja, & Kamalanand, 2017; Raja, Rajinikanth, Fernandes, & Satapathy, 2017). The methods of segmentation of root canal treatment in dental images are performed using clustering techniques and level set method. Segmentation of root canal treatment is a challenging process using FCM clustering whereas level set method with different edge map shows its effectiveness in its result. Three different edge maps are analysed in level set method in which the Gaussian smoothing averages directions instead of orientations. Due to this property, the edge information is blurred which results in detection and localization of edges as difficult. Smoothing operation performed by anisotropic non-linear diffusion process preserves the edges but they do not consider the integration scale. Coherence enhancing diffusion filters are most successful to enhance the edges at different orientations with good scale space representation. The image gradient obtained using coherence enhanced diffusion edge map is found to perform better in the segmentation of root canal treatment compared than other two methods. This is due to the coherence enhancement of diffusion process results in fine scale representation and more edges at diverse orientations. Therefore, the edge map obtained using coherence enhancing diffusion is found to be highly continuously, high contrast and distinct. Subsequently, this method is able to differentiate the region of root canal treatment from the nearby structures through its edge map and facilitating in precise segmentation of root canal treatment. The work flow process which is proposed for dental images confirms its achievement for accurate shape detection of dental root canal.

REFERENCES

Ahmed, S. A., Taib, M. N., Khalid, N. E. A., Ahmad, R., & Taib, H. (2011). *Performance of compound enhancement algorithms on dental radiograph images*. WASET.

Anil, K. (2004). Matching of dental X-ray images for human identification. *Pattern Recognition*, *37*(7), 1519–1532. doi:10.1016/j.patcog.2003.12.016

Bhadauria, H. S., Singh, A., & Dewal, M. L. (2013). An integrated method for haemorrhage segmentation from brain CT imaging. *Computers & Electrical Engineering*, *39*(5), 1527–1536. doi:10.1016/j.compeleceng.2013.04.010

Borchartt, T. B., Conci, A., Lima, R. C., Resmini, R., & Sanchez, A. (2012). Breast thermography from an image processing viewpoint: A survey. *Signal Processing*, *93*(10), 2785–2803. doi:10.1016/j.sigpro.2012.08.012

Caselles, V., Catte, F., Coll, T., & Dibos, F. (1993). A geometric model for active contours in image processing. *Numerische Mathematik*, *66*, 1–31. doi:10.1007/BF01385685

Caselles, V., Ron, K., & Sapiro, G. (1997). Geodesic active contours. *International Journal of Computer Vision*, *22*(1), 61–79. doi:10.1023/A:1007979827043

Chunming, L., Chenyang, X., Changfeng, G., & Martin, D. F. (2005) Level set evolution without re-initialization: A new variational formulation. *Proceedings of the IEEE international conference on computer vision and pattern recognition, 1*, 430-436. 10.1109/CVPR.2005.213

Cremers, D. (2003) A multiphase levelset framework for variational motion segmentation. Proc. In Scale Space Meth. Comput. Vis., 599–614.

Evans, L. (1998). *Partial Differential Equations*. Providence, RI: American Mathematical Society.

Gao, H., & Chae, O. (2010). Individual tooth segmentation from CT images using level set method with shape and intensity prior. *Pattern Recognition, 43*(7), 2406–2417. doi:10.1016/j.patcog.2010.01.010

Jain, K. R., & Chauhan, N. C. (2017). An Automated Level Set Segmentation Approach for Lesion Detection in Dental Radiograph for Endodontic Treatment. *International Journal of Computers and Applications, 172*(5), 17–24. doi:10.5120/ijca2017914092

Jayadevappa, D., Srinivas Kumar, S., & Murty, D. S. (2011). Medical Image Segmentation Algorithms using Deformable Models: A Review. *IETE Technical Review, 28*(3), 248–251. doi:10.4103/0256-4602.81244

Johnson, W. T., & Guttmann, J. L. (2007). Obturation of cleaned and shaped root canal system. In S. Cohen & K. Hargreaves (Eds.), *Pathways of the pulp* (9th ed.). Philadelphia, PA: Elsevier.

Kamalanand, K., & Ramakrishnan, S. (2015). Effect of gadolinium concentration on segmentation of vasculature in cardiopulmonary magnetic resonance angiograms. *Journal of Medical Imaging and Health Informatics, 5*(1), 147–151. doi:10.1166/jmihi.2015.1370

Kang, J., & Ji, Z. (2010) Dental Plaque Quantification using Mean-shift-based Image Segmentation. *IEEE International Symposium on Computer, Communication, Control and Automation*. 10.1109/3CA.2010.5533758

Keshtkar, F., & Gueaieb, W. (2006) Segmentation of Dental Radiographs Using a Swarm Intelligence Approach. *IEEE Canadian Conference on Electrical and Computer Engineering*, 328–331. 10.1109/CCECE.2006.277656

Kuo, H. C., Maryellen, L. G., Ingrid, R., Boone, J. M., Lindfors, K. K., Yang, K., & Edwards, A. (2014). Level Set Segmentation of Breast Masses in Contrast-Enhanced Dedicated Breast CT and Evaluation of Stopping Criteria. *Journal of Digital Imaging, 27*(2), 1–11. doi:10.100710278-013-9652-1 PMID:24162667

Lai & Lin. (2006) Effective Segmentation for Dental X-Ray Images Using Texture-Based Fuzzy Inference System. *LNCS, 5259*, 936–947.

Lai, Y. H., & Lin, P. L. (2008). Effective segmentation for dental x-ray images using texture-based fuzzy inference system. *Lecture Notes in Computer Science, 5259*, 936–947. doi:10.1007/978-3-540-88458-3_85

Lai, Y. H., Lin, P. L., & Huang, P. W. (2010). An effective classification and numbering system for dental bitewing radiographs using teeth region and contour information. *Pattern Recognition, 43*(4), 1380–1392. doi:10.1016/j.patcog.2009.10.005

Laura, F., Rubio, J. L., Ledesma-Carbayo, M. J., Pascau, J., Tellado, J. M., & Ramón, E. (2009) 3D liver segmentation in preoperative CT images using a level sets active surface method. *Proceedings of the 31st Annual International Conference of the IEEE Engineering in Medicine and Biology Society*, 3625-3628.

Li, C., Xu, C., Gui, C., & Fox, M. D. (2010). Distance regularized level set evolution and its application to image segmentation. *IEEE Transactions on Image Processing*, *19*(12), 3243–3254. doi:10.1109/TIP.2010.2069690 PMID:20801742

Li, C., Xu, C., Guiand, C., & Fox, M. D. (2005). Level set evolution without re-initialization: a new variational formulation. *IEEE Computer Society Conference on Computer Vision and Pattern Recognition*, *1*, 430-436.

Li, S., Fevens, T., & Adam, K. S. L. (2006). An automatic variational level set segmentation framework for computer aided dental X-rays analysis in clinical environments. *Computerized Medical Imaging and Graphics*, *30*(2), 65–74. doi:10.1016/j.compmedimag.2005.10.007 PMID:16500077

Li, S., Fevens, T., Krzyzak, A., Jin, C., & Li, S. (2007). Semi-automatic Computer Aided Lesion Detection in Dental X-rays Using Variational Level Set. *Pattern Recognition*, *40*(10), 2861–2873. doi:10.1016/j.patcog.2007.01.012

Machado, D. A., Giraldi, G., Novotny, A. A., Marques, R. S., & Conci, A. (2013) Topological Derivative Applied to Automatic Segmentation of Frontal Breast Thermograms. *IX Workshop de Visão Computacional*.

Malladi, R., Sethian, J. A., & Vemuri, B. C. (1995). Shape modeling with front propagation: A level set approach. *IEEE Transactions on Pattern Analysis and Machine Intelligence*, *17*(2), 158–175. doi:10.1109/34.368173

Manickavasagam, K., Sutha, S., & Kamalanand, K. (2014). An automated system based on 2 d empirical mode decomposition and k-means clustering for classification of Plasmodium species in thin blood smear images. *BMC Infectious Diseases*, *14*(Suppl 3), 13. doi:10.1186/1471-2334-14-S3-P13 PMID:24405683

Manickavasagam, K., Sutha, S., & Kamalanand, K. (2014). Development of systems for classification of different plasmodium species in thin blood smear microscopic images. *Journal of Advanced Microscopy Research*, *9*(2), 86–92. doi:10.1166/jamr.2014.1194

McMichen, F. R. S., Pearson, G., Rahbaran, S., & Gulabilava, K. (2003). A comparative study of selected physical properties of five root-canal sealers. *Journal of Int Endod*, *36*(9), 629–635. doi:10.1046/j.1365-2591.2003.00701.x PMID:12950578

Nomir, O., & Abdel-Mottaleb, M. (2007). Human Identification from Dental X-Ray Images Based on the Shape and Appearance of the Teeth. *IEEE Transactions on Information Forensics and Security*, *2*(2), 188–197. doi:10.1109/TIFS.2007.897245

Oliveira, J., & Proenc, H. (2011). Caries Detection in Panoramic Dental X-ray Images. *Computational Vision and Medical Image Processing: Recent Trends, Computational Methods in Applied Sciences* 19. *Springer Science Business Media BV, 10*. doi:10.1007/978-94-007-0011-6

Osher, S., & Sethian, J. A. (1988). Fronts propagating with curvature-dependent speed: Algorithms based on Hamilton-Jacobi formulations. *Journal of Computational Physics*, *79*(1), 12–49. doi:10.1016/0021-9991(88)90002-2

Perona, P., & Malik, J. (1990). Scale-space and edge detection using anisotropic diffusion. *IEEE Transactions on Pattern Analysis and Machine Intelligence*, *12*(7), 629–639. doi:10.1109/34.56205

Perona, P., Shiota, T., & Malik, J. (1994) Anisotropic diffusion. In Geometry-driven diffusion in computer vision. Springer.

Prabha, S., Anandh, K. R., Sujatha, C. M., & Ramakrishnan, S. (2014). Total variation based edge enhancement for level set segmentation and asymmetry analysis in breast thermograms. *IEEE Engineering in Medicine and Biology Society*, *36*, 6438–6441. PMID:25571470

Prabha, S., Suganthi, S. S., & Sujatha, C. M. (2015). An Approach to Analyze the Breast Tissues in Infrared Images Using Nonlinear Adaptive Level sets and Riesz Transform Features. *Technology and Health Care*, *23*(4), 429–442. doi:10.3233/THC-150915 PMID:26409908

Prabha, S., & Sujatha, C. M. (2014). Analysis of Breast Thermograms using Coherence Enhanced Diffusion Filter and Radon Transform. *Journal of the Instrument Society of India*, *44*(4), 271–274.

Prabha, S., Sujatha, C. M., & Ramakrishnan, S. (2015). Robust Anisotropic Diffusion Based Edge Enhancement for Level Set Segmentation and Asymmetry Analysis Of Breast Thermogramsusing Zernike Moments. *Journal of Biomedical Science Instrumentation*, *51*, 341–348. PMID:25996737

Raja, N., Rajinikanth, V., Fernandes, S. L., & Satapathy, S. C. (2017). Segmentation of Breast Thermal Images Using Kapur's Entropy and Hidden Markov Random Field. *Journal of Medical Imaging and Health Informatics*, *7*(8), 1825–1829. doi:10.1166/jmihi.2017.2267

Rajinikanth, V., Fernandes, S. L., Bhushan, B., & Sunder, N. R. (2018). Segmentation and Analysis of Brain Tumor Using Tsallis Entropy and Regularised Level Set. In *Proceedings of 2nd International Conference on Micro-Electronics, Electromagnetics and Telecommunications* (pp. 313-321). Springer. 10.1007/978-981-10-4280-5_33

Rajinikanth, V., Raja, N. S. M., & Kamalanand, K. (2017). Firefly Algorithm Assisted Segmentation of Tumor from Brain MRI using Tsallis Function and Markov Random Field. *Journal of Control Engineering and Applied Informatics*, *19*(3), 97–106.

Rajinikanth, V., Satapathy, S. C., Dey, N., & Vijayarajan, R. (2018). DWT-PCA Image Fusion Technique to Improve Segmentation Accuracy in Brain Tumor Analysis. In *Microelectronics, Electromagnetics and Telecommunications* (pp. 453–462). Singapore: Springer. doi:10.1007/978-981-10-7329-8_46

Rajinikanth, V., Satapathy, S. C., Fernandes, S. L., & Nachiappan, S. (2017). Entropy based segmentation of tumor from brain MR images–a study with teaching learning based optimization. *Pattern Recognition Letters*, *94*, 87–95. doi:10.1016/j.patrec.2017.05.028

Ravikanth, M., Sethian, A. J., & Vemuri, B. C. (1995). Shape modeling with front propagation: A level set approach. *IEEE Transactions on Pattern Analysis and Machine Intelligence*, *17*(2), 158–175. doi:10.1109/34.368173

Said, Eldin, Nassar, Fahmy, & Ammar. (2006). Teeth Segmentation in Digitized Dental X-Ray Films Using Mathematical Morphology. *IEEE Transactions on Information Forensics and Security*, *1*(2).

Shafer's Tb. (2006). Textbook of Oral Pathology (6th ed.). Academic Press.

Siyal, M. Y., & Yu, L. (2005). An intelligent modified fuzzy c-means based algorithm for bias estimation and segmentation of brainMRI. *Pattern Recognition Letters*, *26*(13), 2052–2062. doi:10.1016/j.patrec.2005.03.019

Srinivasan, S. S., & Swaminathan, R. (2014). Segmentation of Breast Tissues in Infrared Images Using Modified Phase Based Level Sets. In T. D. Pham, K. Ichikawa, M. Oyama-Higa, D. Coomans, & X. Jiang (Eds.), Biomedical Informatics and Technology. Communications in Computer and Information Science. 404. Academic Press. doi:10.1007/978-3-642-54121-6_14

Suganthi, S. S., & Ramakrishnan, S. (2014). Anisotropic diffusion filter based edge enhancement for segmentation of breast thermogram using level sets. *Biomedical Signal Processing and Control*, *10*, 128–136. doi:10.1016/j.bspc.2014.01.008

Szilagyi, L., Szilagyi, S. M., & Benyo, Z. (2007). Efficient Feature Extraction for Fast Segmentation of MR Brain Images. *Lecture Notes in Computer Science*, *4522*, 611–620. doi:10.1007/978-3-540-73040-8_62

Tiwari, R. B., & Yardi, A. R. (2006). Dental radiograph image enhancement based on human visual system and local image statistics. *International Conference on Image Processing, Computer Vision and Pattern Recognition*, 100-108.

Vaishnavi, G. K., Jeevananthan, K., Begum, S. R., & Kamalanand, K. (2014). Geometrical analysis of schistosome egg images using distance regularized level set method for automated species identification. *Journal of Bioinformatics and Intelligent Control*, *3*(2), 147–152. doi:10.1166/jbic.2014.1080

Wang, C. W., Huang, C. T., Lee, J. H., Li, C. H., Chang, S. W., Siao, M. J., ... Fischer, P. (2016). A benchmark for comparison of dental radiography analysis algorithms. *Medical Image Analysis*, *31*, 63–76. doi:10.1016/j.media.2016.02.004 PMID:26974042

Weickert, J. (1999a). Coherence-enhancing diffusion of colour images. *Image and Vision Computing*, *17*(3), 201–212. doi:10.1016/S0262-8856(98)00102-4

Weickert, J. (1999b). Coherence-enhancing diffusion filtering. *International Journal of Computer Vision*, *31*(2-3), 111–127. doi:10.1023/A:1008009714131

Wu, W., Wu, Y., & Huang, Q. (2012) An improved distance regularized level set evolution without re-initialization. *IEEE fifth International Conference on Advanced Computational Intelligence*, 631-636.

Zhou, J., & Abdel-Mottaleb, M. (2005). A content-based system for human identification based on bitewing dental X-ray images. *Pattern Recognition*, *38*(11), 2132–2142. doi:10.1016/j.patcog.2005.01.011

Zhou, Q., Li, Z., & Aggarwal, J. K. (2004). Boundary extraction in thermal images by edge map. *Proceedings of the ACM symposium on Applied computing*, 254-258. 10.1145/967900.967956

Zhuang, A. H., Valentino, D. J., & Toga, A. W. (2006). Skull-stripping magnetic resonance brain images using a model-based level set. *NeuroImage*, *32*(1), 79–92. doi:10.1016/j.neuroimage.2006.03.019 PMID:16697666

ADDITIONAL READING

Caselles, V., Ron, K., & Sapiro, G. (1997). Geodesic active contours. *International Journal of Computer Vision*, *22*(1), 61–79. doi:10.1023/A:1007979827043

Gao, H., & Chae, O. (2010). Individual tooth segmentation from CT images using level set method with shape and intensity prior. *Pattern Recognition*, *43*(7), 2406–2417. doi:10.1016/j.patcog.2010.01.010

Li, C., Xu, C., Gui, C., & Fox, M. D. (2010). Distance regularized level set evolution and its application to image segmentation. *IEEE Transactions on Image Processing*, *19*(12), 3243–3254. doi:10.1109/TIP.2010.2069690 PMID:20801742

Neel, E. A. A., Chrzanowski, W., Salih, V. M., Kim, H.-W., & Knowles, J. C. (2014). Tissue engineering in dentistry. *Journal of Dentistry*, *42*(8), 915–928. doi:10.1016/j.jdent.2014.05.008 PMID:24880036

Perona, P., & Malik, J. (1990). Scale-Space and Edge Detection UsingAnisotropic Diffusion. *IEEE Transactions on Pattern Analysis and Machine Intelligence*, *12*(7), 629–639. doi:10.1109/34.56205

Ravikanth, M., Sethian, A. J., & Vemuri, B. C. (1995). Shape modeling with front propagation: A level set approach. *IEEE Transactions on Pattern Analysis and Machine Intelligence*, *17*(2), 158–175. doi:10.1109/34.368173

Said, E. H., Nassar, D. E. M., Fahmy, G., & Ammar, H. H. (2006). Teeth segmentation in digitized dental X-ray films using mathematical morphology. *IEEE TransactionInformation Forensics and Security*, *1*(2), 178–189. doi:10.1109/TIFS.2006.873606

Wu, W., Wu, Y., & Huang, Q. (2012) An improved distance regularized level set evolution without re-initialization, *IEEE fifth international conference on advanced computational intelligence*, Nanjing, October, 631-636.

Zhou, Q., Li, Z., & Aggarwal, J. K. (2004) Boundary extraction in thermal images by edge map. *Proceedings of the ACM symposium on Applied computing*, Cyprus, March, 254-258. 10.1145/967900.967956

Zhuang, A., Valentino, D. J., & Toga, A. W. (2006). Skull-stripping magnetic resonance brain images using a model-based level set. *NeuroImage*, *32*(1), 79–92. doi:10.1016/j.neuroimage.2006.03.019 PMID:16697666

Chapter 5
Jaya Algorithm-Assisted Evaluation of Tooth Elements Using Digital Bitewing Radiography Images

Kesavan Suresh Manic
Caledonian College of Engineering, Oman

Imad Saud Al Naimi
Caledonian College of Engineering, Oman

Feras N. Hasoon
Caledonian College of Engineering, Oman

V. Rajinikanth
St. Joseph's College of Engineering, India

ABSTRACT

A considerable number of heuristic procedures are widely implemented to evaluate biomedical images. This chapter proposes an evaluation procedure for digital bitewing radiography (DBR) images using the Jaya algorithm. The proposed procedure implements an image processing technique by integrating of the multi-thresholding and segmentation procedure to extract the essential tooth elements recorded with DBR. In this paper, 80 dental x-ray images are considered for the evaluation. The performance of the proposed procedure is confirmed using a relative assessment between the extracted section and its corresponding ground-truth. The results of this study confirm that, for most of the DBR cases, the proposed approach offers better values of picture likeliness measures. Hence, this technique can be considered for the automated detection of tooth elements from the DBR obtained from clinics.

DOI: 10.4018/978-1-5225-6243-6.ch005

INTRODUCTION

Teeth are one of the hardest matters in human body, which play an essential role in the human digestive system. The structure of a tooth is the combination of regions, such as crown, neck and root. The crown is the outermost layer of the tooth which lies on the supporting section called the neck. The neck is the supporting medium surrounded by gum region. Root canal holds the sensitive nerve and blood vessels for an active tooth. The crown is the outermost and visible region of the tooth and it is associated with an outer cover known as the enamel. The major problem in the crown involves in the reasons, such as plaque, erosion of enamel and dentine (Odaira et al., 2011; Weisleder et al., 2009; Jespersen et al., 2014). Tooth decay is one of the chief causes for the damage in outer/interior surface of tooth (Jafarzadeh et al., 2010; Chen and Abbott, 2009; 2011). Tooth decay also commonly known as cavities or caries are the cause of bacterial infection. Bacteria, such as streptococci (mutans & sobrinus) and lactobacilli are the main cause for tooth cavities (Rose and Svec, 2015; Figueiredo et al., 2015; Lin et al., 2012). For these bacteria, food remains or sugar in tooth shell are the principal energy resources. When the bacteria consume sugar and food remains, it will release acid which will affect the outer surface of the tooth wall called enamel. When this process continues, the tooth surface is enclosed with a hard layer called plaque. The bacterial contamination leads to a tiny yellow or black hole on the tooth surface on the tooth called the caries (Wang et al., 2015; 2015a; 2016). If the caries left untreated, it may lead to the trouble, such as pain, sensitivity, infection and tooth loss. To protect the tooth from caries, dental care centers provide various protection methods, like maintaining oral hygiene, avoiding high sugar food, and usage of right amount of fluoride (Son et al., 2018). The dental care procedure also suggest, tooth filling, fitting crown on infected tooth, early stage treatment and root canal treatment (Rad et al., 2013; 2016; 2017). The literature also confirms that, untreated caries will lead to various health problems in human irrespective of their age and gender (Lin et al., 2012; Son et al., 2018). Hence, invasive and non-invasive treatment procedures are implemented by the dentists to treat various dental problems. To plan and implement a suitable treatment procedure, it is necessary to follow a preliminary screening process to identify the cause and the region of tooth infection. The clinical level screening methods involves in; (i) Manual examination of infection by an experienced dentist – this procedure gives the details regarding the type and region of dental infection, (ii) Modern procedures to recognize the severity and intensity of infection – this involves in recording the teeth area using imaging procedure called X-ray (Wang et al., 2015).

Digital Bitewing Radiography (DBR) is a commonly followed imaging procedure to record the entire teeth in order to inspect the various teeth conditions and defects. During this practice, the teeth sections are recorded using the Dental X-ray in a controlled environment under the supervision of a dentist. Later, this DBR is evaluated to identify the infected region using manual, semi-automated and automated procedures to plan for the possible treatment procedures to control or cure the tooth defect by implementing some prescribed treatment procedures. The DBR is extensively considered in dental clinics to record the dental section for the diagnosis and treatment of various dental diseases. It is considered for the examination of teeth orientation, dental infection, early stage dental disease, abnormal teeth growth, and providing the solution for the problems in enamel, dentine, pulp cavity and root canal (Wang et al., 2015).

The proposed work implements a semi-automated practice to evaluate the tooth defects, such as cavities or caries, pulp, Tooth Restoration (TR), and Root Canal Treatment (RCT). In this approach, initially, the image enhancement is implemented to improve the visibility of the DBR picture, a thresholding procedure is implemented with the Tsallis Entropy (TE), and later a segmentation approach based on the Distance-Regularized-Level-Set (DRLS) is implemented to extract the abnormal section. Finally, the extracted section is compared against the Ground-Truth (GT) picture offered by an imaging exert. The results of this chapter confirm that, proposed procedure is efficient in extracting the chosen regions effectively with the greater accuracy.

BACKGROUND

Recently, a considerable number of manual, semi-automated and automated medical image examination procedures are proposed to evaluate a variety of diseases using the digital images recorded with a varied imaging approaches. Most of the procedures will implement a traditional procedure as well as the recent heuristic assisted practices in order to achieve the accuracy in evaluation. Some of the recent works (Rajinikanth et al., 2017; 2018; Vishnupriya et al., 2017; Raja et al., 2018) claims that, heuristic algorithm based evaluation offers better result for most of the medical image cases. These procedures also confirm that, integration of the multi-thresholding and segmentation will help to achieve better outcome compared to the existing approaches. In this chapter, the integration of the Tsallis Entropy (TE) assisted three-level thresholding and the DRLS based segmentation is implemented and tested using the benchmark tooth image datasets. The proposed approach offered enhanced results during the assessment of various tooth sections.

MAIN FOCUS OF THIS CHAPTER

In recent days, a considerable numbers of image assessment methods are proposed to inspect various medical image datasets. The works of Rajinikanth et al. (2018) and Raja et al. (2018) reports that, combination of thresholding and segmentation will help to achieve improved result. Therefore, in this chapter, integration of the multi-thresholding and segmentation is considered to examine the DBR pictures. The multi-thresholding will act as the initial-processing phase and the segmentation acts as the final-processing phase. The steps considered to implement the proposed procedure in this chapter are depicted in Fig 1.

Initially, the digital dental X-ray image is enhanced using the procedures, like the adaptive histogram enhancement (Zuiderveld, 1994; Yadav et al., 2014) or histogram-equalization (Tom and Wolfe, 1982; Stark, 2000) or the image-adjustment (Stark, 2000) procedures existing in the Matlab software. Then, a tri-level thresholding procedure based on the combination of the Jaya algorithm (JA) and Tsallis Entropy (TE) is implemented to enhance the tooth regions to be evaluated, then a segmentation procedure

Figure 1. Block diagram of the tooth evaluation procedure

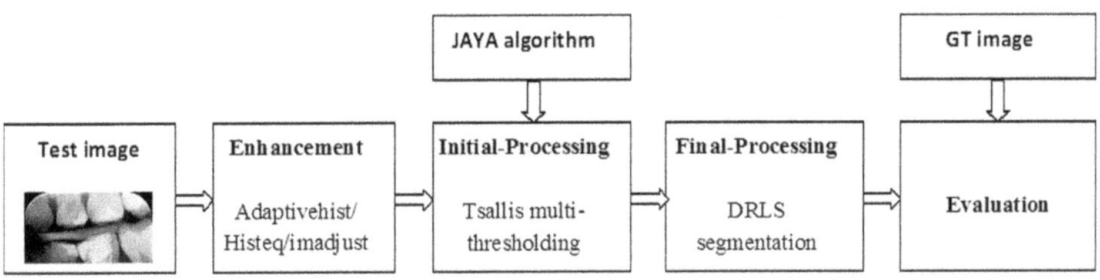

based on DRLS (single as well as multi-well) is implemented to extract the tooth section and finally, a relative examination among the extracted tooth section and GT is executed to confirm the superiority of the proposed procedure.

Implementation steps adopted in this semi-automated tool:

Step 1: Enhance the test picture using the image enhancement procedures like adapthisteq, histeq, and imadjust. This will help to improve the visibility of the tooth element under study.

Step 2: Implementation of JA+TE three-level thresholding to cluster the similar pixels to enhance the abnormal tooth element

Step 3: Execution of the DRLS segmentation with a single or multiple-well in order to extract the tooth region.

Step 4: Performance evaluation based on a relative analysis between the extracted section and GT and computation of the picture likeliness and statistical measures.

Step 5: Computation of the average performance measures to confirm the superiority of the proposed tool.

Overview of Proposed Approach

The major procedure of this chapter is depicted in Fig 1. Initially, the dental X-ray (DXR) image to be examined is considered as the test picture to examine the performance of the proposed semi-automated tooth examination tool. The raw DXR picture requires possible image enhancement procedure in order to improve its visibility. This enhancement includes, adapthisteq (Zuiderveld, 1994; Yadav et al., 2014), histeq (Tom and Wolfe, 1982; Stark, 2000), and regulating the intensity of the test picture (imadjust). The first two functions will provide satisfactory result on the considered gray scale images of DXR and DBR pictures compared to the imadjust function. After the possible enhancement, the picture is the processed using the JA+TE based approach, which helps to enhance the essential tooth region for the assessment. Finally, the tooth section to be examined is separated from the test picture using the DRLS segmentation. Finally, a relative analysis between the extracted section and the GT is performed and the essential picture likeliness measures, such as Jaccard, Dice, Sensitivity (SE), Specificity (SP) and Accuracy (AC) are computed to confirm the superiority of the proposed approach. The outcome of this work confirms that, proposed approach provides better result for the tooth regions, like the pulp, TR, and RCT of benchmark DBR images of ISIB 2015 dataset.

Jaya Algorithm

JA is a newly proposed population based scheme by Rao (2016). The benefit of the JA compared with other approaches is, it needs minimal initial parameters to be assigned.

Let $G(x)_{min}$ is the chosen objective function, $\Re 1 = \Re 2$ are the arbitrary numerals of the choice [0,1], N is the population size (ie. $n=1,2,...,N$) and i and D denotes the number of iterations and dimensions (ie. $j=1,2,...,D$) respectively. At some instance, $G(x)_{best}$ and $G(x)_{worst}$ signifies the best & worst answers attained via the candidate of the population.

The primary equation of JA is presented below;

$$X^1{}_{j,n,i} = X_{j,n,i} + \Re 1_{j,i}(X_{jbest,i} - |X_{j,n,i}|) - \Re 2_{j,i}(X_{jworst,i} - |X_{j,n,i}|) \tag{1}$$

where, $X_{j,n,i}$ represents the jth variable of nth candidate at i^{th} iteration and $X^1_{j,n,i}$ denoted the updated value. More details regarding the JA can be found in (Rao and More, 2017; Rao and Saroj, 2017; 2017a).

Multi-Thresholding With Tsallis Entropy

Thresholding is usually considered to cluster matching pixels in a picture according to user's constraint (Palani et al., 2016; Raja et al., 2014; Rajinikanth and Couceiro, 2015; Rajinikanth et al., 2014). Recently, Tsallis Entropy (TE) supported multi-thresholding proposal is considered to group similar pixels for a class of images (Tsallis, 1988; Rajinikanth et al., 2017).

TE is derived from Shannon's Entropy (Tsallis, 1988) and it can be used to find the optimal thresholds of an image based on assigned threshold value. Chose a given image with L gray levels in the range {0, 1, ..., L-1},with probability distributions $p_i = p_{0,} p_1, ..., p_{L-1}$.

Tsallis multi-level thresholding can then be expressed as:

$$f(T)=[T_1,T_2, ..., T_k] = argmax \left[S_q^A(T) + S_q^B(T) + ... + S_q^K(T) + (1-q).S_q^A(T).S_q^B(T)...S_q^K(T)\right] \quad (2)$$

where

$$S_q^A(T) = \frac{1 - \sum_{i=0}^{t_1-1}\left(\frac{Pi}{P^A}\right)^q}{q-1}, \; P^A = \sum_{i=0}^{t_1-1} Pi$$

$$S_q^B(T) = \frac{1 - \sum_{i=t_1}^{t_2-1}\left(\frac{Pi}{P^B}\right)^q}{q-1}, \; P^B = \sum_{i=t_1}^{t_2-1} Pi$$

$$S_q^K(T) = \frac{1 - \sum_{i=t_k}^{L-1}\left(\frac{Pi}{P^K}\right)^q}{q-1}, \; P^K = \sum_{i=t_k}^{L_2-1} Pi$$

are subject to the following constraints:

$$\left|P^A + P^B\right| - 1 < S < 1 - \left|P^A - P^B\right|$$

$$\left|P^B + P^C\right| - 1 < S < 1 - \left|P^B - P^C\right| \quad (3)$$

$$\left|P^{K}+P^{L-1}\right|-1<S<1-\left|P^{K}-P^{L-1}\right|$$

During multi- thresholding procedure, the aim is to find the best threshold value *T* which maximizes the objective function *f(T)*. In this work, threshold value is chosen as $T=3$; hence the essential likelihood values are P^A, P^B, and P^C. Here, the maximization of function *f(T)*, which deals with the segmentation of a given image, is carried using the JA. Discovering the best threshold based on the manually selected threshold is a time consuming process, Hence, in this work, JA is considered to discover the optimal thresholds based on the chosen *T* value. The thresholding operation enhances the test image by grouping the similar pixel values as described by Eqn. (3).

Distance Regularized Level Set

Segmentation process is employed to extract the required section from the images (Li et al., 2010; 2011). The recent work of Rajinikanth and Satapathy (2018) discussed about the various segmentation scheme existing in the literature for medical image extraction. Even though, varieties of segmentation approaches are available, the Distance-Regularise-Level-Set (DRLS) attracted the researchers due to its superiority. The earlier work on DRLS implements a multiple-well approach to extract more than one section from the test picture. Based on the requirement, the choice of the well size (single/multiple) is fixed. In this chapter, DRLS approach is implemented to extract the required tooth section with single as well as multiple-well based on the tooth section to be examined. For cavity, a single well is used and for other sections, multiple-well procedure is followed. Initially, the well is initialised as a bounding-box and which is allowed to shrink towards the group of pixels existing in the tooth section to be extracted. When the iteration level of the DRLS increases, the bounding-box will shrink and at the end, it will extract the tooth section to be examined. Already, the DRLS procedure is implemented to examine a range of medical pictures and the complete details regarding the DRLS can be found in (Roopini et al., 2018; Rajinikanth and Satapathy, 2018; Vaishnavi et al., 2014; Rajinikanth et al, 2018a).

Assessment

After extracting the required section from tooth image, it is necessary to examine the performance of the proposed procedure. The aim of this section is to compute the features of tooth region for further study. In this work, picture likeness values, such as Jaccard-index, Dice-coefficient, False-Positive-Rate (FPrate), and False-Negative-Rate (FNrate) are computed for extracted tooth section with respect to its GT (Lu et al., 2004; Wang et al., 2004).

The arithmetic expression is shown below;

$$JSC(IGT,IO)=IGT\cap IO\big/IGT\cup IO \tag{4}$$

$$DSC(IGT,IO)=2\left(IGT\cap IO\right)\big/\left|IGT\right|\cup\left|IO\right| \tag{5}$$

$$FPR(IGT, IO) = \left(IGT/IO\right)/(IGT \cup IO) \tag{6}$$

$$FNR(IGT, IO) = \left(IO/IGT\right)/(IGT \cup IO) \tag{7}$$

where, IGT symbolize the GT and IO symbolize mined region.Further, values, like sensitivity (SE), specificity (SP), accuracy (AC), precision (PR), Balanced-Classification-Rate (BCR), and Balanced-Error-Rate (BER) are measured. Mathematical expression for these parameters is given below:

$$SE = T_P/(T_P + F_N) \tag{8}$$

$$SP = T_N/(T_N + F_P) \tag{9}$$

$$AC = (T_P + T_N) / (T_P + T_N + F_P + F_N) \tag{10}$$

$$PR = T_P/(T_P + F_P) \tag{11}$$

where, T_N, T_P, F_N and F_P indicates true-negative, true-positive, false-negative and false-positive; correspondingly.

SOLUTIONS AND RECOMMENDATIONS

The proposed work adopts the well known tooth datasets for the experimental evaluation. Initial investigation is implemented using the Digital X-Ray (DXR) dataset, in which the examination of the tooth cavity is performed. The considered DXR image is initially treated with the contrast enhancement procedures, such as adapthisteq, histeq, and imadjust and the corresponding results are depicted in Fig 2.

Fig 2(a) depicts the test picture under examination and Fig 2(b), (c) and (d) shows the outcome of adapthisteq, histeq and imadjest functions. From this image, it is clear that, imadjust delivers poor result compared to other two enhancement procedures. Hence, in this chapter, the adapthisteq and histeq functions are considered to enhance the DXR pictures.

The tooth of DXR images are evaluated as two cases; (i) Examination of a single tooth and (ii) Examination of the whole DXR picture. During the initial examination, the chosen tooth is cropped and examined using the proposed procedure. Fig 3 presents the result obtained for the single tooth case. In which, Fig 3(a) depicts the chosen test picture and Fig 3(b) presents the threshold outcome. Further, Fig 3(c) and Fig 3(e) represents the enhanced test pictures and Fig (d) and Fig 3(f) shows its correspond-

Figure 2. Picture enhancement results obtained for the sample test picture
(a) Test image, (b) Outcome of (b) Adapthisteq, (c) Histeq and (d) Imadjust

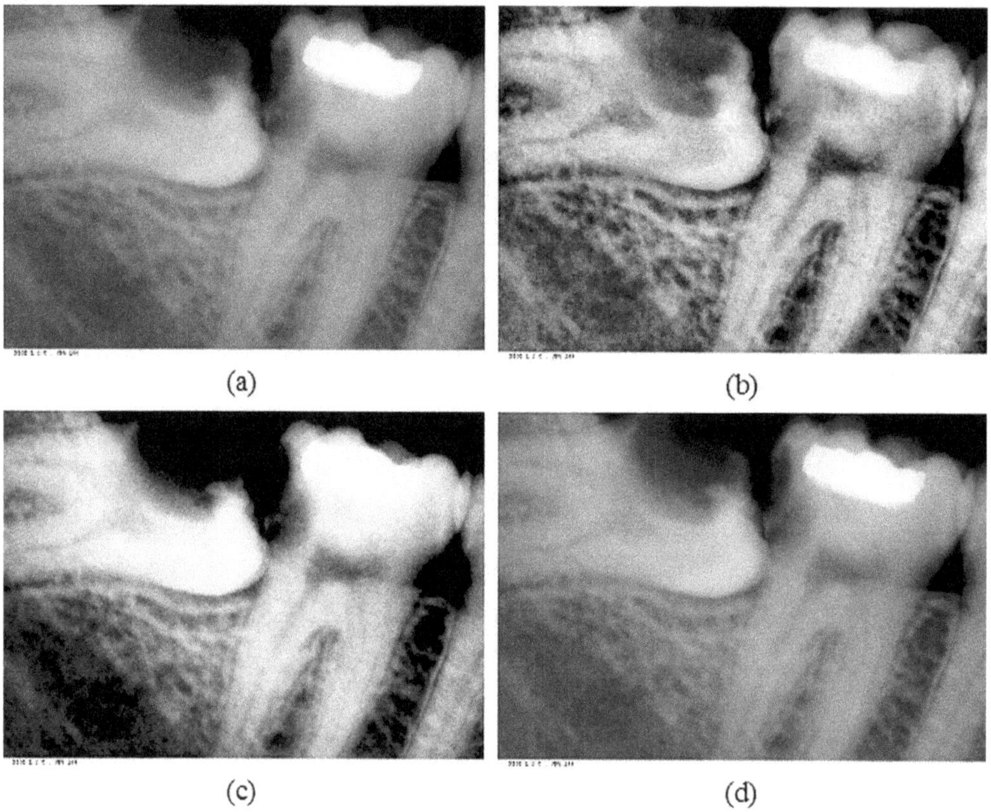

ing results. From this image, it is clear that, the tooth-cavity is more visible in Fig 3(d) and Fig 3(f) compared to Fig 3(b). Hence, from this figure, it is clear that, the enhancement of the DXR is essential before examining the tooth abnormality. Similar procedure is implemented for the whole DXR picture case and its related result is depicted in Fig 4. From Fig 3 and Fig 4, it can be noted that, the adapthisteq and histeq procedures will help to attain better evaluation. Hence, the other images of this chapter are enhanced with adapthisteq/histeq procedures, before implementing the thresholding and segmentation procedures. Fig 3 and Fig 4 also confirms that, single tooth and the whole DXR image cases will help to achieve the similar result on the caries examination process.

The detailed demonstration of the implemented approach on the chosen DXR picture is clearly depicted in Fig 5. Fig 5 (a) and (c) presented the chosen DXR and its related gray histogram respectively. In this histogram image, X-axis represents number of thresholds and the Y-axis denotes the gray pixel levels. The image enhancement procedure can be used to change the pixel and the threshold levels inorder to enhance the test picture. Fig 5(b) and (d) depicts the enhanced image with adapthisteq and its corresponding histogram level. After the enhancement, the three-level thresholding based on the JA+TE is implemented and its result is presented in Fig 5(e). Finally, the caries section from Fig 5(e) is then extracted using the single-well DRLS. Fig 5 (f), (g) and (h) show the initial bounding-box with DRLS, final convergence of the DRLS and the extracted tooth cavity. This tooth cavity section is then further

Figure 3. Results obtained with single tooth case. (a) Test picture, (c) Picture enhanced with Adapthisteq, (e) Picture improved with Histeq, and (b), (d) & (f) Threshold result

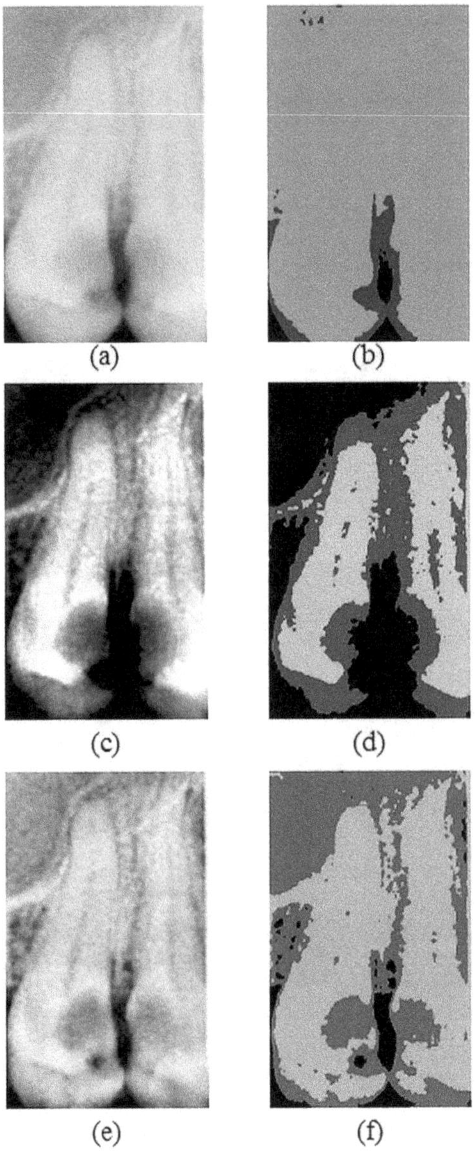

evaluated using the GLCM procedure as discussed in (Haralick et al., 1973). From these images, it is clear that, proposed procedure very efficient in extracting the tooth cavity.

This approach is further tested on the DXR dataset and the sample results are clearly depicted in Table 1. From this table, it can be noted that, proposed approach extracts the tooth cavity from the test images of the DXR dataset. This confirms that, proposed approach is efficient in examining the cavity section using the DXR tooth images. In future, the GLCM values for the extracted section can be computed based on the method discussed in (Rajinikanth et al., 2018), to compute the texture features. This feature can help to examine the tooth cavity in detail.

Figure 4. Results obtained with whole DXR image case. (a) Test picture, (c) Picture enhanced with Adapthisteq, (e) Picture improved with Histeq, and (b), (d) & (f) Threshold result

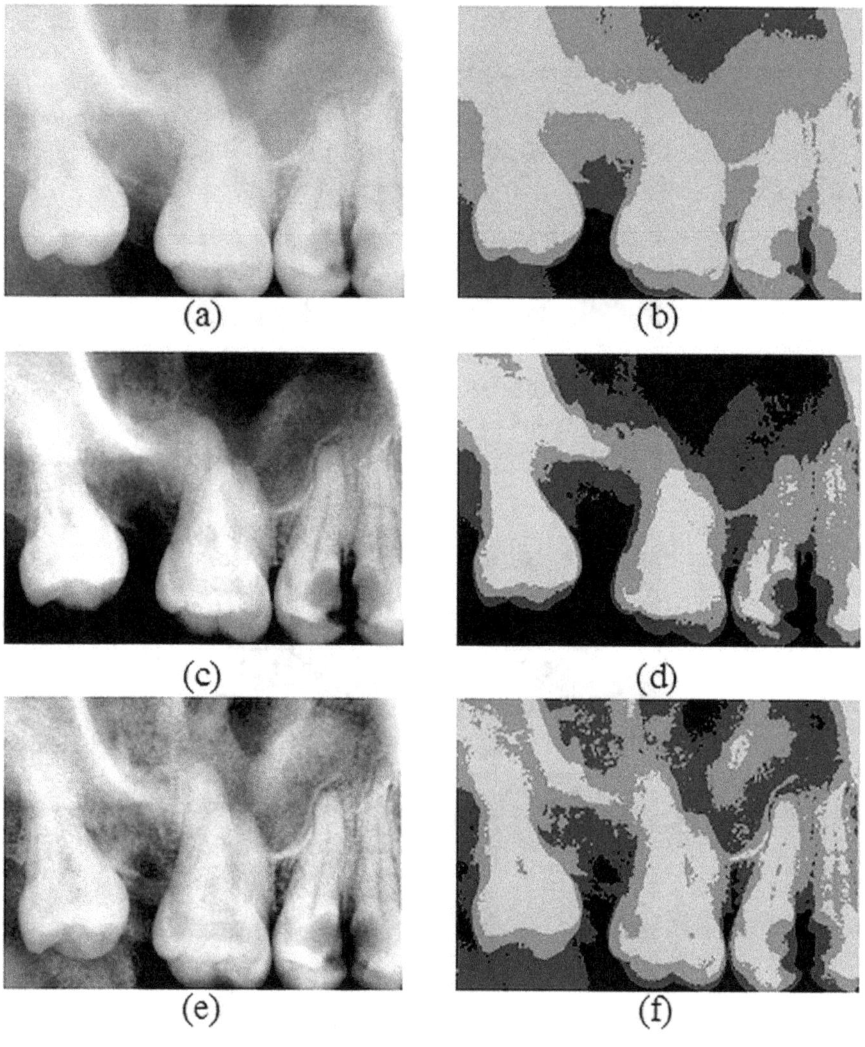

The superiority of the proposed approach is further tested using the benchmark DBR images of ISIB 2015 challenge. This dataset was initially examined by Wang et al. (2016) using an image processing procedure to examine various sections of the tooth. In this chapter, examination of the three vital tooth elements, such as the pulp, tooth restoration and root canal treatment sections are separately analyzed and its outcome is then compared with the GT pictures existing in the ISIB 2015. The major aim of this comparison is to confirm the advantage of the JA+TE and DRLS approach on the ISIB 2015 image datasets.

Initially the test pictures shown in Fig 6 along with its GTs are considered for the examination. The DBR pictures of the ISIB 2015 dataset are available in varied sizes and various illuminations. These pictures will contain all the possible tooth elements and extraction of a particular region is quite complex than the DXR pictures. Fig 6(a) and Fig 6(b) presents the pseudo name and the test pictures respectively.

Figure 5. Outcome of proposed procedure on the DXR tooth dataset

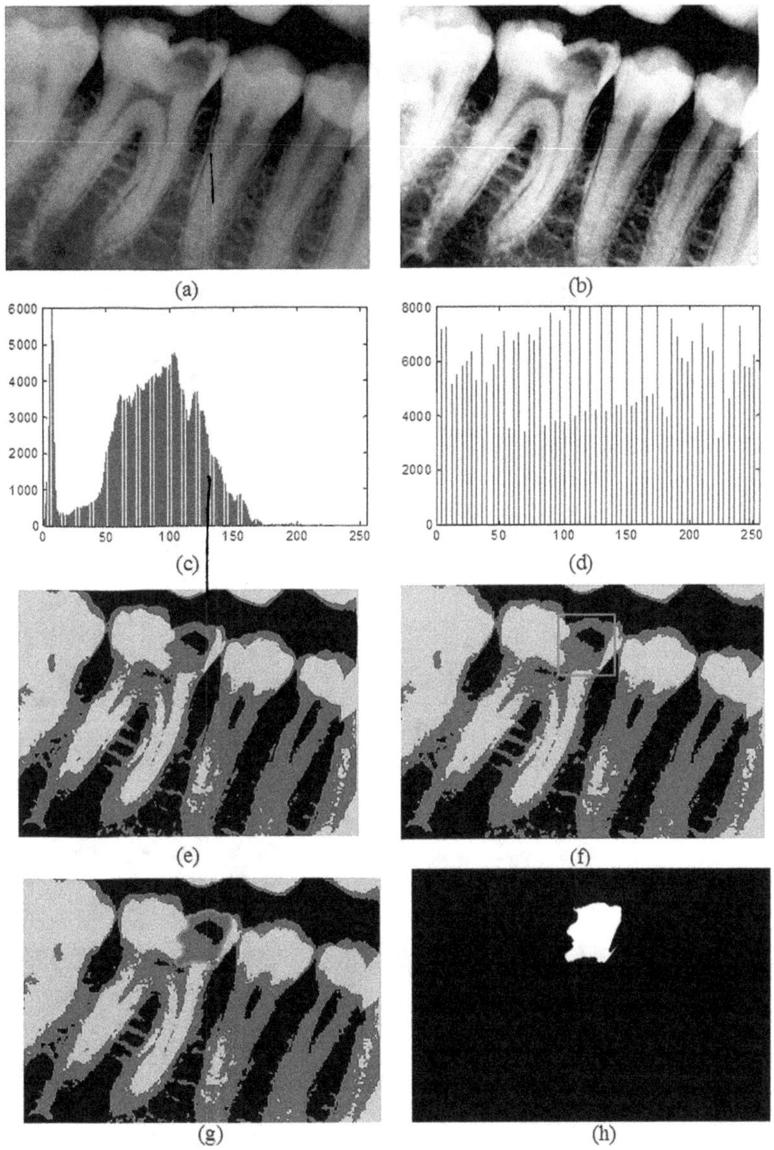

Fig 6(c), (d) and (e) shows the GTs of pulp, tooth restoration and root canal respectively. In this image, there are a number of pulp sections and tooth restoration section and extracting this section will be a complicated task.

Initially, the image enhancement for this image is implemented with the adapthisteq process and which will enhance the visibility of the tooth section to be extracted and evaluated. Later, the thresholding process is implemented with JA+TE by choosing the threshold level as three. The thresholding process further enhances the image region by grouping the identical pixels. The enhanced picture is then undergoing the segmentation operation based on the DRLS. In this, operator assistance is necessary to choose the required tooth element to be examined. The DRLS is initialized with a multiple bounding-

Table 1. Sample DXR pictures and extracted tooth cavity

Pseudo name	Test image	Tooth cavity
12	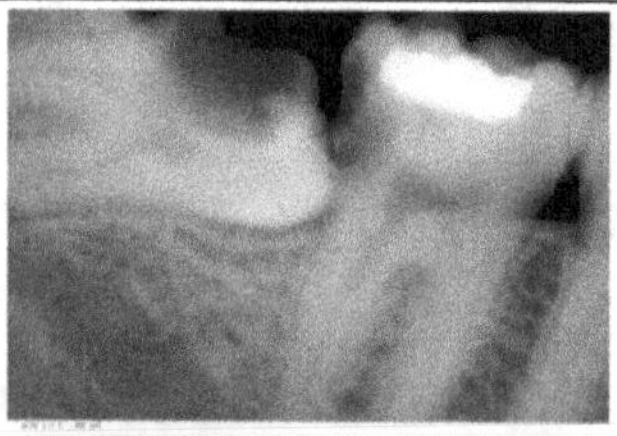	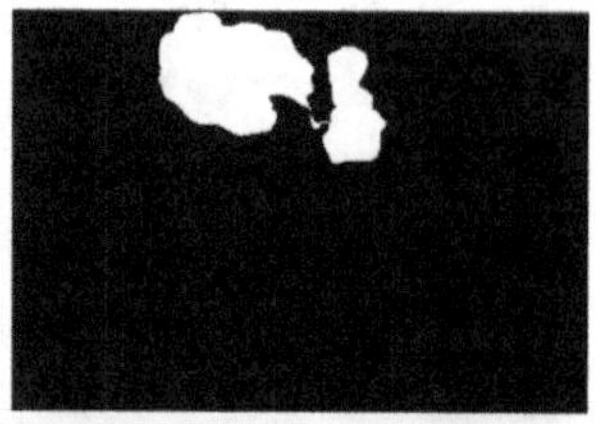
29	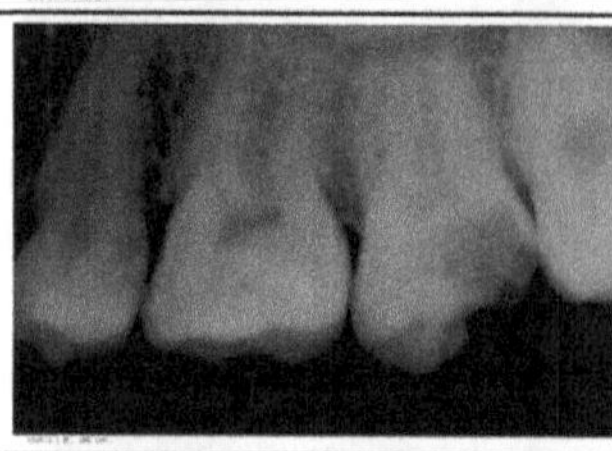	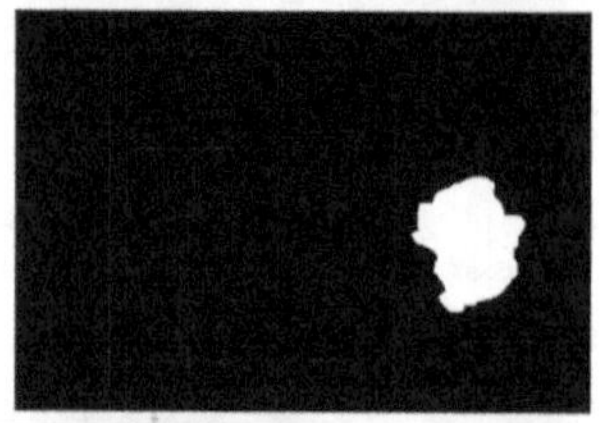
51	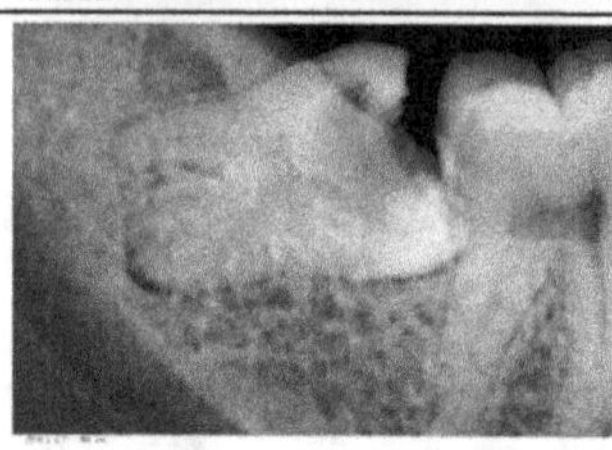	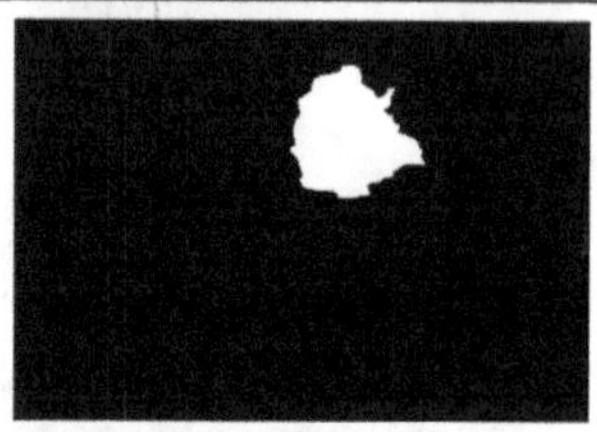
57	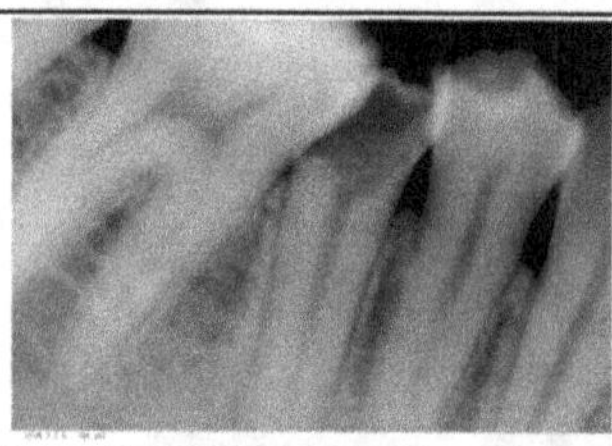	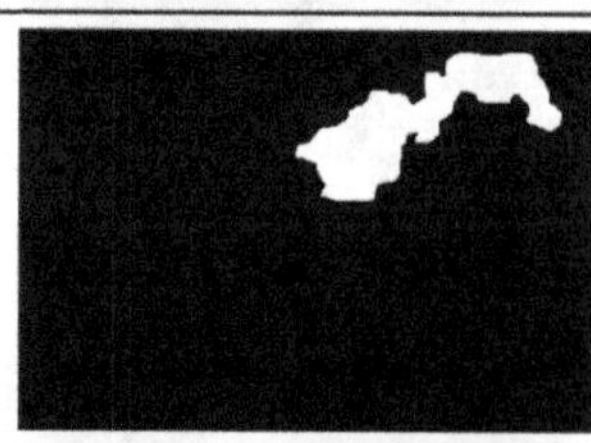
90	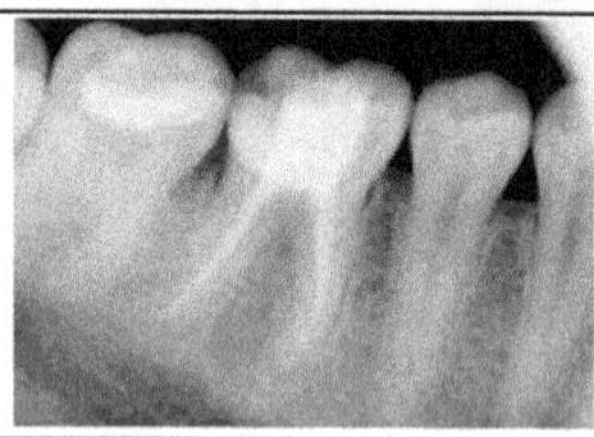	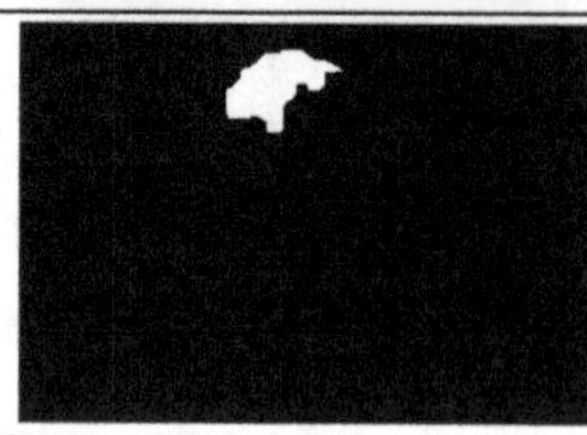

box in order to extract the all the possible elements. This process is initially implemented for the pulp section and similar work is implemented for other sections considered in this chapter.

Fig 7 presents the outcomes obtained using the implemented procedure. Fig 7(a) presents the pseudo names of ISIB 2015, Fig 7(b), (c) and (d) depicts the extracted sections of pulp, restoration and root canal respectively. In order to confirm the accuracy of the implemented procedure, a relative examination among the mined tooth elements and GTs are executed and the essential values like the TPrate, FNrate, TNrate, FPrate, Jaccard, Dice, SE, SP, AC and PR are then computed.

The result of this study also confirms that, TPrate+FNrate = 1 and TNrate+FPrate=1. The computed values are clearly presented in Table 2, Table 3 and Table 4. In these tables, average 1 presents the mean

Figure 6. Chosen test pictures and GT of the ISIB 2015 DBR picture
(a) Pseudo name, (b) Test picture, (c) Pulp, (d) Restoration and (e) Root canal

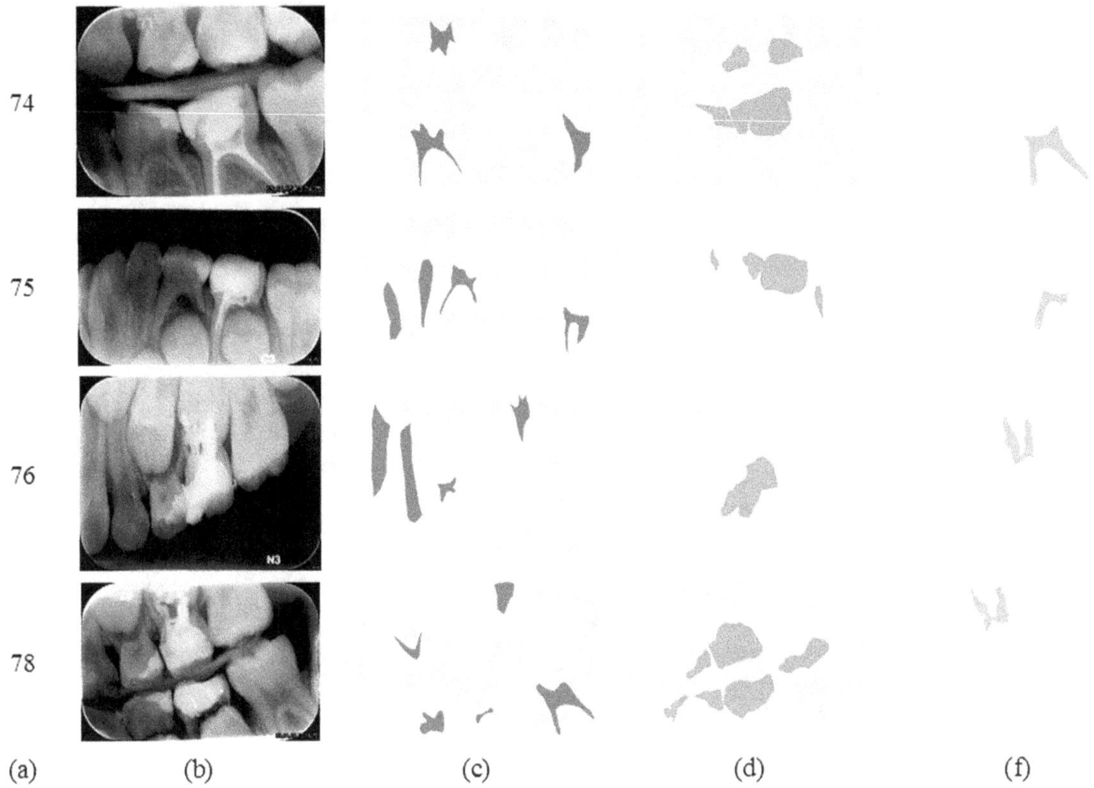

values obtained for the chosen test pictures (four images) and average 2 depicts the mean values obtained for the ISIB 2015 dataset (80 images).

From table 2, it can be noted that, for the pulp section, the proposed procedure helps to achieve more than 84% of mean value for Jaccard, Dice, SE, SP, AC and PR. This confirms that, proposed technique is proficient in extracting the tooth elements (pulp) from the DBR images. Similar results are obtained for other tooth elements, such as the restoration and the root canal.

Proposed approach of this chapter implements a multi-step procedure to extract the tooth elements from the benchmark DXR and DBR pictures. In future, the proposed approach can be tested using the recent image processing procedures such as the machine-learning and deep-leaning approaches.

FUTURE RESEARCH DIRECTIONS

In dental clinics, various manual/automated examination procedures are employed to identify a range of tooth irregularities. If the region and disease severity is identified, then it is possible for the dentist to plan for the further treatment procedures. Mostly, the dentist will always work to restore the dames in tooth elements by implementing all the possible treatment procedures. In order to have a pre-opinion regarding the tooth abnormality in clinical level, it is essential to use a suitable image processing tool to

Figure 7. Extracted tooth elements from the chosen ISIB 2015 DBR images
(a) Pseudo name, (b) Pulp, (c) Restoration and (d) Root canal

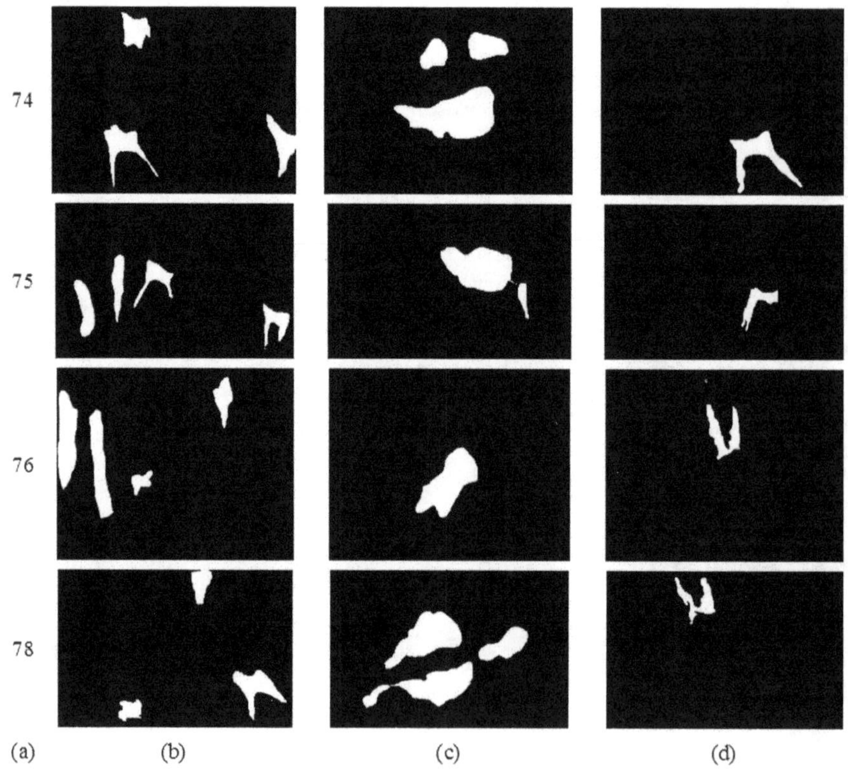

Table 2. Picture similarity measures obtained for the pulp section

Image	TPrate	FNrate	TNrate	FPrate	Jaccard	Dice	SE	SP	AC	PR
74	0.8411	0.1589	0.9977	0.0023	0.7596	0.8634	0.8411	0.9977	0.9945	0.8868
75	0.8926	0.1074	0.9990	0.0010	0.8485	0.9180	0.8926	0.9990	0.9969	0.9450
76	0.9408	0.0592	0.9990	0.0010	0.9137	0.9549	0.9408	0.9990	0.9971	0.9695
78	0.9105	0.0895	0.9958	0.0042	0.8107	0.8426	0.9277	0.9958	0.9918	0.8974
Average1	0.8962	0.1038	0.9979	0.0021	0.8331	0.8947	0.9005	0.9979	0.9951	0.9247
Average2	0.8519	0.1481	0.9828	0.0172	0.8416	0.8602	0.9107	0.9886	0.9910	0.9048

appraise the DXR and DBR images. The main aim of the work discussed in this chapter is to develop a JA assisted procedure to examine the tooth elements. In the proposed work, an approach based on the JA assisted TE and DRLS is proposed to examine tooth sections. The proposed work focuses on extracting and analyzing the chosen elements of tooth.

In future, the proposed work can be directed as follows: i) Implementation of recent heuristic algorithm to improve the pre-processing section, ii) Enhancing the results of the pre-processing by considering other thresholding procedures, such as Otsu's function, Kapur' entropy, Shannon entropy and Fuzzy-Tsallis entropy, iii) Implementing the well known segmentation approaches, like principal component analysis,

Table 3. Image likeness measures acquired for the toot restoration sector

Image	TPrate	FNrate	TNrate	FPrate	Jaccard	Dice	SE	SP	AC	PR
74	0.9561	0.0439	0.9975	0.0025	0.9257	0.9614	0.9561	0.9975	0.9946	0.9668
75	0.9102	0.0898	0.9922	0.0078	0.8264	0.8386	0.9217	0.9922	0.9818	0.9571
76	0.9789	0.0211	0.9976	0.0024	0.9340	0.9659	0.9789	0.9976	0.9968	0.9532
78	0.6147	0.3853	0.9991	0.0009	0.5677	0.7243	0.6147	0.9991	0.9951	0.8814
Average1	0.8650	0.1350	0.9966	0.0034	0.8134	0.8725	0.8678	0.9966	0.9921	0.9396
Average2	0.8819	0.1181	0.9910	0.0090	0.8382	0.8516	0.9011	0.9817	0.9916	0.9158

Table 4. Image similarity values obtained for root canal section

Image	TPrate	FNrate	TNrate	FPrate	Jaccard	Dice	SE	SP	AC	PR
74	0.9613	0.0387	0.9975	0.0025	0.8807	0.9366	0.9613	0.9975	0.9965	0.9131
75	0.9438	0.0562	0.9991	0.0009	0.8831	0.9379	0.9438	0.9991	0.9984	0.9321
76	0.9365	0.0635	0.9969	0.0031	0.7930	0.8846	0.9365	0.9969	0.9959	0.8381
78	0.9008	0.0992	0.9988	0.0012	0.8435	0.9151	0.9008	0.9988	0.9971	0.9298
Average1	0.9356	0.0644	0.9981	0.0019	0.8501	0.9186	0.9356	0.9981	0.9970	0.9033
Average2	0.9018	0.0982	0.9826	0.0174	0.8288	0.8416	0.9281	0.9826	0.9917	0.9319

seed based region growing, and Chan-Vese active contour approaches to enhance the image similarity measures and image statistical measures, iv) Extracting the features of the tooth region to implement the classifier units for the automated detection and diagnose of the tooth abnormalities.

CONCLUSION

This work implements an evaluation tool by integrating Tsallis function with DRLS extraction practice to enhance the performance of tooth abnormality inspection. This work considers two well known tooth benchmark image datasets, like DXR and DBR. Initially, the proposed procedure is tested on the DXR images in order to extract the tooth cavity section. Later, this approach is tested using the benchmark ISIB 2015 DBR dataset and the tooth elements, such as the pulp, restoration and root canal are examined. The considered datasets are initially enhanced using adapthisteq and three-level thresholding with JA + TE. Later, the essential tooth elements are then extracted using the DRLS. The outcome of the DRLS approach is validated against the GT pictures of the ISIB 2015. The results confirm that, proposed procedure help to achieve better values of Jaccard, Dice, SE, SP, AC and PR for the pulp, restoration and root canal regions of the ISIB 2015 dataset. The results of this study confirms that, proposed procedure offers better result on the DXR and DBR pictures and hence the proposed tool can be considered to examine the real time clinical tooth pictures.

REFERENCES

Chen, E., & Abbott, P.V. (2009). Dental pulp testing: A review. *International Journal of Dentistry*. . doi:10.1155/2009/365785

Chen, E., & Abbott, P. V. (2011). Evaluation of accuracy, reliability, and repeatability of five dental pulp tests. *Journal of Endodontics*, *37*(12), 1619–1623. doi:10.1016/j.joen.2011.07.004 PMID:22099893

DXR-database. (n.d.). Retrieved from https://mynotebook.labarchives.com/share/Vahab/MjAuOHw4NTc-2Mi8xNi9UcmVlTm9kZS83NzM5OTk2MDZ8NTIuOA==

Figueiredo, F. E., Martins-Filho, P. R., & Faria-E-Silva, A. L. (2015). Do metal post-retained restorations result in more root fractures than fiber post-retained restorations? A systematic review and meta-analysis. *Journal of Endodontics*, *41*(3), 309–316. doi:10.1016/j.joen.2014.10.006 PMID:25459568

Haralick, R. M., Shanmugam, K., & Dinstein, I. (1973). Textural features for image classification. *IEEE Transactions on Systems, Man, and Cybernetics*, *SMC-3*(6), 610–621. doi:10.1109/TSMC.1973.4309314

ISIB. (2015). *Dataset*. Retrieved from http://www-o.ntust.edu.tw/~cweiwang/ISBI2015/challenge2/

Jafarzadeh, H., & Abbott, P. V. (2010). Review of pulp sensibility tests. Part II: Electric pulp tests and test cavities. *International Endodontic Journal*, *43*(11), 945–958. doi:10.1111/j.1365-2591.2010.01760.x PMID:20726917

Jespersen, J. J., Hellstein, J., Williamson, A., Johnson, W. T., & Qian, F. (2014). Evaluation of dental pulp sensibility tests in a clinical setting. *Journal of Endodontics*, *40*(3), 351–354. doi:10.1016/j.joen.2013.11.009 PMID:24565651

Li, C., Huang, R., Ding, Z., Gatenby, J. C., Metaxas, D. N., & Gore, J. C. (2011). A level set method for image segmentation in the presence of intensity inhomogeneities with application to MRI. *IEEE Transactions on Image Processing*, *20*(7), 2007–2016. doi:10.1109/TIP.2011.2146190 PMID:21518662

Li, C., Xu, C., Gui, C., & Fox, M. D. (2010). Distance regularized level set evolution and its application to image segmentation. *IEEE Transactions on Image Processing*, *19*(12), 3243–3254. doi:10.1109/TIP.2010.2069690 PMID:20801742

Lin, P. L., Lai, Y. H., & Huang, P. W. (2012). Dental biometrics: Human identification based on teeth and dental works in bitewing radiographs. *Pattern Recognition*, *45*(3), 934–946. doi:10.1016/j.patcog.2011.08.027

Lu, H., Kot, A. C., & Shi, Y. Q. (2004). Distance-reciprocal distortion measure for binary document images. *IEEE Signal Processing Letters*, *11*(2), 228–231. doi:10.1109/LSP.2003.821748

Odaira, C., Itoh, S., & Ishibashi, K. (2011). Clinical evaluation of a dental color analysis system: The Crystaleye Spectrophotometer. *Journal of Prosthodontic Research*, *55*(4), 199–205. doi:10.1016/j.jpor.2010.12.005 PMID:21296639

Palani, T. K., Parvathavarthini, B., & Chitra, K. (2016). Segmentation of brain regions by integrating meta heuristic multilevel threshold with Markov random field. *Current Medical Imaging Reviews*, *12*(1), 4–12. doi:10.2174/1573394711666150827203434

Rad, A. E., Rahim, A. S. M., Kolivand, H., & Amin, I. B. M. (2017). Morphological region-based initial contour algorithm for level set methods in image segmentation. *Multimedia Tools and Applications*, *76*(2), 2185–2201. doi:10.100711042-015-3196-y

Rad, A. E., Rahim, M. S. M., Rehman, A., Altameem, A., & Saba, T. (2013). Evaluation of current dental radiographs segmentation approaches in computer-aided applications. *IETE Technical Review*, *30*(3), 210–222. doi:10.4103/0256-4602.113498

Rad, A.E., Rahim, M.S.M., Rehman, A. & Saba, T. (2016). Digital dental x-ray database for caries screening. *3D Research*, *7*(2), 1-5. . doi:10.100713319-016-0096-5

Raja, N. S. M., Fernandes, S. L., Dey, N., Satapathy, S. C., & Rajinikanth, V. (2018). Contrast enhanced medical MRI evaluation using Tsallis entropy and region growing segmentation. *Journal of Ambient Intelligence and Humanized Computing*, 1–12. doi:10.100712652-018-0854-8

Raja, N. S. M., Rajinikanth, V., & Latha, K. (2014). Otsu based optimal multilevel image thresholding using firefly algorithm. *Modelling and Simulation in Engineering*. doi:10.1155/2014/794574

Rajinikanth, V., & Couceiro, M. S. (2015). RGB histogram based color image segmentation using firefly algorithm. *Procedia Computer Science*, *46*, 1449–1457. doi:10.1016/j.procs.2015.02.064

Rajinikanth, V., Dey, N., Satapathy, S. C., & Ashour, A. S. (2018). An approach to examine magnetic resonance angiography based on Tsallis entropy and deformable snake model. *Future Generation Computer Systems*, *85*, 160–172. doi:10.1016/j.future.2018.03.025

Rajinikanth, V., Fernandes, S. L., Bhushan, B., & Sunder, N. R. (2018a). Segmentation and analysis of brain tumor using Tsallis entropy and regularised level set. *Lecture Notes in Electrical Engineering*, *434*, 313–321. doi:10.1007/978-981-10-4280-5_33

Rajinikanth, V., Raja, N. S. M., & Kamalanand, K. (2017). Firefly algorithm assisted segmentation of tumor from brain MRI using Tsallis function and Markov random field. *Journal of Control Engineering and Applied Informatics*, *19*(2), 97–106.

Rajinikanth, V., Raja, N. S. M., & Latha, K. (2014). Optimal multilevel image thresholding: An analysis with PSO and BFO algorithms. *Australian Journal of Basic and Applied Sciences*, *8*, 443–454.

Rajinikanth, V., & Satapathy, S. C. (2018). Segmentation of ischemic stroke lesion in brain MRI based on social group optimization and Fuzzy-Tsallis entropy. *Arabian Journal for Science and Engineering*, 1–14. doi:10.100713369-017-3053-6

Rajinikanth, V., Satapathy, S. C., Fernandes, S. L., & Nachiappan, S. (2017). Entropy based segmentation of tumor from brain MR images–A study with teaching learning based optimization. *Pattern Recognition Letters*, *94*, 87–94. doi:10.1016/j.patrec.2017.05.028

Rao, R. V. (2016). Jaya: A simple and new optimization algorithm for solving constrained and unconstrained optimization problems. *International Journal of Industrial Engineering Computations*, *7*(1), 19–34.

Rao, R. V., & More, K. C. (2017). Design optimization and analysis of selected thermal devices using self-adaptive Jaya algorithm. *Energy Conversion and Management*, *140*, 24–35. doi:10.1016/j.enconman.2017.02.068

Rao, R. V., & Saroj, A. (2017). A self-adaptive multi-population based Jaya algorithm for engineering optimization. *Swarm and Evolutionary Computation*, *37*, 1–26. doi:10.1016/j.swevo.2017.04.008

Rao, R. V., & Saroj, A. (2017a). Constrained economic optimization of shell-and-tube heat exchangers using elitist-Jaya algorithm. *Energy*, *128*, 785–800. doi:10.1016/j.energy.2017.04.059

Roopini, I. T., Vasanthi, M., Rajinikanth, V., Rekha, M., & Sangeetha, M. (2018). Segmentation of tumor from brain MRI using fuzzy entropy and distance regularised level set. *Lecture Notes in Electrical Engineering*, *490*, 297–304. doi:10.1007/978-981-10-8354-9_27

Rose, E., & Svec, T. (2015). An evaluation of apical cracks in teeth undergoing orthograde root canal instrumentation. *Journal of Endodontics*, *41*(12), 2021–2024. doi:10.1016/j.joen.2015.08.023 PMID:26472677

Son, L. H., Tuan, T. M., Fujita, H., Dey, N., Ashour, A. S., Ngoc, V. T. N., ... Chu, D. T. (2018). Dental diagnosis from X-Ray images: An expert system based on fuzzy computing. *Biomedical Signal Processing and Control*, *39*, 64–73. doi:10.1016/j.bspc.2017.07.005

Stark, J. A. (2000). Adaptive image contrast enhancement using generalizations of histogram equalization. *IEEE Transactions on Image Processing*, *9*(5), 889–896. doi:10.1109/83.841534 PMID:18255459

Tom, V. T., & Wolfe, G. J. (1982). Adaptive histogram equalization and its applications. *SPIE Applicat. Dig. Image Process. IV*, *359*, 204–209.

Tsallis, C. (1988). Possible generalization of Boltzmann–Gibbs statistics. *Journal of Statistical Physics*, *52*(1), 479–487. doi:10.1007/BF01016429

Vaishnavi, G. K., Jeevananthan, K., Begum, S. R., & Kamalanand, K. (2014). Geometrical analysis of schistosome egg images using distance regularized level set method for automated species identification. *Journal of Bioinformatics and Intelligent Control*, *3*(2), 147–152. doi:10.1166/jbic.2014.1080

Vishnupriya, R., Raja, N. S. M., & Rajinikanth, V. (2017). An efficient clustering technique and analysis of infrared thermograms. In *Third International Conference on Biosignals, Images and Instrumentation (ICBSII)*. IEEE. doi:10.1109/ICBSII.2017.808227510.1109/ICBSII.2017.8082275

Wang, Huang, C.-T., Lee, J.-H., Li, C.-H., Chang, S.-W., Siao, M.-J., ... Lindner, C. (2016). A benchmark for comparison of dental radiography analysis algorithms. *Medical Image Analysis*, *31*, 63–76. doi:10.1016/j.media.2016.02.004 PMID:26974042

Wang, C. W., Gosno, E., & Li, Y. (2015a). Fully automatic and robust 3D registration of serial-section microscopic images. *Nature-Scientific Reports*, *5*, 15051. doi:10.1038rep15051 PMID:26449756

Wang, C. W., Huang, C.-T., Hsieh, M.-C., Li, C. H., Chang, S. W., Li, W. C., ... Ibragimov, B. (2015). Evaluation and comparison of anatomical landmark detection methods for cephalometric X-ray images: A Grand Challenge. *IEEE Transactions on Medical Imaging*, *34*(9), 1890–1900. doi:10.1109/TMI.2015.2412951 PMID:25794388

Wang, Z., Bovik, A. C., Sheikh, H. R., & Simoncelli, E. P. (2004). Image quality assessment: From error visibility to structural similarity. *IEEE Transactions on Image Processing*, *13*(4), 600–612. doi:10.1109/TIP.2003.819861 PMID:15376593

Weisleder, R., Yamauchi, S., Caplan, D. J., Trope, M., & Teixeira, F. B. (2009). The validity of pulp testing: A clinical study. *The Journal of the American Dental Association*, *140*(8), 1013–1017. doi:10.1109/ICACCI.2014.6968381 doi:10.14219/jada.archive.2009.0312 PMID:19654254

Yadav, G., Maheswari, S., & Agarwal, A. (2014). Contrast limited adaptive histogram equalization based enhancement for real time video system. In *International Conference on Advances in Computing, Communications and Informatics (ICACCI)*. IEEE. 10.1109/ICACCI.2014.6968381

Zuiderveld, K. (1994). *Contrast limited adaptive histogram equalization. In Graphics gems IV* (pp. 474–485). San Diego, CA: Academic Press Professional, Inc.

ADDITIONAL READING

Abhinaya, B., & Raja, N. S. M. (2015). Solving multi-level image thresholding problem—an analysis with cuckoo search algorithm. *Advances In Intelligent Systems And Computing*, *339*, 177–186. doi:10.1007/978-81-322-2250-7_18

Anitha, P., Bindhiya, S., Abinaya, A., Satapathy, S. C., Dey, N., & Rajinikanth, V. (2017). RGB image multi-thresholding based on Kapur's entropy—A study with heuristic algorithms. In. *Second International Conference on Electrical, Computer and Communication Technologies (ICECCT)*. 10.1109/ICECCT.2017.8117823

Ashour, A. S., Samanta, S., Dey, N., Kausar, N., Abdessalemkaraa, W. B., & Hassanien, A. E. (2015). Computed tomography image enhancement using cuckoo search: A log transform based approach. *Journal of Signal and Information Processing*, *6*(3), 244–257. doi:10.4236/jsip.2015.63023

Balan, N. S., Kumar, A. S., Raja, N. S. M., & Rajinikanth, V. (2016). Optimal multilevel image thresholding to improve the visibility of plasmodium sp. in blood smear images. *Advances in Intelligent Systems and Computing*, *397*, 563–571. doi:10.1007/978-81-322-2671-0_54

Dey, N., Ashour, A. S., Beagum, S., Pistola, D. S., Gospodinov, M., Gospodinova, E. P., & Tavares, J. M. R. S. (2015). Parameter optimization for local polynomial approximation based intersection confidence interval filter using genetic algorithm: An application for brain MRI image de-noising. *Journal of Imaging*, *1*(1), 60–84. doi:10.3390/jimaging1010060

Dey, N., Ashour, A. S., Shi, F., Fong, S. J., & Sherratt, R. S. (2017). Developing residential wireless sensor networks for ECG healthcare monitoring. *IEEE Transactions on Consumer Electronics*, *63*(4), 442–449. doi:10.1109/TCE.2017.015063

Dey, N., Ashour, A. S., Shi, F., Fong, S. J., & Tavares, J. M. R. S. (2018). Medical cyber-physical systems: A survey. *Journal of Medical Systems*, *42*(4), 74. doi:10.100710916-018-0921-x PMID:29525900

Dey, N., Rajinikanth, V., Amira, S., Ashour, A. S., & Tavares, J. M. R. S. (2018). Social Group Optimization Supported Segmentation and Evaluation of Skin Melanoma Images. *Symmetry*, *10*(2), 51. doi:10.3390ym10020051

Fernandes, S. L., Gurupur, V. P., Lin, H., & Martis, R. J. (2017). A novel fusion approach for early lung cancer detection using computer aided diagnosis techniques. *Journal of Medical Imaging and Health Informatics*, *7*(8), 1841–1850. doi:10.1166/jmihi.2017.2280

Hore, S., Chakroborty, S., Ashour, A. S., Dey, N., Ashour, A. S., Sifaki-Pistolla, D., ... Chaudhuri, S. R. (2015). Finding contours of hippocampus brain cell using microscopic image analysis. *Journal of Advanced Microscopy Research*, *10*(2), 93–103. doi:10.1166/jamr.2015.1245

Kamalanand, K., & Jawahar, P. M. (2012). Coupled jumping frogs/particle swarm optimization for estimating the parameters of three dimensional HIV model. *BMC Infectious Diseases*, *12*(1Suppl 1), 82. doi:10.1186/1471-2334-12-S1-P82 PMID:22471518

Kamalanand, K., & Jawahar, P. M. (2015). Prediction of Human Immunodeficiency Virus-1 Viral Load from CD4 Cell Count Using Artificial Neural Networks. *Journal of Medical Imaging and Health Informatics*, *5*(3), 641–646. doi:10.1166/jmihi.2015.1430

Kannappan, P. L. (1972). On Shannon's entropy, directed divergence and inaccuracy. *Probability Theory and Related Fields*, *22*, 95–100.

Keerthana, K., Jayasuriya, T. J., Raja, N. S. M., & Rajinikanth, V. (2017). Retinal vessel extraction based on firefly algorithm guided multi-scale matched filter. *International Journal of Modern Science and Technology*, *2*(2), 74–80.

Lakshmi, V.S., Tebby, S.G., Shriranjani, D. & Rajinikanth, V. (2016). Chaotic cuckoo search and Kapur/Tsallis approach in segmentation of T.cruzi from blood smear images. *Int. J. Comp. Sci. Infor. Sec. (IJCSIS)*, 14 (CIC 2016), 51-56.

Mani, M. S., Manisha, S., Thanaraj, K. P., & Rajinikanth, V. (2018). Automated segmentation of Giemsa stained microscopic images based on entropy value. In. International Conference on Intelligent Computing, Instrumentation and Control Technologies (ICICICT). *IEEE*, 1124-1128. Doi: 10.1109/ICICICT1.2017.8342727

Manic, K. S., Priya, R. K., & Rajinikanth, V. (2016). Image multithresholding based on Kapur/Tsallis entropy and firefly algorithm. *Indian Journal of Science and Technology*, *9*(12), 89949. doi:10.17485/ijst/2016/v9i12/89949

Manic, K. S., Rajinikanth, V., Ananthasivam, S., & Suresh, U. (2015). Design of controller in double feedback control loop–an analysis with heuristic algorithms. *Chemical Product and Process Modeling*, *10*(4), 253–262. doi:10.1515/cppm-2015-0005

Manickavasagam, K., Sutha, S., & Kamalanand, K. (2014). An automated system based on 2 d empirical mode decomposition and K-means clustering for classification of plasmodium species in thin blood smear images. *BMC Infectious Diseases*, *14*(3), 1. PMID:24380631

Manickavasagam, K., Sutha, S., & Kamalanand, K. (2014). Development of systems for classification of different plasmodium species in thin blood smear microscopic images. *Journal of Advanced Microscopy Research*, *9*(2), 86–92. doi:10.1166/jamr.2014.1194

Martis, R. J., & Fernandes, S. L. (2017). A special section on early cancer detection and machine vision. *Journal of Medical Imaging and Health Informatics*, *7*(8), 1823–1824. doi:10.1166/jmihi.2017.2274

Palani, T., & Parvathavarthini, B. (2017). Multichannel interictal spike activity detection using time–frequency entropy measure. *Australasian Physical & Engineering Sciences in Medicine*, *40*(2), 413–425. doi:10.100713246-017-0550-6 PMID:28409335

Paramasivam, A., Kamalanand, K., Sundravadivelu, K., & Mannar, J. P. (2017). Effect of electrode contact area on the information content of the recorded electrogastrograms: An analysis based on Rényi entropy and Teager-Kaiser Energy. *Polish Journal of Medical Physics and Engineering*, *23*(2), 37–42.

Paul, S., & Bandyopadhyay, B. (2014). A novel approach for image compression based on multi-level image thresholding using Shannon entropy and differential evolution. Students' Technology Symposium (TechSym), IEEE. 56 – 61. 10.1109/TechSym.2014.6807914

Raja, N. S. M., Kavitha, G., & Ramakrishnan, S. (2012). Analysis of vasculature in human retinal images using particle swarm optimization based Tsallis multi-level thresholding and similarity measures. *Lecture Notes in Computer Science*, *7677*, 380–387. doi:10.1007/978-3-642-35380-2_45

Raja, N. S. M., & Rajinikanth, V. (2014). Brownian distribution guided bacterial foraging algorithm for controller design problem, in ICT and Critical Infrastructure: Proceedings of the 48th Annual Convention of Computer Society of India-Vol I, *Advances in Intelligent Systems and Computing*, 248:141–148. Doi: 10.1007/978-3-319-03107-1_17

Raja, N. S. M., Rajinikanth, V., Fernandes, S. L., & Satapathy, S. C. (2017). Segmentation of breast thermal images using Kapur's entropy and hidden Markov random field. *Journal of Medical Imaging and Health Informatics*, *7*(8), 1825–1829. doi:10.1166/jmihi.2017.2267

Rajinikanth, V., & Couceiro, M. S. (2015). Optimal multilevel image threshold selection using a novel objective function. *Advances in Intelligent Systems and Computing*, *340*, 177–186. doi:10.1007/978-81-322-2247-7_19

Rajinikanth, V., Raja, N. S. M., & Satapathy, S. C. (2016). Robust color image multi-thresholding using between-class variance and cuckoo search algorithm. *Advances in Intelligent Systems and Computing*, *433*, 379–386. doi:10.1007/978-81-322-2755-7_40

Rajinikanth, V., Raja, N. S. M., Satapathy, S. C., Dey, N., & Devadhas, G. G. (2018). Thermogram Assisted Detection and Analysis of Ductal Carcinoma In Situ (DCIS). In. International Conference on Intelligent Computing, Instrumentation and Control Technologies (ICICICT). *IEEE*, 1641-1646. Doi: 10.1109/ICICICT1.2017.8342817

Rajinikanth, V., Raja, N. S. M., Satapathy, S. C., & Fernandes, S. L. (2017). Otsu's multi-thresholding and active contour snake model to segment dermoscopy images. *Journal of Medical Imaging and Health Informatics*, *7*(8), 1837–1840. doi:10.1166/jmihi.2017.2265

Rajinikanth, V., Satapathy, S. C., Dey, N., & Vijayarajan, R. (2018). DWT-PCA image fusion technique to improve segmentation accuracy in brain tumor analysis. *Lecture Notes in Electrical Engineering, 471*, 453–462. doi:10.1007/978-981-10-7329-8_46

Satapathy, S. C., Raja, N. S. M., Rajinikanth, V., Ashour, A. S., & Dey, N. (2016). Multi-level image thresholding using Otsu and chaotic bat algorithm. *Neural Computing & Applications*, 1–23. doi:10.100700521-016-2645-5

Shree, V. T. D., Revanth, K., Raja, N. S. M., & Rajinikanth, V. (2018). A hybrid image processing approach to examine abnormality in retinal optic disc. *Procedia Computer Science, 125*, 157–164. doi:10.1016/j.procs.2017.12.022

Shriranjani, D., Tebby, S. G., Satapathy, S. C., Dey, N., & Rajinikanth, V. (2018). Kapur's entropy and active contour-based segmentation and analysis of retinal optic disc. *Lecture Notes in Electrical Engineering, 490*, 287–295. doi:10.1007/978-981-10-8354-9_26

Shriranjani, D., Tebby, S. G., Satapathy, S. C., Dey, N., & Rajinikanth, V. (2018). Kapur's entropy and active contour-based segmentation and analysis of retinal optic disc. *Lecture Notes in Electrical Engineering, 490*, 287–295. doi:10.1007/978-981-10-8354-9_26

Sudhan, G. H. H., Aravind, R. G., Gowri, K., & Rajinikanth, V. (2017). Optic disc segmentation based on Otsu's thresholding and level set. In. *International Conference on Computer Communication and Informatics (ICCCI)*. 10.1109/ICCCI.2017.8117688

Yang, X. S. (2008). *Nature-Inspired Metaheuristic Algorithms*. Frome, UK: Luniver Press.

Yang, X. S., & Deb, S. (2009). Cuckoo search via Lévy flights. In: Proceedings of World Congress on Nature and Biologically Inspired Computing (NaBIC 2009), IEEE Publications, USA, 210–214. doi:10.1109/NABIC.2009.539369010.1109/NABIC.2009.5393690

Chapter 6
Teeth and Landmarks Detection and Classification Based on Deep Neural Networks

Lyudmila N. Tuzova
Denti.AI, Russia

Dmitry V. Tuzoff
Steklov Institute of Mathematics in St. Petersburg, Russia

Sergey I. Nikolenko
Steklov Institute of Mathematics in St. Petersburg, Russia

Alexey S. Krasnov
Dmitry Rogachev National Research Center of Pediatric Hematology, Oncology, and Immunology, Russia

ABSTRACT

In the recent decade, deep neural networks have enjoyed rapid development in various domains, including medicine. Convolutional neural networks (CNNs), deep neural network structures commonly used for image interpretation, brought the breakthrough in computer vision and became state-of-the-art techniques for various image recognition tasks, such as image classification, object detection, and semantic segmentation. In this chapter, the authors provide an overview of deep learning algorithms and review available literature for dental image analysis with methods based on CNNs. The present study is focused on the problems of landmarks and teeth detection and classification, as these tasks comprise an essential part of dental image interpretation both in clinical dentistry and in human identification systems based on the dental biometrical information.

DOI: 10.4018/978-1-5225-6243-6.ch006

INTRODUCTION

Over the last decade, computer vision models and algorithms based on deep learning models have been successfully applied to various health and medicine domains in a number of medical imaging tasks such as detection and staging of cancer, lung segmentation, diagnosis of colitis, detection and classification of brain diseases, and many others (Lee, et al., 2017; Rezaei, Yang, & Meinel, 2017; Liu, et al., 2017; Litjens, et al., 2017; Shen, Wu, & Suk, 2017). In dentistry, several works have applied deep learning models and algorithms for dental radiograph analysis (Miki, et al., 2016; Lee, Park, & Kim, 2017; Wang, et al., 2016; Ö. Arik, Ibragimov, & Xing, 2017; Tuzoff, et al., 2018). However, deep learning in dentistry still remains an underdeveloped area of research, even though deep neural networks provide state-of-the-art results in many kinds of image recognition tasks (Lee, et al., 2017; Litjens, et al., 2017; LeCun, Bengio, & Hinton, 2015) and problems such as teeth and landmark detection appear to be straightforward object detection problems that could be amenable to modern computer vision approaches based on deep learning methods.

In this chapter, the relevant literature on deep learning methods, specifically convolutional neural networks (CNNs), applied for the tasks of teeth and landmarks detection and classification by type or tooth number is reviewed. These tasks comprise an important part of dental X-ray image analysis. The results can be used to automatically fill a patient's dental records for medical history and treatment planning, preprocess an image for further pathology detection, improve speed and accuracy of postmortem human identification, perform anatomical measurements, and other problems. The deep learning methods have previously been studied for pathology detection purposes (Wang, et al., 2016; Oliveira & Proença, 2011; Imangaliyev, et al., 2016) as well; however, a review of computer-aided disease diagnostics is out of scope of the present study.

CNNs represent state-of-the-art deep learning architectures commonly applied for image recognition tasks. Modern object classification, detection and segmentation approaches based on CNNs have shown promising results, often outperforming methods based on traditional computer vision or other machine learning techniques. An important advantage of deep learning techniques compared with traditional computer vision and other machine learning approaches is that deep learning algorithms do not rely on handcrafted feature extraction and can achieve high performance working with raw input such as pixel values for X-Ray image sources. These methods allow interpreting medical images even if the images are noisy, taken with a different equipment, or in a different setting than that of the data used for training the model. Despite the increasing popularity of the CNNs, the challenges of the application of such architectures still exist. One of the most significant limitation is the amount of annotated data required for the effective model training.

A number of other deep learning models have previously been studied for medical image analysis, including stacked auto encoders (SAEs) and deep belief networks (DBNs) (Shen, Wu, & Suk, 2017; Litjens, et al., 2017). However, CNNs currently represent the state-of-the-art architectures for image recognition tasks. In (Shen, Wu, & Suk, 2017) high performance of CNNs for the images interpretation is explained by the specific properties of the CNNs architectures that better utilize spatial information of images, when most of the other deep learning models process the input in the one-dimensional vector form. Moreover, training of the models, such as DBNs and SAEs, is a complex task combining unsupervised pre-training phase followed by the fine-tuning supervised step. A number of prior studies demonstrated that CNNs models outperformed other deep learning techniques for the image interpretation tasks, when there is annotated data available and end-to-end supervised learning can be performed (Wu, 2015; Song, Zhao, Luo, & Dou, 2017).

In current literature, convolutional neural networks have been studied for the following tasks:

- Landmark detection on cephalometric images (Lee, Park, & Kim, 2017);
- Landmark detection and anatomical type classification on cephalometric images (Ö. Arik, Ibragimov, & Xing, 2017);
- Teeth structures segmentation on bitewing images (Wang, et al., 2016);
- Teeth classification by type on computed tomography (CT) (Miki, et al., 2016);
- Teeth detection and numbering on panoramic views (PV) (Tuzoff, et al., 2018).

In the next sections, a brief history of deep learning and specifically CNNs is provided followed by the review of the methods listed above. Based on the reviewed works, the authors evaluate benefits and limitations of the existing approaches and provide the vision of CNN-based solutions prospects.

BACKGROUND

Deep Learning and Feedforward Neural Networks

Machine learning algorithms aim to learn important features from the set of available data and to apply this knowledge for further interpretation of previously unseen samples. In computer vision, for example, the main problem is to extract features that would be useful for high-level semantic problems such as object recognition.

Deep learning algorithms is one of the most promising classes of machine learning approaches demonstrating the growing popularity in various domains (LeCun, Bengio, & Hinton, Deep learning, 2015; Goodfellow, Bengio, & Courville, 2016; Schmidhuber, 2015). The main limitation of the conventional machine learning algorithms is that they rely on the accuracy of feature engineering when the experts in the specific domain have to design the features by hand. On the contrary, deep learning methods allow computers to learn features from the raw data input.

Many deep learning methods are based on feed-forward neural network architectures. The neural networks have originally been inspired by the desire to imitate the work of a brain. By brain analogy, the core computational unit of the neural network is a neuron. The neurons are grouped together to form the layers of the network. The strengths of the connections (synapses) between the neurons are defined by their weights. At each layer, every neuron computes a weighted sum of inputs from the neurons of the previous layer by the weights of the connections, and passes the result through the non-linear function (activation function). The last output layer of the neural network does not typically use an activation function, but computes the final result, for example, in the form of class scores for classification task or real values for regression task.

There are two main learning approaches used in machine learning methods, both deep and not: supervised, when the annotated data is used to represent the ground truth for training, and unsupervised, when the algorithm aims to reveal clusters of data without using the ground truth annotations. In computer vision, supervised learning is more common.

The supervised learning process can be illustrated using the example of image classification task. To start the training, the model is set with the initial weights. During the training phase, the network processes input images each labeled by a particular class within the known set of possible classes. The output layer of the network produces confidence scores that estimate probability of the image to be any of these classes. The algorithm then measures the error using the special loss function that calculates the difference between the ground truth and predicted scores. To learn from the data, the stochastic gradient descent (SGD) or similar procedure is commonly used. For a batch of images, SGD consists of calculating the average gradients of the loss function, being its local derivatives with respect to the network weights, and adjusting the weights accordingly to decrease the error. The backpropagation procedure based on the chain rule is used to calculate the gradient from the top to the bottom layers. After training, the performance of the model is measured using a different set of images, called test dataset. This step is necessary to evaluate the generalization potential of the model on previously unseen images.

Deep neural networks are represented by multi-layer architectures, where non-linear modules gradually transform the input, producing more and more abstract representations of the data. In computer vision, for example, the first layer can detect simple graphical structures such as lines coming at different angles; the second layer can detect particular arrangements of edges; and so on, with higher layers recognizing complex shapes and patterns. Each next layer detects combinations of features extracted by the previous layer, and the final output layer of a neural network computes the final result. The crucial point for applications is that these layers of features are learned from data but not designed by human experts.

Convolutional Neural Networks

In computer vision, one of the most popular deep network is a convolutional neural network (CNN). The CNN architectures have their roots in the imitation of the visual cortex path organization. In 50th and 60th, studying the brains of the cats, Hubel and Wiesel (Hubel & Wiesel, 1962; Hubel & Wiesel, 1959) discovered that the cats' visual cortex is organized hierarchically, where the simple cells are responsive to simple stimulus such as edges orientation; the complex cells are responsive to both the orientation and the movement; and so on with the growing complexity (Hubel & Wiesel, 1962; Schmidhuber, 2015). In 1982, Fukushima et al. (Fukushima & Miyake, 1982) proposed the neural network model for visual pattern recognition task inspired by the visual cortex structure. This early model was similar to the architecture of modern CNN; however, no supervised learning was used at that moment. Finally, in 1998, Yan LeCun et. al (LeCun, Bottou, Bengio, & Haffner, 1998) introduced the convolutional neural network architecture using gradient-based learning applied for the task of document reading. Since then, a lot of new architectures were proposed; however, the core ideas stay the same. A more detailed history of deep learning is overviewed in (Schmidhuber, 2015).

Similar to traditional feedforward neural networks, CNNs consist of neurons, have weights to be learned, and use both linear and nonlinear transformations to extract features from the data. However, in contrary to the traditional feed-forward neural network, where neurons are fully connected with each other and the number of neurons at each level defines the amount of parameters to be learned, the CNN architecture is organized in a special way where the learnable weights are shared between the neurons of the same feature map. For computer vision, the CNN architectures exploit specific properties of the images allowing to process the raw pixel-level input much more efficiently using significantly lower number of parameters to be learned. In addition to the reduced number of parameters, this structure is also less prone to the overfitting problem.

The architecture of CNNs is represented by the sequence of layers of different types. First layers commonly combine convolutional and pooling layers. Convolutional layers play the key role in the input transformation. Their parameters form a set of learnable filters, where each filter have a fixed size, for example 5x5x3. At the convolutional layer, the neural network slides the filter over the input and computes dot products between the entries of the filter and the input. The results are typically passed through the non-linear function producing the feature map. The intuition behind this process is that the network learns distinctive patterns for each applied filter that are spatially correlated and can be found in different locations of the input. For the image, the network can learn filters that activate when they see some type of a visual pattern, for example a line of a particular orientation.

The pooling layers are used to reduce the spatial size of the input representation. One of the popular ways is to compute maximum value between the local units of the feature map. This procedure allows, on the one hand, to reduce the number of parameters to be learned; and, on the other hand, to control the overfitting. The output of a convolutional layer with or without pooling is a three-dimensional tensor, where the number of filters applied defines the depth. This outputting tensor can serve as an input for subsequent convolutional layers making the network deeper.

The last layers of the CNN are often represented by the fully-connected layers, where each neuron is connected to all neurons of the previous layer. The last fully-connected layer, called the output layer, calculates the output of the network. In Figure 1, examples of a regular feedforward neural network and CNN are presented.

CNNs in Computer Vision

During the last decade, efforts were made to develop architectures based on CNNs to solve various computer vision tasks, including image classification, object localization, object detection and semantic segmentation (Goodfellow, Bengio, & Courville, 2016; LeCun, Bengio, & Hinton, Deep learning, 2015; Schmidhuber, 2015; Huang, et al., 2017).

Figure 1. Regular NN and CNN. a) An example of regular feedforward neural network represented by the sequence of fully-connected layers. It consists of an input layer, two 'hidden' layers, and an output layer. All neurons of adjacent layers are connected with each other; however, neurons are not connected within one layer. b) Convolutional neural network. It consists of the 3D input layer, for example an image, where width and height are the dimensions, and the depth is the number of channels; convolutional layer followed by pooling layer; and two fully-connected layer including the output one.

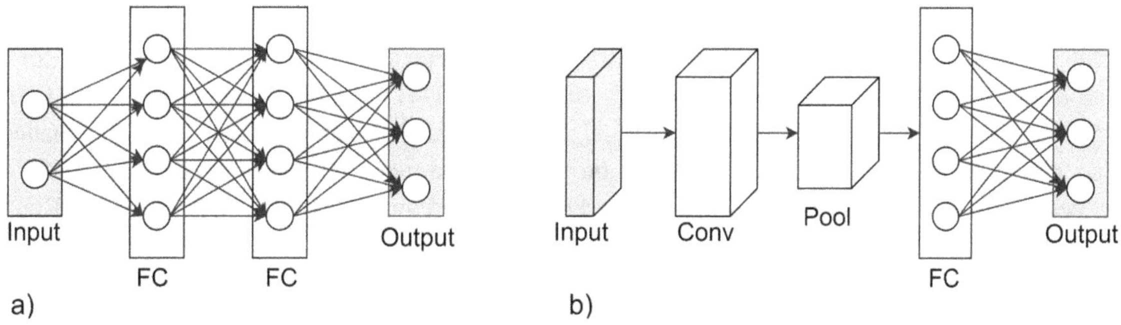

Image classification is one of the most traditional task in computer vision. The purpose of the image classification problem is to label an image with a particular class within a known set of possible classes. In many respects, this task stimulates the current rise of the deep learning and CNNs. The convolutional neural network in its modern form was developed in 1998 by Yan LeCun (LeCun, Bottou, Bengio, & Haffner, 1998); however, for the long time, the CNNs were not actively used in practice. In 2012, the AlexNet CNN architecture (Krizhevsky, Sutskever, & Hinton, 2012) demonstrated spectacular results while classifying a large dataset of hand-annotated images (ImageNet) under the Large Scale Visual Recognition Challenge (ILSVRC). This CNN-based algorithm significantly outperformed other traditional computer vision techniques. Since that moment, the CNNs has been developed rapidly and a number of new architectures was proposed to solve the classification problem, such as ResNet-101 (He, Zhang, Ren, & Sun, 2016), VGG-16 Net (Simonyan & Zisserman, 2015), Inception v3 (Szegedy, Vanhoucke, Ioffe, Shlens, & Wojna, 2016).

In dentistry, the object classification problem is addressed in (Ö. Arik, Ibragimov, & Xing, 2017) for landmark detection, in (Miki, et al., 2016) for teeth classification on CT, and in (Tuzoff, et al., 2018) for teeth classification on panoramic views. Ö. Arik et al. (Ö. Arik, Ibragimov, & Xing, 2017) proposed a custom CNN architecture and used multiple CNNs for binary-class classification. Miki et al. (Miki, et al., 2016) based their work on AlexNet CNN architecture (Krizhevsky, Sutskever, & Hinton, 2012). Tuzoff et al. (Tuzoff, et al., 2018) used VGG-16 (Simonyan & Zisserman, 2015) as a base CNN for teeth classification method.

The localization problem aims not only to classify the singe object on the image, but also to output the bounding box coordinates that enclose the object. This task is commonly addressed as a regression problem, when the network produces both the class scores and the real numbers that can be translated to the coordinates of bounding boxes. The CNNs used for object localization are commonly based on the same architectures as classification CNNs with the difference that the last layer of these networks branch into two heads: regressor and classifier. The joined multitask loss function is commonly used to train the model. In dentistry, the object localization problem was not addressed in prior works. However, Lee et al. (Lee, Park, & Kim, 2017) used a regression approach to define the coordinates of the landmarks on cephalometric images, without predicting class labels. The authors proposed a custom CNN architecture to solve the problem.

Object detection task aims to detect varying number of objects on a single image and mark them with bounding boxes. One of the most popular classes of object detectors uses two-phase approach: first, they find the regions of interests (RoIs); second, they use these region proposals to get class labels and bounding box coordinates for actual objects. The early architectures of this type used external non-learning algorithms for RoIs generation, such as Selective Search (Girshick, Donahue, Darrell, & Malik, 2015). In 2017, a state-of-the-art Faster R-CNN architecture was proposed that combines the region proposal with the object detection in the unified CNN (Ren, He, Girshick, & Sun, 2017). There are also architectures that follow one-phase approach to locate the objects without a separate step for RoI generation, including YOLO (Redmon & Farhadi, 2017) and SSD (Liu, et al., 2016). Huang et al. (Huang, et al., 2017) provided the extended comparison of modern object detectors and their performance. In prior works for dentistry, the object detection problem was addressed in (Tuzoff, et al., 2018) to detect the teeth on panoramic views. The authors based their method on Faster R-CNN architecture (Ren, He, Girshick, & Sun, 2017)

Finally, semantic segmentation task aims to find varying number of objects in the single image on the pixel-level basis. Semantic segmentation is one of the most challenging problem in computer vision, as it requires classifying each pixel to be an object of a particular class or background. Most of the CNN-based methods for pixel-level segmentation use the concept of the fully convolutional networks. The idea of fully convolutional networks is to replace last fully-connected layers of the typical CNN with some kind of upsampling process that allows to output the segmentation map that corresponds to the size of the initial input image. Most well-known architectures for semantic segmentation include Mask R-CNN (He, Gkioxari, Dollár, & Girshick, 2017) and U-Net (Ronneberger, Fischer, & Brox, 2015). In prior works in dentistry, the U-Net (Ronneberger, Fischer, & Brox, 2015) architecture was used to segment teeth structures on the bitewing images (Wang, et al., 2016).

METHODS REVIEW

Landmark Detection and Classification With CNNs

In dentistry, one of the most common types of landmarks to be examined are the landmarks on cephalometric X-ray images. The cephalometric radiograph interpretation is a commonly used tool for human skull examination. This kind of X-rays depicts the patients' head side view and allows to observe dental and skeletal structures of a human skull for further diagnosis, treatment planning and surgery. The essential part of the cephalometric radiograph processing is an accurate localization of the landmarks, special anatomical points used for cephalometric analysis.

By measuring distances and angles between the landmarks, it becomes possible to detect abnormalities of the skull structure. Manual localization of landmarks and an accurate estimation of spatial relationships among them is a time-consuming and subjective procedure (Wang, et al., 2016). An automatic solution of cephalometric landmark detection could eliminate these problems and help improve the quality of the diagnosis and treatment.

Over the recent years, efforts have been made to automate the task of cephalometric landmark detection and classification. In 2015, the Cephalometric X-ray Image Analysis Challenges were performed under the support of IEEE International Symposium on Biomedical Imaging (IEEE ISBI) (Wang, et al., 2016). Wang et al. (Wang, et al., 2016) presented a comparison of different algorithms for automatic cephalometric analysis. In this challenge, the task of cephalometric landmark detection was addressed mostly using the random forests algorithms (RFs). Results of the 2015 challenge demonstrated the progress in this area made over the last years; however, accurate interpretation of cephalometric radiography remains to be a challenging task.

Landmark Detection on Cephalometric Images

In 2017, Lee H. et al. (Lee, Park, & Kim, 2017) proposed a novel method for landmark detection on cephalometric images based on CNNs. Lee H. et al. (Lee, Park, & Kim, 2017) used a dataset of 300 cephalometric images that was randomly divided into 150 training samples and 150 test samples. For each image, two experts provided ground truth annotations for all 19 landmarks.

To train the model, Lee H. et al. (Lee, Park, & Kim, 2017) formulated the task of landmark detection as a regression problem. For each of 19 landmarks, the two target variables were defined: X and Y-coordinates, each normalized by the length of the coordinate axis. Lee H. et al. (Lee, Park, & Kim, 2017) then constructed CNN regression systems for all 38 coordinates with the same underlying architecture and trained CNNs to predict all these coordinate variables from an input X-Ray image. In the proposed system, the authors used a very simple CNN structure with two convolutional layers each followed by the pooling layer, and one fully-connected layer. The output layer of the system produces values that correspond to the normalized coordinate variables of the landmarks (Lee, Park, & Kim, 2017). The schematic representation of the proposed solution is shown on Figure 2.

Lee H. et al. (Lee, Park, & Kim, 2017) did not provide benchmark comparisons with other methods providing the results in form of the box plot of average Euclidean distances between the landmark and ground truths. Lee H. et al. (Lee, Park, & Kim, 2017) reported that the proposed solution demonstrated promising performance locating landmarks within the reasonable distances from the correct annotations; however, the detection accuracy was relatively low. The authors made a conclusion that results of applying CNNs for this new problem is promising considering the simplicity of the method; however, the model performance can be improved with applying of additional techniques: training of deeper networks, using larger input images, and implementing more advanced techniques like data augmentation.

Landmark Detection and Anatomical Type Classification on Cephalometric Images

In 2017, Arik S. et al. (Ö. Arik, Ibragimov, & Xing, 2017) proposed an automatic solution for landmark detection and anatomical type classification on cephalometric images. The authors implemented a framework based on CNNs to locate the landmarks, and used estimated landmark locations to perform a further pathology analysis. The authors used a public dataset of 400 cephalometric images from the IEEE ISBI Challenge (Wang, et al., 2016). In their work, Arik S. et al (Ö. Arik, Ibragimov, & Xing, 2017) provided a comparison with the results of Cephalometric X-ray Image Analysis Challenges 2015 (Wang, et al., 2016). To be consistent with this challenge, Arik S. et al used 150 images for training and 250 for testing.

Figure 2. Lee H. et al. (Lee, Park, & Kim, 2017) method architecture. The input cephalometric image is processed by multiple CNNs each defining one coordinate of a landmark. The resulting set of coordinates is used to place the landmarks on the image.

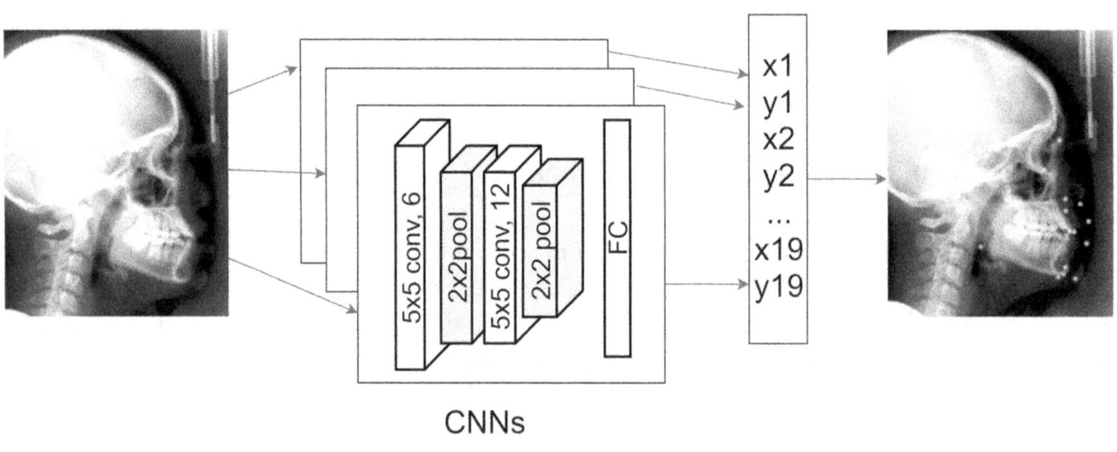

The authors also divided the test set into the two subsets to follow the approach used in this challenge, where Test1 was provided to train and evaluate the models and Test2 was used for on-site competition.

In the proposed method, Arik S. et al. (Ö. Arik, Ibragimov, & Xing, 2017) followed the classification approach to detect landmarks using CNNs. Multiple CNNs were trained to estimate the probability of each pixel being a particular anatomic landmark. A statistical shape model was then used to identify the optimal candidate points. The architecture of the proposed CNN consists of four stages of convolutions each followed by ReLU non-linearity. The max-pooling layers are used to reduce the number of parameters. The last layer uses the sigmoid function to produce the scalar output in the range of (0, 1) estimating the probability of pixel to be a landmark (Ö. Arik, Ibragimov, & Xing, 2017). The simplified architecture is shown on Figure 3.

To evaluate the performance of the algorithms the authors used a success detection rate (SDR) for landmark detection and success classification rate for classification of anatomical types. The success detection rate p_z with precision less then z mm was formulated as:

$$P_z = \frac{\#\{j : \| L_d(j) - L_r(j) \| < z\}}{\#\Omega} \times 100\%, \tag{1}$$

where L_d, L_r are the location of the detected landmark and the ground truth landmark respectively; z is a precision range; $j \in \Omega$, and $\#\Omega$ is the number of detections made. The success classification rate was calculated as:

$$P_{succ} = \frac{\# of\, accurate\, classification}{\# \, of\, classification} \times 100\% \tag{2}$$

Arik S. et al. (Ö. Arik, Ibragimov, & Xing, 2017) reported the following results for their method: SDR of 75.37%, 80.91%, 84.32%, 88.25% in Test1 and 67.8%, 74.16%, 79.11%, 84.63% in Test2 using 2 mm, 2.5 mm, 3 mm, and 4 mm precision ranges; success classification rate of 75.92% and 76.75% for Test1 and Test2 respectively. Wang et al. (Wang, et al., 2016) reported the following results for the winner 2015 method based on RFs: the SDR of 73.68%, 80.21%, 85.19% and 91.47% in Test1 and 66.11%, 72%, 77.63% and 87.43% in Test2 using 2 mm, 2.5 mm, 3 mm, and 4 mm precision ranges; success classification rate of 76.12% and 80.99% for Test1 and Test2 respectively.

Based on these results achieved, the authors concluded that the proposed solution outperformed RF-based methods in the detection accuracy for the 2-, 2.5- and 3-mm ranges. Their method achieved a lower SDR for 4 mm compared with the Lindner et al. method, the best solution of 2015 Challenge (Wang, et al., 2016; Lindner, 2015). The results of classification are slightly worse than the Lindner et al. technique due to the existence of outliers. Arik S. et al. (Ö. Arik, Ibragimov, & Xing, 2017) emphasized the fact that their CNN-based solution demonstrated better results for 2mm range, a clinically accepted detection range.

Arik S. et al. (Ö. Arik, Ibragimov, & Xing, 2017) also reported that while the proposed framework is a synergy of CNNs and a statistical shape model, the overall performance was mostly produced by the CNNs part. The naïve technique such as spatially averaging the highest CNNs output gave the SDR of 67.22%, 79.4%, and 83.20% for the ranges of 2-, 2.5- and 3-mm. The authors also expected that the

Figure 3. Arik S. et al. (Ö. Arik, Ibragimov, & Xing, 2017) method architecture. The CNNs are used to locate the landmarks on cephalometric images. The resulting classification scores are combined together to be refined by statistical shape model.

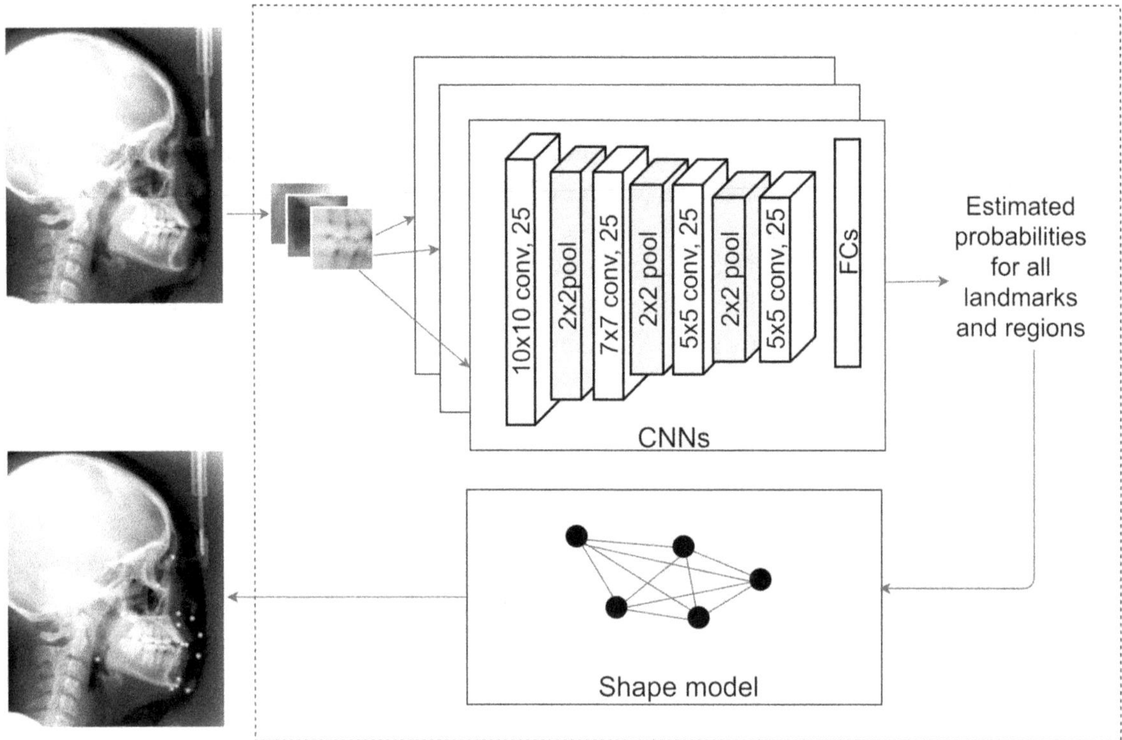

performance of the proposed method can be improved with the increased size and diversity of the training dataset.

Teeth Detection and Classification With CNNs

The task of teeth detection and classification is an essential part of dental radiography interpretation applied for various types of radiographic images. Results of dental radiography interpretation can be used both in clinical practice and in human forensic identification.

In clinical dentistry, results of teeth detection and classification can be used to automatically fill dental charts for monitoring of dental health and treatment planning. Results of teeth detection can also be applied to improve further automated pathology detection. Digitalization of dental X-rays also stimulated the interest in using dental radiographs as a source of biometric information. Automatic dental identification systems (ADIS) use radiographic images to match X-rays for human identification. Accurate detection and classification of teeth on radiographic images can improve the speed and robustness of the ADIS systems (Hosntalab, Zoroofi, Tehrani-Fard, & Shirani, 2010).

During the past decade, a number of methods were proposed for teeth detection and classification: isolation with separation lines or curves (Wanat, 2011; Lin P. L., Huang, Cho, & Kuo, 2013; Zak, et al., 2018; Said, Nassar, Fahmy, & Ammar, 2006), pixel-level teeth segmentation (Hosntalab, Zoroofi,

Tehrani-Fard, & Shirani, 2010; Lin, Lai, & Huang, 2010; Lin P. L., Huang, Huang, Hsu, & Chen, 2014; Oliveira & Proença, 2011; Shah, Baza, Ross, & Ammar, 2006; Tom & Thomas, 2015; Wang, et al., 2016), teeth detection using bounding boxes (Tuzoff, et al., 2018), and teeth classification by numbers or types (Hosntalab, Zoroofi, Tehrani-Fard, & Shirani, 2010; Lin, Lai, & Huang, 2010; Miki, et al., 2016; Tuzoff, et al., 2018).

Most of the methods for teeth detection were based on traditional computer vision techniques, such as thresholding, histogram-based methods, or contour models. The methods for teeth classification (Lin, Lai, & Huang, 2010; Hosntalab, Zoroofi, Tehrani-Fard, & Shirani, 2010) used machine learning approaches, such as SVMs and neural networks; however they were still based on hand-crafted feature extraction algorithms.

Teeth Structures Segmentation on Bitewing Images

In 2015, the method of pixel-level segmentation based on CNN, was introduced under the Computer-Automated Detection of Caries in Bitewing Radiography Challenge (Wang, et al., 2016). Ronneberger et al. (Ronneberger, Fischer, & Brox, 2015; Wang, et al., 2016) proposed a fully-automated solution for segmentation of teeth structures, such as caries, enamel, dentin, pulp, crown, restoration, and root cannel treatment. In this competition, a dataset of 120 bitewing images was used. As with the Cephalometric X-ray Image Analysis Challenges, the bitewing dataset was split into three parts: training set of 40 images, Test1 set of 40 images, and Test2 set of 40 images.

Ronneberger et al. (Ronneberger, Fischer, & Brox, 2015; Wang, et al., 2016) implemented a method based on U-Net (Ronneberger, Fischer, & Brox, 2015), a CNN architecture for image segmentation. The U-Net was initially developed for biomedical image segmentation. This architecture of U-Net consists of two parts: contracting path and expansive path. The standard layers of typical CNN represent the contracting path: the combination of convolutional and pooling layers that compress the original image into a lower-dimensional representation. The expansive path performs upsampling process. At each level, the convolution is applied to halve the number of feature channels. The resulting feature channels are concatenated with the corresponding feature map from the contracting path to add high-resolution features. At the final layer, a 1x1 convolution is used to map each 64-component feature vector to the desired number of classes. In Ronneberger et al. (Ronneberger, Fischer, & Brox, 2015) the seven-class model was trained to segment each of teeth structure types. The architecture of the method is shown on Figure 4.

The following metrics were used to evaluate the methods proposed under the challenge:

$$Sensitivity = \frac{TP}{TP + FN},$$

$$Specificity = \frac{TN}{TN + FP},$$

$$Fscore = \frac{2TP}{2TP + FP + FN}$$

where TP, FP, FN represent the true positives, false positives, and false negatives.

Ronneberger et al. (Ronneberger, Fischer, & Brox, 2015; Wang, et al., 2016) reported the following average results for their method: Sensitivity of 0.578, Specificity of 0.983, and F-score of 0.567 in Test1; and Sensitivity of 0.531, Specificity of 0.982, and F-score of 0.525 in Test 2. The second-best results for the method based on random forests achieved Sensitivity of 0.548, Specificity of 0.912, and F-score of 0.322 in Test1; and Sensitivity of 0.497, Specificity of 0.913, and F-score of 0.287 in Test 2.

Ronneberger et al. (Ronneberger, Fischer, & Brox, 2015; Wang, et al., 2016) method achieved greater than 0.7 F-score for the segmentation of the three fundamental dental structures: enamel, dentin, and pulp. The authors relied on data augmentation because of little available training data. The augmentation techniques allowed them to extend training dataset by transforming the available images in specific ways, such as rotation, zooming, distortion, etc. (Jung, 2015). Wang et al. (Wang, et al., 2016) reported that data augmentation helped to create additional training images for enamel, denting, and pulp; however, the augmentation process was less successful for other types of teeth structures.

Teeth Classification by Type on Computed Tomography

In 2016, the method for teeth classification on CT based on CNN was proposed by Miki et al. (Miki, et al., 2016). This method investigated the application of deep learning for the purpose of teeth classification on CT by seven types: central incisor, lateral incisor, canine, 1st premolar, 2nd premolar, 1st molar, 2nd molar. Miki et al. (Miki, et al., 2016) used dataset of 52 CT images split into two parts: 42 images for training and 10 images for model evaluation.

Figure 4. Ronneberger et al. (Ronneberger, Fischer, & Brox, 2015; Wang, et al., 2016) method architecture. The method is based on the U-Net CNN architecture (Ronneberger, Fischer, & Brox, 2015). The contracting path uses convolutional and pooling layers to extract the features of the image; expansive path upsample the feature maps, concatenate the result with the corresponding feature map of contracting path, and perform convolutions to reduce the number of channels.

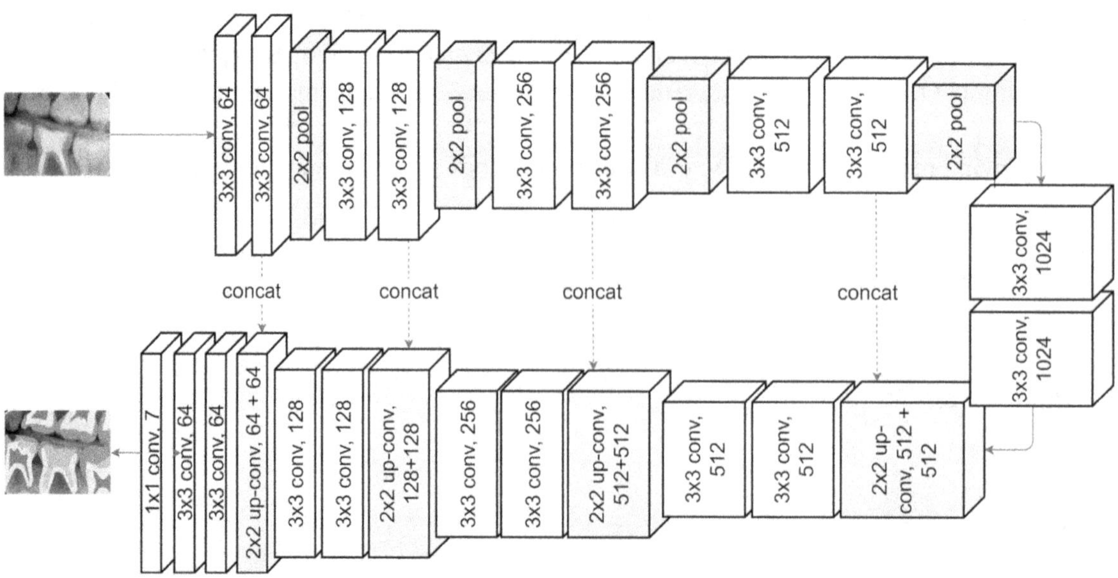

To classify teeth on CT, Miki et al. (Miki, et al., 2016) first manually placed bounding boxes enclosing each tooth on the images. To classify each region into seven teeth types, the authors used a CNN based on classical AlexNet architecture (Krizhevsky, Sutskever, & Hinton, 2012). To increase the variety of the dataset Miki et al. (Miki, et al., 2016) also applied augmentation techniques, such as image rotation and intensity transformation. The architecture of AlexNet consists of five convolutional and three max pooling layers. Each convolutional layer is followed by the rectified linear unit (ReLU) activation function. The network outputs the probabilities for seven teeth types using the softmax function. The architecture of the method is shown on the Figure 5.

To evaluate the model performance, Miki et al. (Miki, et al., 2016) used the accuracy metric. Miki et al. (Miki, et al., 2016) reported 88.8% average accuracy for teeth classification, where the augmentation resulted in an approximately 5% improvement in classification accuracy. Miki et al. (Miki, et al., 2016) also reported that their study has some limitations: the third molars were excluded from the study due to the small number of samples, and teeth effected by metal artifacts were also excluded. Based on the achieved results and a comparison with prior works in the area, the authors of the study concluded that the CNN-based approach can achieve similar classification accuracy based on bounding boxes approach without the precise teeth segmentation. Miki et al. (Miki, et al., 2016) emphasized that the number of samples used to train the network was relatively small and the bigger amount of image available can improve the method performance. Miki et al. (Miki, et al., 2016) also reported that slice images were evaluated independently in this study, and discussed the possible application of 3D convolutions or combining the results for the tooth to improve the accuracy of teeth classification on CT.

Teeth Detection and Numbering on Panoramic Views

In 2018, the method for teeth detection and numbering on panoramic views (PV) was introduced (Tuzoff, et al., 2018). Tuzoff et. al. proposed an end-to-end solution for teeth detection and classification

Figure 5. AlexNet CNN architecture (Krizhevsky, Sutskever, & Hinton, 2012) used in Miki et al. (Miki, et al., 2016)

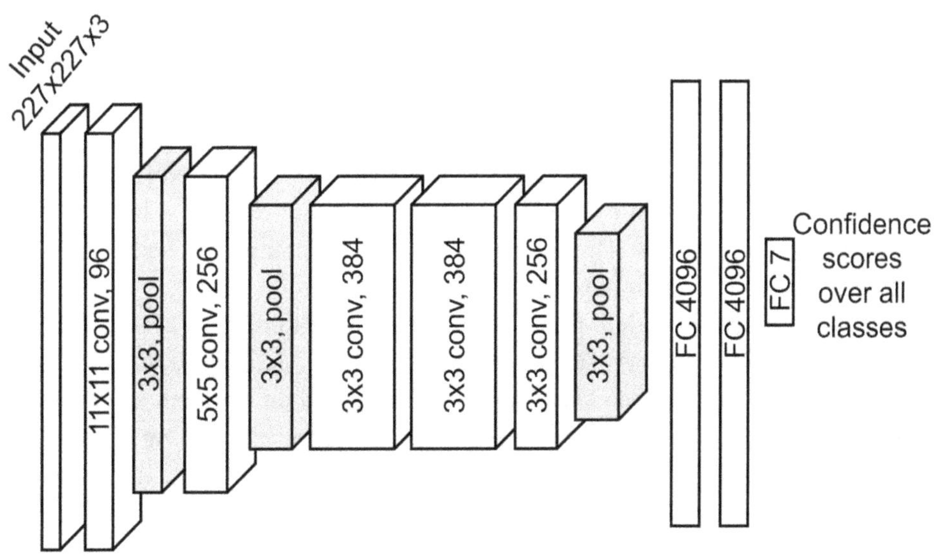

based on CNNs. The teeth were detected on images and numbered with accordance to the two-digit FDI standard (ISO, 2016). The dataset of 1574 images was used to train and test the model. The dataset was split into two groups: training set of 1352 images and test set of 222 images. Five experts provided the ground truth annotations for the images.

To detect and classify teeth on PV, Tuzoff et. al. (Tuzoff, et al., 2018) proposed a system consisting of two modules. The detection module processes the image to detect the teeth in the form of bounding boxes; and the classification module classifies each detected tooth to assign a number among 32 possible teeth numbers. The overall architecture is shown on the Figure 6.

The authors used state-of-the-art Faster R-CNN architecture for the teeth detection module (Ren, He, Girshick, & Sun, 2017). The Faster R-CNN architecture evolved from Fast R-CNN (Girshick, Fast R-CNN, 2015), which, in turn, was based on R-CNN (Girshick, Donahue, Darrell, & Malik, 2015). The basic R-CNN model proposed a solution where RoI proposals were generated separately with the Selective Search algorithm (Uijlings, Sande, Gevers, & Smeulders, 2013), and then each region was evaluated with a CNN, also modifying bounding boxes using special regressors. The Fast R-CNN approach improved performance by simplifying the pipeline and sharing computation between the RoI extraction and object localization. Finally, the Faster R-CNN method introduced a unified network for object detection fully based on convolutional networks. Instead of Selective Search, it added a special region proposal network (RPN) to the CNN. Tuzoff et. al. (Tuzoff, et al., 2018) used the implementation of the Faster R-CNN (Hosang, 2016) with the Tensorflow backend (Abadi, et al., 2015). To initialize the model's parameters, the authors also used the weights pretrained on the conventional ImageNet data source (Deng, et al., 2009).

To generate region proposals the system slides the windows over the convolutional feature map. In (Tuzoff, et al., 2018) the VGG-16 Net was used as a base CNN for both modules. At each window location, Faster R-CNN outputs k potential bounding boxes named "anchors". Features extracted for each anchor are fed to a ROI box-regression layer and a softmax layer. The softmax layer produces objectiveness scores that estimate the probability of a box to be an object or a background. Region proposals generated by RPN were used as input for the Fast R-CNN detector. For each proposal, a region of interest (RoI) pooling layer extracts a feature vector from the feature map. Each feature vector was fed to a sequence of fully connected layers that branch into output layers: softmax layer to produce the class score to be a tooth or a background and a box regression layer to generate the final bounding box coordinates. The network used a combined loss function that consists of RPN and Fast R-CNN parts, jointly training for the classification problem and bounding-box regression. The Faster R-CNN architecture is shown on Figure 7.

The classification module of proposed system was based on VGG-16 Net CNN architecture (Simonyan & Zisserman, 2015). This CNN architecture consists of thirteen convolutional layers, five pooling layers, and three fully-connected layers. The output layer produces confidence scores over all 32-classes corresponding with 32 possible teeth numbers. To classify each tooth, the system cropped the images based on the results of detection module producing the image containing a classified tooth. Additional context was added to improve the classification results. Tuzoff et. al. (Tuzoff, et al., 2018) also implemented a heuristic method to improve classification results based on the teeth arrangement rules. The authors also used simple augmentation techniques, such as zooming and rotating, to increase the variety of the training dataset. Tuzoff et. al. (Tuzoff, et al., 2018) implemented classification method using Keras framework (Chollet, 2015). The VGG-16 Net architecture is shown on the Figure 8.

Tuzoff et. al. (Tuzoff, et al., 2018) used the following metrics to evaluate their method:

$$\text{Precision} = \frac{TP}{TP + FP},$$

$$\text{Recall} = \frac{TP}{TP + FN}, \text{ and}$$

$$\text{F2-score} = \frac{5 * precision * recall}{4 * precision + recall}$$

where TP, FP, FN represent true positives, false positives and false negatives, respectively. For teeth classification, the accuracy metric was used.

The authors also evaluated the human expert performance based on the error analysis using the same metrics. Tuzoff et. al. (Tuzoff, et al., 2018) reported the following results for the system: precision of 0.9944, recall of 0.9941, F2-score of 0.9941 for teeth detection, and 0.98 accuracy for teeth classification. The human experts detected and classified teeth with the following results: precision of 0.9998, recall of 0.9980, F2-score of 0.9994 for teeth detection, and 0.99 accuracy for teeth classification. Based on these results, the authors concluded that their method achieved performance that is comparable with the human level. Tuzoff et. al. (Tuzoff, et al., 2018) also reported that additional context and augmentation allowed improving classification results on 6pp and 2pp, while the heuristic added 0.5pp improvement. The authors emphasized that heuristics had a very modest influence on system performance, and overall results were mostly based on the pure deep-learning techniques.

Figure 6. Tuzoff et. al. (Tuzoff, et al., 2018) method architecture. Teeth detection module is based on Faster-RCNN (Ren, He, Girshick, & Sun, 2017) architecture. Teeth classification module utilizes VGG-16 Net CNN architecture to classify each tooth followed by heuristics to ensure correct teeth arrangement.

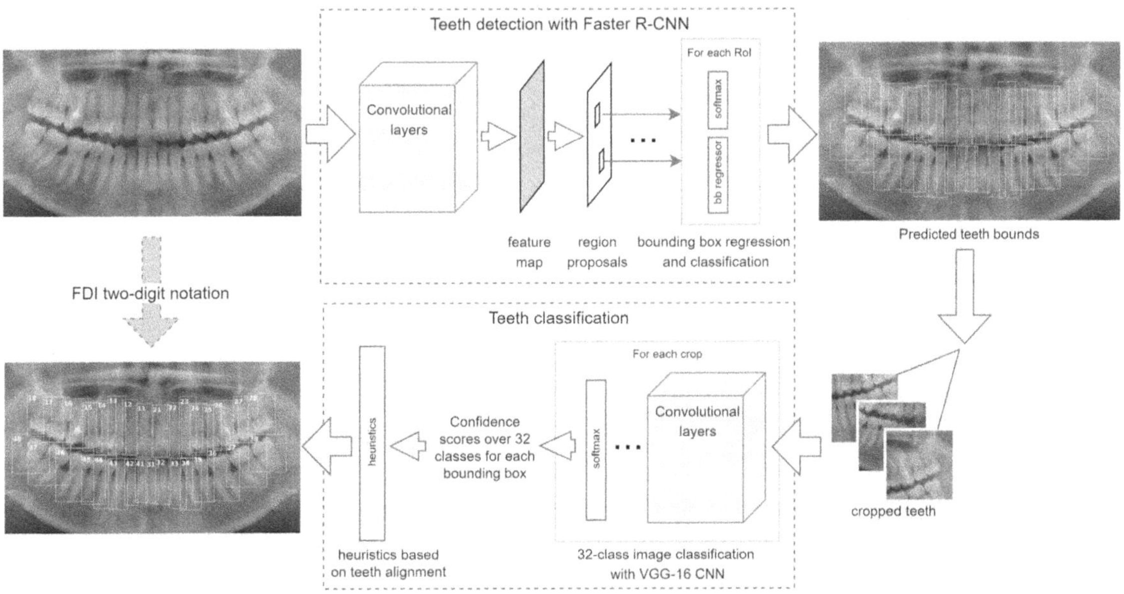

Figure 7. Faster R-CNN architecture used in (Tuzoff, et al., 2018). The RPN generates the region proposals; the object detector use the region proposals to classify each RoI and refine the bounding box coordinates.

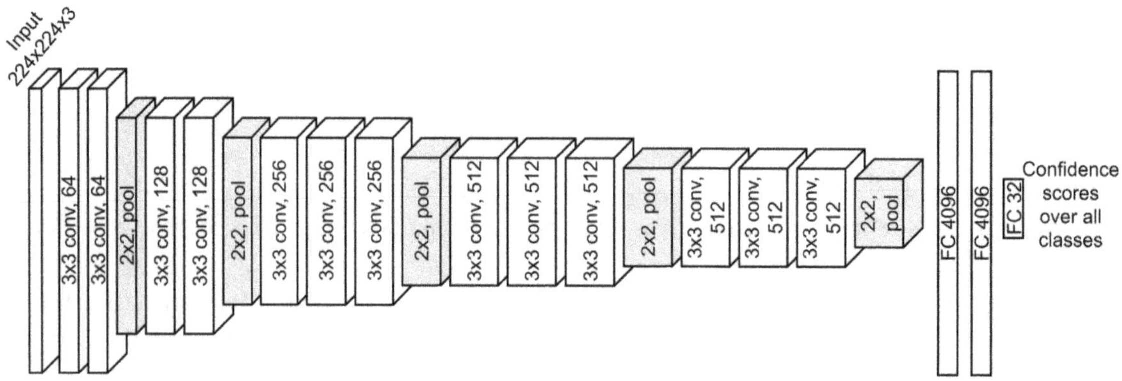

Figure 8. VGG-16 Net architecture used in (Tuzoff, et al., 2018)

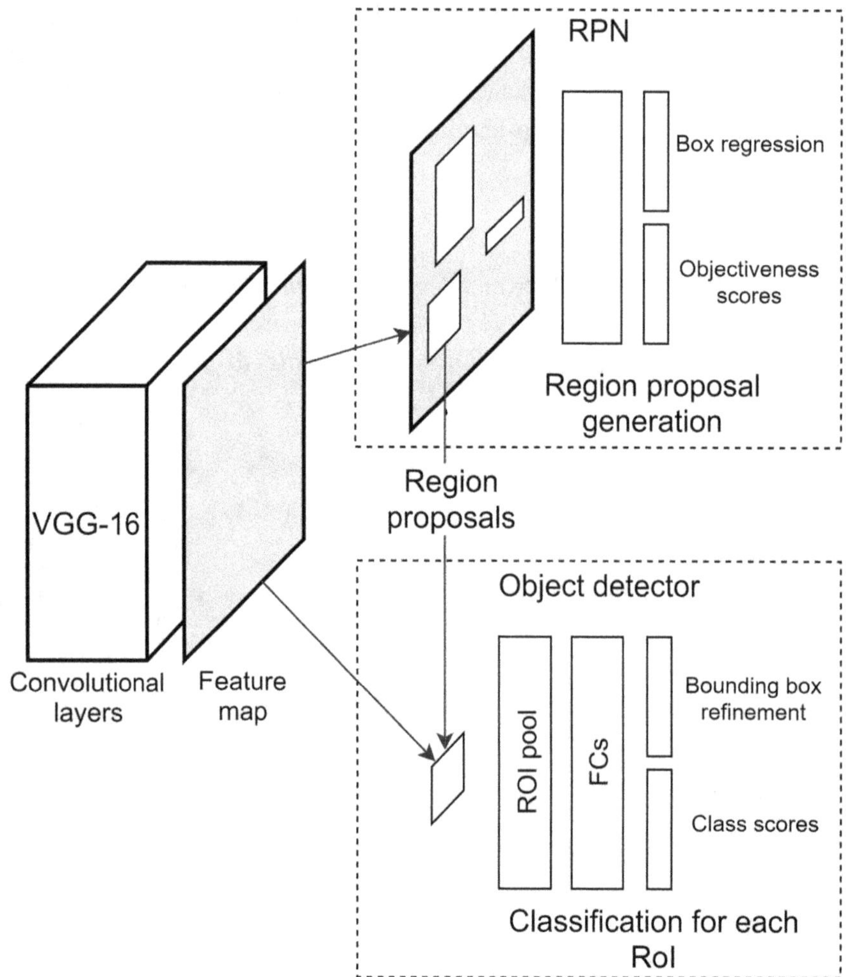

CONCLUSION

In this chapter, the review of deep learning methods in dental radiography interpretation was performed. The present review was mainly focused on the tasks of landmark and teeth detection and classification, as these tasks is an important part of X-ray interpretation process.

Lee H. et al. (Lee, Park, & Kim, 2017) and Arik S. et al. (Ö. Arik, Ibragimov, & Xing, 2017) methods demonstrated the potential of application of CNNs for landmark detection on cephalometric X-Ray images. Lee H. et al. (Lee, Park, & Kim, 2017) used a straightforward approach to experiment with the CNNs. The authors reported that the proposed solution achieved promising results and their method allowed detecting landmarks within reasonable distances from ground truth. Lee H. et al. expected that the results can be improved by implementing more advanced techniques; including using deeper networks, applying image augmentation, training the model on larger training images dataset.

Arik S. et al. (Ö. Arik, Ibragimov, & Xing, 2017) in their work proposed a CNN-based architecture combined with the statistical shape model for cephalometric landmark detection. The authors compared the method performance with the winner solution from the Challenge 2015 (Wang, et al., 2016) that was based on RFs. Arik S. et al. CNN-based method outperformed the RF-based solution for 2-, 2.5- and 3-mm ranges. At the same moment, the method produced slightly lower results for larger ranges. Arik S. et al. concluded that the deep learning approach demonstrated very promising results, especially for the clinically accepted range of 2mm. Arik S. et al. also expected that their method could be further improved using extended training dataset (Ö. Arik, Ibragimov, & Xing, 2017).

Ronneberger et al. (Ronneberger, Fischer, & Brox, 2015; Wang, et al., 2016) introduced a CNN-based solution for teeth structures segmentation on bitewing images under the Challenge 2015 (Wang, et al., 2016). Despite the fact, that the accurate teeth segmentation stayed to be a challenging task, Ronneberger et al. (Ronneberger, Fischer, & Brox, 2015; Wang, et al., 2016) succeeded in demonstrating the high potential of CNN-based architecture for image segmentation. Ronneberger et al. (Ronneberger, Fischer, & Brox, 2015; Wang, et al., 2016) significantly outperformed other methods mostly based on the RFs.

Miki et al. (Miki, et al., 2016) method for teeth classification on CT also showed promising results. The authors emphasized that high accuracy scores were achieved without the precise segmentation of teeth. Miki et al. (Miki, et al., 2016) also expected to improve the accuracy of classification by using more advanced techniques for 3D image processing and extending the training dataset.

Tuzoff et. al. (Tuzoff, et al., 2018) demonstrated the potential of deep learning techniques for the task of teeth detection and classification on panoramic views. The authors reported that the achieved performance results are comparable with the human level. Tuzoff et. al. (Tuzoff, et al., 2018) expected also to improve the results by implementing of more advanced augmentation techniques, using modern CNN architectures, and extending the dataset of the panoramic X-ray images used to train the models.

Based on the reviewed works, it can be concluded that deep learning algorithms, specifically the CNN architectures, have high potential to be applied for the tasks of dental radiography interpretation. At the same time, it can be seen that simple and straightforward approaches cannot compete with the traditional techniques. Effective application of CNN requires both using complex architectures optimized for particular computer vision task and supplementing techniques, such as advanced augmentation. CNN-based methods also significantly benefit from large amounts of annotated data available, which is not always the case.

The important advantage of CNN-based approaches is that they allow to process raw input without by-hand feature engineering. Since these architectures do not rely on hand-crafted features, the CNNs can be studied for a wide range of tasks: various dental and oral structures classification, pathologies detection, post-treatment evaluation. For dental and oral structures, it can be useful to locate implants, bridges, and crowns. For post-treatment procedures, CNNs can be trained to find treatment mistakes, such as root canal overfilling. For pathologies, the CNNs can be studied to detect different diseases, such as caries, periodontitis, and cysts.

Over recent years, the CNNs have become state-of-the-art architectures in all computer vision tasks independent of particular domains. Large community of researches believe in the future of deep learning methods and continue to develop more and more advanced architectures and techniques to meet the growing demand in automation of computer vision problems in various domains, including dentistry.

REFERENCES

Abadi, M., Agarwal, A., Barham, P., Brevdo, E., Chen, Z., Citro, C., . . . Zheng, X. (2015). *TensorFlow: Large-Scale Machine Learning on Heterogeneous Systems*. Retrieved from https://www.tensorflow.org/

Arik, Ö., Ibragimov, B., & Xing, L. (2017). Fully automated quantitative cephalometry using convolutional neural networks. *Journal of Medical Imaging (Bellingham, Wash.)*, *4*(1), 014501. doi:10.1117/1.JMI.4.1.014501 PMID:28097213

Chollet, F. (2015). *keras*. GitHub. Retrieved from https://github.com/fchollet/keras

Deng, J., Dong, W., Socher, R., Li, L. J., Li, K., & Fei-Fei, L. (2009). ImageNet: A large-scale hierarchical image database. *The Conference on Computer Vision and Pattern Recognition (CVPR)*, 248-255.

Fukushima, K., & Miyake, S. (1982). Neocognitron: A new algorithm for pattern recognition tolerant of deformations and shifts in position. *Pattern Recognition*, *15*(6), 455–469. doi:10.1016/0031-3203(82)90024-3

Girshick, R. (2015). Fast R-CNN. *IEEE International Conference on Computer Vision (ICCV)*, 1440-1448.

Girshick, R., Donahue, J., Darrell, T., & Malik, J. (2015). Region-Based Convolutional Networks for Accurate Object Detection and Segmentation. *IEEE Transactions on Pattern Analysis and Machine Intelligence*, *38*(1), 142–158. doi:10.1109/TPAMI.2015.2437384 PMID:26656583

Goodfellow, I., Bengio, Y., & Courville, A. (2016). *Deep Learning*. MIT Press.

He, K., Gkioxari, G., Dollár, P., & Girshick, R. (2017). Mask R-CNN. *IEEE International Conference on Computer Vision (ICCV)*, 2980-2988.

He, K., Zhang, X., Ren, S., & Sun, J. (2016). Deep Residual Learning for Image Recognition. *2016 IEEE Conference on Computer Vision and Pattern Recognition (CVPR)*, 770-778. 10.1109/CVPR.2016.90

Hosang, J. (2016). *Faster RCNN TF*. GitHub. Retrieved from https://github.com/smallcorgi/Faster-RCNN_TF

Hosntalab, M., Zoroofi, R. A., Tehrani-Fard, A. A., & Shirani, G. (2010). Classification and numbering of teeth in multi-slice CT images using wavelet-Fourier descriptor. *International Journal of Computer Assisted Radiology and Surgery*, *5*(3), 237–249. doi:10.100711548-009-0389-8 PMID:20033505

Huang, J., Rathod, V., Sun, C., Zhu, M., Korattikara, A., Fathi, A., ... Murphy, K. (2017, 7). Speed/Accuracy Trade-Offs for Modern Convolutional Object Detectors. *The IEEE Conference on Computer Vision and Pattern Recognition (CVPR)*. 10.1109/CVPR.2017.351

Hubel, D. H., & Wiesel, T. N. (1959). Receptive Fields of Single Neurons in the Cat's Striate Cortex. *The Journal of Physiology*, *148*(3), 574–591. doi:10.1113/jphysiol.1959.sp006308 PMID:14403679

Hubel, D. H., & Wiesel, T. N. (1962). Receptive fields, binocular interaction and functional architecture in the cat's visual cortex. *The Journal of Physiology*, *160*(1), 106–154. doi:10.1113/jphysiol.1962.sp006837 PMID:14449617

Imangaliyev, S., van der Veen, M., Volgenant, C., Keijser, B., Crielaard, W., & Levin, E. (2016). Deep Learning for Classification of Dental Plaque Images. In P. M. Pardalos, P. Conca, G. Giuffrida, & G. Nicosia (Eds.), Lecture Notes in Computer Science: Vol. 10122. Machine Learning, Optimization, and Big Data. MOD 2016 (pp. 407–410). Berlin: Springer. doi:10.1007/978-3-319-51469-7_34

ISO. (2016). *ISO 3950:2016 Dentistry--Designation system for teeth and areas of the oral cavity.*

Jung, A. (2015). *Image augmentation for machine learning experiments.* GitHub. Retrieved from https://github.com/aleju/imgaug

Krizhevsky, A., Sutskever, I., & Hinton, G. E. (2012). ImageNet classification with deep convolutional neural networks. *Annual Conference on Neural Information Processing Systems (NIPS)*, 2, 1097-1105.

LeCun, Y., Bengio, Y., & Hinton, G. (2015). Deep learning. *Nature*, *521*(7553), 436–444. doi:10.1038/nature14539 PMID:26017442

LeCun, Y., Bottou, L., Bengio, Y., & Haffner, P. (1998). Gradient-based learning applied to document recognition. *Proceedings of the IEEE*, *86*(11), 2278–2323. doi:10.1109/5.726791

Lee, H., Park, M., & Kim, J. (2017). Cephalometric Landmark Detection in Dental X-ray Images Using Convolutional Neural Networks. In S. G. Armato & N. A. Petrick (Eds.), SPIE Medical Imaging, 10134 (pp. 1–6). Academic Press.

Lee, J.-G., Jun, S., Cho, Y.-W., Lee, H., Bae Kim, G., Beom Seo, J., & Kim, N. (2017). 7). Deep Learning in Medical Imaging: General Overview. *Korean Journal of Radiology*, *18*(4), 570. doi:10.3348/kjr.2017.18.4.570 PMID:28670152

Lin, P. L., Huang, P. W., Cho, Y. S., & Kuo, C. H. (2013). An automatic and effective tooth isolation method for dental radiographs. *Opto-Electronics Review*, *21*(1), 126–136. doi:10.247811772-012-0051-9

Lin, P. L., Huang, P. Y., Huang, P. W., Hsu, H. C., & Chen, C. C. (2014). Teeth segmentation of dental periapical radiographs based on local singularity analysis. *Computer Methods and Programs in Biomedicine*, *113*(2), 433–445. doi:10.1016/j.cmpb.2013.10.015 PMID:24252317

Lin, P. L., Lai, Y. H., & Huang, P. W. (2010). An effective classification and numbering system for dental bitewing radiographs using teeth region and contour information. *Pattern Recognition*, *43*(4), 1380–1392. doi:10.1016/j.patcog.2009.10.005

Lindner, C., Bromiley, P. A., Ionita, M. C., & Cootes, T. F. (2015). Robust and Accurate Shape Model Matching Using Random Forest Regression-Voting. *IEEE Transactions on Pattern Analysis and Machine Intelligence*, *37*(9), 1862–1874. doi:10.1109/TPAMI.2014.2382106 PMID:26353132

Litjens, G., Kooi, T., Ehteshami Bejnordi, B., Setio, A., Ciompi, F., Ghafoorian, M., & ... Sánchez, C. (2017). A Survey on Deep Learning in Medical Image Analysis. *Medical Image Analysis*, 42. PMID:28778026

Liu, J., Wang, D., Lu, L., Wei, Z., Kim, L., Turkbey, E. B., ... Summers, R. M. (2017). Detection and diagnosis of colitis on computed tomography using deep convolutional neural networks. *Medical Physics*, *44*(9), 4630–4642. doi:10.1002/mp.12399 PMID:28594460

Liu, W., Anguelov, D., Erhan, D., Szegedy, C., Reed, S., & Fu, C.-Y. (2016). SSD: Single Shot MultiBox Detector. In B. Leibe, J. Matas, & N. Sebe (Eds.), Computer Vision -- ECCV 2016 (pp. 21-37). Academic Press.

Miki, Y., Muramatsu, C., Hayashi, T., Zhou, X., Hara, T., Katsumata, A., & Fujita, H. (2016). Classification of teeth in cone-beam CT using deep convolutional neural network. *Computers in Biology and Medicine*, *80*, 24–29. doi:10.1016/j.compbiomed.2016.11.003 PMID:27889430

Oliveira, J., & Proença, H. (2011). Caries Detection in Panoramic Dental X-ray Images. *Computational Methods in Applied Sciences*, *19*, 175–190. doi:10.1007/978-94-007-0011-6_10

Redmon, J., & Farhadi, A. (2017). YOLO9000: Better, Faster, Stronger. *IEEE Conference on Computer Vision and Pattern Recognition (CVPR)*, 6517-6525.

Ren, S., He, K., Girshick, R., & Sun, J. (2017). Faster R-CNN: Towards Real-Time Object Detection with Region Proposal Networks. *IEEE Transactions on Pattern Analysis and Machine Intelligence*, *39*(6), 1137–1149. doi:10.1109/TPAMI.2016.2577031 PMID:27295650

Rezaei, M., Yang, H., & Meinel, C. (2017). Deep Neural Network with l2-norm Unit for Brain Lesions Detection. In Neural Information Processing. ICONIP 2017. Lecture Notes in Computer Science (Vol. 10637, pp. 798-807). Berlin: Springer. doi:10.1007/978-3-319-70093-9_85

Ronneberger, O., Fischer, P., & Brox, T. (2015). U-Net: Convolutional Networks for Biomedical Image Segmentation. *Medical Image Computing and Computer Assisted Interventions Conference (MICCAI)*, *9351*, 234-241. 10.1007/978-3-319-24574-4_28

Said, E. H., Nassar, D. E., Fahmy, G., & Ammar, H. H. (2006). Teeth segmentation in digitized dental x-ray films using mathematical morphology. *IEEE Transactions on Information Forensics and Security*, *1*(2), 178–189. doi:10.1109/TIFS.2006.873606

Schmidhuber, J. (2015). Deep learning in neural networks: An overview. *Neural Networks*, *61*, 85–117. doi:10.1016/j.neunet.2014.09.003 PMID:25462637

Shah, S., Baza, A. A., Ross, A., & Ammar, H. (2006). Automatic tooth segmentation using active contour without edges. *Biometrics Symposium*, 1-6. 10.1109/BCC.2006.4341636

Shen, D., Wu, G., & Suk, H.-I. (2017). Deep Learning in Medical Image Analysis. *Annual Review of Biomedical Engineering*, *19*(1), 221–248. doi:10.1146/annurev-bioeng-071516-044442 PMID:28301734

Simonyan, K., & Zisserman, A. (2015). Very Deep Convolutional Networks for Large-Scale Image Recognition. *International Conference on Learning Representations (ICLR).*

Song, Q., Zhao, L., Luo, X., & Dou, X. (2017). Using Deep Learning for Classification of Lung Nodules on Computed Tomography Images. *Journal of Healthcare Engineering*. PMID:29065651

Szegedy, C., Vanhoucke, V., Ioffe, S., Shlens, J., & Wojna, Z. (2016). Rethinking the Inception Architecture for Computer Vision. *2016 IEEE Conference on Computer Vision and Pattern Recognition (CVPR)*, 2818-2826. 10.1109/CVPR.2016.308

Tom, C. E., & Thomas, J. (2015). Segmentation of Tooth and Pulp from Dental Radiographs. *International Journal of Scientific & Engineering Research, 6.*

Tuzoff, D., Tuzova, L., Bornstein, M. M., Krasnov, A., Kharchenko, M., Nikolenko, S., ... Bednenko, G. (2018). Teeth detection and numbering in panoramic radiographs using convolutional neural networks. *Dento Maxillo Facial Radiology*.

Uijlings, J. R., Sande, K. E., Gevers, T., & Smeulders, A. W. (2013). Selective Search for Object Recognition. *International Journal of Computer Vision*, *104*(2), 154–171. doi:10.100711263-013-0620-5

Wanat, R. (2011). A Problem of Automatic Segmentation of Digital Dental Panoramic X-Ray Images for Forensic Human Identification. *Central European Seminar on Computer Graphics.*

Wang, C. W., Huang, C. T., Lee, J. H., Li, C. H., Chang, S. W., Siao, M. J., ... Lindner, C. (2016). A benchmark for comparison of dental radiography analysis algorithms. *Medical Image Analysis*, *31*, 63–76. doi:10.1016/j.media.2016.02.004 PMID:26974042

Wu, M. a. (2015). Image recognition based on deep learning. *Chinese Automation Congress (CAC).*

Zak, J., Korzynska, A., Roszkowiak, L., Siemion, K., Walerzak, S., Walerzak, M., & Walerzak, K. (2018). The method of teeth region detection in panoramic dental radiographs. In *Advances in Intelligent Systems and Computing* (Vol. 578, pp. 298–307). Cham: Springer. doi:10.1007/978-3-319-59162-9_31

KEY TERMS AND DEFINITIONS

Backpropagation: The algorithm used in in artificial neural networks to calculate the gradient, the vector of partial derivatives, for further update of model parameters. The algorithm is based on the chain rule of derivation.

Convolutional Neural Network (ConvNet or CNN): A special type of feed-forward neural network optimized for image data processing. The key features of CNN architecture include sharing weights, using pooling layers, implementing deep structures with multiple hidden layers.

Feed-Forward Neural Network: The artificial neural network wherein the input data is processed in one direction without any cycle possible. The network gets its input and transforms it with multiple layers of neurons to produce the final result depending on the specific task (e.g., class scores for classification problem or real values for regression problem).

Image Classification: A classical computer vision problem where the task is to label an image with the particular class within a known set of possible classes.

Loss Function: A function used in supervised learning to measure the difference between the prediction and the ground truth.

Object Detection: A computer vision problem that aims to locate a varying number of objects of different classes on a single image.

Semantic Segmentation: A computer vision problem where the task is to identify various objects on a single image on a pixel-level basis.

Stochastic Gradient Descent (SGD): The algorithm that aims to minimize the loss function for the supervised learning algorithms, including neural networks. It calculates the gradient, e.g. using back-propagation, and then changes the parameters of the models in the negative gradient direction. It works iteratively and typically uses a mini-batch of training samples at a moment.

Chapter 7
Dental Cone Beam Computed Tomography for Trabecular Bone Quality Analysis in Maxilla and Mandible

T. Christy Bobby
M. S. Ramaiah University of Applied Sciences, India

Shwetha V.
M. S. Ramaiah University of Applied Sciences, India

Vijaya Madhavi
East Point College of Engineering and Technology, India

ABSTRACT

The stability of a dental implant is one of the most important aspects that decide the success rate of implant treatment. The stability is considerably affected by the strength of trabecular bone present in maxilla and mandible. Thus, finding of trabecular bone strength is a key component for the success of dental implants. The trabecular bone strength is usually assessed by quantity of bone in terms of bone mineral density (BMD). Recently, it has been revealed that along with quantity of bone, strength of the bone also depends on quality features commonly referred as trabecular bone microarchitecture. Since the quality of the trabecular bone is varying across the maxilla and mandible, preoperative assessment of trabecular bone microarchitecture at sub-region of maxilla and mandible are essential for stable implant treatment. Thus, in this chapter, the authors inscribe the quantitative analysis of trabecular bone quality in maxilla and mandible using CBCT images by employing contourlet transform.

DOI: 10.4018/978-1-5225-6243-6.ch007

INTRODUCTION

Noninvasive imaging modalities play a vital role in dental implant analysis such as preoperative, intra-operative and postoperative assessments. In recent years, preoperative diagnosis and treatment plan of implant has been done using Cone Beam Computed Tomography (CBCT). Preoperatively, the quantity and quality of trabecular bone can be calculated from CBCT images for the placement of implant in appropriate position (Shah, Bansal & Logani, 2014). The CBCT imaging system records the images based on a cone-shaped X-ray beam positioned on a 2-D detector. When Compared to conventional MultiSlice CT (MSCT), dental CBCT possess several advantages such as lower radiation doses, fast scanning time, acceptable image resolution (80 to 400 μm) and affordable cost (Ho et al., 2013).

In medical image analysis, surface properties of image are typically analyzed using spatial and spectral domain analysis (Boehm, Lutz, Korner & Mutschler, 2009; Gregory, Stewart, Undrill, Reid & Aspden, 2004; Bullmore et al. 2004, De´fossez et al., 2003). Contourlet transform is a spectral content analysis with distinctive features such as multiresolution exploration and time-frequency localization. Also it explores directionality and anisotropy information of an image. It is a two dimensional multi-scale transform, utilizing fast-iterated directional filter bank algorithm, which captures geometrical and directional information by means of multi-resolution and multi-direction expansion. It can look out for smooth lines, edges, contours and curves and iterated filter banks making it computationally efficient. Hence the resultant transform co-efficient contains multi-resolution, local and directional information of an image. The multiresolution representation allows image to be approximated from coarse to fine resolution in a successive manner. By time-frequency localization the elements of image such as points, lines and curves are localized both in frequency and spatial domains. The directionality feature represents the orientation information of the basic elements oriented at several directions in an image. Anisotropy represents basis elements of elongated shapes with various aspect ratios.

In double filter bank, initially Laplacian Pyramid (LP) is used to capture point discontinuities followed by directional filter banks to link point discontinuities to linear structures. The number of directions are doubled at all finer scales. Hence the final result is image expansion using contour segments and is named as contourlets (Do & Vetterli, 2005). Contourlet transform is used in many medical and other applications due to its simple discrete construction, low complexity and redundancy. Thus Contourlet transform is applied for efficient representation of medical images with contours oriented in different directions (Katsigiannis, Keramidas & Maroulis, 2010).

It is clearly evident from the CBCT images of maxilla and mandible that the trabecular bone structures are represented as curves and straight lines oriented at many directions. Thus in this book chapter I am planning to inscribe about the quantitative analysis of trabecular bone quality in maxilla and mandible using CBCT images by employing contourlet transform.

ANATOMY OF MAXILLA AND MANDIBLE

Jaws are bony sections that hold teeth and it includes two sections namely upper jaw or maxilla and lower jaw or Mandible. Maxilla and mandible are important bone structures in dentistry. Mandible is the largest and strongest facial bone. It is a mobile horseshoe shaped bone in maxillofacial region. Anatomy of mandible is composed of a body and two rami, with their intersection creating the prominent gonion angle. The angle formed may vary between 110 and 140 degrees and usually the mean value is 125 degrees. A rami point upward from the back of the chin, enunciates with the temporal bones and end as coronoid processes. The superior region of the mandible holds 16 lower teeth and is called as alveolar process or alveolar bone while the inferior region, act as a channel for the facial artery. The body and alveolus have central spongy or cancellous bone encapsulated by dense cortical bone. The trabecular patterns in the cancellous bone are oriented in a parallel manner up the ramus to transmit pressures up to the condylar region. The condylar neck of the mandible is the main buttress that in turn transmits pressure to the temporomandibular joint (TMJ).The joint where the mandible and maxilla meets is called the temporo-mandibular joint. The mandible is divided into two halves by a joint that allows some rotation called as mandibular symphysis. Inferior alveolar nerve is a sensory nerve located in mandible that supplies sensation to skin in the chin region (Singh, 2014). The body of the mandible supports the alveolus and dental structures.

Maxilla is the second largest pyramidal shape paired bone of the face forming an upper jaw along with the frontal region of palate of the mouth. It is unification of two large pyramid-shaped maxillary bones at inter-maxillary suture. Each hemi maxilla contains pyramid shape body with anterior surface (facial), posterior surface (infra-temporal), superior surface (orbital), medial surface (nasal) and four prominent projections namely zygomatic, frontal, alveolar and palatine processes. The frontal process is projected upwards and the alveolar process with teeth is projected downwards. Zygomatic process projects laterally whereas palatine process projects medially. The body of the maxilla is hollow and comprises maxillary sinus that opens into nasal fossa (Halim, 2008).

Zygomatic bone is a tiny and quadrangular slice located at the upper and lateral parts of the face. Alveolar bone is an inferior extension with tooth sockets to hold teeth. Alveolar process arises from lower surface of maxilla and contains nerves and blood vessels that support teeth and bone. Palatine bone is a thick horizontal plate projecting medially and palatine process forms part of the roof of mouth. Maxillary tuberosity is a round eminence develops apparently after the growth of third molar and present at lower infra-temporal surface of maxilla. Maxillary sinus is one of the paranasal sinuses located in the body of maxilla. Canine fossa is a cavity on anterior surface of maxilla and canine eminence on the surface of superior maxillary bone. Nasal notch is on the anterior lower border of the frontal bone between the orbits. Infra-orbital foramen is an opening in maxillary bone situated beneath infra-orbital margin of eye socket through which maxillary nerve passes (Mitra, 2005).

DENTAL IMPLANTS

In recent years, implantology has become an imperative division of mainstream dentistry. It helps dentists to rehabilitate the life of huge patient populations by treating functional, anatomical and aesthetic dental related problems. More than 220 types of dental implants are available from 80 manufacturers. The implants are mostly designed and manufactured according to the patient's requirement. Thus implants are manufactured by different process, materials, shapes, lengths, widths and forms (Bashir, Gupta & Ahuja, 2016). Dental implant posts made up of biocompatible materials such as metals or ceramics that are bonded into jawbone surgically so as to provide stable support for artificial teeth. Commonly used metals for implants are pure titanium, titanium alloys or ceramic materials. Implants are classified depending on placement within tissues, materials used, reaction with bone and on treatment options indicated in figure 1 (Yeshwante et al., 2015).

Figure 1. Classification of dental Implants

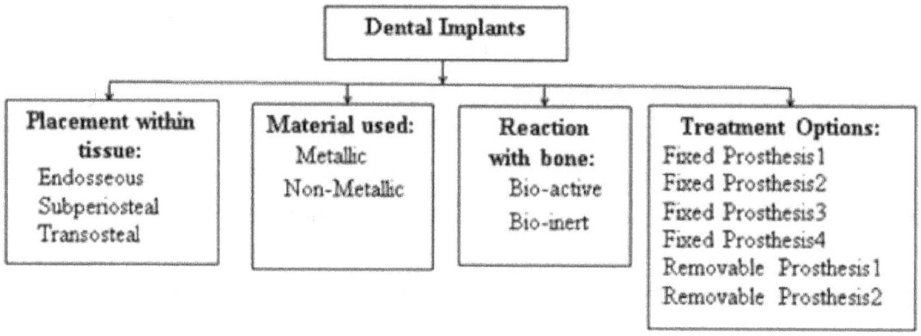

A dental implant works as a substitute for tooth root and its structure has three sections namely implant post fixture, abutment and a crown. Implant post is an artificial tooth root that comes in variety of shapes and sizes. Abutment is a connector made on top of dental implant so as to connect implant to artificial tooth. Crown is an artificial tooth made to match the natural tooth. When both crown and root are lost, dental implant is an appropriate choice. The dental implant is selected and positioned in jaw bone in such a way that it can fuse with natural bone and acts as a substitute for normal tooth. Also the implant surface is corrugated so that it can assimilate with bone tissues in a better way (Bashir et al., 2016).

Dental X-rays or 3D Computed Tomography (CT) scan is usually used to determine bone condition before any implant is being carried out. In dental implant process, a hole is drilled in jawbone creating a cavity. Implant is inserted deep into jawbone that acts as an artificial root. After the jawbone heals, abutment is placed on top of connector that serves as a connector to the crown. Finally crown is attached to abutment to complete dental implant process. The overall implant process usually takes six to twelve months to complete (Haddad, Lauritano, Candotto & Carinci, 2014).

Dental implant stability mainly depends on the quality and quantity of bone in the selected zone. Age related bone loss affects the mandible or the maxilla and serve as diagnostic markers of the disease. Information of jaw bone mineral density along with trabecular bone micro-architecture will aid the surgeon to ascertain the suitable implant locations, in turn improving the success rates of the implant treatment. Also the stability of bone-implant interface determines success of dental implants. Mostly the failure of the implants may be due to improper selection of geometrical and mechanical properties of implant, site of implant and crestal bone resorption (Pisulkar, Pakhan, Godbole & Dahane, 2016).

IMAGING MODALITIES IN DENTAL IMPLANT APPLICATIONS

Commonly used noninvasive imaging modalities for diagnosis in dental treatment are digital X-ray, CT, Magnetic Resonance Imaging (MRI), micro-CT and CBCT. These imaging modalities will help in preoperative preparation, intraoperative and postoperative estimation of dental implants (Gupta, Patil, Solanki, Singh &Laller, 2014). Preoperative dental imaging is essential for accessing quality and quantity of alveolar bone, identifying critical structures at potential implant sites, to plan implant orientation and to determine presence or absence of pathology. Intraoperative assessment involves assessment of bone to implant contact and to improve implant positioning strategy. Postoperative assessment is used to determine if implant shows adequate osseointegration over time (Bornstein, Scarfe & Jacobs, 2014).

Digital radiography is x-ray based imaging technique, where digital x-ray sensors are used for recording and processing of images. When human body is exposed to X-rays, the amount of photons absorbed depends on density of the tissue. The X-ray energy passes through soft tissues and gets absorbed by dense tissues such as bone (Pauwels, Araki, Siewerdsen & Thongvigitmanee, 2015). The X-rays that have passed through soft tissues are converted into digital form and stored as pixel in an image. Intra-oral X-rays are taken inside the mouth and is the common type of radiograph taken in dentistry. The dense tissues such as teeth and bone absorb X-rays, whereas the soft tissues such as gum and cheeks passes X-rays more easily. The radiation exposure is upto 90% less when compared to film X-rays, thus it may not be able to assess the regional three-dimensional anatomical appearance of the jaw bones. Also dental radiographs are typically affected with random distribution of X-ray photons termed as quantum noise (Pauwels et al., 2015).

MRI uses magnets, varying gradient fields and radio frequencies for acquiring sectional images of body in any imaging plane. MRI provides contrast between soft components like soft tissues and solid components like teeth, bone at the same time. MRI does not use ionizing radiation and provides 2D cross-sectional views in sagittal, coronal, axial and oblique planes. (Gupta et al., 2014).These information are not enough for dental implant planning.

CT scan is an established radiology based imaging modality for acquiring jaw bone images before performing dental implant surgery. It records precise 3D anatomic structures of jaw bones and Bone Mineral Density (BMD) value in Hounsfield Units (HU). CT comprises of X-ray source with fan beam geometry and a detector producing images representing sections of body. X-ray source and detectors are moved synchronously during exposure focusing on objects in the focal plane. In dentistry, the detector measures X-ray absorption of tissues present in bone traversing the slice at all angles. These attenuation values are converted into CT numbers expressed in Hounsfield units. During CT scans, patients are exposed to radiation of the order of 10 mSv per scan that is higher compared to conventional X-ray imaging. A high-resolution 3D micro-CT can describe detailed bone morphology and BMD more accurately than CT. It comprises of a low energy micro-focus X-ray source that illuminates objects wherein the object is rotated to acquire series of projection images. Very fine scale internal structure of objects such as metals, bones and soft tissues are imaged, and the acquired 3D images have increased resolution compared to CT. Though CT and Micro-CT appear to be precise methods to analyze bone micro structure and BMD, its high radiation doses, cost and longer scanning time associated with CT imaging limits its usage (Silva, Freitas, Ambrosano, Bóscolo & Almeida, 2012; Kim et al., 2013).

CBCT is a relatively new technology comprising of an X-ray source and a detector, fixed with rotating framework. From the source, a pyramid shaped X-ray beam excited by 100-120 kV potential with 1.5-10 mA current is directed through middle of two-dimensional (2D) detector (Kim et al., 2013). The X-ray source and detector positions are fixed beforehand and the revolving framework performs one spin around the patient, producing 2D images in various planes such as axial, sagittal and coronal planes. These 2D images obtained in various planes are used for the reconstruction of 3D skull image. In dentistry, CBCT is replacing expensive and high radiation producing medical CT scans to examine oral and maxillofacial regions. Currently, CBCT is frequently used for the investigation of dental bone related pathologic conditions such as jaw bone fracture, structural maxillofacial deformity, preoperative planning of wedged teeth, temporomandibular joint related problems and in the analysis of bone density for implant placement (Shah et al., 2014; Yilmaz, Misirlioglu & Adisen, 2014). It has been reported that CBCT has few drawbacks such as higher noise and lower contrast than conventional CT systems. However, these artefacts can be corrected by image processing algorithms. Otherwise, it has many potential advantages such as lower cost, lower radiation and higher spatial resolution output images (Kim, 2014).

IMPORTANCE OF TRABECULAR BONE QUALITY ANALYSIS

Trabecular bone structure comprises network of trabeculae (plates and rods) that has highly anisotropic distribution. It is one of the important components in bone quality analysis. Interruption of trabecular continuity by trabecular gap results in reduced connectivity of trabecular bone structure thereby increasing bone brittleness. Reduction of trabecular bone density and structural variation are directly related to variation in bone strength. The structural changes related to trabecular bone loss include decrease of bone density, transformation of plate like trabecular structure into rod like trabecular structure, increase of anisotropy and decrease of trabecular thickness (Chappard, Baslé, Legrand & Audran, 2008).

The residual alveolar bones are classified into four main types depending on the proportion and structure of cortical and trabecular bone. Type I classification denotes dense cortical bone found at anterior mandible region; type II classification indicates dense to porous cortical bone encapsulating dense trabecular bone, found at anterior mandible, posterior mandible and anterior maxilla region; type III classification represents thin porous cortical layer encapsulating fine trabecular bone, found at posterior mandible, anterior maxilla and posterior maxilla region; type IV classification signifies fine trabecular bone found at posterior maxilla region (Gulsahi, 2011). The quality of the bone influences choice of implant with regard to its number, diameter, length and type. Hence when planning implant treatment, it is important to know the bone quality. Implant positioned in low density trabeculae has a higher chance of implant failure. Also higher survival rate of dental implants is reported in mandible region compared to maxilla region. This is predominantly due to its higher bone density in the anterior region of mandible (Gulsahi, 2011). Hence preoperative assessment of trabecular bone microarchitecture at sub-regions of maxilla and mandible is essential for a stable and successful dental implant. Commonly used morphometric or morphological measurements associated with trabecular bone micro architecture include trabecular thickness, number and separation. Trabecular number is measured as mean number of trabeculae per unit of length, trabecular thickness is calculated as mean thickness of the trabeculae present in the specific region and trabecular separation is defined as mean distance between each trabeculae. In previous research, it has been reported that the bone loss lead to reduction in the quantity of trabeculae without affecting the trabeculae thickness (Ho et al., 2013; Kim et al., 2013).

TEXTURE ANALYSIS OF TRABECULAR BONE

Texture is an important feature, used to characterise images in computer based processing techniques. Texture analysis uses various image processing procedures that assist in quantification of gray-level patterns, pixel distribution and its interrelationships, sub-patterns in an image and the spectral components of an image (Backes, Gonçalves, Martinez, & Bruno, 2010). They also describe the image characteristics such as pixel size, brightness, colour variation, surface intensity variation, patterns that are indiscernible to the human visual system. In medical image analysis, inspecting the nature of gray-level transitions helps to extract textural features that characterizes the pathology or disease. Also, texture features assists in disease progress monitoring, diseases severity analysis and evaluation of emerging therapies. Texture analysis is widely employed for medical images recorded using various imaging modalities (Kassner & Thornhill, 2010).

Different existing methods for extracting texture features are generally classified into the spatial and spectral or transform based methods. Spectral methods have the advantage of being less sensitive to noise and have been commonly used to represent image textures. The transform methods such as Fourier, Wavelets, Gabor filters and Radon transforms are most repeatedly used to extract texture information from an image recorded with different modalities (Boehm et al., 2009; Gregory et al., 2004; Bullmore et al. 2004). Also spatial methods such as co-occurence matrix, run length matrix and Laws masks are often used to characterise bone micro architecture at different anatomic locations.

Three groups of image analysis techniques that are applicable to describe the trabecular structure from radiographic images. In the first group, matrix analyses employed to differentiate highly spaced trabecular pattern from normal pattern. The Run-length matrix computation helps to find information of trabecular orientation, whereas co-occurrence matrix provides trabecular texture as well as the trabecular direction information. The second group of methodologies is based on spatial-frequency analyses of an image, it allows understanding of the dominant trabecular direction. The power-spectrum computation provide details about roughness of the trabeculae. Further the results of spatial-frequency analyses are associated to mechanical properties of the bone such as elasticity and strength. Lastly third group of approaches is resulting from multifractal analyses, where features of the image such as texture description and trabecular direction are defined at different scales (Dé fossez et al., 2003).

Anisotropy computation from texture images provides information about extent of directional component in an Image. The anisotropy property of trabecular bone influenced by the skeletal site, anatomy and the main direction of loading applied to the bone. The mean intercept length method, fractal analysis and fast Fourier transform are the commonly used methods for connectivity and architectural anisotropy analysis (Chappard et al. 2005). Commonly used statistical measures for texture analysis are mean, standard deviation, energy, entropy, contrast, homogeneity, variance, correlation, maximum probability, sum mean, cluster tendency and inverse difference moment (Mokji & AbuBakar, 2007).

The two important components of texture analysis are the proper selection of the best suitable texture analysis method and selection of appropriate Regions of Interest (ROI). The choice of ROI in trabecular strength analysis is important because different region in bone are related differently to the bone's biomechanical strength and the power to predict fracture risks varies considerably for different regions (Huber et al., 2009).

SPECTRAL METHODS

Two-dimensional transforms based texture analysis have been used extensively in digital image processing for image depiction and enhancement. Fourier transform is one of the transforms that is most extensively used. Fourier analysis can be used to study the properties of textured images, wherein the power spectrum divulges information on the texture periodicity and directionality (Gonzalez & Woods, 2001). Spatial granularity and repetitiveness are the aspects of texture that can be determined by the frequency content of the image.

The Discrete Wavelet Transforms (DWT) is well established and effective method in the area of image de-noising, enhancement, compression, and image fusion. But the limitation of DWT is, they do not produce an optimal discrete-time basis from the perspective of time-localization. Hence Slantlet has been projected as an improvement over the conventional DWT in time-localization. The Slantlet transform filter bank employs filters of shorter supports, thus it has an octave-band characteristic, orthogonality with two zero moments and a scale-dilation factor of two. It has been employed for multispectral image fusion and texture analysis of medical images (Maitra & Chatterjee 2006; Selesnick 1999).The wavelet transform based multi-resolution analysis plays vital role in image processing applications such as image enhancement, compression, noise removal and feature extraction. It is efficient in localizing point singularities spatially. As the wavelets are non-geometrical, in 2D space they fail to localize highly anisotropic geometrical elements such as lines, curves and edges that present in most of the medical images. Discrete wavelet transform possess directional wavelets in horizontal, vertical and diagonal direction to capture the image texture information. Thus medical images which contain a high level of directionality elements will not be completely characterized by wavelet spectral domain analysis. Due to the above mentioned flaws of discrete wavelet transform, other spectral approaches such as discrete ridgelet, curvelet, and contourlet transforms capture multiscale, multi-resolution and multi directional information from an image are used for texture representation (Starck, Candès & Donoho, 2002; Starck & Fadili, 2007).

The ridgelet transform is multiscale and multi-orientation analysis tool capable to represent an image in a local and sparse manner. It represents objects with line singularities and the prominence of ridgelet is that it can compress energy of the image into smaller number of ridgelet coefficients (Cui & Zhang, 2010). Ridgelet transform is a two-step procedure; in the first step, radon transform of the image is computed to map line singularity into a point singularity and the in next step one dimensional wavelet transform is applied to the slices of radon transform to provide sparse representation. Thus, ridgelet transform efficiently represents line singularities present in the image by means of compact representation when compare to wavelet transform. (Arivazhagan, Ganesan & Kumar, 2006).Discrete form of ridgelet transform is known as Finite RIdgelet Transform (FRIT), which involves Discrete Finite Radon Transform (DFRT). DFRT of an image is computed as summations of pixels over a defined set of lines and its DWT. The important properties of FRIT are, it is reversible, orthogonal and achieves perfect reconstruction. Thus FRIT provides efficient representation of images with line discontinuities and straight edges oriented in different directions (Do &Vetterli, 2003; Ratnaparkhe, Manthalkar & Joshi, 2009; Chen & Kegl, 2010).

In most of the medical images the edges are formed by curves, thus ridgelets alone cannot represent it in an efficient way. At fine scales, the curved edge is almost appears like straight line, and hence to capture curved edges, ridgelets are applied in a localized manner at sufficiently fine scales is known as curvelet transform. Curvelet transform is a multiscale transforms with frame elements indexed by multi scale, orientation and position. It have the capability of apprehending image intensity information along curves in the form of radial segments in frequency domain. Also, from high frequency components more directional information of lines and curves can be fetched. In the first step of the curvelet transform computation approach, the input image is decomposed into a set of sub bands. In the second step, each of sub bands are partitioned into several blocks by fine-tuning to different scales and orientations. The decomposition procedure by means of spatial partitioning includes overlapping of windows to evade blocking effects leading to huge amount of redundancy and is a slow procedure making it less feasible for complicate texture patterns and large image database analysis (Dettori & Lindsay, 2007; Candès & Donoho, 1999; Starck et al., 2002). In digital implementation of cuvelet transform, the transform coefficients are derived using Fourier samples split into Unequally-Spaced Fast Fourier Transform (USFFT). The wrapping based curvelet transform comprises wrapping of specifically selected Fourier samples and the advantages of this method is computation time very less and is more robust than ridgelet transform. Both these approaches tabulate curvelet coefficients at various scale, orientation and location for an image (Candes, Demanet, Donoho, & Ying, 2006).

Contourlet transform is a 2D representation of images that can capture the geometry of edges from an image. Contourlet transform has localization, orientation, and parabolic scaling similar to curvelet transform whereas it has certain differences such as lower redundancy, direct implementation in discrete domain and simple construction. The dominant features in the image such as contours can be effectively captured with few contourlet coefficients. The Contourlet transform's multiscale implementation is based on dual filter bank structure comprising of pyramidal band-pass image decomposition followed by a directional filtering stage so as to get smooth contours of image. Contourlets hold few important features namely, multiresolution capability, time-frequency localization, directionality and anisotropy (Wang, Liang, Cheng & Tang, 2008; Chen & Kegl, 2010). The contourlets have stretched support at high number scales, orientations or directions and aspect ratios allowing it to efficiently approximate smooth the geometrical structures such as curves and contours at several resolutions. (Do & Vetterli, 2005; Rubinstein, Zibulevsky & Elad, 2010).

TRABECULAR MICRO ARCHITECTURE ANALYSIS OF MAXILLA AND MANDIBLE CBCT IMAGES

CBCT images used for the analysis (N = 21) are recorded at M S Ramaiah Dental College and Hospital using Care Stream 9300 unit with CS3D acquisition interface. The exposures are made with 6.3 mA at 90 kV for 17 seconds. Three views such as coronal, sagittal and axial views recorded are shown in figure 2, figure 3 and figure 4 respectively.

In coronal view, the trabecular structures are perceptible evidently. Thus in the proposed work trabecular structure analysis has been done in first and second molar region of maxilla and mandible in

Figure 2. Coronal view

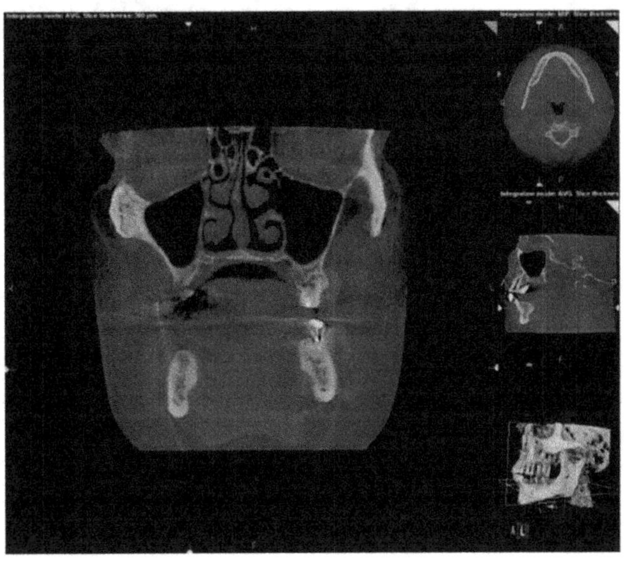

Figure 3. Sagittal view

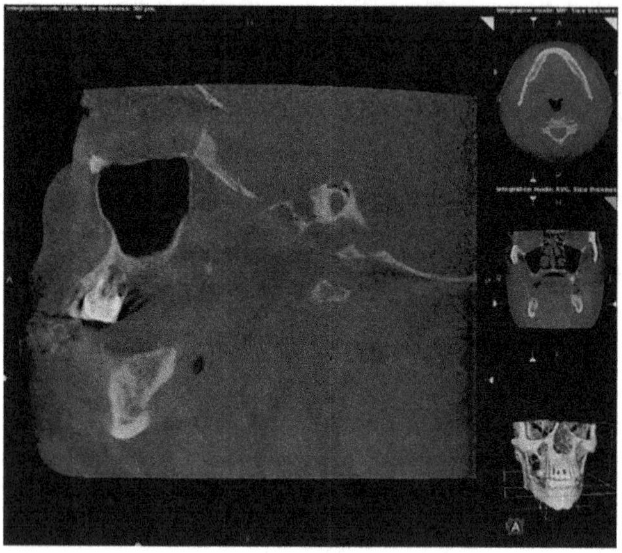

Figure 4. Axial view

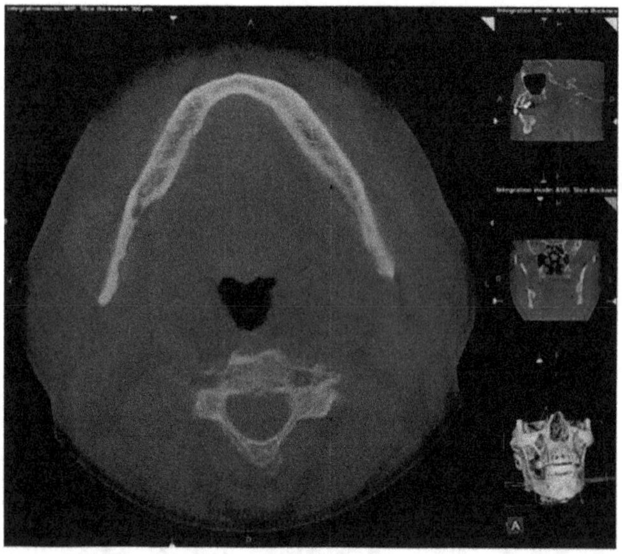

coronal view. Three slices of first molar namely slice1, slice2 and slice3 are represented in figure 5 (a), figure 5 (b) and figure 5 (c) respectively. Similarly three slices of second molar namely slice1, slice2 and slice3 are represented in figure 5 (d), figure 5 (e) and figure 5 (f) respectively.

Trabecular bone microarchitecture analysis has been performed on CBCT coronal view images of 21 patients (male and female) belonging to an age group of 41 to 67 years. Representative coronal CBCT image employed for trabecular analysis is shown in figure 6.

Figure 5a. Slice 1 of first molar CBCT coronal dental images

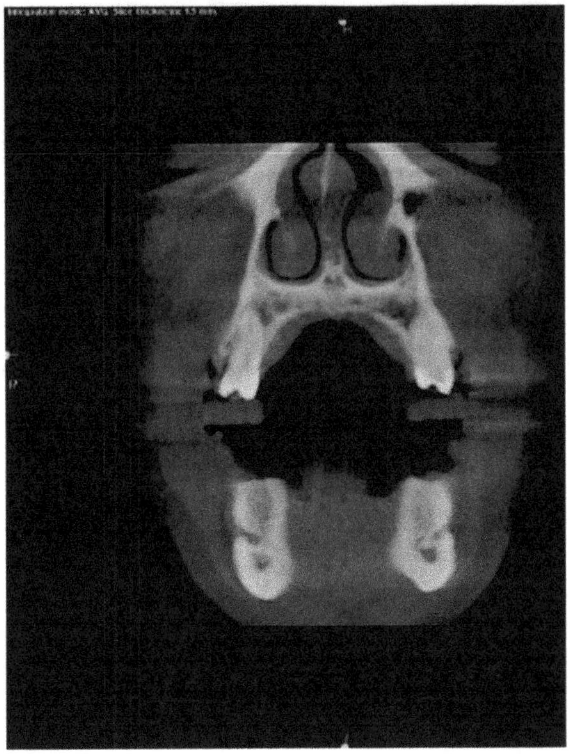

The quality of CBCT images is affected by artefacts, noise and poor soft tissue contrast, thus bilateral filtering method is employed to enhance the contrast and smoothening of the image while preserving edges. Bilateral filter parameters chosen are half size of Gaussian bilateral filter window, w=5, standard deviation of bilateral filter is $\sigma_r = 2.25$ and $\sigma_d = 0.05$. The enhanced image is shown in figure 7.

ROI is the section of image to be segmented in order to perform analysis in that specific image region. In this analysis, ROI is chosen to encompass trabecular structure in first and second molar region of maxilla and mandible. The ROI of size 40x40 for maxilla region and 40x60 for mandible region is manually delineated and is considered for the further analysis. Representative ROI regions marked in right maxilla, left maxilla, right mandible and left mandible are shown in figure 8 (a), figure 8 (b), figure 8 (c) and figure 8 (d) respectively.

The ROI's extracted from right maxilla, left maxilla and right mandible, left mandible are shown in figure 9 (a), figure 9 (b), figure 9 (c) and figure 9 (d) respectively. It is clearly evident from the ROI images of maxilla and mandible that the trabecular bone structures are represented as curves and lines oriented at many directions. Thus contourlet transform is applied and coefficients are extracted at four levels using 9/7 biorthogonal pyramidal and 9/7 biorthogonal directional filter.

In frequency domain, contourlet transform delivers multiscale and directional information of an image. The capability of this transform to capture directional features from heterogeneous images in a multi-resolution manner makes it suitable to recognize unknown pattern in CBCT bone images. Contourlet transform is a two-step process which acquires sub-bands decomposition using LP and directional information using directional filter bank. The spectrum of input image comprises of frequency components

Figure 5b. Slice 2 of first molar CBCT coronal dental images

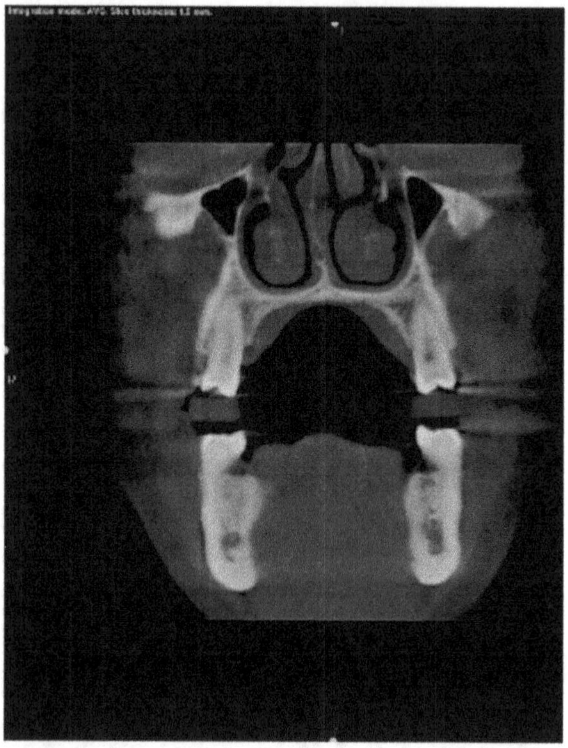

Figure 5c. Slice 3 of first molar CBCT coronal dental images

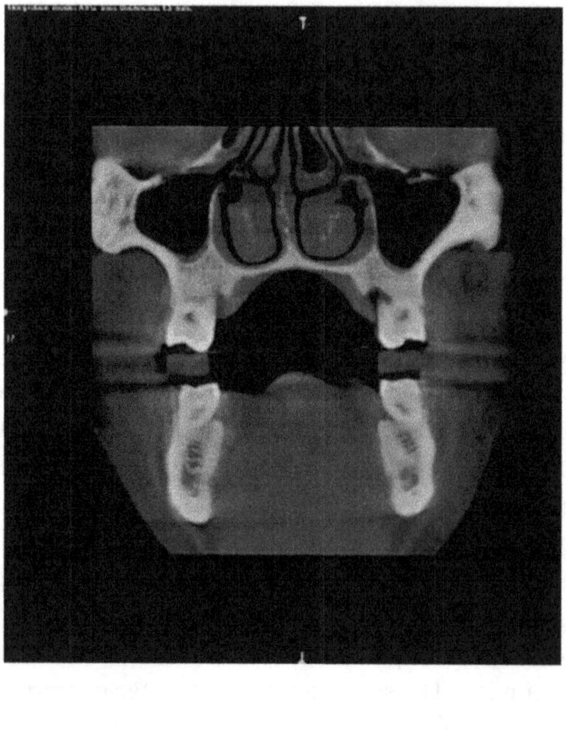

Figure 5d. Slice 1 of second molar CBCT coronal dental images

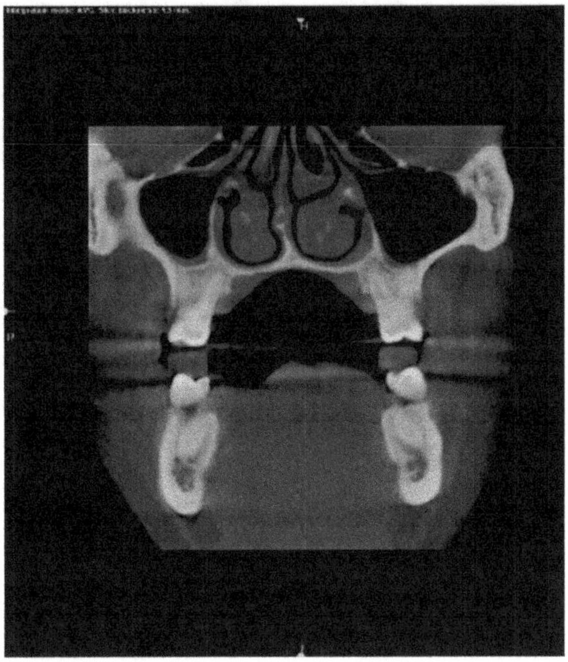

Figure 5e. Slice 2 of second molar CBCT coronal dental images

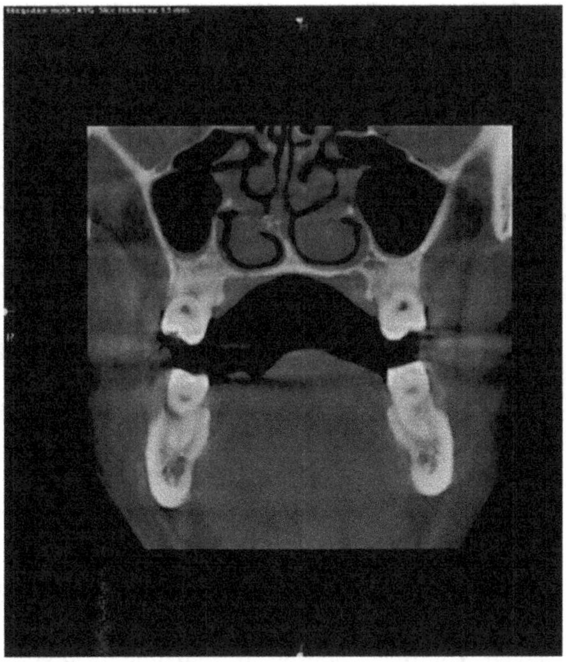

Figure 5f. Slice 3 of second molar CBCT coronal dental images

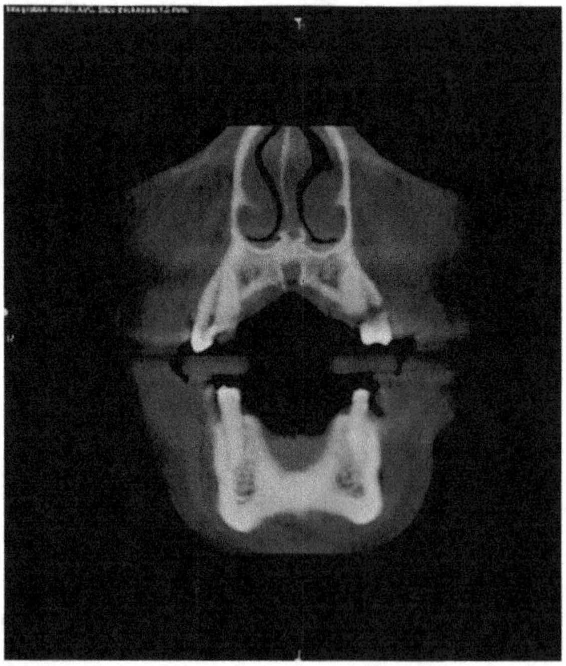

Figure 6. Representative coronal CBCT image

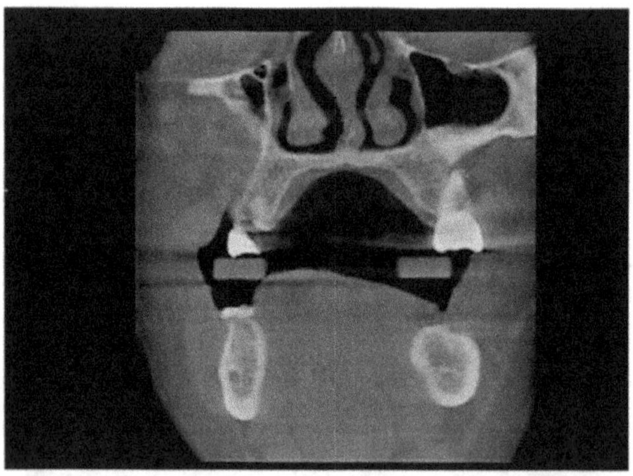

such as low pass output (LL) and bandpass output (LH, HL and HH). In the first level of computation, the bandpass output is passed into directional filter bank which results in contourlet coefficients. In the successive level LL is passed through the LP to obtain more coefficients and this step is repeated till fine details of image are achieved. Totally 32 contourlet coefficients are tabulated at various scales and orientations are fed to feature selection module and the most significant orientations are derived. Trabecular micro architecture pattern is highly anisotropic and consists of lines, edges, curves and contours.

Figure 7. Bilateral filtered coronal CBCT image

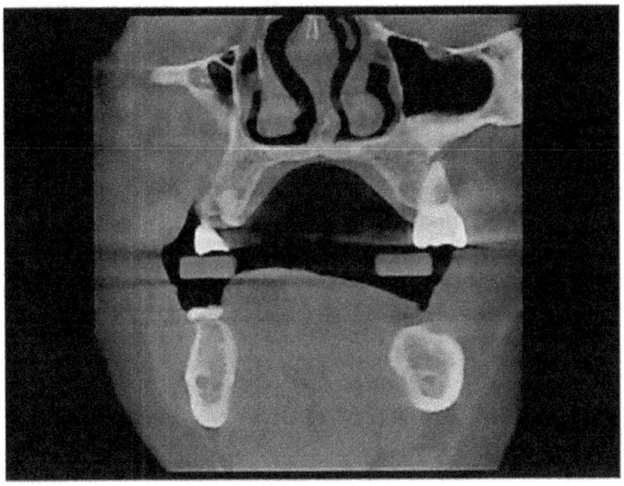

Figure 8a. Representative highlighted ROI of right maxilla

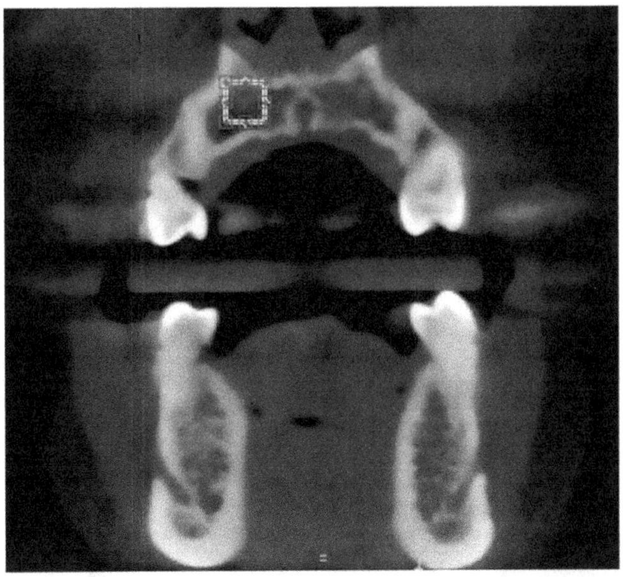

Statistical features such as energy and kurtosis (Murakami, Tan, Kim, Ishikawa & Aoki, 2012) are extracted from three slices at various orientations. Further, mean of three slices is computed and normalized values are obtained for each ROI. Figure 10 (a), figure 10 (b), figure 10 (c), figure 10 (d) represents scatter plot of normalized energy at significant orientation derived for maxilla and mandible regions. In the maxilla region, energy values are found to be low and lie in the same range except for few images. The low value of energy in maxilla region represents the heterogeneity or less density of trabecular structures. Also the larger value of energy in mandible represents the homogeneity and densely packed nature of the trabecular structure. Figure 10 (e), figure 10 (f), figure 10 (g), figure 10 (h) represents scatter plot of normalized kurtosis at significant orientation derived for maxilla and mandible regions. On

Figure 8b. Representative highlighted ROI of left maxilla

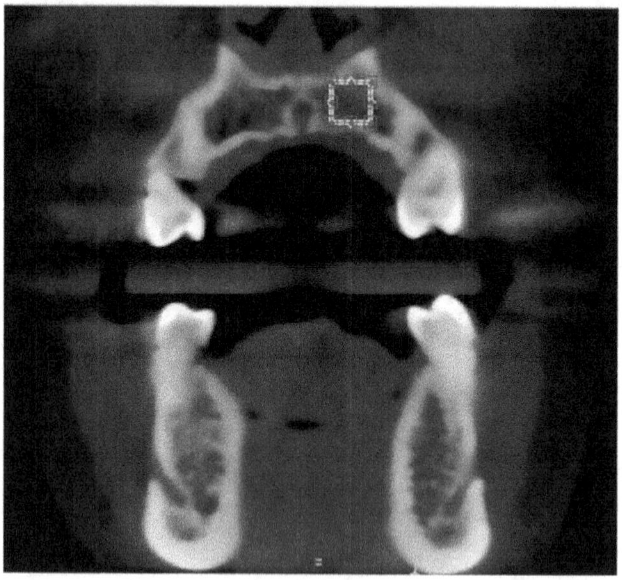

Figure 8c. Representative highlighted ROI of right mandible

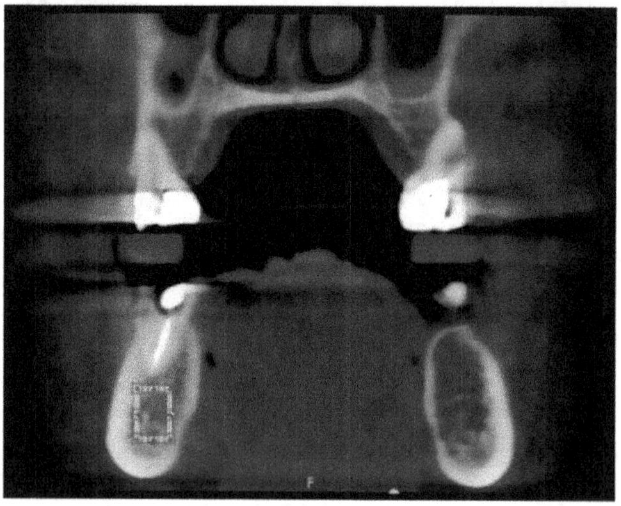

comparing the kurtosis values of maxilla and mandible, the values are high in maxilla for orientation1 and 4. In contrast to orientation1 and 4, in orientation2 and 3 the kurtosis values are high in mandible; this indicates that the discontinuities of the trabecular structure are different for different orientations. Thus from the result it is evident that strength of the trabecular bone in mandible may be high when compare to maxilla, also implant success rate will be high in mandible.

Figure 8d. Representative highlighted ROI of left mandible

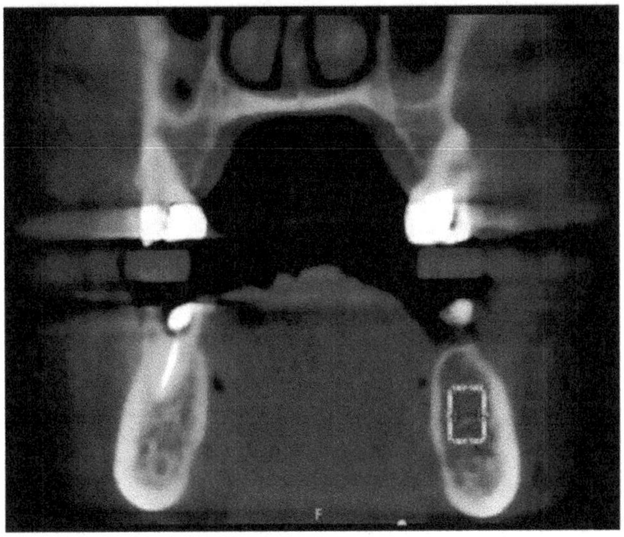

Figure 9a. ROI extraction from right maxilla

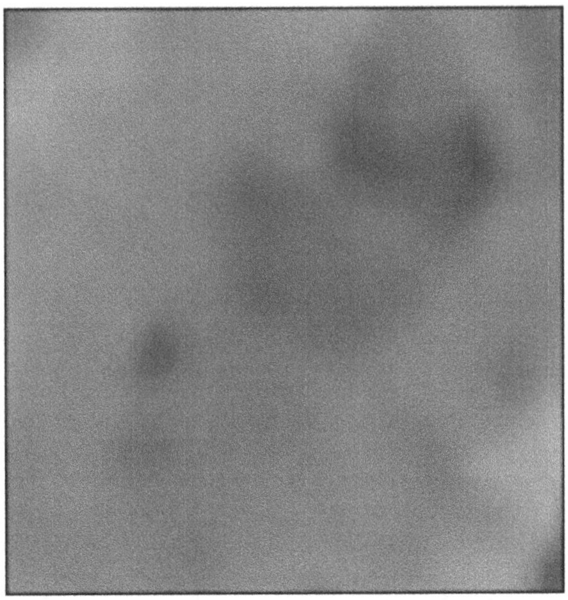

CONCLUSION

In bone biomechanics, estimation of bone strength is an important component for assessment of various pathological conditions such as osteoporosis and fracture. Analysis of trabecular micro architecture is to be analysed in addition to BMD recording for better understanding of bone strength. In trabecular

Figure 9b. ROI extraction from left maxilla

Figure 9c. ROI extraction from right mandible

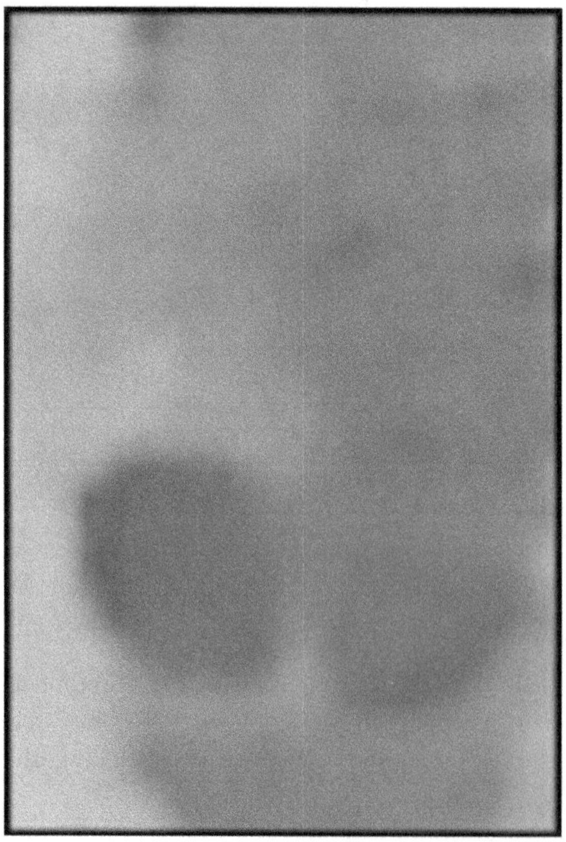

Figure 9d. ROI extraction from left mandible

Figure 10a. Scatter plot of normalized energy for maxilla and mandible region at Orientation 1

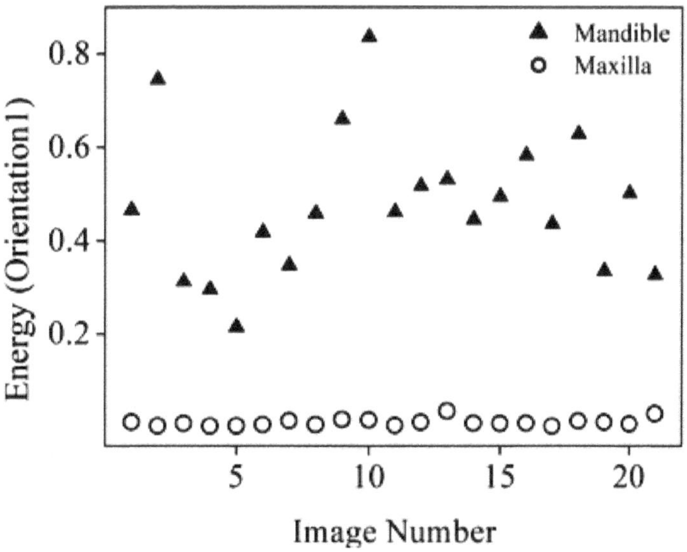

Figure 10b. Scatter plot of normalized energy for maxilla and mandible region at Orientation 2

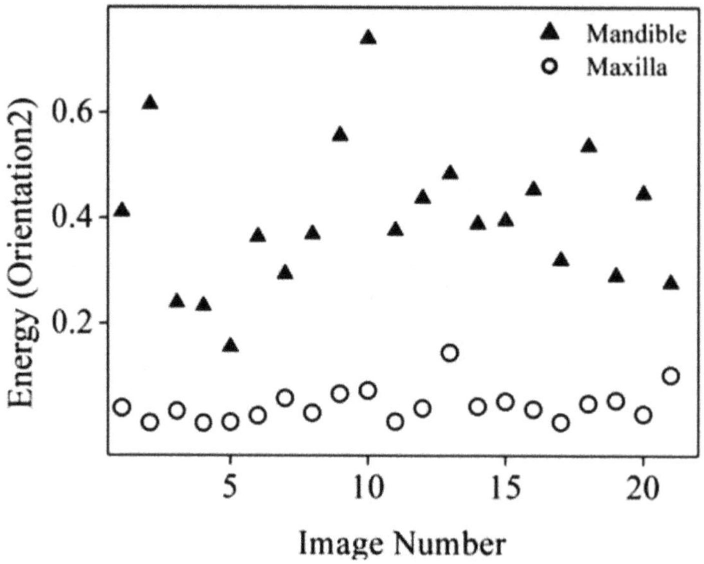

Figure 10c. Scatter plot of normalized energy for maxilla and mandible region at Orientation 3

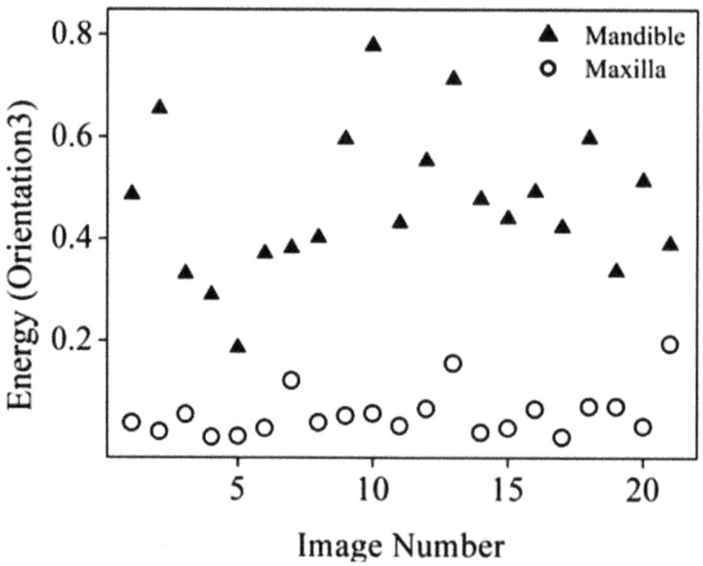

micro architecture analysis, orientations and geometrical features of the trabecular bone remain the significant determinants of bone strength. Analyses of such factors provide valuable additional information regarding trabecular bone fracture risk, strength and response to therapy (Linwei, Guangwei, He, Dong, & Weimin, 2012; Sun et al., 2006). It has also been revealed that there is an association between morphometric alterations and bone loss (Pulkkinen, Partanen, Jalovaara & Jamsa, 2004). Also analyses

Figure 10d. Scatter plot of normalized energy for maxilla and mandible region at Orientation 4

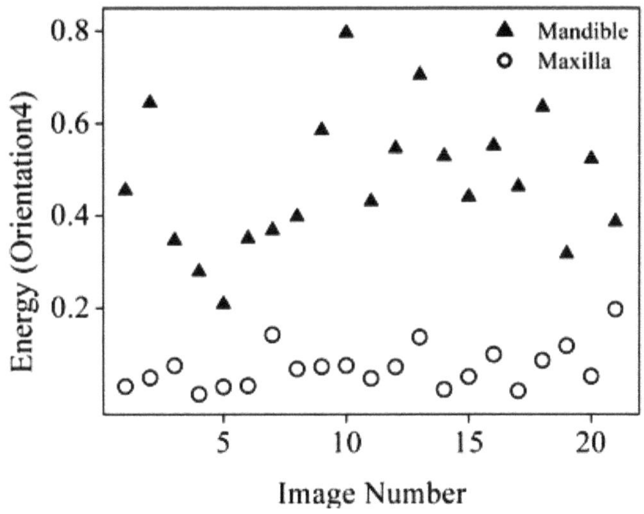

Figure 10e. Scatter plot of normalized kurtosis for maxilla and mandible region at Orientation 1

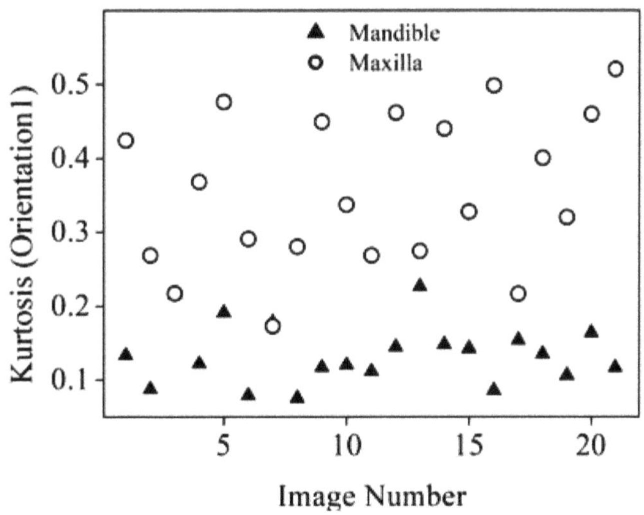

of various medical images using different transform based methods have shown good results. Thus transform based approach is applied in many diagnostic equipment for various clinical inferences. Especially, they are used to extract and compare the variations in the pattern manifested in images due to diseases conditions. (Shi, Sherry, Wang, Edward & Niebur, 2010). Jaw bone comprising of maxilla and mandible are rich in trabecular bone in the anterior region. Dental implant success depends on trabecular bone microarchitecture. In CBCT, patient is exposed to less radiation compared to CT. Also texture analysis of CBCT trabecular jaw bone images provides information about trabecular micro architecture and aids in determining its suitability for dental implant. Contourlet which captures the directional information

Figure 10f. Scatter plot of normalized kurtosis for maxilla and mandible region at Orientation 2

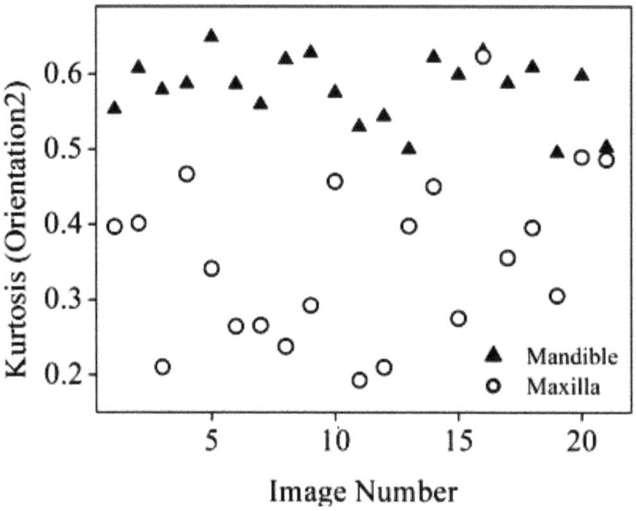

Figure 10g. Scatter plot of normalized kurtosis for maxilla and mandible region at Orientation 3

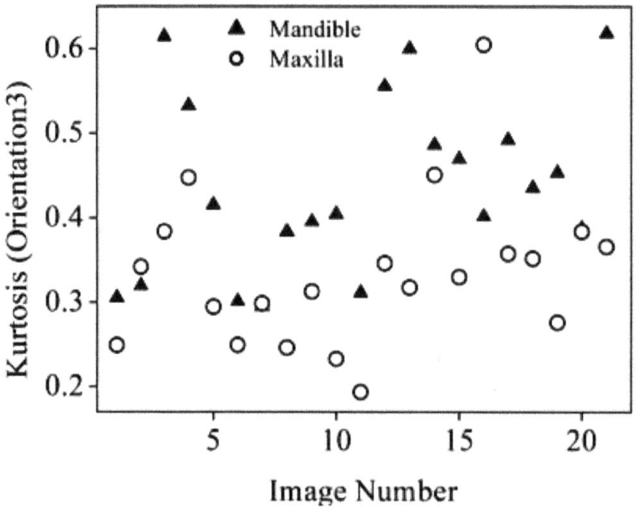

of trabecular structure at various scale and orientations show clear demarcation between maxilla and mandible images in selected four orientations. Further, combining BMD and micro architectural information of the trabecular bone derived from the CBCT images could be used to find the strength index is conceivable.

Figure 10h. Scatter plot of normalized kurtosis for maxilla and mandible region at Orientation 4

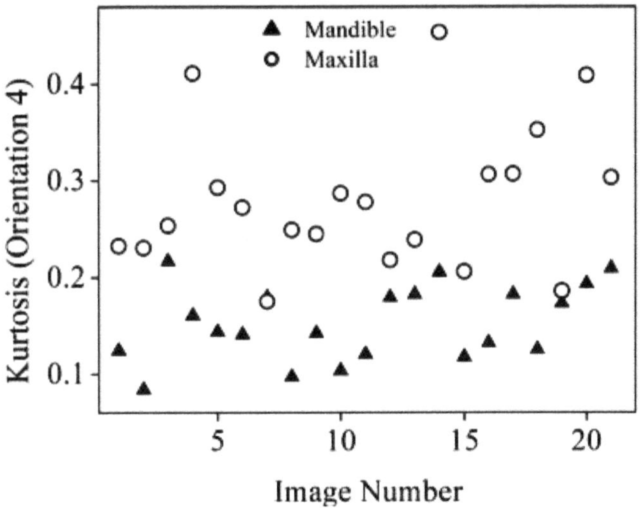

ACKNOWLEDGMENT

I take this opportunity to thank Dr. Sujatha.S, Professor and Head, Department of Oral Medicine and Radiology, MSRUAS, Bangalore, for providing support to record CBCT images. I extend my thanks to Dr. A. Kavitha, Professor and Head, Department of Biomedical Engineering, SSN College of Engineering, Chennai, for the constant support and encouragement.

REFERENCES

Arivazhagan, S., Ganesan, L., & Kumar, T. G. S. (2006). Texture classification using ridgelet transform. *Pattern Recognition Letters*, *27*(16), 1875–1883. doi:10.1016/j.patrec.2006.04.013

Backes, A. R., Gonçalves, W. N., Martinez, A. S., & Bruno, O. M. (2010). Texture analysis and classification using deterministic tourist walk. *Pattern Recognition*, *43*(3), 685–694. doi:10.1016/j.patcog.2009.07.017

Bashir, U., Gupta, M., & Ahuja, R. (2016). Implant systems. *International Journal of Applied Dental Sciences*, *2*(2), 35–41. PMID:26783652

Boehm, H. F., Lutz, M., Korner, M., Mutschler, W., Reiser, M., & Pfeifer, K.-J. (2009). Using radon transform of standard radiographs of the hip to differentiate between post-menopausal women with and without fracture of the proximal femur. *Osteoporosis International*, *20*(2), 323–333. doi:10.100700198-008-0663-6 PMID:18560746

Bornstein, M., M, Scarfe, W. C., & Jacobs, R. (2014). Cone beam computed tomography in implant dentistry: A systematic review focusing on guidelines, indications, and radiation dose risks. *The International Journal of Oral and Maxillofacial Implants, 29*, 55-77.

Bullmore, E., Fadili, J., Maxim, V., Sendur, L., Whitcher, B., Suckling, J., ... Breakspear, M. (2004). Wavelets and functional magnetic resonance imaging of the human brain. *NeuroImage*, *23*, 234–249. doi:10.1016/j.neuroimage.2004.07.012 PMID:15501094

Candes, E., Demanet, L., Donoho, D., & Ying, L. (2006). Fast discrete curvelettransforms. *Multiscale Modeling & Simulation*, *5*(3), 861–899. doi:10.1137/05064182X

Candes, E. J., & Donoho, D. L. (1999). Ridgelets: A key to higher-dimensional Intermittency. *Philosophical Transactions of Royal Society London*, *357*(1760), 2495-2509.

Chappard, C., Brunet-Imbault, B., Lemineur, G., Giraudeau, B., Basillais, A., Harba, R., & Claude, L. B. (2005). Anisotropy changes in post-menopausal osteoporosis: Characterization by a new index applied to trabecular bone radiographic images. *Osteoporosis International*, *16*(10), 1193–1202. doi:10.100700198-004-1829-5 PMID:15685395

Chappard, D., Baslé, M.-F., Legrand, E., & Audran, M. (2008). Trabecular bone microarchitecture: A review. *Morphologie*, *92*(299), 162–170. doi:10.1016/j.morpho.2008.10.003 PMID:19019718

Chen, G. Y., & Kegl, B. (2010). Invariant pattern recognition using contourlets and AdaBoost. *Pattern Recognition*, *43*(3), 579–583. doi:10.1016/j.patcog.2009.08.020

Cui, Z., & Zhang, G. (2010). A novel medical image dynamic fuzzy classification model based on ridgelet transform. *Journal of Software*, *5*(5), 458–465.

Dé fossez, H., Hall, R. M., Walker, P. G., Wroblewski, B. M., Siney, P. D., & Purbach, B. (2003). Determination of the trabecular bone direction from digitised radiographs. *Medical Engineering & Physics Journal*, *25*(9), 719-729.

Dettori, L., & Lindsay, S. (2007). A comparison of wavelet, ridgelet, and curvelet-based texture classification algorithms in computed tomography. *Computers in Biology and Medicine*, *37*(4), 486–498. doi:10.1016/j.compbiomed.2006.08.002 PMID:17054933

Do, M. N., & Vetterli, M. (2003). The finite ridgelet transform for image representation. *IEEE Transactions on Image Processing*, *12*(1), 16–28. doi:10.1109/TIP.2002.806252 PMID:18237876

Do, M. N., & Vetterli, M. (2005). The contourlet transform: An efficient directional multiresolution image representation. *IEEE Transactions on Image Processing*, *14*(12), 2091–2106. doi:10.1109/TIP.2005.859376 PMID:16370462

Gonzalez, R. C., & Woods, R. E. (2001). Digital Image Processing. Upper Saddle River, NJ: Prentice Hall.

Gregory, J. S., Stewart, A., Undrill, P. E., Reid, D. M., & Aspden, R. M. (2004). Identification of hip fracture patients from radiographs using Fourier analysis of the trabecular structure: A cross-sectional study. *BMC Medical Imaging*, *4*(4), 1–11. PMID:15469614

Gulsahi, A. (2011). Bone quality assessment for dental implants. In *Implant Dentistry* (pp. 437–452). Ankara, Turkey: The Most Promising Discipline of Dentistry.

Gupta, S., Patil, N., Solanki, J., Singh, R., & Laller, S. (2014). Oral implant imaging: A review. *The Malaysian Journal of Medical Sciences*, *22*(3), 7–17. PMID:26715891

Haddad, E. E., Lauritano, D., Candotto, V., &Carinci, F. (2014). Guided bone regeneration is a reliable technique in implant dentistry: An overview and a case report. *OA Dentistry*, *2*(1).

Halim, A. (2008). *Human Anatomy: Volume 3. Head, Neck and Brain*. New Delhi, India: I. K. International Publisher.

Ho, J. T., Wu, J., Huang, H. L., Chen, M. Y. C., Fuh, L. J., & Hsu, J. T. (2013). Trabecular bone structural parameters evaluated using dental cone-beam computed tomography: Cellular synthetic bones. *Biomedical Engineering Online*, *12*(1), 115. doi:10.1186/1475-925X-12-115 PMID:24207062

Huber, M. B., Carballido, G. J., Fritscher, K., Schubert, R., Haenni, M., Hengg, C., ... Link, T. M. (2009). Development and testing of texture discriminators for the analysis of trabecular bone in proximal femur radiographs. *Medical Physics*, *36*(11), 5089–5098. doi:10.1118/1.3215535 PMID:19994519

Kassner, A., & Thornhill, R. E. (2010). Texture analysis: A review of neurologic MR imaging applications. *AJNR. American Journal of Neuroradiology*, *31*(5), 809–816. doi:10.3174/ajnr.A2061 PMID:20395383

Katsigiannis, S., Keramidas, E. G., & Maroulis, G. (2010). Contourlet transform for texture representation of ultrasound thyroid images. *IFIP Advances in Information and Communication Technology,* 138-145.

Kim, D.-G. (2014). Can dental cone beam computed tomography assess bone mineral density? *Journal of Bone Metabolism*, *21*(2), 117–126. doi:10.11005/jbm.2014.21.2.117 PMID:25006568

Kim, J. E., Shin, J. M., Oh, S. O., Yi, W. J., Heo, M. S., Lee, S. S., ... Huh, K. H. (2013). The three-dimensional micro structure of trabecular bone: Analysis of site-specific variation in human jaw bone. *Imaging Science in Dentistry*, *43*(4), 227–233. doi:10.5624/isd.2013.43.4.227 PMID:24380061

Linwei, L., Guangwei, M., He, G., Dong, Z., & Weimin, Z. (2012). A new method for the measurement and analysis of three-dimensional morphological parameters of proximal male femur. *Biomedical Research*, *23*(2), 219–226.

Maitra, M., & Chatterjee, A. (2006). A slantlet transform based intelligent system for magnetic resonance brain image classification. *Biomedical Signal Processing and Control*, *1*(4), 299–306. doi:10.1016/j.bspc.2006.12.001

Mitra, S. (2005). Anatomy: Part 3. In A. Tapadar (Ed.), *Anatomy: osteology, important landmarks and surface markings* (pp. 43–53). Kolkata, India: Academic Publishers.

Mokji, M. M., & AbuBakar, S. A. R. (2007). Gray level co-occurrence matrix computation based on Haar wavelet. Proceeding of Computer Graphics Imaging and Visualisation, 273-279. doi:10.1109/CGIV.2007.45

Murakami, S., Tan, J. K., Kim, H., Ishikawa, S., & Aoki, T. (2012). Development of a quantitative method for the detection of periarticular osteoporosis using density features within ROIs from computed radiography images of the hand. In Technical Advancements in Biomedicine for Healthcare Applications (pp. 55-67). IGI Global.

Pauwels, R., Araki, K., Siewerdsen, J. H., & Thongvigitmanee, S. S. (2015). Technical aspects of dental CBCT: State of the art. *Dento Maxillo Facial Radiology*, *44*(1). doi:10.1259/dmfr.20140224 PMID:25263643

Pisulkar, S. G., Pakhan, A. J., Godbole, S. R., & Dahane, T. M. (2016). Correlation of bone mineral density on primary implant stability in elderly edentulous patients: A literature review. *Scholars Academic Journals of Biosciences*, *4*(10A), 822–826.

Pulkkinen, P., Partanen, J., Jalovaara, P., & Jamsa, T. (2004). Combination of bone mineral density and upper femur geometry improves the prediction of hip fracture. *Osteoporosis International*, *15*(4), 274–280. doi:10.100700198-003-1556-3 PMID:14760516

Ratnaparkhe, V. R., Manthalkar, R. R., & Joshi, Y. V. (2009). Texture characterization of CT images based on ridgelet transform. *International Journal Graphics Vision Image Processing*, *8*, 43–50.

Rubinstein, R., Zibulevsky, M., & Elad, M. (2010). Double sparsity: Learning sparse dictionaries for sparse signal approximation. *IEEE Transactions on Signal Processing*, *58*(3), 1553–1564. doi:10.1109/TSP.2009.2036477

Selesnick, I. W. (1999). The slantlet transform. *IEEE Transactions on Signal Processing*, *47*(5), 1304–1313. doi:10.1109/78.757218

Shah, N., Bansal, N., & Logani, A. (2014). Recent advances in imaging technologies in dentistry. *World Journal of Radiology*, *6*(10), 794–807. doi:10.4329/wjr.v6.i10.794 PMID:25349663

Shi, X., Sherry, L. X., Wang, X., Edward, G. X., & Niebur, G. L. (2010). Type and orientation of yielded trabeculae during overloading of trabecular bone along orthogonal directions. *Journal of Biomechanics*, *43*(13), 2460–2466. doi:10.1016/j.jbiomech.2010.05.032 PMID:20554282

Silva, I. M., Freitas, D. Q., Ambrosano, G. M., Bóscolo, F. N., & Almeida, S. M. (2012). Bone Density: Comparative evaluation of Hounsfield units in multislice and cone-beam computed tomography. *Brazilian Oral Research*, *26*(6), 550–556. doi:10.1590/S1806-83242012000600011 PMID:23184166

Singh, V. (2014). Osteology of Head and Neck. In Anatomy: Head, Neck and Brain (vol. 3, pp. 24-40). Elsevier Publishers.

Starck, J. L., Candès, E. J., & Donoho, D. L. (2002). The curvelet transform for image denoising. *IEEE Transactions on Image Processing*, *11*(6), 670–684. doi:10.1109/TIP.2002.1014998 PMID:18244665

Starck, J. L., Fadili, J., & Murtagh, F. (2007). The undecimated wavelet decomposition and its reconstruction. *IEEE Transactions on Signal Processing*, *16*(2), 297–309. PMID:17269625

Sun, X., Shu, F. L., Fei, Y. D., Shan, W., Christopher, P., James, H., ... Hong, W. D. (2006). Genetic and environmental correlations between bone geometric parameters and body compositions. *Calcified Tissue International*, *79*(1), 43–49. doi:10.100700223-006-0041-3 PMID:16868663

Wang, F. N., Liang, D., Cheng, Z. Y., & Tang, J. (2008). Nonsubsampled pyramid contourlet transform and its application, In *Proceedings of International Symposium on Information Science and Engineering* (*vol. 2*, pp. 567-570). Academic Press. 10.1109/ISISE.2008.220

Yeshwante, B., Patil, S., Baig, N., Gayakwad, S., Swami, A., & Doiphode, M. (2015). Dental implant-Classification, success and failure-An overview. *IOSR Journal of Dental and Medical Sciences*, *14*(5), 1–8.

Yilmaz, S. Y. Y., Misirlioglu, M., & Adisen, M. Z. (2014). A diagnosis of maxillary sinus fracture with Cone-Beam CT: Case report and literature review. *Craniomaxillofacial Trauma & Reconstruction*, *7*(2), 85–91. doi:10.1055-0034-1371550 PMID:25045417

ADDITIONAL READING

Amer, M. E., Heo, M. S., Brooks, S. L., & Benavides, E. (2012). Anatomical variations of trabecular bone structure in intraoral radiographs using fractal and particles count analyses. *Imaging Science in Dentistry*, *42*(1), 5–12. doi:10.5624/isd.2012.42.1.5 PMID:22474642

Angelopoulous, C. (2014). Anatomy of maxillofacial region in three planes of section. *Dental Clinics*, *58*(3), 497–521. doi:10.1016/j.cden.2014.03.001 PMID:24993921

Berkovitz, B. K., & Moxham, B. J. (1988). *A Textbook of Head & Neck Anatomy* (1st ed.). Mosby-Year Book.

Couture, R. A, Whiting, B. R., Hildebolt, C. F., Dixon, D. A., Mo, S. L.,& Ill, A. (2003). Visibility of trabecular structures in oral radiographs. *oral surgery, oral medicine oral pathology*, 96(6) 764-771.

David, O. T., Leretter, M., & Neagu, A. (2014). The quality of trabecular bone assessed using cone-beam computed tomography. *Romanian Journal of Biophysics*, *24*(4), 227–241.

Gonzalez, S. M. (2013). *Interpretation basics of cone beam computed tomography*. Wiley Blackwell Publisher. doi:10.1002/9781119421177

Klintstrom, E. (2017). *Image analysis for trabecular bone properties on Cone-Beam CT Data*. Sweden: LinköpingUniversity. doi:10.3384/diss.diva-142066

Kreider, J. M., & Goldstein, S. A. (2009). Trabecular bone mechanical properties in patients with fragility fractures. *Clinical Orthopaedics and Related Research*, *467*(8), 1955–1963. doi:10.100711999-009-0751-8 PMID:19247731

Scarfe, W. C., & Farman, A. G. (2008). What is cone beam CT and how does it works? *Dental Clinics of North America*, *52*(4), 707–730. doi:10.1016/j.cden.2008.05.005 PMID:18805225

Vasovic, M., Jovanovic, L., & Djordjevic, A. (2015). Bone quality assessment of dental implant recipient sites. *Serbian Journal of Experimental and Clinical Research*, *16*(4), 327–331. doi:10.1515jecr-2015-0050

Section 2

Dental Materials, Mechanics, and Instrumentation

Chapter 8
Comparison and Analysis of Dental Imaging Techniques

Najumnissa D.
B. S. A. Crescent Institute of Science and Technology, India

ABSTRACT

Fluoride dental decay is the second most common disease around the world. Detection methods for early disease are very crude. Precise oral diagnosis and treatment are very strongly connected to the quality of dental imaging techniques which advances the diagnostic procedure. To study the external appearance of the teeth arches, 2D images are used. CBCT images were used to locate the bone at dental implant sites. Fiberoptic transillumination, fluorescence imaging detects caries. For qualitative and quantitative analysis of dental applications, laser-induced breakdown spectroscopy (LIBS) is used. Electron caries monitor (ECM), fiberoptic transillumination (FOTI), digital fiberoptic transillumination (DIFOTI), quantitative light-induced fluorescence (QLF) are also some of the detection methods used. Hence, in this chapter, the methodologies are analyzed and compared for easy use of the dentist.

INTRODUCTION

As dental lesions develop they can be noticed using X-rays. Radiopacity is a useful property of implants used in dental analysis. They alternate energy differently than surrounding body tissues. Computational models of dental image analysis address several problems. Image modalities rely on X-ray images, even though Ultrasound, CT and MRI provide more accurate interpretations. For investigation and visualization of teeth, 2D images are used which do not provide adequate information. Photogrammetric approach gives solutions with required accuracy and for treatment planning with 3D models. CBCT images are used to locate the bone at dental implant sites. Optical methods were developed for Caries detection based on quality modification of tooth configurations which comprise Fiber Optic Trans illumination (FOTI), fluorescence, photo thermal radiometry, multi photon imaging and optical coherent tomography. Some investigations with laser and light induced fluorescence have demonstrated lesion detection. For qualitative and quantitative analysis of dental applications, Laser-induced breakdown spectroscopy (LIBS) is used. Some recent methods used for detection of dental infections are Electron Caries Monitor, Fiber optic trans-illumination, Digital Fiber optic Trans-illumination, Quantitative Light Induced Fluorescence, etc.

DOI: 10.4018/978-1-5225-6243-6.ch008

In this chapter the dental imaging techniques will be discussed, analyzed and compared as the detection methods will have their own advantages and disadvantages. Hence it will be useful for the dentist for identification and selection of technology to make the dentist feel more confident and comfortable.

DENTAL IMAGING USING X-RAYS

One of the best biomedical field accomplishments with the examination of vacuum tubes in the nineteenth century, prompted the disclosure of x-rays by Wilhelm Conrad Roentgen. Analysts doing research on x - rays experienced repeated radiation which harmed the tissues. Today the innovation has enhanced on the x-ray hardware, photographic film and its advanced strategies are utilized for different medical applications.

In 1895, Professor Wilhelm Conrad Roentgen made an attempt to investigate the cathode beams. He found that the x-rays could enter thick materials to varying degrees. Combining the information that specific materials would assimilate the light emission beams when put in its way, Roentgen exhibited that a picture of dense object could be caught on the photographic plate (Davis, 1981; Dhaliwal, Singh, Kapila & Rajput, 2012).

The revelation of the unknown beams in the year 1895 denoted a novel period in the field of dental surgery. In dental surgery, significant advancement was accomplished through the new potential outcomes of a radiological examination. The primary skiagraph (a photographic picture created on a radiosensitive surface by radiation other than unmistakable light - particularly by X-rays or gamma beams) of the tooth was introduced by Professor Koenig in the month of February, 1896, to the Frankfurt Society of Physics during. During the month April, of that year, Dr. Walkhoff exhibited numerous skiagraphs of the teeth in the human subject at the Congress of Erlangen. Dr. W. J. Morton has written a research article which became visible in June 1896 at the Dental Cosmos, on "The X-rays and their Application in Dentistry" (Friedman & Friedland, 1998). In the next few years, radiographs emerged and were utilized for the therapeutic analysis and dental conditions, for treatment using x-rays and for technical research (Dhaliwal, Singh, Kapila & Rajput, 2012).

BASICS OF DENTAL RADIOLOGY

Radiography taken from the mouth called as oral radiography is the recording of the patients oral structures recorded on an x-ray film using x- rays. Like visible light energy, X-rays are electromagnetic beams that travel in a wave movement. The measurement of this wave movement is known as a wavelength. X-rays contain an amazingly small wavelength that empowers the beam to go through materials that typically attract or return the light energy or other electromagnetic beams with longer wavelengths. In spite of the fact that X-rays contributes to the properties of other electromagnetic beams, their activities are extensively unique. A portion of the properties and characteristic features of X-rays are:

- They take a straight line path.
- When they penetrate an image is produced in the x-ray film.
- It causes fluorescence of certain substances.
- When the cells are exposed to x-rays they degenerate.

For diagnosis of dental issues, treatment planning and monitoring of the development of lesions, dentists usually go for radiographs. But its exposure to patients and to the clinical staff should be considered to ensure appropriate protection. Selecting the suitable radiographic examination depends upon the prevalence of the disease, the progression rates and the imaging technique used for accurate diagnosis. Radiographic Referral Criteria have been characterized as "descriptions of clinical conditions resulting from patient warning signs, indications and patient history that identify them who are expected to get benefited from a particular radiographic technique". Therefore the dentist should ensure that sufficient patient medical information is made available to the person taking responsibility for the exposure of the x-rays.

Whilst patients experience X-ray assessment, a large number of photons penetrate through the bodies of the subject and may harm any particle inside the system; especially the DNA in the chromosomes. DNA is repaired instantly; however a small portion of a chromosome might change permanently. While dosages and dangers for dental radiology are little, various epidemiological investigations have given indication of an augmented threat to the salivary gland, brain and thyroid tumors. Risks of the individual person in dental radiography are very mild yet they become more prominent below the age group of 30 in which dental radiography is performed recurrently.

Application of x-rays in dentistry can be classified into two types as, intraoral and extraoral. In Intraoral case, x-ray is taken inside the mouth and in extra oral case, the x-ray is taken exterior to the mouth. Intraoral X-rays are the widely known technique used in dentistry. They present an abnormal state of the teeth, jaw bone and supporting tissues of the mouth. These X-rays enable dental specialists to:

- Locate the different cavity
- Observe infections at the root of the tooth
- Verify the hard region strength in the region of the tooth
- Decide if periodontal illness is an issue of oral care
- Examine the condition of growing teeth
- Observe excellent tooth health throughout by prevention

The advantages of x-rays are tremendous: They enable the dental specialists and the clinicians to analyze regular issues, for example, cavities, illness of the gum and a few types of tooth infections. Radiographs enable dental practitioners to analyze the inside of a tooth and underneath the gums to evaluate the effectiveness of the jaw bone and the tissues that support that grasp the teeth set up.

In this section, the absolute and most regular sorts of X-rays performed are discussed in detail:

- **Periapical:** A periapical (PA) X-ray denotes to an X-ray that is taken to reveal a particular region of concern. In the event that a tooth ache occurs, the dentist will prescribe a PA film to examine that entire tooth including the root. It gives a perspective of whole tooth, from the bone that helps to carry the tooth to the crown.
- **Bite-Wing:** Bitewings are common amongst the most widely recognized arrangements of X-rays. They show teeth over the gum line and the stature of the bone height between teeth. Also it helps to analyze the gum infection and cavities between teeth. The bitewing X-ray is placed on the tongue side of the teeth and kept in place by biting down on a cardboard tab. Four bitewings are usually taken as frequently as every six months for people with cavities or once in two or three years for people with excellent oral hygiene and no cavities. This kind of x-ray demonstrates the

dental practitioner on how these teeth get in touch with each other (or impede) and will help them to decide whether decay is available between the back of the teeth.

- **Panoramic/ Panorex:** A panorex X-ray that is taken by using a X-ray film into the mouth. As the person sits still, the X-ray head turns around, giving one extensive picture of the jaws and teeth. This kind of X-ray is especially useful for examining the upper and lower jaws at the same time and can give an idea about the impacted teeth or other hidden structures that could be difficult to see on the film utilized for a full-set. Therefore it demonstrates a perspective of the teeth, area of the jaws, region of the nose, sinuses and the joints of the jaw, and is typically taken when a patient might require orthodontic treatment or implant.
- **Occlusal/Full Set:** The"full-set" X-rays reveals all of the teeth and the majority of the surrounding bone, helping to analyze cavities, ulcer or tumors, abscesses, affected teeth, and gum infections. Usually as a rule occlusal X-ray comprises of 14-20 singular X-rays and is usually recommended during the initial visit with the dentist to help in suitable finding and planning for treatment. It offers an apparent examination of the bottom of the mouth to demonstrate the bite of the lower or upper jaw.

Thus, dental specialists have to give the most ideal dental care to their patients; however a visual examination doesn't let them know all that they have to know. Dental specialists can precisely analyze and treat dental issues utilizing dental X-rays, before they turn out to become severe.

ADVANTAGES OF X-RAYS

In spite of the fact that x-rays are utilized as a major aspect of a normal examination to prevent dental infection, they additionally help dental practitioner in diagnosing a particular or inaccessible dental issues. Radiographs are utilized to check for cavities and assess the degree of decay. Also, some X-rays demonstrate the foundation of the tooth, as indicated by the Academy of General Dentistry (AGD), nearness of any growths; abscesses and different masses can be analyzed. Inherently absent or affected teeth, wisdom teeth are regularly identified along these lines, and the degree of bone loss because of periodontal problem is effectively observed through dental X-rays also.

DENTAL X-RAYS SAFETY

Since X-ray machines and other causes of dental radiographs are intended to limit radiation, these procedures are safe and the exposure is insignificant. Current issues have been raised that x-ray based diagnostic systems triggers thyroid cancer. The American Dental Association (ADA) suggest patients to have additional safety by wearing a leaded apron to cover the abdomen and a leaded collar to shield the thyroid and also keep informed the dental practitioner regarding pregnancy or breastfeeding. An increased danger of thyroid malignancy has been accounted for in dental practitioners and dental assistants and employees working with x-rays (Singh & Mathur, 2013). There are various issues of the selection of the radiation level received by patient while undergoing radiographic measurement. They comprise of:

- Selection of the x-ray machine
- The result in low patient exposure by the usage of technique
- The combinational use of fast films and screen/film
- Observance to correct film processing methods
- The use of digital sensors
- The use of collimators and filtration
- The use of lead aprons and thyroid collars to protect the patient from unnecessary radiation exposure

The kilo voltage setting is factor that set up the image contrast, along with the quantity of dose to the patient. In the 60-80 kVp range, natural hazard estimation from dental radiology is the same. Settings beneath 60 kVp are not suggested for regular dental radiography as a result of high patient exposures. Table 1.shows the typical doses from dental x-rays.

CONE BEAM CT IMAGES

Two-dimensional (2D) imaging modalities have been utilized as a part of dentistry from the inception of x-rays. Tomography isolated the areas of infections using radiographic examinations while panoramic procedures used principles of tomography, making it promising to envisage the maxillofacial structures in the complete tomographic image. Modern enhancements in computerized or digital imaging have implied lesser radiation dosages and quicker response devoid of disturbing the excellence of diagnosis of the intraoral or panoramic images. On the other hand, 2D pictures have unique characteristic restriction (deformation, intensification and superimposition) that can create to get the wrong idea about structures. CBCT is able to create three-dimensional (3D) images that can be used for diagnosing, providing treatment, and also follow up (William, Martin, Gane & Farman, 2009).

Established in 1998 for dento-alveolar imaging, CBCT produces 3D information at a lesser cost and with lesser dosages of radiation than conventional CT. CBCT's imaging method depends on a funnel shaped X-ray beam that is focused on detection that is two dimensional. In CBCT an equivalent shift of detection is not required during rotation as it uses the tube power. Hence it gives a higher rate of acquisition than the conventional CT. The cone-shaped beam rotates, which produces as many of 2D images of a characterized anatomical volume rather than the slice to slice imaging found in ordinary CT. (Alamri, Sadrameli, Alshalhoob, Sadrameli & Alshehri, 2012). The images are then reconstructed in 3D by means of an algorithm. CBCT offers various advantages compared with conventional 2D data, including an absence of superimposition, 1:1 estimations, the nonappearance of distortions and 3D output. It's essential to note down that by using moderately low dosage of ionizing radiation, CBCT provides 3 dimensional

Table 1. Typical patient doses from dental x-ray exams

Type	Skin Dose (mR)	Effective Dose in Millirem
Full mouth	2,300-3,100	1.5
Bite-wing	200	0.4
Panoramic	700-950	1.1

information of hard tissues with less information of the soft tissues compared to the conventional CT which acquires images with a large amount of ionizing radiation and extended scanning time.

The utilization of progressive imaging have been restricted for most dental specialists in view of rate, accessibility and contemplation of dosage measurements. In addition, the presentation of CBCT for the maxillofacial environment gives chances to dental professionals to bring multiplanar imaging.

Advantages of CBCT

The craniofacial area is best analyzed using CBCT. It gives apparent descriptions of extremely dissimilar arrangement and will be helpful for assessing jaw bone. Even though restrictions persists in the utilization of the equipment for imaging soft tissues, numerous attempts are carried out towards advancement of the methods and algorithms, to enhance ratio of signal to noise and amplify the contrast. The employment of CBCT technology in medical observation gives various potential importances for maxillofacial imaging contrasted and ordinary CT. Frequently, CBCT units may be tuned to examine minute areas for particular task of diagnosis. Sometimes, when necessary they are used to scan the entire craniofacial complex. The advantages are:

- Quick Scan time
- Reduction of Dose
- Imaging of maxillofacial exclusive display modes
- Reduced Image artifact

Importance of CBCT to Dental Imaging

Like traditional CT scanners, which are costly to sustain, CBCT is appropriate for use in dental practice wherever price and dosage observations are essential, provided the scanning is limited to the head. The cost of imaging in dental implants by using CBCT, evaluation of pathology in surgery, assessment of TMJ and pre and postoperative evaluation of craniofacial breaks is reported. In orthodontics, imaging using CBCT is helpful in evaluating development and growth. In maxillofacial imaging CBCT interacts with the data and produce images which are useful in clinical practice. In addition, all exclusive programming is associated with real-time data (Hechler, 2008). These systems and their clinical relevance include:

- **Oblique Planar Reorganization:** In this mode particular structure (e.g., TMJ, affected third molars) is in particular are assessed as significant features may not be prominently evident on Multiplanar Reconstructed (MPR) images.
- **Curved Planar Renewal:** This mode is used in showing the dental curve, producing usual thin-slice images and the images are accurate with least errors.
- **Serial Transplanar Reformation:** Here the images are commonly slices of the thickness of 1mm. Resulting images are important in the estimation of particular morphologic feature, for instance, bone height of the alveolar and evaluation of location of implant width, the inter relation to the affected mandibular molars and the inferior alveolar canal and the symptomatic TMJ shape or pathological circumstances assessment of affected the jaws.
- **Multiple Plane Renewal of Volume:** In this form of approach any image of multiple plane dimensions could be "condensed" by escalating the nearby voxels quantity. This image character-

izes a particular volume of the patient. The easiest method is including its assimilation also. This approach may be utilized to produce more panoramic images by expanding reformatted images thickness of curved planar along with the dental curve to 25– 30 mm. This is practically identical to the image layer of all panoramic radiographs.

- **Cone Beam Artefacts:** An artifact is an unwanted disturbance or inaccurate image that is not associated to the subject under assessment (Scarfe & Farman, 2008). Unwanted disturbances can be categorized as scientific-based, subject-related or equipment-based (Barrett & Keat, 2004; Scarfe & Farman, 2008). Besides, cone-beam associated artifacts are a feature of CBCT (Scarfe & Farman, 2008).

Patient movement creates contamination or noise in reconstructed images (Scarfe &Farman, 2008). In imaging of dental and maxillofacial, firm position of head and small imaging time reduce the difficulty of artifact (Scarfe & Farman, 2008). The activity of the head of the patient, motion of the Jaw or swallowing motion change the quality of the whole amount of data in CBCT. Because of long acquired motion, artefacts are common in medical images. When scanning an organ that is moving like the lungs or heart, movement artifacts frequently corrupt the image quality and confine the usage of CBCT (Leng, Zambelli, Tolakanahalli, Nett, Munro, Star-Lack & Chen, 2008). To lessen the motion artifacts, 4D CBCT procedure is incorporated (Leng, Zambelli, Tolakanahalli, Nett, Munro, Star-Lack & Chen, 2008)

Scanning Equipment-related artifacts arise due to Fault in scanning apparatus, exposure or meager adjustment cause circular or loop fashioned artifacts (Mozzo Procacci, Tacconi, Martini & Andreis 1998; Barrett & Keat 2004; Sijbers & Postnovz 2004, Scarfe & Farman, 2008). To minimize these artefacts detector calibration and software corrections are carried out (Barrett & Keat 2004; Sijbers & Postnovz, 2004). On the other hand, while final spatial decision is required, it is difficult to avoid ring artifacts (Sijbers & Postnovz, 2004).

There is a possibility of the impact of cone beam as a source of artifact, particularly in the marginal segment of scan volume (Scarfe & Farman, 2008). The entire quantity of data acquired for marginal structure is decreased since the external row pixels confirm a partially unusual degree of the entity when axial projection angle varies along the scan rotation, which results in image deformation, streaking artefacts and greater peripheral noise (Scarfe & Farman, 2008).

The advancement and quick promotion of CBCT equipment contribute imaging the maxillofacial area that will certainly boost dental specialist use 3D radiographic measurement. CBCT images furnishes dentists with very sensitive of the order of mm spatial resolution of high investigative excellence with comparatively small scan times (10–70 seconds) and a discharge amount corresponding to that required for 4 to 15 panoramic radiographs.

FIBEROPTIC TRANSILLUMINATION (FOTI)

FOTI is a basic, easily suitable, non hazardous technique, which supplements the medical assessment (Davies, 2001; Cortes, Ellwood & Ekstrand, 2000; Vaarkamp, Ten Bosch & Verdonschot, 1997). Thin beams of intense light pass transversely through the area of connection between approximal planes. The variation of the white light transmission results as light pass through the teeth and create an image for analysis. Research demonstrating utilization of this procedure lead to a point of variability in examination and led to the progress of DIFOTI (Stookey & González-Cabezas, 1999). FOTI is a simple, quick and inexpensive technique of imaging teeth in the presence of various scattering. Depending on the variations in scattering and assimilation phenomenon of photons of the light, it improves the difference involving enamel caries and sound.

FOTI Applications: It is used as an alternative diagnostic support for frontal and latter inter proximal and occlusal caries analysis; assessment of the stained limits of composite resins; Detection of cusp fractures and cracked teeth; to light up endodontic admittance and root canal orifices inside the chamber of teeth during endodontic treatment; as a device for enhanced assessment of lesions in the soft tissues; for assessment of all-ceramic renovation to cancel out any rupture before cementation; for clinical valuation of crack and trend in all-ceramic renovation and ordinary teeth; and for assessment of strength of extrinsic coloration to decide on the suitable treatment suggestions.

It is a risk free, modestly persistent, simple procedure which could be used throughout the dental assessment. By means of a narrow beam light focussed along the facial and interproximal planes, the desiccated tooth can be diagnosed for colour and texture variations, appearance of the surface of the tooth variations and the existence or deficiency of shadows inside the tooth.

FOTI have been demonstrated throughout the research to be a suitable marker of the existence or nonexistence of formation of infected toot (Pine, 1996; Peers & Hill, 1993; Reddy & Sugandhan, 1994; Côrtes, Ekstrand, Elias-Boneta & Ellwood, 2000). The sensitivity or true positive rate is the percentage of subjects having the condition. Specificity measures, the percentage of subjects who do not have the condition. The investigation strongly recommends that diagnosis of proximal caries have to be supported with a blend of tests, and not with one technique of diagnosis alone, and hence x-ray clinical examination and FOTI collectively offers the precise inference (Hintze, Wenzel, Danielsen & Nyvad, 1998; Choksi, Brady, Dang & Rao, 1994; Mialhe, Pereira, Pardi & de Castro, 2003; Strassler & Sensi, 2008).

At times dentists make use of bright lights to brighten the teeth region, which has displayed a massive exposure associated through "blue light hazard" which could cause eye damage and macular tissue disintegration (Wielgus & Roberts, 2012; Wielgus, Collier & Martin, 2010). Caries, which appears as a shadow inside the tooth, has a lesser index of light transmission. Using FOTI, Calculus can be diagnosed as a dark area on the tooth surface. Transparent tooth soothing resources can be simply recognised from normal tooth.

Thus FOTI is a suitable method for making a diagnosis of caries with dentists, technicians skilled and qualified in the system procedure can detect additional carious lesions than normal examination alone. In quite few cases considerably additional lesions could be able to detect with FOTI. The result was obvious both with and without the use of the x-ray measurement.

DIGITAL IMAGING FIBER-OPTIC TRANSILLUMINATION (DIFOTI)

The inconsistency problem in FOTI is succeeded by, a new method called DIFOTI (Ripa, 1985) has been designed and examined (Newbrun, 1992). This method makes use of digital image processing for diagnosing quantitatively in dentistry. In DIFOTI the images are recorded instantly with a CCD imaging camera. This diagnostic device is used to detect caries using light scattering in sound enamel and light absorption in decalcific enamel by non ionizing radiation. DIFOTI could detect occlusal, proximal, smooth surface caries and recurrent caries around restoration. Further, tooth fracture or success of restoration can be diagnosed.

Images from DIFOTI may perhaps be obtained in frequent way by preserving tuning of several control values. Further, it makes use of digital image processing techniques to make most of the difference among typical and carious tissues and to compute parameters of caries lesions for analysis. It has a number of benefits, like no exposure to ionizing radiation, instant diagnosis and more sensitive in finding the lesions near the beginning. Especially, its detection of selected lesions over an extent of time period is an exclusive benefit of DIFOTI any caries detection method. Nevertheless, the distribution of spatial resolution in images of DIFOTI is responsive to variation of lighting of source illumination. Hence, when modifications are identified at different time intervals of the image of the same tooth, it might be difficult to make a decision whether they are because of the variations in the internal tooth structure or because of a failure in setting parameters for acquiring the image.

By analyzing digital image of a tooth researchers made an attempt to increase the efficiency of FOTI (Schneiderman, Elbaum, Shultz, Keem, Greenebaum & Driller, 1997; Vaarkamp, ten Bosch, Verdonschot & Tranæus, 1997; Zero, Mol, Sá Roriz, Spoon, Jacobs, Keem & Elbaum, 2000). Digital Fiber Optic Transillumination, DIFOTI, which is a quantitative technique of measurement have been appreciated in some of the research work and the initial outcome indicate that the sensitivity and the specificity are very high. DIFOTI is a procedure for hard tissue diagnostics. It is observed that the researchers were able to find more sound surfaces on radiography images and were able to detect more caries lesions and smaller lesions using DIFOTI images. As it is easy and fast to operate, DIFOTI does not create any pain for the patient and can be easily used on children. The disadvantage of DIFOTI is that it cannot detect features covered with gingival tissue.

ELECTRICAL CARIES MONITOR (ECM)

The association with the amount of caries in teeth and resistance has been studied in the past decades. By considering different features like contact area of the electrode, enamel depth and dentin tissue thickness, porosity, temperature, enamel hydration, the dental tissue fluid ionic nature and the growth of the tooth, it is likely to review caries lesions. (Neuhaus, Longbottom; Ellwood & Lussi, 2009). The research on ECM has evaluated the above features and has exposed the outcome of the strength and the ability to reproduce. (Longbottom & Huysmans, 2005; Huysmans, Kühnisch & Ten Bosch, 2006). Research works have revealed that the existence of stain affects the ECM measurements.

ECM employs alternating current to determine the resistance of the tooth tissue. ECM is based on the opinion that porosity is proportional to electrical conductivity. When there are a number of pores the enamel demineralization occurs. The pores are filled with saliva which forms the conductive pathways for electrical transmission. As the conductivity of saliva is more than enamel, it becomes demineralized and hence the conductivity increases. High conductivity indicates healthy enamel tissue while low conductivity indicates unhealthy enamel tissue.

Fluids that are highly concentrated with ions from the oral environment (ie., tooth paste) fill the pores of caries lesions to decrease the electrical resistance or impedance. A probe is applied on to the site of caries lesion directly and converts the output into an electrical resistance. When the electrical resistance value is high the caries lesions are very deep. As the quantitative value of electrical resistance is proportional to the amount of caries lesions developed this instrument is used for measuring caries lesions. The most important benefit of ECM is to provide data, that have potential for examining the lesion growth, arrest, or remineralization. In regular clinical practice, ECM is not used alone as it cannot be applied to proximal surfaces and also it takes time when used to examine full mouth.

Research has been done using ECMs to diagnose small occlusal caries lesions and it was found that ECM better predicted lesions than visual assessment. (Le, Verdonschot, Schaeken & Van't Hof, 1995). Some studies contrast the efficiency of visual examination which was used to identify signs of fissure decalcification and discoloration. ECMs are also used to predict the onset of occlusal caries in children and are a better predictor of occlusal caries than fissure discoloration and FOTI. Another study, the posterior permanent teeth with occlusal cavitation was examined and it was observed that the ECM gave accurate diagnosis for the *in vitro* diagnosis of begining stage of non-cavitated occlusal lesions on posterior teeth (Sengun, Özbay, Akdemir, Öztürk, Özer & Bağlar, 2013).

The conductance values (Mital, Mehta, Saini, Raisingani & Sharma, 2014) obtained from ECM can be interpreted as shown in table2

LASER INDUCED BREAKDOWN SPECTROSCOPY (LIBS)

Laser induced breakdown spectroscopy (LIBS) is a powerful method which is fast and precise equipment for detection and the investigation of germs and human teeth samples. LIBS is a form of optical emission spectroscopy (El-Sherbini, Hegazy Cristoforretti & Legnaioli, 2005). The high power laser beam interacts with matter (solid, liquid and gas). While interaction it uses the light that is discharged from the plasma. The emitted light analyzed using atomic spectroscope shows significant data about the essential formation and the fundamental physical process in the plasmas (Sturm, Vrenegor, Noll, & Hemmerlin, 2004). LIBS have its relevance in industries, surroundings, medical and forensic sciences (Capitelli, Colao, Provenzano, Fantoni, Brunetti & Senesi, 2002) and give a proper method for study that is better than other traditional methods. Also it is a convenient and flexible technique for rapid determination of sample composition. In the dentistry field, researchers used this method to analyze the teeth affected by caries and to determine the accumulation of trace elements in teeth, quantity of mercury in the silver dental amalgam (Rehse, 2014). It has been reported that the fundamental analysis of element of teeth section can give essential decision about the causes of caries in teeth, which is the main oral health problem (El-Sherbini, 2016).

Table 2. ECM Interpretation

Range of Conductance Values	ECM Analysis
-1 – 3	initial phase of caries
3 - 6	caries till the Dento enamel junction (DEJ)
6 to 8	dentinal caries
8 to 13	extending half of dentine thickness

QUANTITATIVE LIGHT-INDUCED FLUORESCENCE (QLF)

A number of studies done previously have given a conclusion that QLF arrangement are realistic procedure for identifying and diagnosing beginning stage of caries lesions (Ellwood, Goma & Pretty, 2012). Finding of plaque of final stages in oral cavity have discrete benefits to care besides oral infection. Mature plaque might be a sign to alert danger of tooth infection. On the other hand, young plaque does not directly affect subject's oral infection. To distinguish severe and minute plaque, a two-tone disclosing substance (Ismail, Sohn, Tellez, Amaya, Sen, Hasson, & Pitts, 2007) has been frequently used by dentists for plaque assessment. Yet, identifying plaque has some restrictions that they not only confirm plaque, but also dye soft debris and pellicle. In addition, to remove the plaque it needs time. A different method of measurement for old plaque is Silness-Löe plaque indicator (Hee Eun Kima, Baek-Il Kim, 2015). These indicators have been devised for evaluating the amount of mineralized deposits and are also time consuming. QLF illuminator is a new dental diagnostic tool which is based on the auto fluorescence of the teeth. Earlier research work has revealed that full-grown plaque may produce red color fluorescence and they are related with microbe metabolism substances that are named Porphyrins. These organic compounds are known to be produced from delayed settling oral bacteria, such as Porphyromonas gingivalis (an inflammatory disease) and Prevotella intermedia (periodontal infections) which are normally recognized in heavily accumulated plaque. A current research gave their findings that the red fluorescence intensity of accumulated plaque in varied sucrose concentrations is connected with pH of low value and cariogenic plaque (Chalmers, 2008).

In current research, the Quantitative light-induced fluorescence – digital (QLF-D) technology, improved edition of the QLF technology, is developed that uses an optical device that contains a modified filter, an increased blue light of 405 nanometer and visible source of light and a digital single-lens reflex (DSLR) camera with high quality. The QLF-D technology shows the teeth in usual color as well as identifies the fluorescence variations of developing lesions. In addition, it is promising to observe dynamic variations of net de mineralization in the lesions which is the best advantage of the QLF-D technology. In clinical practice the beginning stage of caries lesions appear in white color becomes fairly prominent with larger porosity from sound teeth and clear plaque image in red in colour (Marthaler, 2004). Therefore, it can be used as a identifying tool to differentiate pathological area and healthy surface of the teeth by observing the red fluorescence. Thus QLF is capable as an alternative for the conventional chemical agent method to detect old plaque.

Due to its versatility the QLF technology is an extremely promising method to the dental investigator and the dental clinician. It has the prospective to become a typical investigation of dentistry and may continue glowing to set a trend toward preventive dentistry. Hence the QLF-D would enable dentists to properly detect demineralization at various phase in a comparatively less period of time, decide suitable treatment and observe, and assist patients understand their oral health conditions.

CONCLUSION

In conclusion, we can observe and appreciate the different fundamental principles that are employed for dental imaging. Each of the principles presented and the various techniques based on the principles give the reader an opportunity to understand different possibilities and appreciate the options available. This will help in diagnostics planning for different patients. During diagnostics planning, given this choice, it is important that the practitioner studies, along with the clinical indications of interest, the advantages and disadvantages of the various technologies involved and the cost of the equipment used in the context of the indications to be diagnosed and proceed planning. It is important to observe that some techniques offer observation data that may be not be helpful for clinical diagnosis but nevertheless be useful for research and study. The general recommendation here is to mindfully opt for one or more technologies by validating the purpose of reason for opting into a dental imaging based cariogenic study of the subject. A methodical selection approach may also be adopted for sophisticated study reason, especially for clinical diagnosis of patients with specific indications and for research.

REFERENCES

Alamri, H. M., Sadrameli, M., Alshalhoob, M. A., & Alshehri, M. A. (2012). Applications of CBCT in dental practice: A review of the literature. *General Dentistry*, *60*(5), 390–400. PMID:23032226

Campbell, D. J. (1995). A brief history of dental radiography. *The New Zealand Dental Journal*, *91*(406), 127–133. PMID:8602286

Capitelli, F., Colao, F., Provenzano, M. R., Fantoni, R., Brunetti, G., & Senesi, N. (2002). Determination of heavy metals in soils by laser induced breakdown spectroscopy. *Geoderma*, *106*(1-2), 45–62. doi:10.1016/S0016-7061(01)00115-X

Chalmers, J. M. (2008). Minimal intervention dentistry: A new focus for dental hygiene. *Dentistry Today*, *27*(4), 132–134. PMID:18497207

Choksi, S. K., Brady, J. M., Dang, D. H., & Rao, M. S. (1994). Detecting approximal dental caries with transillumination: a clinical evaluation. *Journal of the American Dental Association, 125*(8), 1098-1102.

Côrtes, D. F., Ekstrand, K. R., Elias-Boneta, A. R., & Ellwood, R. P. (2000). An in vitro Comparison of the Ability of Fibre–Optic Transillumination, Visual Inspection and Radiographs to Detect Occlusal Caries and Evaluate Lesion Depth. *Caries Research*, *34*(6), 443–447. doi:10.1159/000016621 PMID:11093016

Davies, G. M., Worthington, H. V., Clarkson, J. E., Thomas, P., & Davies, R. M. (2001). The use of fibre-optic transillumination in general dental practice. *British Dental Journal*, *191*(3), 145–147. doi:10.1038j.bdj.4801123 PMID:11523886

Davis, A. B. (1981). *Medicine and its technology: an introduction to the history of medical instrumentation*. Academic Press.

Dental Radiographs (X-Rays). (n.d.). Retrieved from California Dental Association: http://www.cda.org

El Sherbini, A. M., El Sherbini, T. M., Hegazy, H., Cristoforetti, G., Legnaioli, S., Palleschi, V., ... Tognoni, E. (2005). Evaluation of self-absorption coefficients of aluminum emission lines in laser-induced breakdown spectroscopy measurements. *Spectrochimica Acta. Part B, Atomic Spectroscopy*, *60*(12), 1573–1579. doi:10.1016/j.sab.2005.10.011

El-Sherbini, T. M. (2016). Impact of Physics on Medical Sciences and Applications: Lasers and Nanotechnology. *Journal of Medical Physics and Applied Sciences*, *1*(1).

Ellwood, R. P., Goma, J., & Pretty, I. A. (2012). Caries clinical trial methods for the assessment of oral care products in the 21st century. *Advances in Dental Research*, *24*(2), 32–35. doi:10.1177/0022034512449464 PMID:22899676

Evens, R. G. (1995). Roentgen retrospective: One hundred years of a revolutionary technology. *Journal of the American Medical Association*, *274*(11), 912–916. doi:10.1001/jama.1995.03530110074039 PMID:7674507

Fennis-Ie, Y. L., Verdonschot, E. H., & Van't Hof, M. A. (1998). Performance of some diagnostic systems in the prediction of occlusal caries in permanent molars in 6-and 11-year-old children. *Journal of Dentistry*, *26*(5-6), 403–408. doi:10.1016/S0300-5712(97)00060-2 PMID:9699429

Forrai, J. (2005). *Culture history of dentistry*. Academic Press.

Harrell, W. E. Jr. (2007). Three-dimensional diagnosis and treatment planning: The use of 3D facial imaging and 3D cone beam CT in orthodontics and dentistry. *Australian Dental Practice*, *18*, 102–113.

Hechler, S. L. (2008). Cone-beam CT: Applications in orthodontics. *Dental Clinics*, *52*(4), 809–823. doi:10.1016/j.cden.2008.05.001 PMID:18805230

Hintze, H., Wenzel, A., Danielsen, B., & Nyvad, B. (1998). Reliability of visual examination, fibre-optic transillumination, and bite-wing radiography, and reproducibility of direct visual examination following tooth separation for the identification of cavitated carious lesions in contacting approximal surfaces. *Caries Research*, *32*(3), 204–209. doi:10.1159/000016454 PMID:9577986

Ie, Y. L., Verdonschot, E. H., Schaeken, M. J. M., & Van't Hof, M. A. (1995). Electrical conductance of fissure enamel in recently erupted molar teeth as related to caries status. *Caries Research*, *29*(2), 94–99. doi:10.1159/000262048 PMID:7728835

Ismail, A. I., Sohn, W., Tellez, M., Amaya, A., Sen, A., Hasson, H., & Pitts, N. B. (2007). The International Caries Detection and Assessment System (ICDAS): An integrated system for measuring dental caries. *Community Dentistry and Oral Epidemiology*, *35*(3), 170–178. doi:10.1111/j.1600-0528.2007.00347.x PMID:17518963

Kapila, R., Dhaliwal, A., & Singh, N. (2014). History of X-rays in dentistry. *Annals of Dental Research*, *2*(1), 22–25.

Kim, H. E., & Kim, B. I. (2015). An in vitro comparison of quantitative light-induced fluorescence-digital and spectrophotometer on monitoring artificial white spot lesions. *Photodiagnosis and Photodynamic Therapy*, *12*(3), 378–384. doi:10.1016/j.pdpdt.2015.06.006 PMID:26117198

Levinson, M. H. (1999). Medicine's 10 Greatest Discoveries. *et Cetera, 56*(2), 232.

Luiz Mialhe, F., Carlos Pereira, A., Pardi, V., & de Castro Meneghim, M. (2004). Comparison of three methods for detection of carious lesions in proximal surfaces versus direct visual examination after tooth separation. *The Journal of Clinical Pediatric Dentistry*, *28*(1), 59–62. doi:10.17796/jcpd.28.1.g121387868676514 PMID:14604144

Marthaler, T. M. (2004). Changes in dental caries 1953–2003. *Caries Research*, *38*(3), 173–181. doi:10.1159/000077752 PMID:15153686

Mital, P., Mehta, N., Saini, A., Raisingani, D., & Sharma, M. (2014). Recent advances in detection and diagnosis of dental caries. *Journal of Evolution of Medical and Dental Sciences*, *3*(1), 177–192. doi:10.14260/jemds/1807

Miziolek, A. W., Palleschi, V., & Schechter, I. (Eds.). (2006). *Laser induced breakdown spectroscopy*. Cambridge University Press.

Newbrun, E. (1992). Preventing dental caries: current and prospective strategies. *Journal of the American Dental Association, 123*(5), 68-73.

Peers, A., Hill, F. J., Mitropoulos, C. M., & Holloway, P. J. (1993). Validity and reproducibility of clinical examination, fibre-optic transillumination, and bite-wing radiology for the diagnosis of small approximal carious lesions: An in vitro study. *Caries Research*, *27*(4), 307–311. doi:10.1159/000261556 PMID:8402807

Pine, C. M. (1996). Fiber-optic transillumination (FOTI) in caries diagnosis. *Early Detection of Dental Caries*, *I*, 51–65.

Reddy, V. V., & Sugandhan, S. (1994). A comparison of bitewing radiography and fibreoptic illumination as adjuncts to the clinical identification of approximal caries in primary and permanent molars. *Indian Journal of Dental Research, 5*(2), 59-64.

Rehse, S. J. (2014). Biomedical applications of LIBS. In *Laser-Induced Breakdown Spectroscopy* (pp. 457–488). Berlin: Springer. doi:10.1007/978-3-642-45085-3_17

Ripa, L. W. (1985). The current status of pit and fissure sealants. *Journal - Canadian Dental Association, 5*, 367–380. PMID:3160450

Scarfe, W. C., Levin, M. D., Gane, D., & Farman, A. G. (2009). Use of cone beam computed tomography in endodontics. *International Journal of Dentistry*. PMID:20379362

Schneiderman, A., Elbaum, M., Shultz, T., Keem, S., Greenebaum, M., & Driller, J. (1997). Assessment of dental caries with digital imaging fiber-optic translllumination (DIFOTITM): In vitro Study. *Caries Research, 31*(2), 103–110. doi:10.1159/000262384 PMID:9118181

Sengun, A., Özbay, Y., Akdemir, B., Öztürk, B., Özer, F., & Baglar, S. (2013). Reliability of electronically detection of fissure caries (by using a prototype device): An alternative diagnostic electronic caries monitor device. *Journal of Restorative Dentistry*, *1*(1), 26. doi:10.4103/2321-4619.111230

Singh, M. M., & Mathur, R. (2013). *Thyroid Cancer Risk from Dental X-ray Diagnostics*. Academic Press.

St-Onge, L., Kwong, E., Sabsabi, M., & Vadas, E. B. (2004). Rapid analysis of liquid formulations containing sodium chloride using laser-induced breakdown spectroscopy. *Journal of Pharmaceutical and Biomedical Analysis*, *36*(2), 277–284. doi:10.1016/j.jpba.2004.06.004 PMID:15496320

Stookey, G. K., Jackson, R. D., Zandona, A. G., & Analoui, M. (1999). Dental caries diagnosis. *Dental Clinics of North America*, *43*(4), 665–677. PMID:10553249

Strassler, H. E., & Sensi, L. G. (2008). Technology-enhanced caries detection and diagnosis. *Compendium of Continuing Education in Dentistry, 29*(8), 464-5.

Thomas, A. (1998). The Radiology History & Heritage Charitable Trust. *Newsletter of the Radiology History and Heritage Charitable Trust Winter, 10*, 5-24.

Tognoni, E., Palleschi, V., Corsi, M., & Cristoforetti, G. (2002). Quantitative micro-analysis by laser-induced breakdown spectroscopy: A review of the experimental approaches. *Spectrochimica Acta. Part B, Atomic Spectroscopy*, *57*(7), 1115–1130. doi:10.1016/S0584-8547(02)00053-8

Vaarkamp, J., Ten Bosch, J. J., & Verdonschot, E. H. (1995). *Propagation of light through human dental enamel and dentin*. Academic Press.

Vaarkamp, J., Ten Bosch, J. J., Verdonschot, E. H., & Tranaeus, S. (1997). Quantitative diagnosis of small approximal caries lesions utilizing wavelength-dependent fiber-optic transillumination. *Journal of Dental Research*, *76*(4), 875–882. doi:10.1177/00220345970760040901 PMID:9126184

Wielgus, A. R., Collier, R. J., Martin, E., Lih, F. B., Tomer, K. B., Chignell, C. F., & Roberts, J. E. (2010). Blue light induced A2E oxidation in rat eyes–experimental animal model of dry AMD. *Photochemical & Photobiological Sciences*, *9*(11), 1505–1512. doi:10.1039/c0pp00133c PMID:20922251

Wielgus, A. R., & Roberts, J. E. (2012). Retinal photodamage by endogenous and xenobiotic agents. *Photochemistry and Photobiology*, *88*(6), 1320–1345. doi:10.1111/j.1751-1097.2012.01174.x PMID:22582903

Zero, D., Mol, A., Roriz, C. S., Spoon, M., Jacobs, A., Keem, S., & Elbaum, M. (2000). Caries detection using digital imaging fibre-optic transillumination (DIFOTITM): A preliminary evaluation. *Early Detection of Dental Caries*, *2*, 169–183.

FURTHER READING

Ashley, P. F., Blinkhorn, A. S., & Davies, R. M. (1998). Occlusal caries diagnosis: An in vitro histological validation of the Electronic Caries Monitor (ECM) and other methods. *Journal of Dentistry*, *26*(2), 83–88. doi:10.1016/S0300-5712(97)00007-9 PMID:9540303

Ashley, P. F., Ellwood, R. P., Worthington, H. V., & Davies, R. M. (2000). Predicting occlusal caries using the electronic caries monitor. *Caries Research*, *34*(2), 201–203. doi:10.1159/000016590 PMID:10773640

Fennis-Ie, Y. L., Verdonschot, E. H., & Van't Hof, M. A. (1998). Performance of some diagnostic systems in the prediction of occlusal caries in permanent molars in 6-and 11-year-old children. *Journal of Dentistry*, *26*(5-6), 403–408. doi:10.1016/S0300-5712(97)00060-2 PMID:9699429

Huysmans, M. C. D., Longbottom, C. H., Hintze, H., & Verdonschot, E. H. (1998). Surface–specific electrical occlusal caries diagnosis: Reproducibility, correlation with histological lesion depth, and tooth type dependence. *Caries Research*, *32*(5), 330–336. doi:10.1159/000016468 PMID:9701657

Huysmans, M. C. D. N. J. M., Longbottom, C., Christie, A. M., Bruce, P. G., & Shellis, R. P. (2000). Temperature dependence of the electrical resistance of sound and carious teeth. *Journal of Dental Research*, *79*(7), 1464–1468. doi:10.1177/00220345000790070601 PMID:11005729

Ie, Y. L., Verdonschot, E. H., Schaeken, M. J. M., & Van't Hof, M. A. (1995). Electrical conductance of fissure enamel in recently erupted molar teeth as related to caries status. *Caries Research*, *29*(2), 94–99. doi:10.1159/000262048 PMID:7728835

Pretty, I. A. (2006). Caries detection and diagnosis: Novel technologies. *Journal of Dentistry*, *34*(10), 727–739. doi:10.1016/j.jdent.2006.06.001 PMID:16901606

Rock, W. P., & Kidd, E. A. (1988). The electronic detection of demineralisation in occlusal fissures. *British Dental Journal*, *164*(8), 243–247. doi:10.1038j.bdj.4806415 PMID:3164192

Verdonschot, E. H., Bronkhorst, E. M., Burgersdijk, R. C. W., König, K. G., Schaeken, M. J. M., & Truin, G. J. (1992). Performance of some diagnostic systems in examinations for small occlusal carious lesions. *Caries Research*, *26*(1), 59–64. doi:10.1159/000261429 PMID:1568239

Yukizaki, H., Kawaguchi, M., Egashira, S., & Hayashi, Y. (1998). Relationship between the electrical resistivity of enamel and the relative humidity. *Connective Tissue Research*, *38*(1-4), 53–57. doi:10.3109/03008209809017020 PMID:11063015

KEY TERMS AND DEFINITIONS

Beam Limiting Devices: Tube cover for motionless and movable investigative x-ray units that is supplied with light beam collimators. They conform with the seepage radiation level prescribed for tube housing.

Cone: A system by which x-ray beam is restricted to a particular region, volume, or dimension.

Radiation: Rays consisting of alpha particles, beta particles, neutrons, protons of Gamma rays, x-rays, and other nuclear subatomic particles, but not sound or radio waves, or visible, infrared, or ultraviolet light.

Chapter 9
Thermal Analysis of Artificial Dental Materials Using Numerical Simulation

Thanigaiarasu Subramanian
Anna University, India

Chitimada Narendra Kumar
Madras Institute of Technology, India

ABSTRACT

A numerical investigation of steady state heat transfer phenomenon in human tooth is carried out in this present study. The materials generally used for repairing human tooth are ceramics, gold, structural steel, and copper. These materials are considered for the three-dimensional heat transfer analysis. The shape of the teeth is considered as perfect cubic of dimensions 5mm X 5mm X 1.5mm. The simulation results show that the teeth are subjected to temperatures of varying nature in both X and Y directions whereas Z direction is same. It is found that the temperature is maximum at the top of the teeth and minimum at bottom of the teeth (i.e., root). The contours of the heat transfer analysis for various teeth materials shows that the layers of the teeth are subjected to varying temperature. It is also found that the heat transfer characteristics depend not only on shape and boundary conditions but also on the materials used for the teeth selection.

INTRODUCTION

Investigations on the material properties of soft and hard tissues such as its electrical properties, mechanical properties, optical properties and thermal properties is highly useful for understanding is structure, functions and cellular interactions (Kamalanand, Sridhar, Rajeshwari, & Ramakrishnan, 2010; Apparsamy, & Krishnamurthy, 2018; Krishnamurthy, Sridhar, Rajeshwari, & Swaminathan, 2009; Sahana, Paramasivam, & Kamalanand, 2017). Heat transfer phenomenon plays a major role in most of the fields of science and engineering including medicine (Kandlikar, Garimella, Li, Colin, & King, 2005). The analysis of temperature distribution in human teeth would be highly useful for the diagnosis of many dental

DOI: 10.4018/978-1-5225-6243-6.ch009

diseases and also would provide useful information on the life of the teeth and its replacement (Wilkins & McCullough, 1989). Many materialswhich are subjected to heat transfer over a long period of time, would change the material characteristics, including its strength and lifetime. In particular, human teeth are made up of different materials and are subjected to thermal loads in day to day life at regular intervals (Brown, Dewey, & Jacobs, 1970). Proper understanding of the heat transfer phenomenon in human tooth will greatly help in the protection of human teeth for its increased lifetime and also to have a better understanding on artificial teeth. Hence, there is a need for the analysis of heat transfer phenomenon in dental materials (Lin, Liu, Niu, Xu, & Lu, 2011). A knowledge of the heat transfer phenomenonin teeth mayprovide more efficient ways to performsurgical procedures and toprevent certain tooth problems to some extent (Lin, Xu, Lu, & Bai, 2010).

Based on different requirements, the Dentist can utilize materials with different compositions to fill a cavity in a patient's teeth. For example, among different varieties of amalgam, metalamalgam and cast gold can be chosen as a material for filling (Torabinejad, & White, 1995). Due to recent advancements, dentists prefer certain advanced materials for use in dentistry (Moorthy, Hogg, Dowling, Grufferty, Benetti, & Fleming, 2012). As all the basic materials belong to same family most of the properties like strength, melting ranges, thermal conductivity, ductility, and electrical conductivity, are similar. The material properties such as brittleness in ceramics and flexibility in polymers are often utilized in dentistry for the development of dental materials (Lin, Liu, Niu, Xu, & Lu, 2011). Out of several materials, only three basic materials are more convenient for use. Theclassification of basic materials is presented in Figure 1.

Figure 1 explains the general behaviour of the three basic materials such as metals, ceramics, and polymers. The properties of materials are important factors for selecting the materials to be used in the field of dentistry (Craig, & Powers, 1989). These materials are significantly tough and possess high stiffness or Young's modulus. With these desirable properties, these materials are also used as repairing materials. But, metals are not usually preferred as repairing materials due to its high conductivity (Roberts, Berzins, Moore, & Charlton, 2009). On the other hand, ceramics and polymers are semi-transparent and exhibit lower heat transfer through the material. However, these materials are less strong and exhibit lower

Figure 1. Classification of basic materials

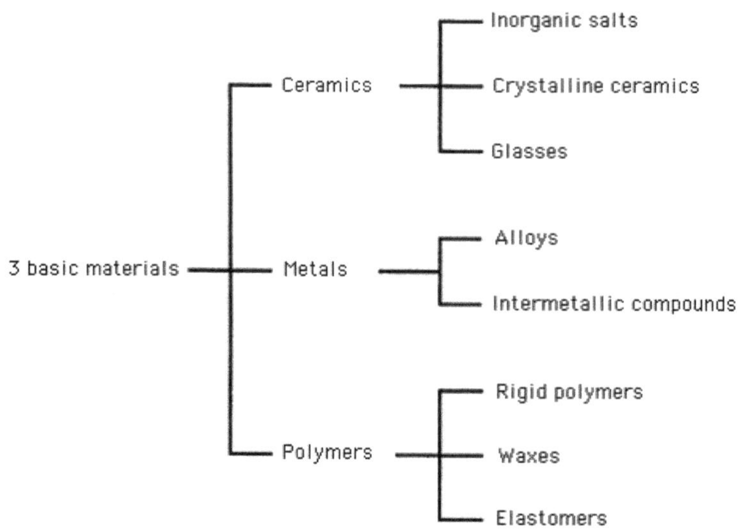

hardness when compared to metals (Anusavice, Shen, & Rawls, 2013). Since, a single material cannot have all the required properties, we go for composite materials for use in dentistry (Braden, & Clarke, 1984). A Dental resin composite filler material is formed by reinforcement of high thermal-expansion, low-strength, and low-elastic-modulus polymer with a low thermally expansion, high-strength, and high-elastic-modulus material (Blum, Lynch, & Wilson, 2014). Taking the advantages and disadvantages into account, the best material can be selected by analysing the material properties such as the thermal properties.

Filling Materials

There are many choices for the selection of dental fillings (Yengopal, Harneker, Patel & Siegfried, 2009; Johnson, 1999; Van Meerbeek, Vanherle, Lesaffre, Braem, & Lambrechts, 1991). Recent advances in dentistry has also led to achange in priorities and to havepreferences inthe dental materials for various dental applications (Ferracane, 1995). Ideally, a dental filling material should have the following important qualities (1) Dimensional solidity (doesn't change shape over time), (2) Physical strength (is able to resist wear and breakage) and (3) Cosmetically appealing (looks like a tooth) (Kirsch, Tchorz, Hellwig, Tauböck, Attin, & Hannig, 2016). It is important to know the differences between the types of filling materials. When a restoration is made, it is important to understand how each material will behave, how long it will last, and how it will look.

Dental Amalgam

The mercury in the amalgam material may enter into the bloodstream (although the mercury levels detected have never been proven to cause illness and other problems) Bharti, Wadhwani, Tikku, & Chandra, 2010). Most of the patients don't like the way it looks either, as its dark colour contrasts strongly with the natural colour of the tooth. The main problem of dental amalgam isthat it is not suitable for the teeth becausethe material is not dimensionally stable andcontinually expands with time (Arenholt-Bindslev, 1992). Research indicates that, in practice, nearly all large amalgam restored teeth fracture over time, and sometimes to the point where the tooth is split beyond repair. This is why doctors don't useamalgam. Its compositions consist of a mixture of silver, tin, zinc, copper and mercury with mercury contributing to 50% of the total composition. It is mainly usedas a filler in back teeth and has aminimum lifespan of10 years (Thomas, 2013).

- **Merits:** The forces of chewing can be easily resisted by the Amalgam fillers and it is also less expensive when compared to other materials. Amalgam fillings require single seating only andthey are delicate to moisture when compared to composite resin.
- **Demerits**: Amalgam doesn't blend with the shade of teeth. Amalgam fillings can also weaken gradually over a period of time. For proper fixing of cavity, the dentist has to remove surface of the other tooth to create a proper holding of material (Hanson & Pleva, 1991).

Composite Resin

Composite resin material is generally unstable since it doesn't maintain its shape. When a dentist places a composite resin filling in patient's mouth, it's in a semi-liquid form. It is then cured with a high-intensity blue light as it solidifies and hardens. This process is called polymerisation. With itcomes shrinkage the filling becomes smaller than the cavity. Researchers have worked hard to reduce this shrinkage and it also causes the decay to come backdespite theprecautions. The ultimate goal of the filling is compromised over time and bacteria gets back in. A refilling might be required around 2-7 years later for the same tooth. So composite resins are normally used in special cases. Some of the composite resin properties and its merits and demerits are discussed in this section. It is composed of a blend of plastic and fine glass particles. A direct filling is placed as such by using a blue light that tightens the soft material whereas an indirect filling is one which is prepared separately from an impression (Ferracane, 2011; Brunthaler, König, Lucas, Sperr, & Schedle, 2003; Douglas, 2000; Peutzfeldt, 1997).

- **Merits:** The filling color blends quite well with the color of the teeth and it also takes a singleseating only to fit into teeth cavity. It is also stronger than the amalgam filling. Making holes in the teeth is not required. Indirect composite fillings, if used, increases the strength of filling.
- **Demerits:** It is expensive when compared to other materials. Though material is stronger, life span is less compared to amalgam materials. Shrinkage in material is also seen after filling the tooth cavityand this can be avoided if it is filled in thin layers. Filling these materials is a time-consuming process as ithas to be filled layer by layer (Jandt & Sigusch, 2009).

Ceramics

In researcher's opinion, porcelain (sometimes called dental ceramics), is the best filling material that can be used for restoration. It is cosmetically appealing and is also durable for a long-term. As it is designed exactly to fit into the tooth, there is no trouble of shrinkage as with plastic fillings. The downside is, it is not cheap and these fillings take longer time to prepare. Ceramics fillings are prepared separatelyby a dental technician or CAD-CAM (Computer Aided Design – Computer Aided Manufacturing) system. This new development or technology means that porcelain fixings or fittings can be made and cemented during the same day anditreduces the visiting time for the patients. Some of the ceramic's properties and its merits and demerits are discussed in this part. Generally, Porcelain is used for the purpose ofindirect restoration (filling) of crowns, teeth whitening, implants, and braces. Its lifespan is greater than 7 years (Denry & Holloway, 2010; Yamauchi, Masuhara, Nakabayashi, Shibatani & Wada, 1981; Cummings, Rolf, Rosenflanz, Rusin & Swanson, 2006).

- **Merits:** It is the same color as of the tooth and areerosion resistantwhen compared to composite resin.
- **Demerits:** Ceramic fillings are easily breakable and also the tooth size should be decreased for the extra fitting on the surface.

Gold

Before the introduction of dental porcelain the best filling material was gold. It has similar qualities to porcelain and is very stable and very strong. However, as the price of gold increases each year, it is expensive to manufacture and because it looks 'gold' in colour it lacks the cosmetic appeal. Unlike porcelain, gold fillings used to be done over two appointments and are made by a dental technicianor using CAD-CAM process. For these reasons, it's rarely used today, but occasionally it is used when really high strength is required and to withstand high stress on the tooth. It is composed of Gold alloy and it is used mainly as a filling material and as a crown material. It has a very long lifespan (15 years) (Tobias, 1988; Wataha, 2000).

- **Merits:** Gold will not be damaged easily and its life span is more.
- **Demerits:** It is highly expensive. Galvanic shock may also be seen inside the mouth due the interactions between Gold and amalgam if used one after other.

Glass Ionomer

Glass ionomer is a white filling just like ceramics and composite resin. These fillings are prepared from a natural looking cream coloured substance designed to fill gaps in teeth structure. Glass Ionomer is basically made from a combination of glass particles, or fluoroalumino silicate, acrylic and a setting agent. Unlike traditional silver amalgam fillings, this materialactuallyforms a chemical bond with the teeth, whichreduces the chance of damage or decay of the tooth. Resin-modified or hybrid composite glass ionomer fillings are much stronger than the generally used glass ionomers. Glass ionomer is used as a filler for inlay, front teeth fillings and also the most preferred filler for babies, most damaged part of the tooth and also used as a liner in some cases (Wilson, & Kent, 1971; Lohbauer, 2009; Croll, & Nicholson, 2002).

- **Merits:** It blends with teeth color. It releases fluoride that helps fight tooth decay and it also has a strong bonding character around the tooth in general.
- **Demerits:** It easily gets damaged when compared to composite resins. Long duration is also required for resin modified ionomer type.

BACKGROUND

Over the past 60 years, research studies have been carried out in the selection of materials for teeth restoration. For the present investigation, several research papers ondental materialsand its thermal effectswere examined. There are many reasons and causes to study thermal and heat transfer changes in an artificial tooth. Heat transfer analysis is of crucial importance for modern dental industry. The heat transfer analysis of tooth can help in diagnosis of many diseases and also the temperature and temperature related factors are major factors fordentalmaterial selection. The main focus of present study is the tooth behavior when it is subjected to different temperatures under specified heat transfer conditions. In recent years, scientists and dentistshave been investigating the heat transfer treatment over artificial tooth and its behaviour with respect toall the natural conditionsexperienced by the tooth. Considering major dental heat treatments, one should keep in mind that the artificial teeth are limited to certain temperatures only. In the study of dental materials, material characteristics are important since there are several materials to choose from (Brown, Dewey, & Jacobs, 1970;Jakubinek, O'Neill, Felix, Price, & White, 2008).A number of procedures and productsused in dental practisesand treatments, namelycavity preparation, rotary instruments, light-curing of dental restorations and high-intensity lasers, can cause harmful thermal effects on dental tissues.The strength of the tooth may weaken by the dental tooth fixing procedures and may lead topermanent tooth damage.

In addition, heat transfer in a tooth can be affected by factorssuch as the geometry of tooth components, thickness of enamel and dentin, type of restorativematerials, blood perfusion and pulp (Lin, Xu, Lu, &Bai, 2010; Liang, Sa, Sun, Ma, Wang, Xing, Jiang& Wang, 2012; Liang, Sa, Jiang, Ma, Xing, Wang & Wang, 2013; Davis, Gluskin, Livingood, & Chambers, 2010).There are very strict limitations in vivo studies due to health care regulations and for this reason most experiments need to be carried out in vitro or using simulation studies. There are some research reports on temperature measurements on tooth (Louer Kraft, Rothlauf & Zwigers, 1990). The investigation of heat transfer intooth requires both mathematical modelling and experimental measurements. Generally, to check the validity of modelling predictions, experimental measurements are required (Hannig, &Bott, 1999; Linsuwanont, Palamara, & Messer, 2007; Linsuwanont, Versluis, Palamara, & Messer, 2008). The objective of these studies areto investigate the thermal variations in different layers of permanent teeth (Lipski, Woźniak, Lichota, & Nowicka, 2011).

In 2013, a technique was implemented to use lasers in medical science for dental restorations (Secilmis, Bulbul, Sari, & Usumez, 2013). Numerical techniques and finite element analysis methods (Oskui, Ashtiani, Hashemi, & Jafarzadeh, 2013) were used to obtain the temperature distribution data through human teeth.The wide range of testing conditions in numerical studies allow better results whencompared to the experimental results.A three-dimensional finite element thermal analysis of dental tissues was conducted by Ana, Velloso, & Zezell, 2008.The heat transfer phenomenon over the tooth was extensively studiedusing finite element analysis by dividing single tooth section into multiple nodes and elements (Preiskorn, Zmuda, Trykowski, Panas, & Preiskorn, 2003).

A method for estimation of thermal diffusivity of the teeth and for identification of the damage in the teeth, was studied by Onisor, Asmussen, & Krejci, 2011; Saghlatoon, Soleimani, Moghimi, & Talebi, 2012. For the present investigation, dental materials have been considered in the perfect cube shape and are studied numerically at three different boundary conditions. Adopting numerical techniques instead of experimental investigation enables the testing of the dental materials in thermal related problems without causing any actual damage.

HEAT TRANSFER

- **Definition:** Heat transfer is the science that deals with the flowof heat from a higher temperature body to a lower temperature body. In other words, thermal energy travels from high thermal body to a low thermal body. Heat is defined as energy movement due to the variation in temperature between cold and hot bodies (Rohsenow, & Cho, 1998). Heat transfer tells us how and in what mode and rate the heat is transferred between two bodies as shown in Figure 2.

Figure 2. Heat transfer process

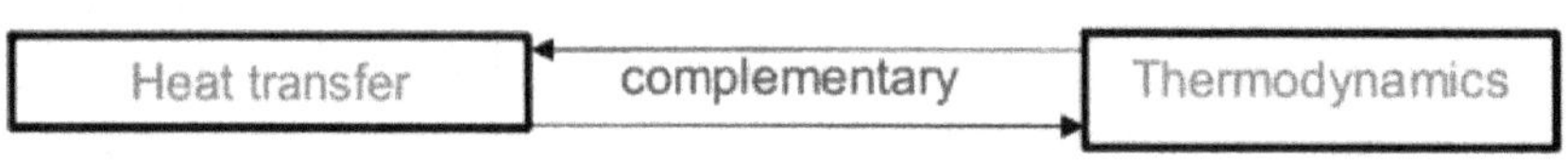

Modes of Heat Transfer

Generally there are three different types of heat transfer modes in which heat passes from a hot surface to a cold surface, namely Conduction, Convention, andRadiation (Bergman, &Incropera, 2011; Hewitt, Shires, & Bott, 1994; Rohsenow, & Cho, 1998).

- **Conduction:** Conduction is the heat transfer in a continuous substance without any observable movement of matter. Thus, heat conduction is basically the transmission of energy by inter-molecular movement. Some of the examples of conduction heat transfer are, (1) Heat transfer through brick, solid wall, and plate, (2) The heat from warm liquid makes the cup itself hot, (3) In summer, sand can transfer the heat into feet while walking, (4) A heat exchanger uses a hot fluid to pass heat to a cooler fluid without the interaction of each other (Rohsenow, & Cho, 1998).

Laws of Conduction

- **Fourier Law of Heat Conduction:** Fourier law states that the heat flow rate 'q' through a congruent solid is directly proportional to the surface area A, of the section at right angles to the heat flow, and to the temperature difference ΔT along the path of the heat flow. Fourier law of Heat Conduction states that therate of transfer of heat via a material is proportional to the negative gradient in temperature to the area and mathematically given as,

$$q = -K\Delta T \tag{1}$$

where, q = Heat flux (W/m^2), K = Thermal conductivity of material (W/m K) and ΔT = Temperature gradient(K). The property of a material to conduct heat is called as its Thermal Conductivity. It is expressed in terms of the Fourier's law of heat conductionas shown in equation (1). Heat transfer is low in materials having low thermal conductivity as compared to the materials having high thermal conductivity. Heat sink applications are largely made from high thermal conductivity materials, whereas insulation applications are made from materials having low thermal conductivity. Temperature dependency is an important parameter in determining the thermal conductivity of a material. The inverse of thermal conductivity gives the thermal resistivity_(Rohsenow, & Cho, 1998).

- **Convection:** The fluid always moves from a high temperature zone to a low temperature zone of a macroscopic particle. During this process, some amount of enthalpy is carried. This enthalpy flow is known as Convection. Convection could be either natural convection or forced convection. The Convection equation is given by:

$$q = h\Delta T \tag{2}$$

where, q = Heat transferred per unit area (W/m^2), h= Convective heat transfer coefficient (W/m^2 K), and ΔT = Difference in temperature (K) (Rohsenow, & Cho, 1998).

- **Natural Convection:** In natural convection, the movement of the fluid substances is obtained by the buoyancy forces where the transfer of heat is from the high thermal region density to the low thermal region density. Here the mobility is triggered by the warm fluid. The hot fluids have less density and are lighter when compared with the cold fluids and therefore circulation of the fluids happens between them constantly between them based on the temperature difference. Some of the examples for free convection in daily life are (1) Boiling of water - The heat transfer from the burner into the pot. The cycle movement of the hot and cold water due to the changes in densities happens continuously as the bottom water gets heated first and goes up as the low-density attainment and the upper cold water gets down as its density is more, (2) Boiling cup of hot tea – The heat transfer is due to the vapours when boiling of tea, (4) Hot air balloon – Air is heated inside the balloon by a heater, due to this air moves upward. The hot air is released to descend causing the balloon to move downward (Rohsenow, & Cho, 1998).

- **Forced Convection:** In forced convection, mobility of fluid particles is triggered usuallyfrom a high thermal region to low thermal region. The triggering of fluid mobility is caused by an external source. The most commonly used sources are fans. Some of the examples for forced convection in daily life are (1) Using a fan to reduce the temperature of boiled water, (2) A convection oven, (3) Stirring the hot milk to cool it to normal temperature, (4) Water heater to heat the cool water, (5) Climate conditioning (Air conditioners), (6) Radiators and boilers, (7) Wind or Steam turbines, (8) Blood circulation in animals with moderately high temperature of blood (Rohsenow, & Cho, 1998).
- **Radiation**: The process in which heat energy is transferred from one system to another in the form of electromagnetic waves with no need of material medium is called as Radiation. The formula which relates the radiation heat transfer is given by Stefan Boltzmann law which states that the total radiant heat energy emitted from a surface is proportional to the fourth power of its absolute temperature.

$$q = \sigma T^4 \tag{3}$$

where, q = Heat flux per unit area in W/m^2,σ = Stefan Boltzmann constant (5.67 x 10^{-8} $W/m^2\ K^4$), and T = Absolute temperature in K (Rohsenow, & Cho, 1998).

TYPES OF BOUNDARY CONDITIONS

- **Conduction Boundary Condition:** In this boundary condition, the temperature has been specified. For the present study, the top And bottom surfaces are maintained at higher temperature and lower temperature or vice versa. Generally, the bottom surface has been maintained at atmospheric conditions for all heat transfer applications like slabs, plates and heat exchangers. In this, the flow of heat takes places from solid to solid. The formula associated with this kind of boundary condition is given by (Lienhard, 2013)

$$Q = -KA(dt/dx) \tag{4}$$

where, Q= Heat Transfer(W/m^2), K = Thermal Conductivity of material (W/m K), A = Surface Area (m^2), and (dt/dx) = Temperature gradient (K).

Table 1. Material with Thermal Properties

S.NO	Material	ThermalConductivity (W/mK)	Heat Transfer Coefficient (W/m^2 K)	Specific heat (J/g °c)
1	Copper	406	500	0.385
2	Ceramics	325	300	0.210
3	Structural Steel	50.2	150	0.09
4	Gold	314	285	0.168

Convection Boundary Condition: In this boundary condition the temperature has been specified at one side and the other side is exposed to convection fluid through which heat transfer takes places into the body.Also the body may be subjected to convection on both sides if required, by maintaining the difference in temperature as in the heat transfer applied in industrial heat exchanger applications from hot fluid to cold fluid.The formula that describes this kind of boundary condition is given by (Lienhard, 2013):

$$-K\,(dt/dx) = h\Delta T \qquad (5)$$

where, K = Thermal Conductivity of material (W/m K), (dt/dx) = Temperature Gradient(K), h = Heat Transfer Coefficient in (W/m^2 K), and ΔT = Temperature difference between the surface and the fluid in K.

- **Internal Heat Generation Boundary Condition:** In this boundary condition the heat generates inside the body and it passes throughout the structure. This type of boundary conditions can be seen in the industrial sectors where large amount of heat is generated from inside the body which passes to the surface. For this case, the temperature has been specified at one surface andthe internal heat generated condition acts on the other (Cengel, 2003). Throughout the heat transfer, internal heat generation is kept constant in the present investigation.
- **Numerical Calculations of Heat Transfer in Tooth:** Heat transfer in tooth primarily depends upon its geometry and material properties. Due to the complex structure of tooth, experimental investigations are tough. Numerical methods software such as ANSYS offers an alternative approach to this problem. The numerical study was executed using three-dimensional steady state thermal model of tooth. The present investigation was carried out using the analysis software ANSYS. The tooth is designed by using design software CATIA using part design and the shape of the tooth is assumed as a cubewith the dimensions of 5mm x 5mm x 1.5mm. The materials that has been chosen for the present study are copper, Ceramics, Structural Steel and Gold. The details of these materials are given in Table 1. The given values such as thermal conductivity (K), Heat transfer coefficient (h) and specific heat (C_p) are taken from literature. To formulate the numerical problem, an assumption has been considered that the heat transfer inside the tooth was conductive in nature with no internal heat sources. It is assumed that the thermal conductivity of the material is same in all directions. The governing equation which describes the heat transfer in tooth in three dimensional is as follows. The below given equation is solved by using finite difference method and uses time dependent schemes such as implicit, explicit and Crank Nicholsonschemes.

$$\rho c_p \frac{\partial T}{\partial t} = K\left[\frac{\partial^2 T}{\partial x^2} + \frac{\partial^2 T}{\partial y^2} + \frac{\partial^2 T}{\partial z^2}\right] \quad (6)$$

where, ρ is the density(kg/m³), C_p is the Specific Heat (J/kg K), T is the Temperature(K), t is Time (sec), and K is Thermal Conductivity (W/m K). Boundary conditions considered for the present problem are conduction boundary condition, convective boundary condition and both side convection boundary condition. The mentioned boundary conditions are differentiated with respect to space and time respectively. For the top tooth surface, the condition imposed is second order boundary condition of heat flux i.e. given below and bottom surface is maintained at ambient condition.

$$q = -K \frac{\partial T}{\partial x} \quad (7)$$

where, q is heat flux (W/m²), K is Thermal Conductivity (W/m K) and $\frac{\partial T}{\partial x}$ is the Temperature Gradient. During the next boundary condition, a convective heat transfer has been considered. The heat flux from this condition was assumed to be,

$$q = h\,(T-T_a) \quad (8)$$

where, q is heat flux (W/m2), h is Heat Transfer Coefficient (W/m2 K), T is the Surface Temperature(K) and T_a is the Ambient Temperature(K). For both side convection boundary conditions, the convective heat flux equation is consideredas above. The heat transfer coefficient values are calculated by using standard heat transfer formulas.

RESULTS AND DISCUSSION

In the present study, single teeth section has been considered in the shape of a perfect cube with a size of (5mm X 5mm X 1.5mm). The teeth have been studied with three boundary conditions which are mentioned in the previous section and the variation in temperature has been studied in one direction i.e. XZ. In other directions, the change in temperature is symmetric and so the graphical results are plotted only for XZ direction. Numerical package ANSYS (Steady State Thermal) has been used to analyse the temperature distribution over the teeth with predefined boundary conditions. For the present study, four different types of materials were chosen, i.e. Copper, Ceramics, Gold and Structural Steel. The selection of material was donebased on a detailed study of dental materials. For this investigation, the top of the teeth surface was maintained ata temperature of 353 K and the bottom surface of tooth i.e. root was maintained at ambient temperature of 288 K for all the investigated boundary conditions.

Conduction Boundary Condition

Figures 3 to 6 shows the temperature distribution for variousmaterials in conduction mode. Figure 7 shows the variation of temperature for different materials with conduction boundary condition in Y direction. In this, the upper surface of the teeth is assigned with 353K and lower surface is assigned with a temperature of 288K.The above graph shows that the temperature distribution for copper material is more when compared to other materials because the thermal conductivity of copper is very high

Figure 3.Temperature distribution for copper Material in conduction

**For a more accurate representation see the electronic version.*

Figure 4. Temperature distribution for ceramics Material in conduction

**For a more accurate representation see the electronic version.*

Figure 5. Temperature distribution for gold Material in conduction

**For a more accurate representation see the electronic version.*

Figure 6. Temperature distribution for structural steel Material in conduction

**For a more accurate representation see the electronic version.*

Figure 7. Variation of temperature for conduction boundary condition

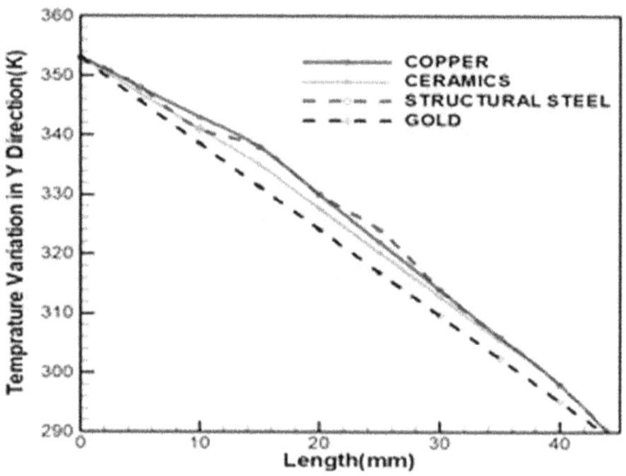

and it is a good conductor of heat.The maximum temperature 353K was observed at the top surface of the teeth i.e. upper surface and later it dropped to the ambient temperature at the fixed surface of the teeth. Ceramics and Gold gave lesser temperature values when compared to copper. The variation in temperature is linear in both the cases. Structural steel showed fluctuations in the temperature due to its material properties. From the above graphical results, it is noticed that the variation in temperature is minimum for gold. It can be taken as a better choice compared to other materials for conduction type of boundary conditions.

Convection Boundary Condition

Figures 8 to 11 shows the temperature distribution for various materials in convection mode. Figure 12 shows the variation of temperature for different materials with convection boundary condition in Y direction. In this, the upper surface of the teeth is assigned with a heat transfer coefficient at a particular temperature and lower surface is assigned with a temperature of 288K.The graph shows the temperature variation for copper and ceramics are almost the same as both had a linear variation from upper surface to teeth fixed surface i.e. lower surface and also the maximum in temperature is 334K for both the materials. The structural steel maximum temperature is also attained at upper surface but the magnitude of temperature has been reduced to 323K due to its low thermal conductivity. When compared to copper,

Figure 8. Temperature distribution for copper Material in convection

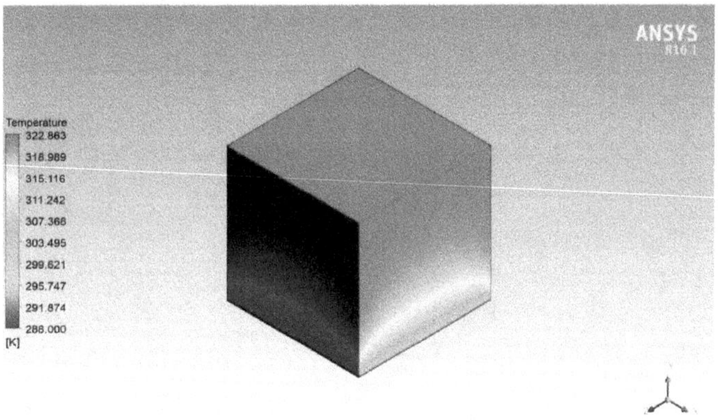

**For a more accurate representation see the electronic version.*

Figure 9. Temperature distribution for ceramics Material in convection

**For a more accurate representation see the electronic version.*

Figure 10. Temperature distribution for gold Material in convection

**For a more accurate representation see the electronic version.*

Figure 11. Temperature distribution for structural steel Material in convection

**For a more accurate representation see the electronic version.*

Figure 12. Variation of temperature for convection boundary condition.

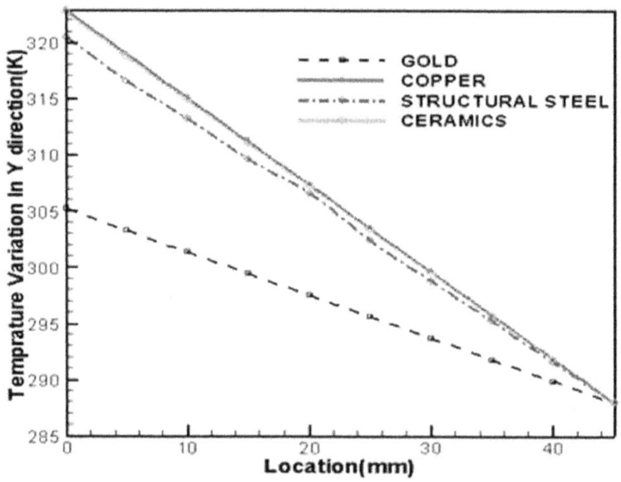

**For a more accurate representation see the electronic version.*

Figure 13. Temperature distribution for structural steel Material in both side convection

**For a more accurate representation see the electronic version.*

Figure 14. Temperature distribution for structural steel Material in both side convection

*For a more accurate representation see the electronic version.

Figure 15. Temperature distribution for structural steel Material in both side convection

*For a more accurate representation see the electronic version.

Figure 16. Temperature distribution for structural steel Material in both side convection

*For a more accurate representation see the electronic version.

ceramics and structural steel, gold has the less magnitudes of temperature variation because of its low conductivity of heat, i.e. the amount of heat transfer from upper surface to lower surface with convection type of boundary condition is very poor and thermal conductivity is also very less compared to cooper.

Both Side Convection

Figures 13 to 16 shows the temperature distribution for differentmaterials in both side convection mode. Figure 17 shows the variation of temperature for different materials with both side convection boundary condition in Y direction. In this the upper surface and lower surface of the teeth is assigned with a heat transfer coefficient, maintaining a difference in temperature between upper surface and lower surfaces

Figure 17. Variation of temperature for both side convection boundary condition

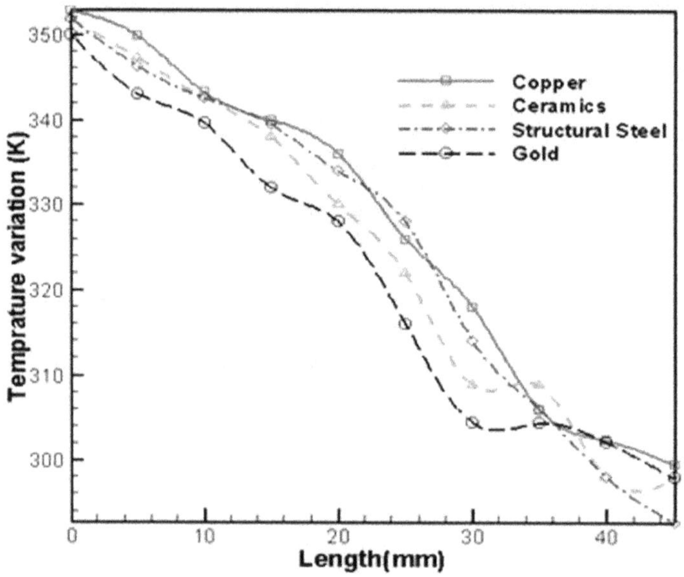

of teeth. When the teeth are exposed to both side convection, the temperature is more for copper because of its higher value of thermal conductivity as compared to other materials. For gold and ceramics, the temperature value has been limited to 306K and 310K from length 30mm to 45mm because the maximum amount of heat transfer has already taken place due to convection. Later it drops to a lower surface temperature. At this specified boundary condition, gold material has low heat transfer rate compared to the other materials.

Mathematical modelling is an important technique for analyzing complex medical scenarios (Krishnamurthy, 2016; Kumar, Anantharaj, Sakthioli, Vinnakota, & Krishnamurthy, 2016;). In recent years, the tools derived from thermodynamics such as energy and entropy have been widely used for solving biomedical problems, even bio signal processing and image processing (Alagumariappan, Rajagopal & Krishnamurthy, 2016; Alagumariappan, Krishnamurthy, Kandiah & Ponnuswamy, 2017; Alagumariappan & Krishnamurthy, 2018; Alagumariappan & Krishnamurthy, 2017; Ambikapathy & Krishnamurthy, 2018; Rajinikanth, Raja & Kamalanand, 2017; Manic, Priya & Rajinikanth, 2016; Lakshmi, Tebby, Shriranjani & Rajinikanth, 2016; Rajinikanth, Satapathy, Fernandes & Nachiappan, 2017; Rajinikanth & Satapathy, 2018; Kamalanand, & Ramakrishnan, 2015). Further, the numerical analysis techniques such as the finite element methods have been utilized for the development of biosensors for medical diagnostics (Arunachalam, Jacob, & Kamalanand, 2014; Krishnamurthy, & Jacob, 2014), and also for biomedical analysis (Kamalanand, & Srinivasan, 2011a,b; Kamalanand, Srinivasan, & Ramakrishnan, 2011).

From the present numerical investigation of steady state heat transfer phenomenon, it is observed that human tooth exhibits high heat transfer rate for the materials having higher thermal conductivity i.e. copper. From the investigation, it is found that the temperature is maximum at the top of the teeth and minimum at bottom of the teeth i.e. root. From the convection boundary condition, the chosen materials are limited to a particular temperature at a location and from there onwards there is no heat transfer. The contours of the heat transfer analysis for various teeth materials shows that the layers of the teeth are subjected to varying temperature. From the computational analysis, it is understandable that copper material shows better transfer rate when compared to other materials such as ceramics, structural steel and gold. Material selection of teeth is also important besides the shape and boundary conditions of teeth. Computational analysis reduces the time as compared to the experimental analysis.

REFERENCES

Alagumariappan, P., & Krishnamurthy, K. (2017). Analysis of Normal and Abnormal Electrogastrograms using Teager-Kaiser Energy. *Proceedings of the 2nd World Research Journals Congress*, 134-136.

Alagumariappan, P., & Krishnamurthy, K. (2018). An Approach Based on Information Theory for Selection of Systems for Efficient Recording of Electrogastrograms. In *Proceedings of the International Conference on Computing and Communication Systems* (pp. 463-471). Springer. 10.1007/978-981-10-6890-4_45

Alagumariappan, P., Krishnamurthy, K., Kandiah, S., & Ponnuswamy, M. J. (2017). Effect of electrode contact area on the information content of the recorded electrogastrograms: An analysis based on Rényi entropy and Teager-Kaiser Energy. *Polish Journal of Medical Physics and Engineering*, *23*(2), 37–42. doi:10.1515/pjmpe-2017-0007

Alagumariappan, P., Rajagopal, A., & Krishnamurthy, K. (2016). Complexity Analysis on Normal and Abnormal Electrogastrograms Using Tsallis Entropy. In *3rd International Electronic and Flipped Conference on Entropy and Its Applications*. Multidisciplinary Digital Publishing Institute. 10.3390/ecea-3-A003

Ambikapathy, B., & Krishnamurthy, K. (2018). Analysis of electromyograms recorded using invasive and noninvasive electrodes: A study based on entropy and Lyapunov exponents estimated using artificial neural networks. *Journal of Ambient Intelligence and Humanized Computing*, 1–9.

Ana, P. A., Velloso, W. F. Jr, & Zezell, D. M. (2008). Three-dimensional finite element thermal analysis of dental tissues irradiated with Er, Cr: YSGG laser. *The Review of Scientific Instruments*, *79*(9), 093910. doi:10.1063/1.2953526 PMID:19044431

Anusavice, K. J., Shen, C., & Rawls, H. R. (2013). *Phillips' science of dental materials*. Elsevier Health Sciences.

Apparsamy, S., & Krishnamurthy, K. (2018). Design and Development of Capacitance Sensor Array for Analyzing the Non-Homogeneity in Biological Materials: A Smart Sensing Approach. In Expert System Techniques in Biomedical Science Practice (pp. 70-96). IGI Global. doi:10.4018/978-1-5225-5149-2.ch004

Arenholt-Bindslev, D. (1992). Dental amalgam—environmental aspects. *Advances in Dental Research*, *6*(1), 125–130. doi:10.1177/08959374920060010501 PMID:1292452

Arunachalam, K., Jacob, L. V., & Kamalanand, K. (2014). Design and analysis of finite element based sensors for diagnosis of liver disorders using biocompatible metals. *Technology and Health Care*, *22*(6), 867–875. PMID:25391529

Bergman, T. L., & Incropera, F. P. (2011). *Introduction to heat transfer*. John Wiley & Sons.

Bharti, R., Wadhwani, K. K., Tikku, A. P., & Chandra, A. (2010). Dental amalgam: An update. *Journal of conservative dentistry. JCD, 13*(4), 204. PMID:21217947

Blum, I. R., Lynch, C. D., & Wilson, N. H. (2014). Factors influencing repair of dental restorations with resin composite. *Clinical, Cosmetic and Investigational Dentistry, 6*, 81. doi:10.2147/CCIDE.S53461 PMID:25378952

Braden, M., & Clarke, R. L. (1984). Water absorption characteristics of dental microfine composite filling materials: I. Proprietary materials. *Biomaterials, 5*(6), 369–372. doi:10.1016/0142-9612(84)90038-3 PMID:6525397

Brown, W. S., Dewey, W. A., & Jacobs, H. R. (1970). Thermal properties of teeth. *Journal of Dental Research, 49*(4), 752–755. doi:10.1177/00220345700490040701 PMID:5269374

Brunthaler, A., König, F., Lucas, T., Sperr, W., & Schedle, A. (2003). Longevity of direct resin composite restorations in posterior teeth: A review. *Clinical Oral Investigations, 7*(2), 63–70. doi:10.100700784-003-0206-7 PMID:12768463

Craig, R. G., & Powers, J. M. (Eds.). (1989). Restorative dental materials. Academic Press.

Croll, T. P., & Nicholson, J. W. (2002). Glass ionomer cements in pediatric dentistry: Review of the literature. *Pediatric Dentistry, 24*(5), 423–429. PMID:12412956

Cummings, K. M., Rolf, J. C., Rosenflanz, A. Z., Rusin, R. P., & Swanson, J. E. (2006). *U.S. Patent No. 6,984,261*. Washington, DC: U.S. Patent and Trademark Office.

Davis, S., Gluskin, A. H., Livingood, P. M., & Chambers, D. W. (2010). Analysis of temperature rise and the use of coolants in the dissipation of ultrasonic heat buildup during post removal. *Journal of Endodontics, 36*(11), 1892–1896. doi:10.1016/j.joen.2010.08.027 PMID:20951308

Denry, I., & Holloway, J. A. (2010). Ceramics for dental applications: A review. *Materials (Basel), 3*(1), 351–368. doi:10.3390/ma3010351

Douglas, R. D. (2000). Color stability of new-generation indirect resins for prosthodontic application. *The Journal of Prosthetic Dentistry, 83*(2), 166–170. doi:10.1016/S0022-3913(00)80008-6 PMID:10668028

Ferracane, J. L. (1995). Current trends in dental composites. *Critical Reviews in Oral Biology and Medicine, 6*(4), 302–318. doi:10.1177/10454411950060040301 PMID:8664421

Ferracane, J. L. (2011). Resin composite—state of the art. *Dental Materials, 27*(1), 29–38. doi:10.1016/j.dental.2010.10.020 PMID:21093034

Hannig, M., & Bott, B. (1999). In-vitro pulp chamber temperature rise during composite resin polymerization with various light-curing sources. *Dental Materials, 15*(4), 275–281. doi:10.1016/S0109-5641(99)00047-0 PMID:10551096

Hanson, M., & Pleva, J. (1991). The dental amalgam issue. A review. *Experientia, 47*(1), 9–22. doi:10.1007/BF02041243 PMID:1999251

Hewitt, G. F., Shires, G. L., & Bott, T. R. (1994). *Process heat transfer* (Vol. 113). Boca Raton, FL: CRC Press.

Jakubinek, M. B., O'Neill, C., Felix, C., Price, R. B., & White, M. A. (2008). Temperature excursions at the pulp–dentin junction during the curing of light-activated dental restorations. *Dental Materials*, *24*(11), 1468–1476. doi:10.1016/j.dental.2008.03.012 PMID:18448161

Jandt, K. D., & Sigusch, B. W. (2009). Future perspectives of resin-based dental materials. *Dental Materials*, *25*(8), 1001–1006. doi:10.1016/j.dental.2009.02.009 PMID:19332352

Johnson, B. R. (1999). Considerations in the selection of a root-end filling material. *Oral Surgery, Oral Medicine, Oral Pathology, Oral Radiology, and Endodontics*, *87*(4), 398–404. doi:10.1016/S1079-2104(99)70237-4 PMID:10225620

Kamalanand, K., & Ramakrishnan, S. (2015). Effect of gadolinium concentration on segmentation of vasculature in cardiopulmonary magnetic resonance angiograms. *Journal of Medical Imaging and Health Informatics*, *5*(1), 147–151. doi:10.1166/jmihi.2015.1370

Kamalanand, K., Sridhar, B. T. N., Rajeshwari, P. M., & Ramakrishnan, S. (2010). Correlation of Dielectric Permittivity with Mechanical Properties in Soft Tissue-Mimicking Polyacrylamide Phantoms. *Journal of Mechanics in Medicine and Biology*, *10*(02), 353–360. doi:10.1142/S0219519410003411

Kamalanand, K., & Srinivasan, S. (2011a). Modeling of Normal and Atherosclerotic Blood Vessels using Finite Element Methods and Artificial Neural Networks. *World Academy of Science, Engineering and Technology*, *60*, 1314.

Kamalanand, K., & Srinivasan, S. (2011b). Modelling and analysis of normal and atherosclerotic blood vessel mechanics using 3D finite element models. *ICTACT Journal on Soft Computing: Special Issue on Fuzzy in Industrial and Process Automation*, *2*(1), 261–264. doi:10.21917/ijsc.2011.0040

Kamalanand, K., Srinivasan, S., & Ramakrishnan, S. (2011). Analysis of Normal and Atherosclerotic Blood Vessels Using 2D Finite Element Models. In *5th Kuala Lumpur International Conference on Biomedical Engineering 2011* (pp. 411-414). Springer. 10.1007/978-3-642-21729-6_105

Kandlikar, S., Garimella, S., Li, D., Colin, S., & King, M. R. (2005). *Heat transfer and fluid flow in minichannels and microchannels*. Elsevier.

Kirsch, J., Tchorz, J., Hellwig, E., Tauböck, T. T., Attin, T., & Hannig, C. (2016). Decision criteria for replacement of fillings: A retrospective study. *Clinical and Experimental Dental Research*, *2*(2), 121–128. doi:10.1002/cre2.30 PMID:29744158

Krishnamurthy, K. (2016). Parameter Estimation of Nonlinear Biomedical Systems Using Extended Kalman Filter Algorithm: Development of Patient Specific Models. *Computational Tools and Techniques for Biomedical Signal Processing*, 76.

Krishnamurthy, K., & Jacob, L. (2014). Finite element based design of a capacitive sensor for diagnosis of cirrhotic and malignant liver. *Journal of Clinical and Experimental Hepatology*, *4*, S65. doi:10.1016/j.jceh.2014.02.126

Krishnamurthy, K., Sridhar, B. T. N., Rajeshwari, P. M., & Swaminathan, R. (2009). Correlation of Electrical Impedance with Mechanical Properties in Models of Tissue Mimicking Phantoms. In *13th International Conference on Biomedical Engineering* (pp. 1708-1711). Springer. 10.1007/978-3-540-92841-6_424

Kumar, V. S., Anantharaj, U. J., Sakthioli, M., Vinnakota, R., & Krishnamurthy, K. (2016). Mathematical modelling of the effects of prebiotic concentration on lactobacillus casei growth. *International Journal of Infectious Diseases*, *45*, 205. doi:10.1016/j.ijid.2016.02.470

Lakshmi, V. S., Tebby, S. G., Shriranjani, D., & Rajinikanth, V. (2016). Chaotic cuckoo search and Kapur/Tsallis approach in segmentation of T. cruzi from blood smear images. *International Journal of Computer Science and Information Security*, *14*, 51–56.

Lauer, H. C., Kraft, E., Rothlauf, W., & Zwingers, T. (1990). Effects of the temperature of cooling water during high-speed and ultrahigh-speed tooth preparation. *The Journal of Prosthetic Dentistry*, *63*(4), 407–414. doi:10.1016/0022-3913(90)90228-5 PMID:2184223

Liang, S., Sa, Y., Jiang, T., Ma, X., Xing, W., Wang, Z., & Wang, Y. (2013). In vitro evaluation of halogen light-activated vs chemically activated in-office bleaching systems. *Acta Odontologica Scandinavica*, *71*(5), 1149–1155. doi:10.3109/00016357.2012.757355 PMID:23294115

Liang, S., Sa, Y., Sun, L., Ma, X., Wang, Z., Xing, W., ... Wang, Y. (2012). Effect of halogen light irradiation on hydrogen peroxide bleaching: An in vitro study. *Australian Dental Journal*, *57*(3), 277–283. doi:10.1111/j.1834-7819.2012.01702.x PMID:22924349

Lienhard, J. H. (2013). *A heat transfer textbook*. Courier Corporation.

Lin, M., Liu, S., Niu, L., Xu, F., & Lu, T. J. (2011). Analysis of thermal-induced dentinal fluid flow and its implications in dental thermal pain. *Archives of Oral Biology*, *56*(9), 846–854. doi:10.1016/j.archoralbio.2011.02.011 PMID:21411060

Lin, M., Xu, F., Lu, T. J., &Bai, B. F. (2010). A review of heat transfer in human tooth—experimental characterization and mathematical modeling. *Dental Materials, 26*(6), 501-513.

Linsuwanont, P., Palamara, J. E. A., & Messer, H. H. (2007). An investigation of thermal stimulation in intact teeth. *Archives of Oral Biology*, *52*(3), 218–227. doi:10.1016/j.archoralbio.2006.10.009 PMID:17109811

Linsuwanont, P., Versluis, A., Palamara, J. E., & Messer, H. H. (2008). Thermal stimulation causes tooth deformation: A possible alternative to the hydrodynamic theory? *Archives of Oral Biology*, *53*(3), 261–272. doi:10.1016/j.archoralbio.2007.10.006 PMID:18037388

Lipski, M., Woźniak, K., Lichota, D., & Nowicka, A. (2011). Root surface temperature rise of mandibular first molar during root canal filling with high-temperature thermoplasticized Gutta-Percha in the dog. *Polish Journal of Veterinary Sciences*, *14*(4), 591–595. doi:10.2478/v10181-011-0088-6 PMID:22439330

Lohbauer, U. (2009). Dental glass ionomer cements as permanent filling materials? – Properties, limitations and future trends. *Materials (Basel)*, *3*(1), 76–96. doi:10.3390/ma3010076

Manic, K. S., Priya, R. K., & Rajinikanth, V. (2016). Image multithresholding based on Kapur/Tsallis entropy and firefly algorithm. *Indian Journal of Science and Technology*, *9*(12). doi:10.17485/ijst/2016/v9i12/89949

Moorthy, A., Hogg, C. H., Dowling, A. H., Grufferty, B. F., Benetti, A. R., & Fleming, G. J. P. (2012). Cuspal deflection and microleakage in premolar teeth restored with bulk-fill flowable resin-based composite base materials. *Journal of Dentistry*, *40*(6), 500–505. doi:10.1016/j.jdent.2012.02.015 PMID:22390980

Onisor, I., Asmussen, E., & Krejci, I. (2011). Temperature rise during photo-polymerization for onlay luting. *American Journal of Dentistry*, *24*(4), 250. PMID:22016921

Oskui, I. Z., Ashtiani, M. N., Hashemi, A., & Jafarzadeh, H. (2013). Thermal analysis of the intact mandibular premolar: A finite element analysis. *International Endodontic Journal*, *46*(9), 841–846. doi:10.1111/iej.12069 PMID:23480124

Peutzfeldt, A. (1997). Resin composites in dentistry: The monomer systems. *European Journal of Oral Sciences*, *105*(2), 97–116. doi:10.1111/j.1600-0722.1997.tb00188.x PMID:9151062

Preiskorn, M., Zmuda, S., Trykowski, J., Panas, A., & Preiskorn, M. (2003). In vitro investigations of the heat transfer phenomena in human tooth. *Acta of Bioengineering and Biomechanics*, *5*(2), 23–36.

Rajinikanth, V., Raja, N. S. M., & Kamalanand, K. (2017). Firefly Algorithm Assisted Segmentation of Tumor from Brain MRI using Tsallis Function and Markov Random Field. *Journal of Control Engineering and Applied Informatics*, *19*(3), 97–106.

Rajinikanth, V., & Satapathy, S. C. (2018). Segmentation of Ischemic Stroke Lesion in Brain MRI Based on Social Group Optimization and Fuzzy-Tsallis Entropy. *Arabian Journal for Science and Engineering*, 1–14.

Rajinikanth, V., Satapathy, S. C., Fernandes, S. L., & Nachiappan, S. (2017). Entropy based segmentation of tumor from brain MR images–a study with teaching learning based optimization. *Pattern Recognition Letters*, *94*, 87–95. doi:10.1016/j.patrec.2017.05.028

Roberts, H. W., Berzins, D. W., Moore, B. K., & Charlton, D. G. (2009). Metal-Ceramic Alloys in Dentistry: A Review. *Journal of Prosthodontics*, *18*(2), 188–194. doi:10.1111/j.1532-849X.2008.00377.x PMID:19178620

Rohsenow, W. M., & Cho, Y. I. (1998). *Handbook of heat transfer* (J. P. Hartnett, Ed.; Vol. 3). New York: McGraw-Hill.

Saghlatoon, H., Soleimani, M., Moghimi, S., & Talebi, M. (2012, May). An experimental investigation about the heat transfer phenomenon in human teeth. In *Electrical Engineering (ICEE), 2012 20th Iranian Conference on* (pp. 1598-1601). IEEE. 10.1109/IranianCEE.2012.6292616

Sahana, A. S., Paramasivam, A., & Kamalanand, K. (2017, March). Experimental investigations on capacitive imaging of biological materials. In *Biosignals, Images and Instrumentation (ICBSII), 2017 Third International Conference on* (pp. 1-3). IEEE. 10.1109/ICBSII.2017.8082289

Secilmis, A., Bulbul, M., Sari, T., & Usumez, A. (2013). Effects of different dentin thicknesses and air cooling on pulpal temperature rise during laser welding. *Lasers in Medical Science*, *28*(1), 167–170. doi:10.100710103-012-1108-1 PMID:22562450

Thomas, G. P. (2013). *Amalgam-chemical composition, mechanical properties, and common applications*. Academic Press.

Tobias, R. S. (1988). Antibacterial properties of dental restorative materials: A review. *International Endodontic Journal*, *21*(2), 155–160. doi:10.1111/j.1365-2591.1988.tb00969.x PMID:3151895

Torabinejad, M., & White, D. J. (1995). *U.S. Patent No. 5,415,547*. Washington, DC: U.S. Patent and Trademark Office.

Van Meerbeek, B., Vanherle, G., Lesaffre, E., Braem, M., & Lambrechts, P. (1991). Trends in the selection of dental filling materials. *Journal of Dentistry*, *19*(4), 207–213. doi:10.1016/0300-5712(91)90118-I PMID:1787208

Wataha, J. C. (2000). Biocompatibility of dental casting alloys: A review. *The Journal of Prosthetic Dentistry*, *83*(2), 223–234. doi:10.1016/S0022-3913(00)80016-5 PMID:10668036

Wilkins, E. M., & McCullough, P. A. (1989). *Clinical practice of the dental hygienist* (Vol. 235). Lea &Febiger.

Wilson, A. D., & Kent, B. E. (1971). The glass-ionomer cement, a new translucent dental filling material. *Journal of Chemical Technology and Biotechnology (Oxford, Oxfordshire)*, *21*(11), 313–313.

Yamauchi, J., Masuhara, E., Nakabayashi, N., Shibatani, K., & Wada, T. (1981). *U.S. Patent No. 4,259,117*. Washington, DC: U.S. Patent and Trademark Office.

Yengopal, V., Harneker, S. Y., Patel, N., & Siegfried, N. (2009). Dental fillings for the treatment of caries in the primary dentition. *Cochrane Database of Systematic Reviews*, 2. PMID:19370602

KEY TERMS AND DEFINITIONS

Anisotropy: Changes of thermal conductivity in all directions.

CATIA: Computer-aided three-dimensional interactive application.

Heat Flux (q): Flow of heat energy per unit of area per unit of time.

Specific Heat (C_p): It is the amount of heat per unit mass is required to rise the temperature by one degree Celsius.

Thermal Conductivity (K): Property of material to conduct heat.

Tooth Cavity: Caused by a combination of factors including bacteria in the mouth.

Chapter 10
Recent Trends in 3D Printing of Dental Models:
Rapid Prototyping in Dental Implants

Kayalvizhi Mohan
Agni College of Technology, India

ABSTRACT

This chapter introduces the recent trend in 3D printing (3DP) in dentistry. The advantage and disadvantages of 3DP are discussed. It elaborates on different types of 3DP techniques involved and their significance. The chapter further discuss about the biomaterial used. It also describes the complete steps involved in 3DP such as image acquisition, modeling, segmentation, and printing techniques. The merits and demerits of the different methodologies pertaining to steps involved in 3DP are illustrated. Rapid prototyping in dental implants is discussed in detail. It ends with review of a case study in implementing the technique.

INTRODUCTION

This chapter reviews the incipient role of 3D printing in dentistry, a topic of growing relevancy (Schubert, Van Langeveld, & Donoso, 2014). Further this chapter gives an insight in bioprinting that aids medical community to design structurally complex 3D constructs by comprehensive positioning and aiding complex patterns that are formed by different structures of cells, biomaterials and bioactive molecules within a single make. With the evolution of several bioprinting strategies during recent period, 3D bioprinting is applied in several research areas that include dental implant (Klein, Lu & Wang, 2013).

Bioprinting has emerged over the past decade as a prominent technology in the field of dentistry (Banks, 2013). Implementation of 3D printer helps in reducing the cost, it enhances the speed of production and is easy method for implementation in dental rehabilitation. It supports the surgeons in planning for surgery in resection and reconstruction. It represents as an exceptional aid for the academicians to train their students (Bertassoni, Cecconi, Manoharan, Nikkhah, Hjortnaes, Cristino, & Khademhosseini, 2014). Medical Rapid Prototype based pre-bended reconstruction helps in decreasing the resection dura-

DOI: 10.4018/978-1-5225-6243-6.ch010

tion, rate and anaesthesia menace bared by the subjects (Mertz, 2013). The need for reconstruction for dental rehabilitation is continuous challenge.

The very purpose of reconstruction is the repair of appropriate aesthetics and regularty of facet. Reconstruction also helps in stabilising arrangements and the strengthening the jaw (Salgueiro, & Stevens, 2010). The defects in ideal model in reconstruction is obtained by using titanium plate during resection. For mandibular reconstruction, is not easy to shape the titanium plates and the whole procedure is time consuming (Cohen, Laviv, Berman, Nashef, & Abu-Tair, 2009). The intraoperative prearrangement leads to extensive surgical operation and upsurges the budget (Yovchev, Stanimirov, & Mihaylova, 2014). The appropriate alteration of the plate to match the skeletal structure is indispensable to get a positive outcome.

BACKGROUND

The dental prosthesis fabrication includes many of manual laboratory processes and the quality of restorations depends on the dentists and dental technician's skill and expertise. The dental restoration is also a very individual and complex process. Computer aided manufacturing (CAM) is methodology is introduced to overcome the disadvantage in manual process. Introducing CAM for the processing procedure will highly rise the manufactured quality of dental restorations. The manufacturing process includes: data acquisition, data processing and model fabrication (Nayar, Bhuminathan, & Bhat, 2015).

3D Printing Intial Steps

The first step often used towards 3d printing involves, Computer Tomography scan with field of view within twenty cm, the slice thickness usually selected is slightly more than one cm (equal to scan spacing) and the gantry tilt of zero degree is set. During the whole procedure the patients are not allowed to move and the occlusal planes are set parallel to the gantry. The whole arrangement helps in reducing the artefacts in the anatomic model. The acquired images are stored in DICOM file format.

The open source 3D imaging software is used to process CT DICOM images. The commonly used software version is Vesalius 3.0.0 version. The bone mask is created using the tool for mask creation in Hounsfield unit range. The 3D surfaces are crated using the bone mask as reference. Finally, the data is transferred in stereolithography (STL) file. The process takes few minutes depending the CPU configuration.

The Autodesk Meshmixer 2.9.1 version is used for editing STL file. The selection/analysis tools are used to isolate and repair the mandible. Further the compatible model for printing is produced by transforming the complex geometry of the bone. Finally mesh of the mandible is prepared for the slicing using open source Matter Control and printing is performed using the ROBO 3D R1 with 1.75mm polylactid acid filament.

Imaging Modalities

The first step important step towards 3D printing is imaging. The imaging techniques such as CT, MRI, digital photography is commonly used.

The objectives of diagnostic imaging depend on

- The extent and nature of data required (combinations of conventional dental images)
    - Facility to envisage the implant site in the mesiodistal, buccolingual, and superioinferior dimensions
    - Facility to allow consistent, precise estimation
    - Normal anatomical structures (incisive canal, nasal floor, maxillary sinus, mandibular canal)
    - Capability to estimate trabecular bone density and cortical thickness (height, width)
- The imaging should have least radiation risk to the patient
- It should include appropriate clinical examination and patient's need.
- The reasonable access and cost-effective

Dental implant is a promptly intensifying area of dentistry. The imaging modalities range from 2D projections to complex 3D imaging. The 2D modalities are voluntarily available, less expensive with very less radiation exposure, but have restrictions like exaggerations and overlapping of planes and so clinician are not able to develop a 3D perspective of patient's anatomy with a single image. However, in complex cases, more extensive and advanced radiographic evaluation is needed. Hence, cross-sectional capturing is more and more considered crucial for optimal implant placement, important roles in pre- and postoperative evaluation of the implant patient especially in complex reconstructions and multiple implants.

The feat of endodontic method mostly depends on the detection of all root canals so that are opened, cleaned, shaped, and kept immovable (Vertucci, 1984). Conventional radiographic method is able to identify only 55% of these structures (Ramamurthy, Scheetz, Clark, & Farman, 2006). The occurrence of a second mesiobuccal canal in maxillary first molars is testified to fluctuate from 69% to 93% subject to the study method involved. This inconsistency happens in the buccolingual plane where overlapping of anatomic arrangements hampers the identification of small structural density variations (Pineada,1973, Nance, Tyndall, Levin, & Trope, 2000).

Ramamurthy et al. (2006) established that evaluators using different 2D film modalities were seldom capable of detecting more than 50% presence of MBC canals. They also established the dissimilarities in perception rates with complementary metal oxide semiconductors (CMOSs), analog film, and photostimulable phosphor plates (PSP) detecting 55%, 44%, and 39% of MBC canals, respectively. Matherne et al. (2008) detected the number of root canals on intraoral digital (both charged-couple device and photostimulable phosphor) plate images with CBCT in 72 extracted teeth and related the capability of three board-certified endodontists. It was established that on average the contributors did not detect even one root canal in 40% of teeth using intraoral radiographs.

CBCT assessments recognized an average of 56% root canals (RCS) per maxillary molar, 19% per mandibular premolar, and 24% per mandibular incisor (Gepreel, & Niinomi, 2005). Estimation of CCD images revealed an average number of 58% per maxillary molar, 19% RCS per mandibular incisor and 19% per mandibular first premolar. Assessment of PSP images signified an average number of 23% RCS per mandibular incisor, 18% per mandibular first premolar, and 54% per maxillary molar. In Taiwanese individuals the high occurrence of the distolingual canal was described using CBCT imaging (Alamri, Sadrameli, Alshalhoob, & Alshehri,2012; Alamri, Sadrameli, Alshalhoob, & Alshehri,2009), it also was helpful in determination of root curvature (Estreala, Bueno, Sousa-Neto, & Pécora, 2008) and acme glitches in the root canal system of mandibular premolars (Cleghorn, Christie, & & Fu-yuan,, 2008). Baratto Filho et al. (2009) studied the internal change in shapes of the extracted maxillary first molars by matching finding rates attained by means of an operating microscope and CBCT to ex vivo

Table 1. Modalities in dental implants

	Common Imaging Types	Advantage	Disadvantage
1	**Periapical radiography (Bagchi,& Joshi, 2012)**	• Long cone paralleling technique. • Best resolution (line pairs/mm). • Area of interest are inspected for types of trabecular patterns, the changes in residual root and periodontiumn and how the angulation of adjacent teeth is formed. • It is readily available, inexpensive and involves less radiation dose	• 2D perspective of 3D anatomy are inadequate for inspecting the amount of available bone (facio-lingual dimension) in the edentulous area. • The valuation is inadequate due to the limited size of mandibular, large edentulous areas and maxillary structures • The maxillary and mandibular structures face image distortion due to inadequate size and anatomic limitations in image receptor inflexibility
2	**Digital intraoral radiographic images (Lingeshwar, Dhanasekar & Aparna, 2010)**	• This gives less exposure compare to conventional periapical radiography, instant results, contrast is manipulated.	• But its limitation is, it will provide the dentist the limited area for imaging, and no facio-lingual dimension is assessed.
3	**Bitewing (Siu, Chu, Li, Chow, Deng, 2010, Kalra, Jain, Deoghare & Lambade, 2010)**	• This modality helps in the assessment of the coronal view of tooth and in relation to the adjacent edentulous space of alveolar bone	• Periapical area cannot be assessed so approximation with adjacent vital structures is a major limitation
4	**Occlusal radiograph (DelBalso, Greiner & Licata, 1994, Siu, Chu, Li, Chow, Deng, 2010, Kalra, Jain, Deoghare & Lambade, 2010)**	• Provide us with facio-lingual dimension technique is readily available, high image definition, inexpensive, less radiation exposure	• Due to distortion, it records only the widest portion of mandible, with limited reproducibility and makes it unusable in maxillary arch capturing
5	**Panoramic radiograph (Siu, Chu, Li, Chow, Deng, 2010, Kalra, Jain, Deoghare & Lambade, 2010)**	• It is commonly used for diagnosis of gross pathoses within the jaws in the implant site • It helps in assessing the anatomic structures such as foramen, sinuses, canals, fossa, and foramen • This helps in better visualization of bony anatomy clearly	• The main disadvantage is the sharpness and resolution is very much reduced resolution and does not give cross-sectional imaging • The magnification is not stable and unpredictable specifically in the vertical aspect and is extra prominent in posterior than in anterior regions. Thus, can contribute a wrong sense that extra bone subsists between the crest of the alveolar process and the inferior alveolar canal, nasal fossa or maxillary sinuses. • Incorrect subject placing can additionally contribute to image misrepresentation. Even appropriately placed and exposed panoramic radiographs are not used for direct bony measurements.
6	**Lateral cephalogram (Siu, Chu, Li, Chow, Deng, 2010, Kalra, Jain, Deoghare & Lambade, 2010)**	• Along with soft tissue profile of the patient, axial tooth inclination and dentoalveolar ridge relationship in the midline can be assessed.	• Provide dentist with cross-sectional dimension only at midline, image magnification and also contralateral side superimposition evident.
7	**Dentascan imaging (Siu, Chu, Li, Chow, Deng, 2010, Kalra, Jain, Deoghare & Lambade, 2010, Matherne, Angelopoulos, Kuliid, & Tira, 2008)**	• This imaging modality provides with programmed formation, recollection, and display of the image, mandibular arch, the maxillary arch and cross sectional and at a tangent panoramic image of the alveolus that enables accurate preprosthetic treatment planning. • Diagnostic template is necessary for determining the Location and alignment of dental implants.	• Magnification of images, grayscale values, is limited, and the inclination of the patient's head all through the check-up is necessary.
8	**Interactive CT (ICT) (Tyndall & Rathore, 2008)**	• It overcomes limitations of CT, also enables radiologist to transfer the image to the computer, so clinician can view • Hence, 3D treatment is planned according to patient's anatomy and are seen before surgery thus helping in determining the number and size of implants correctly according to density of bone at suggested sites. ICT is more relevant diagnostic technique when compared to conventional techniques for implant planning.	• Determining the Exact position of implants is difficult, tedious and time consuming
9	**MRI (Manisundar, Saravanakumar, Hemalatha, Manigandan, & Amudhan, 2014)**	• In MRI, cortical bone is clearly delineated from the cancellous bone. • Vital structures such as nerves, vessels, and the floor and mucosa of the maxillary sinus are readily identified by the implant surgeon, exceptionally safe due to the absence of ionizing radiation, high image quality.	• High capital and running costs with lack of availability have a significant barrier to the use of MRI for implant assessment
10	**CBCT and multidetector CT (Alamri, Sadrameli, Alshalhoob, & Alshehri, 2012; Kumar & Satheesh, 2013, Arai, Tammisalo, Iwai, Hashimoto, & Shinoda, 1999)**	• It allows the surgeon to have an accurate information of implantsite, in edentulous patient or when multiple implants are considered thus aiding in diagnosis and provides the dentist with controlled surgical plan, is used when conventional radiograph fails to provide the needed information. • Evaluation are done in all three surfaces of space without image misrepresentation, overlapping of structures, and differential enlargement of the image based on geometry, also accurate position of anatomic markers as well as the altitude, thickness, inclination, and quality of alveolar bone for implant site. • Also, for case selection and a postsurgical evaluation to assess implant's position in the alveolus.	• Increased susceptibility to movement artefacts and metallic restoration often leads to streak artefact, in CBCT no standard grayscale system present, so universality is questionable

sections. It was stated that an ex vivo incidence of a fourth canal in 67.14% of teeth and additional root canals in 92.85% of mesiobuccal roots. CBCT showed lowest overall (37.05%) detection rate. Clinical valuation suggested slightly lower overall (53.26%) but higher (95.63%) MBC detection rates. It was also indicated that CBCT delivered a worthy method for the initial assessment of maxillary first molar internal morphology but that the practice of operating microscopes was optimal. Once the images are captured, they have to be processed.

IMAGE PROCESSING TECHNIQUES

During the past few decades numerous image-processing techniques have been proposed to delineate tooth images. Interpolation method was used to obtain an N dimensional object from a set of (N $_{2-1}$) dimensional objects by Bors et al (2002). They also applied their proposed method for reconstruction of the three-dimensional tooth shape with 2D CT images, but the tooth regions were segmented manually. Edge detection was proposed by Mol and van der Stelt (1991) in his work to detect the modification in bone to the morbid state, from an X-ray image. Gao et al (2006) reconstructed three-dimensional shapes using shadow speckles. Region growing method were applied for extracting the region of all the teeth by & Fu-yuan, -ri and Fu-yuan (2010).

A semiautomatic method was proposed by Omachi et al (2007) for extracting the three-dimensional shapes of teeth from dental CT images. The coutour method was really challenging because the adjacent teeth or the surrounding tissues is also extracted (Jayadevappa, Kodhandarama, Santosh, & Rashid 2010). A three-dimensional statistical model by Buchaillard et al (2007) was proposed with the objective of obtaining a good estimate of the entire tooth when only partial information of a tooth is provided. A method for segmenting and envisioning all of the teeth with CT images was recommended by Akhoondali et al. (2009). Hirogaki et al. (2001) projected a technique for rebuilding a three-dimensional dental cast model with a line laser. Carter et al. (2010) analyzed tooth reconstruction techniques. Hoy et al. (2013) suggested improved methods for extracting the contours of teeth in dental radiographs. The region growing method is used as an interactive tool for segmentation and visualization of the teeth in CT scans.

Commonly Used Image Processing Techniques

The commonly used step in reconstruction of three-dimensional shape from the CT image is region growing method (Grenier, Revol-Muller, Costes, Janier, M & Gimenez, 2005). Normalization, contrast enhancement and smoothing are introduced as pre-processing techniques for attaining precise delineation of the tooth. Normalization is used to reduce the variance of the brightness values along the elevation. It is followed by contrast enhancement that enhances the boundary between the tooth region and the surrounding tissues. Further the noise in the CT image is reduced by using a commonly used smoothing bilateral filter (Akhoondali, Zoroofi, & Shirani, 2009). The three main constraints in delineating the tooth structures are the continuity of the brightness values, the maximum brightness value and isobrightness contour information. The constraints results in over-extraction, which is a major issue when using the existing method (Bors, Kechagias, & Pitas, 2002). To overcome such constraints, region competition is accomplished by growing regions initially selecting multiple seeds. This helps in determining the boundary among the tooth to be delinated and the neighboring teeth (Bors, Kechagias, & Pitas, -Nielsan, 1996). The irregular points are eliminated by applying morphological operations (Dong & Fu-yuan, 2010).

Pre-Processing

In dental CT images, tooth regions do not have same intensity value, it mainly varies in the enamel and dental pulp (Omachi, Saito, Aso, Kasahara & Yamada Kimura, 2007). The region to be delineated contains different intensity value compared to the other regions. Normalization is the process in which the target to be extracted is processed to have uniform distribution of the intensity value. It is achieved by Normalization based on slice energy. This is achieved by taking summation of intensity values of each slice on x-y plane as normalization factor (Omachi,Saito, Aso, Kasahara & Yamada Kimura, 2008).

The precise delineation is achieved when the intensity levels of tooth to be extract is different from the surrounding regions and it is achieved by contrast enhancement (Juergens, Krol, Zeilhofer, Beinemann, Schicho, Ewers & Klug, 2009). The real challenge in extraction of tooth is the similarity in the intensity levels of regions such as alveolar bone, gum and periodontal ligament, to the desired tooth region (Stavropoulos, & Wenzel, 2007; Kalra,2010). To delineate the desired tooth with overlapping similar intensity region, the intensity value of each voxel is substituted by a square as suggested by Pratt (2007). The final step in the preprocessing is removing the noise inherent and specific in CT. The bilateral filter, which is a boundary preserving smoothing filters is preferred to remove the noise. The smoothing is achieved when the distance from one pixel to the other is minimal or these pixels have similar intensity values. This method helps in noise reduction and upholding the boundary information are accomplished instantaneously (Yau, Lin, Tsou, & Lee, 2008).

Region Extraction

On the CT image, the intensity values are distributed like the dental pulp with low intensity value, the dentine with high intensity value and the enamel with highest intensity value in a tooth (Kumar, Moni, & Rajeesh, 2011).

As suggested by R Yanagisawa et al (2011) the pixel with minimum intensity value is selected as the seed because the dental pulp has a low intensity value (Rahimov, & Farzaliyev, 2011; Roser, Ramachandra, Blair, Grist, Carlson, Christensen, & Steed, 2010). Further the pixels with a lower intensity value are explored iteratively from the adjoining region of the speculative initialized seed to determine the pixel that has minimum intensity value (Mupparapu, & Singer,, 2000). After identifying the seed, the region-growing method is applied by using this seed as the starting point (Scarfe,2006). The dental pulp, dentine and enamel of each delineated tooth are the tissues that are anticipated to be delineated. The major typical feature of these tissues is that the intensity value of a pixel increases as it approaches the surface of a tooth (Palomera, Martinez, Benitez & Ortega, 2010).

In spite of all the restrictions applied in the algorithm, adjacent teeth around the crown of the tooth are over extracted (Grenier, Revol-Muller, Costes, Janier, & Gimenez,2005; Yanagisawa, & Omachi,2011; Palomera, Martinez, Benitez & Ortega, 2010). The major disadvantage of the region growing method is the presence of disjointed points, due to the left-over noise on the image. These disjoint are often removed using morphological operation. This is achieved by opening and closing process with a structure element for n iteration (Grenier, Revol-Muller, Costes, Janier, & Gimenez, 2005; Yanagisawa, & Omachi, 2011; Palomera, Martinez, Benitez & Ortega, 2010).

Data Processing

The quality of dental or medical model is depend on data processing which is very complex step. The acquired data file format is converted into a file format that are compatible with engineering and industrial procedure. Sterolithographic file format (STL) is suitable in most situations for Rapid manufacturing.

Model Fabrication

Many types of RP technologies were used such as stereolithography, selective laser sintering, fused deposition modelling (FDM), and Ink Jet printing techniques to create dental models. The type of dental restoration will determine the chosen technology, it depended on exacting precision, finishing surface, visible manifestation of inner structures, number of colors which needed for each model, strength and properties of the materials etc. The model is ready for implementation if the team decided that there are no errors after the RP medical model is created (Klein, Lu, & Wang, 2013) .

3D MEDICAL RAPID PROTOTYPE DESIGN PROCESS

Stereolithography

Stereolithography is the most popular from the advanced RP technologies. It creates three dimensional models by computer-controlled laser beam that moves in order to build up the required objects from a liquid in a layer by layer manner. To fabricate prototypes, models, patterns, the material has to be dipped in photosensitive liquid resin, a model-construction platform, and an ultraviolet (UV) or laser for curing the resin are used in 3D printing technology. The layers sequentially bind together to form the solid that begins from the bottom to the top. A distinct visualization of anatomical structures is used for educational purposes and it preferable for surgical planning before surgery as suggested by Azariet et al (2009). It is used to produce the model for reconstructive surgical procedure and sub-periosteal dental implant surgical procedure. Currently, the main purpose for to use these models is to fabricate surgical drilling templates during the insertion of dental implant.

Advantages

- It has Good speed.
- The needed time depends on the model volume and complexity.
- The accuracy of producing the details is increased.
- It has close tolerance.
- Good finished surface and smooth.
- It is possible to make it transparent.
- The possible density is 100percent.
- High mechanical strength, its Prototypes are used as master models for injection moulding, thermoforming, blow moulding.

Disadvantages

- The equipments and materials are highly Expensive.
- It used with polymers only.
- Handling of wet materials.
- It needs Post-processing procedure.

Inkjet-Based System (3D Printing - 3DP)

3DP technique got a great importance because it is precise, cost effective and speedy. It is an additive technique that differs in duration of design and print a wax pattern of a restoration. The machine constructs wax patterns of full crowns and frameworks similar to that of manual waxed restorations.

Similar to ink-jet printing, the pattern fabricated over the surface of a powder bed by spraying binder materials on a thin distribution of powder and it spreads over the complete bed. An amount of powder is measured and added from a source compartment by movement of piston ascendant, then a roller spreads and squeezes the powder at the upper manufacturing compartment. After that a head of multi- channel jetting will deposit a fluid bonding agent in a 2D pattern on the coating of powder. After a finishing of a layer, the piston which supports the powder bed 3DP is a coating - by- coating technique which continued progressively till the sample is totally manufactured. After heat- treating, unbounded powder is flounced up and leaves the created part intact as implemented by Azari et al. (2009).

3DP technology has high tolerance and hence not useful for molding intents as SLA, this tolerance considers as a problem to make flimsy dental prototypes, to form configurations to create educational models. Due to this high tolerance, 3DP technology is not useful for impression objectives as SLA. 3DP is inaccurate technology because it breaks off easily which requires extra period for accurate fitting of medical implant throughout the surgical procedure, However, it is supportive to work-out the surgical processes.

Advantages

- They have short manufacturing time.
- The used material is inexpensive
- Their colouring ability.
- Models are used in casting purposes with low ash burnout.
- The toxicity is Low.
- Variety of materials are used.

Disadvantages

- They have High tolerance.
- The models with low strength.
- The surface finish is rough.
- The resolutions are Limited.

Selective Laser Sintering

Selective laser sintering (SLS) creates the desired three-dimensional shape solid mass by fusing small particles powdered materials like plastic, metal, ceramic or glass powders with a high-power laser (CO2 laser). A wide variety of materials are used by coating particles with thermal binders. The powders have many higher properties than resin-based technologies with higher yields and faster post-finishing. Fusing of the material is done by the laser selectively after scanning of the cross- sections that created from a 3-D digital description of the model on surface of a powder bed.

By using a roller, the powdered material will spread on the surface of a build cylinder. Then moving the piston cylinder down just one-layer thickness, to give sufficient space to the new layer of powder. The distribution system of powder is similar in action to the build cylinder, in which a piston moves upward gradually to give a measured amount of powder for each layer. Then the surface of this tightly compacted powder is exposed to laser beam. An interaction will occur between laser beam and the powder which elevates its temperature to the melting point, leading to fusing of the powder particles and solid mass forming. Laser beam energy is adjusted to melt only the powder in areas which determined by the object's form over that cross section. The temperature of manufacturing chamber is kept lower than powder's melting point and the laser heat will raised the temperature a little only to cause sintering. SLS machine keeps the bulk powder material temperature in the powder bed less than its melting point by using infrared heating to reduce thermal deformation ″curling″ and simplify binding to the prior layer. This will reduce a lot of processing time. After finishing of the first layer, extra powder layer will be added by a roller technique over the layer which scanned previously. This process is repeated until the whole object is finished, then the object is removed from the building chamber and the powder which is not scanned and fused are reused. Post-processing is needed, depending on the desired application. SLS technology is used to fabricate removable partial denture (RPD) frameworks with cobalt- chromium alloy spherical powder that has maximum particle size of 0.045mm (particle size range 0.005- 0.045mm), the mean particle size approximately 0.030mm. A complete cobalt- chrome RPD framework is fabricated with a demonstrate successful.

Advantages

- Extensive variety of materials are used especially in dentistry.
- Excellent accuracy.
- Simple post- processing.
- Un-sintered powder remains at the sites of support structure. So, support structures are not needed, after completing the model it can be recycled.

Disadvantages

- Material and equipment are costly.
- Large tolerance is required for dental procedures (0.5 ± 0.2).

Fused Deposition Modeling

FDM is the other vastly used rapid prototyping technology, following stereo lithography. It is a technique in which a thermoplastic material was threw in a layer by layer manner through a temperature-controlled head. In this system, supplied materials is released from a coil with a plastic filament in to an extrusion nozzle head. The plastic melts in to a semi liquid form by the heated nozzle, the melted plastic then permitted to flow in order to be rotated on and off (Ventola,2014).

This procedure functions in three axes, essentially, portrayal of the model in one layer at the same time. When machine nozzle is progressed over the stand table in the essential path, it drops a slim bead of extruded plastic that forms the first layer. After ejection of it from the nozzle, the plastic will solidify directly and unites with the layer below.

The procedure is composed of a compartment that is fixed at a temperature lesser than temperature that melts the plastic. This technology permit designing of a wide range of materials and colors like wax of investment casting and medical grade ABS (Kumar, Moni, & Rajeesh, SS,2011).

FDM is suitable for fabrication of bone imprints. It uses build materials which manufacture strong and tough models in one step. Furthermore, it gives excellent visualization by highlighting a chosen feature with different color. For overhanging geometries, supporting material is required which can be broken away from the object after finishing it, or can be easily washed away when a water-soluble support material is used (Kumar, Moni, & Rajeesh, 2013). The system is slow in fabricating models have wide cross sections, while in fabricating small models that have tall, thin form-factors it will be faster.

Advantages

- Wax pattern is directly fabricated.
- Constructed in multi-color.
- Relatively, fast and speed technique.

Disadvantages

- Removing the support structure after model fabrication is completed.
- Finishing surface is rough.
- Only thermoplastic material can be used.
- Its dense is less than 100 percent.

For designing frameworks of RPD, STL is the best choice. STL model is often recommended to recognize the existing undercut areas on the computer-generated CAD model which prepared for the teeth and surrounding tissues of the patient. Undesirable undercuts should be blocked out to ensure a path way for replacement and removal of the RPD into the patient mouth. Soft tissues will be relieved, then carving will be done virtually with special software tools making it similar to conventional carving by using haptic device which enabling models′ movement, rotation and translation in all axes, as well as hand moving.

3D PRINTING IN DENTISTRY

Velasco et al. (2017), in his paper discussed evaluation of right perimandibular swelling in the case study a multilocular lesion of few cms in size in the right mandible was disclosed in the study of Panorex and maxillofacial CT scan of a 40-year-old patient It also spread commencing from root of the canine to the mid-ramus. Apart from this the incisional tissue removal of the lesion resulted in solid ameloblastoma.

For the first stage a titanium reconstruction plate was planned on an mandibular resection along with safeguarding right condyle. The second stage involved delayed restoration of the flaw with iliac crest bone graft. The restoration plate is placed without modifications and hovels are pierced in healthy bone. Plate elimination and segmental mandibulectomy is accomplished, preceded by relocation of the plate and fixation to the predrilled hoveles. The whole procedure took around 4 hrs. A skull x-ray taken after the completion of the operation showed the precise placing of the restoration plate and mandibular symmetry.

Velasco et al. (2017) in his paper another a case of 34 old patient with a large anterior mandibular solid ameloblastoma formerly biopsied. A multilocular lesion around few cms were found predominantly in the symphysis region, spreading from the root of the right canine to the left first molar in the panoramic x-ray and maxillofacial CT scan. The treatment planning of mandibular resection with titanium 2.7-mm mandible reconstruction plate was suggested. The complete surgery was planned using rapid prototype Modelling.

Benefits of 3D Printing in Dentistry

The utmost benefit that 3D printers offers is the liberty to fabricate custom-made medical tools, devices and equipment. 3D printing is preferably used to tailor prosthetics and implants that bestows unlimited value for both patients and general practitioner (Ursan, Chiu, & Pierce 2013). In addition, 3D printing aids to fabricate jigs and fixtures for utility in surgical rooms. Custom-made implants, fixtures, and surgical tools have a constructive effect in terms of the time required for surgery, patient recovery time, and the success of the surgery or implant (Erkal, Lockwood, Chen, & Spence 2014).

The significant assistance offered by 3D printing is, the ability to manufacture items at minimal cost (Lipson, 2013, Li, 2014). It is realized with standard implants or prosthetics, used for craniofacial disorders. The cost to custom-print 3D object is nominal, with the first item being low-priced. This technique is beneficial for corporations that have low fabrication sizes or that assemble parts or models that are multifarious or necessitate regular changes, economical to supply to patients (Chae, Rozen, McMenamin, Findlay, Spychal, & Hunter-Smith,2015, Cui, Boland, D'Lima, & K Lotz, 2012).

The main positive attribute offered by 3D printing is the patient specific personalized design and manufacturing of medical tools and devices. An increasing group of biomaterials are accessible for use in 3D printing, and hence it decreases the cost. 3D printer allows flexibility to produce novel products for individual and marketable usage. The nature of 3D printing data files also offers an extraordinary opportunity for sharing among researchers (Cunningham, Madsen, & Peterson, 2005). Getting approval from regulatory bodies is another major obstacle in use of medical application of 3D printing. (De Vos, Casselman, J & Swennen, 2009).

BIOMATERIALS USED IN DENTAL IMPLANTS

Titanium and its alloys are the main optimal choice for implantation application. Titanium readily passives itself to form a thin oxide layer. This characteristic makes it biocompatible as well as forming a direct bond between bone and implant leading to permanent fixation of the implant. Titanium is commonly used in fracture plates, dental implants, etc.

Although 3D bioprinting schemes are laser-based, inkjet based, or extrusion-based, inkjet-based bioprinting is most common. This method deposits "bioink," or droplets of biomaterials, onto a substrate according to digital instructions to replicate dental implants. Multiple print heads are to deposit different cell types (region specific, blood vessel), a necessary feature for fabricating whole heterocellular tissues. A process for bioprinting involves to create a blueprint of a region with its vascular architecture. Then generate a bioprinting process plan. Further isolates stem cells and then differentiate the stem cells into tissue-specific cells (Tyndall DA & Brooks,2000). It is followed by preparing bioink reservoirs with tissue-specific cells, blood vessel cells, support medium and load them into the printer. Laser printers have also been employed in the cell printing process, in which laser energy is used to stimulate the cells in a particular pattern, providing spatial control of the cellular environment (Derand & Hirsch,2009).

PLANNING FOR 3D PRINTING

Dental rehabilitation planning has progressed promptly in the past two decades with the developments in medical image processing. Medical rapid prototyping (MRP) is defined as the manufacture of dimensionally accurate physical models of human anatomy derived from medical image data using a variety of rapid prototyping technology. The construction of the physical models is based on CT scans and there are many techniques for production of MRP models including stereolithography, selective laser sintering, and 3D printing (Eckardt & Swennen, 2005). There are several 3D printing processes for building 3D objects in any shape (Estrela, Bueno, Sousa-Neto & Pécora, 2008). MRP models are generally used in the perioperative period for improving the likelihood of treatment (Frederiksen,1995). Such technology is becoming increasingly acceable (Fullerton, Frodsham, G. C., & Day,2014). Recently rapid prototyping is used to produce solid object of computer models. It can form functional structures in a direct way such as metal parts, as well it is used in nano manufacturing and bio manufacturing, it is powerful manufacturing method.

Like other medical branches, dentistry also uses rapid prototyping as surgical guide, maxillofacial repair or physical model of dental implantology implemented by Yan et al (2008) and Prosthodontics implemented by Hoang et al. (2015), Sun et al. (2011) and Bibb et al. (2011). From the definition, metal-free ceramic restorations that made by CAD/CAM system falls into this group of technology.

One of the considerable other benefits of RP when used in dentistry is medical modeling fabrication, it is quite useful when used in mass production of patterns for casting purposes as analyzed by Toro (2007)). In this method, it is easy to create the difficult parts of restorations even without human involvement in a short time as experimented by Bibb (2010). Clinically, rapid prototyping technology reduces the time and possible injury which may occur during implantation process as reported by Hoang (2015).

By RP technology, dental prosthesis is constructed layer by layer directly from digital model simply and rapidly without special tooling and human interference, it makes a revolutionary progressing in the manufacturing of dental prosthesis as proposed by Tyndall (2008).

After the evolution of RP technologies there is a possibility to create many types of dental prostheses for many various implementations, Such as dental prosthesis wax pattern, dental (facial) prosthesis mold (shell), dental metallic prosthesis, and zirconia prostheses. RP thought to play an important role in prosthodontics and it will be most acceptable technique for digital manufacturing of dental prostheses as reported by Sun et al (2011). Some of RP implementations are: crowns and RPD wax pattern manufacturing, complete dentures and casting molds, maxillofacial prosthesis and dental metallic prosthesis manufacturing.

RP technologies will be an alternative method instead of usual method which depends on the skills of dentists and technicians in the manufacturing of dental.

Advantages of RP Systems

RP is fast, it needs 1to 2 days to construction and insertion of the prosthesis. The design defects are detected early and repaired with RP technique. The confidence of the design integrity is one of the properties of RP technique. It also has good management for the tissues with less trauma. It further enables accurate assessment of anatomical landmarks such as the size of the maxillary sinus in the upper jaw and location of the alveolar nerve in the lower jaw. It is well known to have accurate Osseous Topography analysis and precise location of implants. The data is got without direct patient, to avoid tissue tension. Also, can be used with patients who can't cooperate when preparing usual full face moulage impression. CAD and RP technique can be reduce lab work due to creation of positive models which invested for casting later. An object with definite accurate dimensions can be created by using RP approach of stereo lithography. If primary model is wasted or damaged during fabrication process, digital one can be used. The fabricated model has the internal detail as well as outer surface.

Disadvantages

Rp technique is unsupported in soft tissue. A two-steps impression may be required to have proper orientation of the soft tissue. Undercuts can be minimized by positioning the patient correctly, also it will be helpful to create any additional undercuts by using digital manipulation of the STL file, as proposed by Sun et al (2011).

CONCLUSION

This review chapter discusses about different technologies that were developed in past two decades in dentistry. Rapid prototyping manufacturing (RPM) is one of recent technologies, it is also known as an advanced and resilient fabrication technique as suggested by Mertz et al. (2013). There are continuous attempts to improve them by: amending its speed and accuracy, reducing the system and items cost, and developing more obtainable popular items. RP technologies plays a very important role in dental applications and it is noticeable that using of its models in dentistry will be widened in the future with continuous evolution, as mentioned by Yau et al. (2008). It will be useful in many fields of dentistry such as planning for surgery and designing for prosthodontics as reported by Winder et al, (2005).

REFERENCES

Akhoondali, H., Zoroofi, R. A., & Shirani, G. (2009). Rapid automatic segmentation and visualization of teeth in CT-scan data. *Journal of Applied Sciences (Faisalabad)*, *9*(11), 2031–2044. doi:10.3923/jas.2009.2031.2044

Alamri, H. M., Sadrameli, M., Alshalhoob, M. A., & Alshehri, M. A. (2012). Applications of CBCT in dental practice: A review of the literature. *General Dentistry*, *60*(5), 390–400. PMID:23032226

Arai, Y., Tammisalo, E., Iwai, K., Hashimoto, K., & Shinoda, K. (1999). Development of a compact computed tomographic apparatus for dental use. *Dento Maxillo Facial Radiology*, *28*(4), 245–248. doi:10.1038j.dmfr.4600448 PMID:10455389

Bagchi, P., & Joshi, N. (2012). Role of radiographic evaluation in treatment planning for dental implants: A review. *J Dent Allied Sci*, *1*(1), 21–25. doi:10.4103/2277-4696.159112

Banks, J. (2013). Adding value in additive manufacturing: Researchers in the United Kingdom and Europe look to 3D printing for customization. *IEEE Pulse*, *4*(6), 22–26. doi:10.1109/MPUL.2013.2279617 PMID:24233187

Baratto Filho, F., Zaitter, S., Haragushiku, G. A., de Campos, E. A., Abuabara, A., & Correr, G. M. (2009). Analysis of the internal anatomy of maxillary first molars by using different methods. *Journal of Endodontics*, *35*(3), 337–342. doi:10.1016/j.joen.2008.11.022 PMID:19249591

Barker, T. M., Earwaker, W. J. S., & Lisle, D. A. (1994). Accuracy of stereolithographic models of human anatomy. *Journal of Medical Imaging and Radiation Oncology*, *38*(2), 106–111. PMID:8024501

Bartlett, S. (2013). Printing organs on demand. *The Lancet. Respiratory Medicine*, *1*(9), 684. doi:10.1016/S2213-2600(13)70239-X PMID:24429271

Bertassoni, L. E., Cecconi, M., Manoharan, V., Nikkhah, M., Hjortnaes, J., Cristino, A. L., ... Khademhosseini, A. (2014). Hydrogel bioprinted microchannel networks for vascularization of tissue engineering constructs. *Lab on a Chip*, *14*(13), 2202–2211. doi:10.1039/C4LC00030G PMID:24860845

Bornstein, M. M., Scarfe, W. C., Vaughn, V. M., & Jacobs, R. (2014). Cone beam computed tomography in implant dentistry: A systematic review focusing on guidelines, indications, and radiation dose risks. *International journal of oral & maxillofacial implants*, 29. PMID:24660190

Bors, A. G., Kechagias, L., & Pitas, I. (2002). Binary morphological shape-based interpolation applied to 3-D tooth reconstruction. *IEEE Transactions on Medical Imaging*, *21*(2), 100–108. doi:10.1109/42.993129 PMID:11929098

Bro-Nielsen, M., Larsen, P., & Kreiborg, S. (1996). Virtual teeth: a 3D method for editing and visualizing small structures in CT scans. In Proc. Computer Assisted Radiology (CAR'96) (pp. 921-924). Academic Press.

Buchaillard, S. I., Ong, S. H., Payan, Y., & Foong, K. (2007). 3D statistical models for tooth surface reconstruction. *Computers in Biology and Medicine*, *37*(10), 1461–1471. doi:10.1016/j.compbiomed.2007.01.003 PMID:17336957

Carter, C. N., Pusateri, R. J., Chen, D., Ahmed, A. H., & Farag, A. A. (2010, September). Shape from shading for hybrid surfaces as applied to tooth reconstruction. In *Image Processing (ICIP), 2010 17th IEEE International Conference on* (pp. 4049-4052). IEEE. 10.1109/ICIP.2010.5653040

Chae, M. P., Rozen, W. M., McMenamin, P. G., Findlay, M. W., Spychal, R. T., & Hunter-Smith, D. J. (2015). Emerging applications of bedside 3D printing in plastic surgery. *Frontiers in Surgery, 2*, 25.

Chen, H., & Jain, A. K. (2005, January). Dental biometrics: Alignment and matching of dental radiographs. In Application of Computer Vision, 2005. WACV/MOTIONS'05 Volume 1. Seventh IEEE Workshops on (Vol. 1, pp. 316-321). IEEE.

Cleghorn, B. M., Christie, W. H., & Dong, C. C. S. (2008). Anomalous mandibular premolars: A mandibular first premolar with three roots and a mandibular second premolar with a C-shaped canal system. *International Endodontic Journal*, *41*(11), 1005–1014. doi:10.1111/j.1365-2591.2008.01451.x PMID:19133090

Cohen, A., Laviv, A., Berman, P., Nashef, R., & Abu-Tair, J. (2009). Mandibular reconstruction using stereolithographic 3-dimensional printing modeling technology. *Oral Surgery, Oral Medicine, Oral Pathology, Oral Radiology, and Endodontics*, *108*(5), 661–666. doi:10.1016/j.tripleo.2009.05.023 PMID:19716728

Cui, X., Boland, T., & D'Lima, D., & Lotz, M. (2012). Thermal inkjet printing in tissue engineering and regenerative medicine. *Recent Patents on Drug Delivery & Formulation*, *6*(2), 149–155. doi:10.2174/187221112800672949 PMID:22436025

Cunningham, L. L. Jr, Madsen, M. J., & Peterson, G. (2005). Stereolithographic modeling technology applied to tumor resection. *Journal of Oral and Maxillofacial Surgery*, *63*(6), 873–878. doi:10.1016/j.joms.2005.02.027 PMID:15944992

De Vos, W., Casselman, J., & Swennen, G. R. J. (2009). Cone-beam computerized tomography (CBCT) imaging of the oral and maxillofacial region: A systematic review of the literature. *International Journal of Oral and Maxillofacial Surgery*, *38*(6), 609–625. doi:10.1016/j.ijom.2009.02.028 PMID:19464146

DelBalso, A. M., Greiner, F. G., & Licata, M. (1994). Role of diagnostic imaging in evaluation of the dental implant patient. *Radiographics*, *14*(4), 699–719. doi:10.1148/radiographics.14.4.7938761 PMID:7938761

Dérand, P., & Hirsch, J. M. (2009). Virtual bending of mandibular reconstruction plates using a computer-aided design. *Journal of Oral and Maxillofacial Surgery*, *67*(8), 1640–1643. doi:10.1016/j.joms.2009.03.039 PMID:19615575

Dong-ri, S., & Fu-yuan, G. (2010, April). The segmentation algorithm of dental CT images based on fuzzy maximum entropy and region growing. In *Bioinformatics and Biomedical Technology (ICBBT), 2010 International Conference on* (pp. 74-78). IEEE. 10.1109/ICBBT.2010.5479006

Eckardt, A., & Swennen, G. R. (2005). Virtual planning of composite mandibular reconstruction with free fibula bone graft. *The Journal of Craniofacial Surgery*, *16*(6), 1137–1140. doi:10.1097/01.scs.0000186306.32042.96 PMID:16327572

Estrela, C., Bueno, M. R., Azevedo, B. C., Azevedo, J. R., & Pécora, J. D. (2008). A new periapical index based on cone beam computed tomography. *Journal of Endodontics*, *34*(11), 1325–1331. doi:10.1016/j.joen.2008.08.013 PMID:18928840

Estrela, C., Bueno, M. R., Leles, C. R., Azevedo, B., & Azevedo, J. R. (2008). Accuracy of cone beam computed tomography and panoramic and periapical radiography for detection of apical periodontitis. *Journal of Endodontics*, *34*(3), 273–279. doi:10.1016/j.joen.2007.11.023 PMID:18291274

Estrela, C., Bueno, M. R., Sousa-Neto, M. D., & Pécora, J. D. (2008). Method for determination of root curvature radius using cone-beam computed tomography images. *Brazilian Dental Journal*, *19*(2), 114–118. doi:10.1590/S0103-64402008000200005 PMID:18568224

Frederiksen, N. L. (1995). Diagnostic imaging in dental implantology. *Oral Surgery, Oral Medicine, Oral Pathology, Oral Radiology, and Endodontology*, *80*(5), 540–554. doi:10.1016/S1079-2104(05)80153-2 PMID:8556464

Fullerton, J. N., Frodsham, G. C., & Day, R. M. (2014). 3D printing for the many, not the few. *Nature Biotechnology*, *32*(11), 1086–1087. doi:10.1038/nbt.3056 PMID:25380438

Gahleitner, A., Watzek, G., & Imhof, H. (2003). Dental CT: Imaging technique, anatomy, and pathologic conditions of the jaws. *European Radiology*, *13*(2), 366–376. PMID:12599003

Gao, J., Xu, W., & Geng, J. (2006). 3D shape reconstruction of teeth by shadow speckle correlation method. *Optics and Lasers in Engineering*, *44*(5), 455–465. doi:10.1016/j.optlaseng.2005.04.013

Gepreel, M. A. H., & Niinomi, M. (2013). Biocompatibility of Ti-alloys for long-term implantation. *Journal of the Mechanical Behavior of Biomedical Materials*, *20*, 407–415. doi:10.1016/j.jmbbm.2012.11.014 PMID:23507261

Grenier, T., Revol-Muller, C., Costes, N., Janier, M., & Gimenez, G. (2005). Automated seeds location for whole body NaF PET segmentation. *IEEE Transactions on Nuclear Science*, *52*(5), 1401–1405. doi:10.1109/TNS.2005.859354

Gross, B. C., Erkal, J. L., Lockwood, S. Y., Chen, C., & Spence, D. M. (2014). *Evaluation of 3D printing and its potential impact on biotechnology and the chemical sciences*. Academic Press.

Hirogaki, Y., Sohmura, T., Satoh, H., Takahashi, J., & Takada, K. (2001). Complete 3-D reconstruction of dental cast shape using perceptual grouping. *IEEE Transactions on Medical Imaging*, *20*(10), 1093–1101. doi:10.1109/42.959306 PMID:11686444

Hoy, M. B. (2013). 3D printing: Making things at the library. *Medical Reference Services Quarterly*, *32*(1), 93–99. doi:10.1080/02763869.2013.749139 PMID:23394423

Jain, A. K., & Chen, H. (2004). Matching of dental X-ray images for human identification. *Pattern Recognition*, *37*(7), 1519–1532. doi:10.1016/j.patcog.2003.12.016

Jayadevappa, B. S., Kodhandarama, G. S., Santosh, S. V., & Rashid, W. T. (2010). Imaging of dental implants. *Journal of Oral Health Research*.

Juergens, P., Krol, Z., Zeilhofer, H. F., Beinemann, J., Schicho, K., Ewers, R., & Klug, C. (2009). Computer simulation and rapid prototyping for the reconstruction of the mandible. *Journal of Oral and Maxillofacial Surgery*, *67*(10), 2167–2170. doi:10.1016/j.joms.2009.04.104 PMID:19761910

Kalra, D., Jain, G., Deoghare, A., & Lambade, P. (2010). Role of imaging in dental implants. *Journal of Indian Academy of Oral Medicine and Radiology*, *22*(1), 34–38. doi:10.5005/jp-journals-10011-1007

Khaled, S. A., Burley, J. C., Alexander, M. R., & Roberts, C. J. (2014). Desktop 3D printing of controlled release pharmaceutical bilayer tablets. *International Journal of Pharmaceutics*, *461*(1-2), 105–111. doi:10.1016/j.ijpharm.2013.11.021 PMID:24280018

Klein, G. T., Lu, Y., & Wang, M. Y. (2013). 3D printing and neurosurgery—ready for prime time? *World Neurosurgery*, *80*(3), 233–235. doi:10.1016/j.wneu.2013.07.009 PMID:23871811

Krol, Z., Chlebiej, M., Zerfass, P., Sader, R., Zeilhofer, H. R., Mikołajczak, P., & Keeve, E. (2002). Surgery planning tools for the osseous grafting treatment. *Biomedizinische Technik/Biomedical Engineering, 47*(s1a), 97-100.

Kumar, S. S., Moni, R. S., & Rajeesh, J. (2011). Automatic segmentation of liver and tumor for CAD of liver. *Journal of Advances in Information Technology, 2*(1), 63-70.

Kumar, V., & Satheesh, K. (2013). Applications of cone beam computed tomography (CBCT) in implant treatment planning. *JSM, 1*(1008).

Li, Y., Yang, C., Zhao, H., Qu, S., Li, X., & Li, Y. (2014). New developments of Ti-based alloys for biomedical applications. *Materials (Basel)*, *7*(3), 1709–1800. doi:10.3390/ma7031709 PMID:28788539

Lingeshwar, D., Dhanasekar, B., & Aparna, I. N. (2010). Diagnostic imaging in implant dentistry. *International Journal of Oral Implantology & Clinical Research*, *1*(3), 147–153.

Lipson, H. (2013). New world of 3-D printing offers" completely new ways of thinking": Q&A with author, engineer, and 3-D printing expert Hod Lipson. *IEEE Pulse*, *4*(6), 12–14. doi:10.1109/MPUL.2013.2279615 PMID:24215725

Lofthag-Hansen, S., Huumonen, S., Gröndahl, K., & Gröndahl, H. G. (2007). Limited cone-beam CT and intraoral radiography for the diagnosis of periapical pathology. *Oral Surgery, Oral Medicine, Oral Pathology, Oral Radiology, and Endodontics*, *103*(1), 114–119. doi:10.1016/j.tripleo.2006.01.001 PMID:17178504

Manisundar, N., Saravanakumar, B., Hemalatha, V. T., Manigandan, T., & Amudhan, A. (2014). Implant imaging-A literature review. *Biosciences Biotechnology Research Asia*, *11*(1), 179–187. doi:10.13005/bbra/1251

Matherne, R. P., Angelopoulos, C., Kulild, J. C., & Tira, D. (2008). Use of cone-beam computed tomography to identify root canal systems in vitro. *Journal of Endodontics*, *34*(1), 87–89. doi:10.1016/j.joen.2007.10.016 PMID:18155501

Mehra, P., Miner, J., D'Innocenzo, R., & Nadershah, M. (2011). Use of 3-d stereolithographic models in oral and maxillofacial surgery. *Journal of Maxillofacial and Oral Surgery*, *10*(1), 6–13. doi:10.100712663-011-0183-3 PMID:22379314

Mertz, L. (2013). Dream it, design it, print it in 3-D: What can 3-D printing do for you? *IEEE Pulse*, *4*(6), 15–21. doi:10.1109/MPUL.2013.2279616 PMID:24233186

Mol, A., & van der Stelt, P. F. (1991). Application of digital image analysis in dental radiography for the description of periapical bone lesions: A preliminary study. *IEEE Transactions on Biomedical Engineering*, *38*(4), 357–359. doi:10.1109/10.133231 PMID:1855798

Mupparapu, M., & Singer, S. R. (2004). Implant imaging for the dentist. *Journal - Canadian Dental Association*, *70*(1), 32–32. PMID:14709253

Nance, R., Tyndall, D., Levin, L. G., & Trope, M. (2000). Identification of root canals in molars by tuned-aperture computed tomography. *International Endodontic Journal*, *33*(4), 392–396. doi:10.1046/j.1365-2591.2000.00330.x PMID:11307216

Nayar, S., Bhuminathan, S., & Bhat, W. M. (2015). Rapid prototyping and stereolithography in dentistry. *Journal of Pharmacy & Bioallied Sciences*, *7*(5Suppl 1), S216. doi:10.4103/0975-7406.155913 PMID:26015715

Omachi, S., Saito, K., Aso, H., Kasahara, S., Yamada, S., & Kimura, K. (2008). Semi-automatic reconstruction of tooth shape from CT images by contour propagation. [in Japanese.]. *IEICE Transactions on Information and Systems*, *J91-D*, 2426–2429.

Omachi, S., Saito, K., Aso, H., Kasahara, S., Yamada, S., & Kimura, K. (2007). Tooth shape reconstruction from CT images using spline curves. *Proceedings of the International Conference on Wavelet Analysis and Pattern Recognition*, 393–6.

Ozbolat, I. T., & Yu, Y. (2013). Bioprinting toward organ fabrication: Challenges and future trends. *IEEE Transactions on Biomedical Engineering*, *60*(3), 691–699. doi:10.1109/TBME.2013.2243912 PMID:23372076

Özen, T., Kamburoğlu, K., Cebeci, A. R. I., Yüksel, S. P., & Paksoy, C. S. (2009). Interpretation of chemically created periapical lesions using 2 different dental cone-beam computerized tomography units, an intraoral digital sensor, and conventional film. *Oral Surgery, Oral Medicine, Oral Pathology, Oral Radiology, and Endodontics*, *107*(3), 426–432. doi:10.1016/j.tripleo.2008.08.017 PMID:18996725

Palomera-P'erez, M. A., Martinez-Perez, M. E., Benitez-P'erez, H., & Ortega-Arjona, J. L. (2010). Parallel multiscale feature extraction and region growing: Application in retinal blood vessel detection. *IEEE Transactions on Information Technology in Biomedicine*, *14*(2), 500–506. doi:10.1109/TITB.2009.2036604 PMID:20007040

Patel, S., & Kumar, N. A. (2009). Implant radiology. *Journal of the Indian Dental Association*, 3.

Pineda, F. (1973). Roentgenographic investigation of the mesiobuccal root of the maxillary first molar. *Oral Surgery, Oral Medicine, and Oral Pathology*, *36*(2), 253–260.

Pratt, W. K. (2007). *Digital image processing* (4th ed.). Los Altos, CA: Wiley-Interscience. doi:10.1002/0470097434

Rahimov, C., & Farzaliyev, I. (2011). Virtual bending of titanium reconstructive plates for mandibular defect bridging: Review of three clinical cases. *Craniomaxillofacial Trauma & Reconstruction*, *4*(04), 223–234. doi:10.1055-0031-1293523 PMID:23205175

Ramamurthy, R., Scheetz, J. P., Clark, S. J., & Farman, A. G. (2006). Effects of imaging system and exposure on accurate detection of the second mesio-buccal canal in maxillary molar teeth. *Oral Surgery, Oral Medicine, Oral Pathology, Oral Radiology, and Endodontics*, *102*(6), 796–802. doi:10.1016/j.tripleo.2006.02.009 PMID:17138184

Roser, S. M., Ramachandra, S., Blair, H., Grist, W., Carlson, G. W., Christensen, A. M., & Steed, M. B. (2010). The accuracy of virtual surgical planning in free fibula mandibular reconstruction: Comparison of planned and final results. *Journal of Oral and Maxillofacial Surgery*, *68*(11), 2824–2832. doi:10.1016/j.joms.2010.06.177 PMID:20828910

Salgueiro, M. I., & Stevens, M. R. (2010). Experience with the use of prebent plates for the reconstruction of mandibular defects. *Craniomaxillofacial Trauma & Reconstruction*, *3*(04), 201–208. doi:10.1055-0030-1268520 PMID:22132258

Scarfe, W. C., Farman, A. G., & Sukovic, P. (2006). Clinical applications of cone-beam computed tomography in dental practice. *Journal - Canadian Dental Association*, *72*(1), 75. PMID:16480609

Schubert, C., Van Langeveld, M. C., & Donoso, L. A. (2014). Innovations in 3D printing: A 3D overview from optics to organs. *The British Journal of Ophthalmology*, *98*(2), 159–161. doi:10.1136/bjophthalmol-2013-304446 PMID:24288392

Siu, A. S. C., Chu, F. C. S., Li, T. K. L., Chow, T. W., & Deng, F. (2010). *Imaging modalities for pre-operative assessment in dental implant therapy: an overview. Hong Kong Dental Journal.*

Stavropoulos, A., & Wenzel, A. (2007). Accuracy of cone beam dental CT, intraoral digital and conventional film radiography for the detection of periapical lesions. An ex vivo study in pig jaws. *Clinical Oral Investigations*, *11*(1), 101–106. doi:10.100700784-006-0078-8 PMID:17048029

Tomasi, C., & Manduchi, R. (1998, January). Bilateral filtering for gray and color images. In *Computer Vision, 1998. Sixth International Conference on* (pp. 839-846). IEEE. 10.1109/ICCV.1998.710815

Toro, C., Robiony, M., Costa, F., Zerman, N., & Politi, M. (2007). Feasibility of preoperative planning using anatomical facsimile models for mandibular reconstruction. *Head & Face Medicine*, *3*(1), 5. doi:10.1186/1746-160X-3-5 PMID:17224060

Tyndall, D. A., & Brooks, S. L. (2000). Selection criteria for dental implant site imaging: A position paper of the American Academy of Oral and Maxillofacial Radiology. *Oral Surgery, Oral Medicine, Oral Pathology, Oral Radiology, and Endodontics*, *89*(5), 630–637. doi:10.1067/moe.2000.106336 PMID:10807723

Tyndall, D. A., & Rathore, S. (2008). Cone-beam CT diagnostic applications: Caries, periodontal bone assessment, and endodontic applications. *Dental Clinics*, *52*(4), 825–841. doi:10.1016/j.cden.2008.05.002 PMID:18805231

Ursan, I. D., Chiu, L., & Pierce, A. (2013). Three-dimensional drug printing: A structured review. *Journal of the American Pharmacists Association*, *53*(2), 136–144. doi:10.1331/JAPhA.2013.12217 PMID:23571620

Ventola, C. L. (2014). Medical applications for 3D printing: Current and projected uses. *P&T*, *39*(10), 704. PMID:25336867

Velasco, I., Vahdani, S., & Ramos, H. (2017). Low-cost method for obtaining medical rapid prototyping using desktop 3d printing: A novel technique for mandibular reconstruction planning. *Journal of Clinical and Experimental Dentistry*, *9*(9), e1103. PMID:29075412

Vertucci, F. J. (1984). Root canal anatomy of the human permanent teeth. *Oral Surgery, Oral Medicine, and Oral Pathology*, *58*(5), 589–599. doi:10.1016/0030-4220(84)90085-9 PMID:6595621

Winder, J., & Bibb, R. (2005). Medical rapid prototyping technologies: State of the art and current limitations for application in oral and maxillofacial surgery. *Journal of Oral and Maxillofacial Surgery*, *63*(7), 1006–1015. doi:10.1016/j.joms.2005.03.016 PMID:16003630

Yanagisawa, R., & Omachi, S. (2010, November). Extraction of 3D shape of a tooth from dental CT images with region growing method. In *International Workshop on Computational Forensics* (pp. 68-77). Springer.

Yau, H. T., Lin, Y. K., Tsou, L. S., & Lee, C. Y. (2008). An adaptive region growing method to segment inferior alveolar nerve canal from 3D medical images for dental implant surgery. *Computer-Aided Design and Applications*, *5*(5), 743–752. doi:10.3722/cadaps.2008.743-752

Yovchev, D., Stanimirov, P., & Mihaylova, H. (2014). Lower jaw reconstruction using prototype from cone-beam computed tomography data. *Int Journal of Sciences and Research*, *3*, 57–59.

KEY TERMS AND DEFINITIONS

Fused Deposition Modeling (FDM): It is an additive manufacturing (AM) technology commonly used for modeling, prototyping, and production. It is one of the techniques used for 3D printing. Thus, FDM is also known as a solid-based AM technology.

Implantology: A dental implant (also known as an endosseous implant or fixture) is a surgical component that interfaces with the bone of the jaw or skull to support a dental prosthesis such as a crown, bridge, denture, facial prosthesis or to act as an orthodontic anchor.

Prosthesis: An artificial body part, such as a limb, a heart, or a dental implant.

Rapid Prototyping: Rapid prototyping is the idea of quickly assembling a physical part, piece or model of a product. This is often done using sophisticated computer-aided design or other assembly software, and physically implemented using 3-D printers.

Selective Laser Sintering: It is an additive manufacturing (AM) technique that uses a laser as the power source to sinter powdered material (typically nylon/polyamide), aiming the laser automatically at points in space defined by a 3D model, binding the material together to create a solid structure.

Stereolithography: Stereolithography is a process for making models and parts in which a solid is built up a layer at a time by hardening parts of a container of resin using light. In stereolithography, parts are produced by successive solidification of thin resin layers using a UV laser beam.

Chapter 11
Phytochemicals:
Their Therapeutic Potential Against Dental Caries

Karthikeyan Ramalingam
B. S. Abdur Rahman Crescent Institute of Science and Technology, India

Bennett T. Amaechi
University of Texas Health Science Center at San Antonio, USA

ABSTRACT

The chapter gives a picture of the current data on the available anticariogenic natural products and their mechanism of action. Different phytochemicals such as phenols, flavanoids, alkaloids, terpenoids, tannins, lectins, etc. and their anticariogenic efficacy have been discussed in detail. All the data emphasise the fact that the use of natural products is emerging as an effective strategy in the prevention and treatment of dental caries. Consequently, these natural products could be incorporated in toothpastes and other oral hygiene products to promote oral health.

INTRODUCTION

Dental caries can be defined as a multistep process involving demineralization and remineralisation rather than unidirectional demineralization (Robinson et al., 1998). In spite of the global claims of success in treating this disease, it still haunts most of the world's population (Mandel, 1993; Stephen, 1993). Although according to Glass (1982), a decline in its prevalence has been witnessed in some populations of the developed countries, the level has been exaggerated (NIH, 1987, 1989). An NIH survey (1989) claims that 50% of 12 year old children of US are free from tooth decay. However, according to the survey, deciduous teeth were not examined and 85% of the 17 year old children had one or more carious teeth. The mean value which gives the percentage of caries free children itself is not reliable since the data were not normally distributed (Edelstein and Douglass, 1995). Even if it is considered true, 50% of the population have dental caries.

DOI: 10.4018/978-1-5225-6243-6.ch011

The most important treatment strategy for dental caries is the reduction and/or elimination of bacterial accumulations on the tooth surface and between teeth by brushing teeth regularly and frequent dental cleanings (Loesch and Grossman, 2001). Antibacterial mouth rinses which generally contain fluorides, alcohols, detergents and other antimicrobial substances effectively reduce the plaque formation. Synthetic antimicrobials used in tooth pastes and mouth rinses include povidone iodine products, chlorhexidine, cetylpyridinium chloride, triclosan and zinc citrate. However, these substances cause unwarranted undesirable effects like vomiting, diarrhea and teeth staining. There is also a problem of development of antimicrobial resistant strains with the use of these synthetic antimicrobials. These adverse consequences have led to the development of natural antimicrobial agents which are being incorporated in tooth pastes and mouth rinses (Cai and Wu, 1996). Some popular examples include miswak, tea tree oil, peppermint, green tea and manuka honey.

WORLD SCENARIO

Around 250,000 to 500,000 species of plants have been reported on earth of which relatively a small percentage (1 to 10%) is used as food by humans and other animals Borris, 1996). It is possible that even more are used for medicinal purposes (Moerman, 1996). Natural products are being used to treat many diseases since time immemorial wing to their anti bacterial, anti inflammatory, cytostatic (Wu-Yuan et al., 1990), antifungal and anti viral activities. According to WHO, 80% of the population of developing countries exclusively use traditional medicine (Eloff, 1998; Nascimento et al., 2000).

It has been well documented that medicinal plants confer considerable antibacterial activity against various microorganisms including bacteria responsible for dental caries (Jonathan et al., 2000). Compelling evidences show that a variety of plant products have the potential to control dental caries (Onishi et al., 1981; Kashket et al., 1985; Kakiuchi et al., 1986; Hada et al., 1989; Sakanaka et al., 1989; Kubo et al., 1992; Nakahara et al., 1993). The anticariogenic activity of many of the plant products can be attributed to their rich polyphenolic content (Ito et al., 1984; Kakiuchi et al., 1986; Sawamura et al., 1992; Tagashira et al., 1997; Mitsunaga and Abe, 1997). Drinking green tea extracts after every meal is a well known Japanese custom. The anticaries effect of this polyphenol rich source was investigated by several researchers (Hattori et al., 1990; Sakanaka et al., 1989, 1990, 1992; Otake et al., 1991; Nakahara et al., 1993; Ooshima et al., 1993, 1994, 1998). The polyphenols in tea extract comprise catechin, epicatechin, gallocatechin, epigallocatechin, epicatechin gallate, epigallocatechin gallate, etc. which are found to inhibit the primary cariogenic agent, *S. mutans* (Otake et al., 1991). They also prevent caries formation by inhibition of bacterial adherence, acid production, and glucosyl transferase activity (which is involved in extracellular polysaccharide formation). Some combinations of Oolong tea monomeric polyphenols showed the highest level of antibacterial activity. These results suggest that the antibacterial activity of oolong tea extract is caused by a synergistic effect of monomeric polyphenols, which can easily bind to proteins (Sasaki et al., 2004). The growth of many oral bacteria was found to be inhibited by spice extracts like C*innamon bark* oil, *Papua mace* extract and C*love bud* oil (Saeki et al., 1989). *Sanguinaria,* an alkaloid extract from the rhizome of *Sanguinaria canadensis* has been found to show broad antibacterial activity against cariogenic bacteria (Joann et al., 1985). The antibacterial activity of some plant species like *Melia azadirachta, Calotropis gigantean, Leucas aspera, Vitex negundo* and others have been extensively investigated (Rao, 2000). Manuka (honey derived from the flowers of the manuka tree in NewZealand), tea tree, eucalyptus, lavandula and rosmarinus (rosemary) oils have been shown to inhibit cariogenic bacteria, of which, manuka was the most effective. All the oils except lavandula have shown immediate bactericidal activity. Oral bacterial adhesion was inhibited by tea tree and manuka oils, which are less toxic to eukaryotic cells (Takarada et al., 2004). Essential oils from *Syzygium aromaticum* (clove) and *Zanthoxylum limonella* (makaen) are used to treat dental caries in Thailand and other countries (Trongtokit et al., 2004). Essential oils like Australian tea tree oil, peppermint oil and sage oil were found to be bactericidal to anaerobic oral bacterial, the potent components being thymol and eugenol (Shapiro et al., 1994). Tea tree oil comprising mono terpenes, sesquiterpenes and associated alcohols has shown antagonistic activity against a wide range of oral pathogens. The widely used essential oil rinse, Listerine™ contains thymol, eucalyptol, methyl salicylate and menthol as its active ingredients. Thymol and eucalyptol are antimicrobial, while methyl salicylate and menthol act as a cleaning agent and local anaesthetic respectively. Essential oils are considered far more superior that chlorhexidine as they do not stain, alter taste perception or promote calculus formation and most of all, they are not impaired in terms of efficacy when incorporated in the toothpaste (Santos, 2003).

Potent antibacterial activity against *S. mutans* has been reported from magnold and honokiol isolated from the bark of *Magnoliae cortex* (Namba et al., 1986). Oleonolic acid and ursolic acid were found to be active principles in *Zizyphi fructus* (Kohda et al., 1958). These compounds inhibit the formation of insoluble glucan by the glucosyl transferase of cariogenic bacteria. Polyphenolic 5'-nucleotidase inhibitors showing growth inhibitory activities against *S. mutans* were identified from the seeds and skin of the wine grape, 'Koshu' (Toukairin et al., 1991). They were also found to prevent caries by inhibiting glucan formation from sucrose. Chewing sticks (miswak) prepared from the roots of *Salvadora persica* have found their use as a potential oral hygiene tool a few centuries back. These sticks possess antiplaque, anticaries and antibacterial activities (Gazi et al., 1992). Phytoalexins from *Sophora exigua* have been shown to exhibit anticaries and antibacterial activities (Tsuchiya et al., 1994). The leaves of *Eucalyptus globules* were found to contain eucalyptone which showed potential cariostatic activity (Osawa et al., 1995). Isoprenyl flavones from *Artocarpus heterophyllus* were identified as anticariogenic compounds by Sato et al. (2003).

Allium cepa shows antibacterial activity against *S. mutans* and *S. sobrinus* (Kim, 1997). Plumbagin isolated from the aerial parts of *Drosera peltata* also shows growth inhibitory properties against cariogenic bacteria (Didry et al., 1998). Aqueous and ethanolic extracts of *Juglandaceae regia*, used in chewing sticks were found to affect both growth and physiological functions of *S. mutans* (Jagtap and Karkera, 2000). Bakuchiol which has a profound use in traditional oriental medicine was isolated from the seeds of *Psoralea corylifolia* and found to be a potential anticariogenic agent (Katsura, 2001). The solvent extracts of *Streblus asper* were also found to control dental caries (Wongkham et al., 2001).

Some of the cariogenic bacteria *viz.*, *S. mutans*, *Actinomyces viscosus*, and *P. gingivalis* were found to be inhibited by ceanothic acid and ceanothetric acid from *Ceanothus americanus*, a plant native to America (Katsura, et al., 2001). The antibacterial activity of Perilla seed extract against oral cariogenic *Streptococci* was reported by Yamamoto and Ogawa (2002). The anticariogenic activity of *Garcinia manii* (stick chewing) can be attributed to their ability to reverse acidogenic challenge to teeth (Addai et al., 2002). Kuwanon-G is an antibacterial agent isolated from the ethyl acetate fraction of methanol extract of *Morus alba* showing potential activity against cariogenic bacteria like *S. mutans, S. sobrins, S. sanguis* and *Porpyromonas gingivalis* (Park et al., 2003). Cranberry, a native American fruit is a rich source of phenolic compounds including flavonoids, phenolic acids and complex phenolic polymers and is reported to play a beneficial role in human health (Puupponen-Pimia et al., 2005). A nondialysable material (NDM) from cranberry juice is found to inhibit coaggregation of oral bacteria and streptococci biofilm formation (Steinberg et al., 2004, 2005 and Yamanaka et al., 2004).

Kaempferia pandurata commonly called as temu kunci in Indonesia and krachai in Thailand is a traditional medicinal plant whose compounds were shown to inhibit *S. mutans* in less than a minute, thus making it a potential anticaries agent (Kwan Hwang et al., 2004). Extracts of several medicinal plants are incorporated in commercial tooth pastes as antigingivitis and anticaries agents in Thailand (Pongpan et al., 1982; Avirutnant and Pongpan, 1983). *Gymnema sylvestre* is an yet another traditional medicinal plant widely used in Australia, Japanese, Vietnamese and India. This traditional plant has found its place in the treatment of diabetes mellitus, obesity and caries. Erycristagallin, a compound isolated from *Erythrina variegate* showed potential activity against dental caries (Sato et al. 2003). Around 430 species of *Mikania* are distributed in the tropical areas of Africa, Asia and America (King and Robinson, 1987) and 200 species were identified in Brazil (Barroso, 1986). However, the most widely studied species include *Mikania laevigata* and *Mikania glomerata*. The active principles of these plants were found to be Cupressenic, Kurenoic and diterpenic acids which showed potential activity against *S. mutans, S. sobrinus* and *Actinomyces naeslundii* (Yatsuda et al., 2005).

Chewing sticks made from selected medicinal plants are used in many African and Middle Eastern countries (Wu et al., 2001). This practice has been advocated by WHO for oral hygiene. A single chewing stick may contain heterogeneous antimicrobial agents. For example, various alkaloids are present in the Nigerian chewing stick (*Fagara zanthoxyloides*). *Serindeiawernecki* extracts were also found to inhibit the growth of oral pathogens (Rotimi et al., 1988). Assessment of bacterial taxa in chewing stick users' saliva by DNA-DNA hybridization technique revealed that the use of miswak (twigs of the *Salvadora persica* tree) inhibited certain bacteria especially oral streptococci (Darout et al, 2002). Anticariogenic activities are more pronounced in several African plants like *Piper guineense*, *Xylopia aethiopica*, *Citrus aurantifolia* and *Aframomum melegueta*. In general, in most African homes, teeth are cleaned by chewing the roots of certain plants (El-Said *et al*, 1971). Chewing sticks are commonly used in certain parts of West Africa. *Zanthoxylum zanthoxyloides* (Lam.) Waterm. roots impart tingling and peppery taste while *Masularia acuminata* (G. Don) Bullock ex Hoyle stem produces a bitter taste and frothing and *Vernonia amygdalina* Del. Root, an initial bitterness and a later sweet taste (Tella, 1976). The root of *Terminalia glaucescens* Planch. is used in the discolouration of mouth (Sofowora, 1993). All the chewing sticks exhibit a varying degree of antimicrobial activity. The benzoic acid derivatives confer antibacterial activity to *Zanthoxylun zanthoxyloides* (Odebiyi and Sofowora, 1979). The phenolics are active at the acidic pH and the alkaloids such as Canthin–6–one, berberine and chelerythrine are active at the alkaline pH which makes them active at both the ends of the pH spectrum. Some African chewing sticks were found to possess fluoride ions (Odusanya et al., 1979), silicon, tannic acid, sodium bicarbonate as well as other natural substances that inhibit bacterial colonoisation and plaque formation (Akpata *et al* 1977). The role of chewing sticks in preventive oral hygiene was reported by Bruneton (1995) and Kocry (1983).

In the Indian subcontinent, there is a rich ethnobotanical diversity greatly exploited for their therapeutic potential and medicinal efficacy. Several herbal extracts from plants like *Melia azadirachta*, *Moringa pterygosperma* and *Balsamodendron mukul* are being used in the prevention of dental caries (Chopra et al., 1958). Many parts of *Melia azadirachta* are used as tonic, antiseptic, astringent and antibacterials and they have found a place in indigenous tooth powders, toothpastes, toilet soaps, etc. Dentists have recommended the use of neem twigs as tooth brushes and neem mouth wash was found to inhibit growth of *S. mutans* and carious lesions (Vanka et al., 2001). The medicinal virtues of *Moringa pterygosperma* (*Moringaceae*) have long been known and appreciated in India. Almost all parts of this plant were found to possess medicinal properties with roots showing potential activity against dental caries as recommended by the Ayurvedic practitioner, Hakims (Chopra et al., 1958). The oleogum resin from the plant, *Balsamodendron mukul* shows activity against a wide spectrum of diseases like leprosy, rheumatism, syphilis, nervous disorders and skin diseases. They are also being used as a gargle to treat dental caries, weak and spongy gums, pyorrhea alveolaris, chronic tonsillitis, pharyngitis and ulcerated throat (Chopra et al., 1958). Powder formulations from different medicinal plants *viz., Calotropis gigantean, Cassia hirsute, Cassia tora* and *Leucas aspera* are used in western ghat region of Dharwad district of Karnataka (India), to massage teeth and gums for plaque and caries (Hebber et al., 2004).

The native chinese medicines such as Radix et *Rhizoma* rhei, *Semen areca, Rhizoma* ligustici chuanxiong and Catechu were shown to control dental caries by preventing the adherence of *S. mutans* (Xiao et al., 2004). Isopanduratin A, isolated from *Kaempferia pandurata* is reported to be a potent anticariogenic agent by virtue of their marked antibacterial activity against cariogenic *S. mutans* (Hwang et al., 2004). The growth of cariogenic microorganisms like *S. mutans, S. sobrinus* and *C. albicans* was inhibited by propolis, a resinous hive product (Uzel et al., 2005). Brazilian propolis samples were found to be rich sources of flavonoids, diterpenes, triterpenic, lignans, and acetophenones, as well as prenylated *p*-coumaric acids. Prenylated phenylpropanoids were found in *B. dracunculifolia* and other South American *Baccharis* species as secondary metabolite phenolic compounds (Banskota et al., 1998; Marcucci et al., 2001). The effect of propolis from different sources on glucan synthesis and acidogenic potential of *S. mutans* were extensively investigated by Leitao et al. (2004).

Phytoalexins are plant derived compounds showing antimicrobial activities. Many edible plants such as soybean, carrot, tomato, potato, eggplant and rice synthesize phytoalexins which exhibit a broad range of antimicrobial activity (Dixon et al., 1983; Ebel and Grisebach, 1988). More than 200 phytoalexins have been characterized from over 20 plant families, most of which are flavonoids (Pathak et al., 1991). The antimicrobial potential of flavanoids make effective therapeutic tools to treat bacterial infections (Pathak et al., 1991; Havsteen, 1983; Biswas,et al., 2002). Phytoalexins purfied from *Sophora exigue* were found to be potent antimicrobial agents. The presence of such antibacterial phytoalexins in edible foods make them effective source of anticaries agents (Tsuchiya et al., 1994).

Xylitol-sweetened chewing gums are found to be cariostatic as cariogenic microorganisms do not metabolize xylitol and xylitol does not decrease plaque pH (Beckers, 1988; Grenby et al., 1989). There is a drop in the count of mutans streptococci with the use of xylitol-sweetened chewing gums. This might be due to the fact that it starves the cariogenic bacteria (Hayes, 2001; Gales and Nguyen, 2000). There are also reports of reduction in plaque accumulation by the xylitol gum (Mäkinen et al., 1996a,b). Therefore, xylitol sweetened chewing gums are being investigated for any anticariogenic activity.

Palatinose which is α-D glucopyranosyl-1,6-fructose, is found in honey and sugarcane extract (Barea-Alvarez et al., 2014). Most of the oral streptococci cannot ferment palatinose and produce acid *in vitro*. There was also an inhibition of insoluble glucan synthesis from sucrose by the streptococci GTase. These results show that palatinose is anticariogenic (Ooshima et al., 1983). However, certain serotypes of *S. mutans* were found to ferment palatinose after repeated sub culture in palatinose containing media. This might attributed to the adaptation of some oral bacteria resulting in the metabolism of palatinose by alternate pathways.

Plant derived monoclonal antibodies were also produced to prevent adherence of cariogenic bacteria to teeth (Lehner et al., 1985, 1986; Ma et al., 1987, 1989 and Ma and Lehner, 1990). The murine IgG1 (Guy's 13) was directed against the 185-kDa streptococcal cell surface adhesin (SA I/II) protein which showed broad serotype specificity and ability to prevent recolonization of teeth by MS (Smith and Lehner, 1989). The antigen is responsible for the attachment of the oral streptococci to the salivary pellicle of the teeth. The gene coding for the antibody is cloned into a plant which secreted chimeric murine secretory IgA/G (SIgA/ G) (Ma et al., 1994, 1995). The purified plant derived antibodies were also used in clinical trials to test its efficiency against dental caries (Weintraub et al., 2005) following a preliminary test on four volunteers (Ma et al., 1998).

CLASSIFICATION OF ANTIMICROBIAL PHYTOCHEMICALS

Phytochemicals which are non-nutritive plant components having protective properties are used as drugs since time immemorial (Mtunzi et al., 2017) and most of them were found to be antimicrobial agents (Mathias et al., 2000; Ganguly et al., 2001; Martino et al., 2002). More than thousand phytochemicals have identified which are mostly plant secondary metabolites. They can be classified into alkaloids, steroids, tannins, and phenol compounds. These compounds are complex and are specific to certain taxa. Therefore, the medicinal property of each plant species is unique. These phytochemicals are homologous to endogenous metabolites, ligands, hormones, signal transduction molecules or neurotransmitters which accounts for their biological activities. Hence, a random screening of all plants is required for targeting active principles.

Simple Phenols and Phenolic Acids

The simplest phytochemicals belong to the group of phenyl propanes which have a single substituted phenolic ring like cinnamic acid, caffeic acid, catechol and pyrogallol. These phytochemicals are present in the highest oxidation states and show a broad spectrum of antimicrobial activity. Essential oils are those possessing a C3 side chain with low oxidation states and with no oxygen. These phenolics exhibit a diverse range of bioactivity including anti-inflammatory, antiallergic, antimicrobial, anticarcinogenic, and antiviral activities (Medina et al., 2007).

Quinones

Quinones are those compounds with aromatic rings with two ketone substitution. They are highly reactive compounds and the conversion between diphenol (hydroxyl quinine) and diketone (quinine) is via the oxidation reduction reactions. This high redox potential of this class of compounds is responsible for the browning reaction in cut fruits or vegetables, dying of henna and they are also found to intermediates in the melanin synthesis pathway (Schmidt, 1988). They act as a source of stable free radicals and bind with nucleophilic aminoacids thereby inhibiting the protein action (Stern et al., 1996).

Flavones, Flavanoids and Flavanols

Flavones are hydroxylated phenolics structures containing one carbonyl group (two in quinines), while the addition of a 3-hydroxyl group yields a flavonol (Fessenden and Fessenden, 1982). Flavonoids occurs as a C6-C3 unit is linked to an aromatic ring and are synthesized in response to microbial infection of plants (Dixon et al., 1983). This group of phenolic compounds exhibits a broad range of antimicrobial activity by virtue of their interactions with extra cellular and soluble protein, interference and microbe-host interactions and their ability to inhibit microbial enzymes like protease, integrase and reverse transcriptase.

Tannins

Tannins are amorphous, astringent, polymeric phenolic compounds of plant origin most of which are glucosides. They produce blue or green colour, precipitate with iron salts and form insoluble and imputrescible leather by interacting with animal membrane. Molecular weights of tannins range between 500 and 3000 (Haslam, 1996) and they occur in wide range of plants and in all parts of the plant body (Apisariyakul et al., 1995; Afolayan and Meyer, 1997). Tannins are formed by condensation of flavan derivatives in woody tissues or by the polymerisation of quinine units and they can be classified into hydrolysable and condensed tannins (Geissman, 1963).

Coumarins

Coumarins are benzopyrone compounds possessing benzene ring attached to a pyrone (Keating and O'Kennedy, 1997). They occur as glucosides in combination with sugars and are responsible for the odor of hay (Murray *et al.*, 1982). Around 1300 coumarins had been identified and are found to exhibit wide range of biological activities like antithrombotic (Thastrup et al., 1985), anti-inflammatory (Piller, 1975), vasodilatory (Namba et al., 1988), anticoagulant and rodenticide properties (Keating and O'Kennedy, 1997). They also possess antimicrobial potential.

Terpenoids and Essential Oils

The fragrance in plants is bestowed by essential oils which are complex compounds with an isoprene structure. They consist of two groups; terpenes and those with an additional oxygen called terpenoids. These two classes of compounds show broad range of activities against bacteria, fungi, viruses and protozoa. Some common examples include menthol and camphor which are monoterpenoids and farnesol and artemisin which are sesquiterpenoids (Dhifi et al., 2016).

Alkaloids

Alkaloids are the most important phytochemicals in terms of bioactivity. They are chemically diverse organic nitrogen compounds and are found to be extremely toxic in spite of their therapeutic potential. Hence, compounds containing alkaloids are generally used externally in folk medicine. Pure and synthetic derivatives of alkaloids show analgesic, antispasmodic, and bactericidal effects (Stary, 1996).

Lectins/Antimicrobial Peptides

Lectins are derived from both plants and animals containing protein and sugar moieties. They bind to the cell membrane and effect changes in the signaling pathways. Peptides showing antimicrobial property contain disulphide bonds and are positively charged.

Other Compounds

Compounds like polyamines, isothiocyanates, thiosulfinates, glucosides and polyacetylenes are also found to exert antimicrobial properties.

MECHANISM OF ACTION OF ANTICARIOGENIC PHYTOCHEMICALS

Homeostasis of oral microorganisms is an important phenomenon in healthy individuals. The oral microbiota cohabit different sites of mouth depending upon the pH, nutrient availability and mucous surface (Marsh and Bradsha, 1997). Adhesion of cariogenic microorganisms mediated by adhesion factors like adhesions and salivary glycoproteins disrupt this homeostasis resulting in microbial colonization, biofilm formation and ultimately dental caries (Gibbons, 1984). The primary pathogenic organism in this common disease is identified as *Streptococcus mutans* (Bowen, 1999). Adhesion, acidogenicity and acid tolerance of cariogenic bacteria are the key factors involved in dental caries (Ooshima et al., 2000). The GBP and GTF of mutans streptococci bind to glucans and result in bacterial aggregation (Gibbons and Van Houte, 1975). Aggregation or biofilm formation results in production and secretion of considerable amounts of lactic acid, which can cause demineralization of the tooth structure when present in sufficient amounts in close proximity to the tooth surface and it eventually results in a carious lesion (Loesche, 1987) (Figure 1). Some of the bacteria associated with dental caries include *Actinomyces, Bacteroides, Campylobacter, Capnocytophaga, Corynebacterium, Eikenella, Eubacterium, Fusobacterium, Gemella, Haemophilus, Lactobacillus, Neisseria, Peptostreptococcus, Porphyromonas, Prevotella, Propionibacterium, Rothia, Selenomonas, Treponema, Veillonella and Bacteroides forsythus Actinomyces naeslundii* and *Actinomyces visosus* (Whittaker et al., 1996).

Biofilm formation is a multistep process involving salivary proteins and enzymes (Marsh and Bradshaw, 1998; Bowden and Hamilton, 1998). The GTF emzymes of cariogenic bacteria provide conducive sites on tooth surfaces by fermenting glucose on which oral bacteria adhere, accumulate and propagate resulting in the formation of a biofilm (Schilling and Bowen, 1992; Rozen et al., 2001; Steinberg et al., 2004). Hence, prevention of dental caries is one of the important strategies in the prevention of dental caries.

Antibiotics penicillin and erythromycin effectively prevent dental caries however, their use is limited due to adverse effects like hypersensitivity reaction, suprainfections, teeth staining and most of all, the development of resistant strains of viridans group *streptococci* including *S. mitis, S. sanguis* and *S. mutans* (Leclercq et al., 1988; Venditti et al., 1989). These adverse consequences warrant development of natural antibiotics as an alternative treatment for the treatment of dental caries and other oral infections. Although, plant products are greatly exploited for therapeutic potential to cure various oral ailments, this article describes natural products beneficial only in dental caries.

Inhibition of Adherence

One of the most important processes in dental caries is the adherence of cariogenic bacteria to the tooth surface (Matsumoto et al., 1999). Hence, inhibition of adherence is essential for the prevention of dental caries. Some of the model systems used to study adherence include metal slabs or wires, glass, hydroxyapatite, pulverized enamel, and sections of teeth (Schilling and Doyle, 1995).

Figure 1. Mechanisms of carious lesion formation by Streptococcus mutans and their prevention by phytochemicals

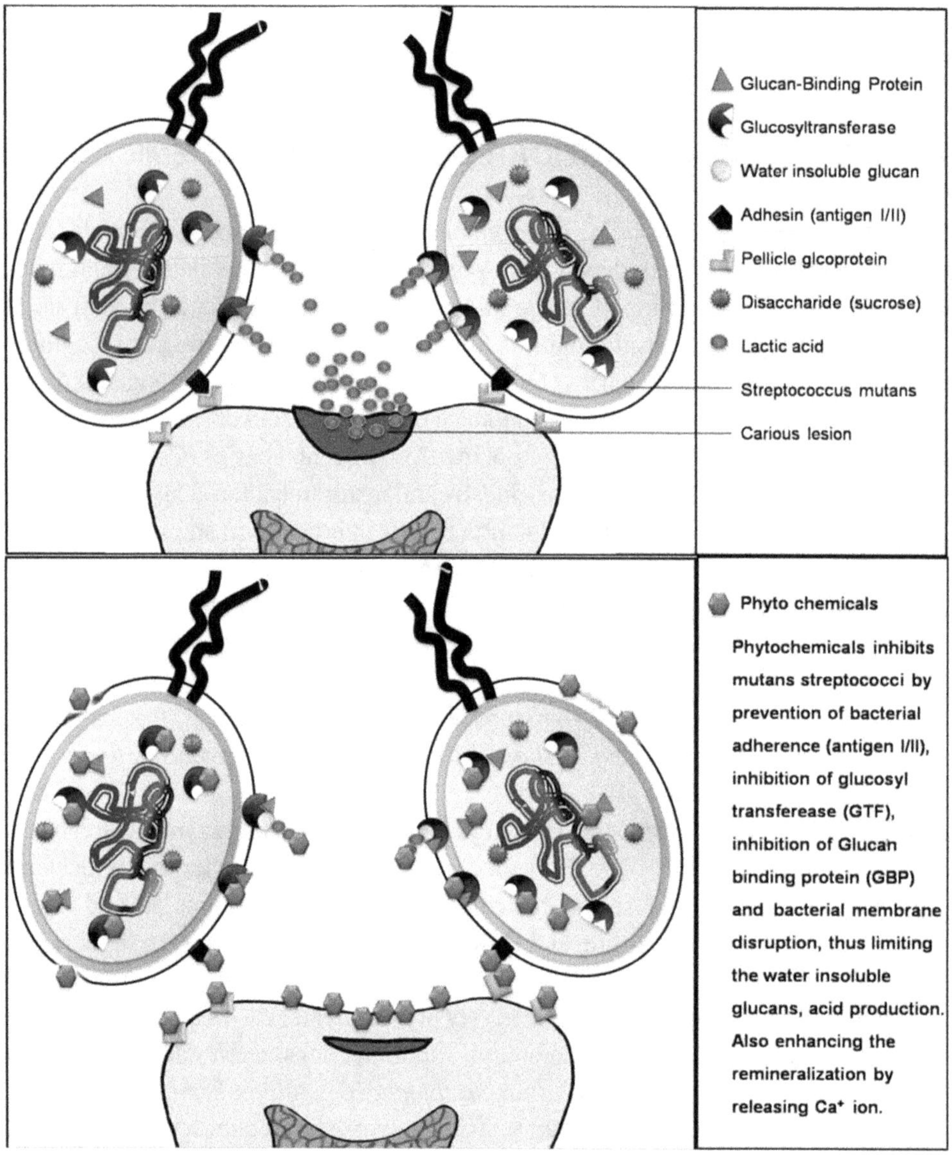

The anticariogenic potential of tannins can be attributed to their ability to inhibit the attachment of bacteria to proline rich proteins of salivary pellicle or with cell surface lipoteichoic acid (Hogg and Embery, 1982; Wolinsky and Sote, 1984). They are also found to interact with insoluble enzyme, binding to an active site or changing enzyme structure. It also prevents the adsorption of GTF to bacterial cells (Jagtap and Karkera, 2000). Catechins act by damaging the bacterial membranes and disrupting the function of cell surface proteins (Ikigai et al., 1993; Hamilton-Miller, 2001). The polyphenols of oolong tea extracts that are structurally different from their green and black tea counterparts affect the bacterial adherence and GTF activity (Nakahara et al., 1993; Ooshima et al., 1993; Matsumoto et al., 1999).

Evidences show pellicle adsorbed enzymes are resistant to antimicrobial agents than thoe in solution (Wunder and Bowen, 1999). Several other researchers have also reported on the effect of antimicrobials on bacterial adherence (Yu et al., 2007; Ooshima et al., 2000; Weiss et al. 2002; Steinberg et al. 2004; Yamanaka et al. 2004; Duarte et al. 2006; Koo et al. 2006; Yatsuda et al., 2005; Limsong et al., 2004). Some phytochemicals that inhibit the bacterial adherence include neoandrographolide in *Andrographis paniculata* whole aerial parts (Chan et al., 1968), anthraquinones in *Cassia alata* leaves (Hauptmann and Lacerda Nazario, 1950), tannins in guava leaves (Osman et al., 1974), limonoid and chromone in *Harrisonia perforate* (Mitsunaga et al., 1995).

Glass adherence assay is an effective in vitro model o assess sucrose dependent bacterial adhesion (Koo et al. 2000; Carter et al. 2001; Nostro et al. 2004). Guaijaverin isolated from Psidium guajava inhibited the adherence of *S. mutans* to glass surface. However, the effect differs between different bacterial strains (Limsong et al., 2004). P. guajava and *Piper betle* extracts significantly inhibited the adhesion of early plaque colonizers (Fathilah and Rahim, 2003). The bioactivities of these two extracts are conferred by flavanoids and tannins (Osman et al. 1974). Flavanoids act by inhibiting the GTase activity (Iio et al. 1984). Thereby controlling the conversion of sucrose to sticky insoluble glucans and inhibiting the adherence of *S. mutans* to tooth surface.

Cell surface hydrophobicity affects the bacterial adhesion to tooth surface. This is evidenced by the reports of Westergren and Olsson (1983) which showed that *Streptococcus sanguis* and *Streptococcus salivarius* failed to adhere as they lacked cell surface hydrophobicity. The bioactive principles from Nidus vespae were shown to inhibit *S. mutans* adherence due to their high chemical diversity (Xiao et al., 2007). The oolong tea extracts were found to inhibit bacterial adherence by decreasing the hydrophobicity (Matsumoto et al., 1999). High molecular weight polyphenols of apple interfered with the glucan synthesis by glucosyl transferases of *S. mutans* thus inhibiting the sucrose dependent bacterial adherence (Yanagida et al., 2000).

The anticariogenic activity of active ingredients of coffee is due to their adsorption to host surfaces thereby preventing both reversible and irreversible interaction between tooth receptors and bacterial adhesions (Daglia et al., 2002). The propolis extracts inhibit bacterial adhesion by inhibiting glucan synthesis (*Hayacibara et al.,* 2005) which is similar to the mechanism of palatinose (Ooshima et al., 1983).

Inhibition of Glucosyl Transferases (GTF)

The main virulence factor in dental caries is the glucosyl transferases (GTF) produced by mutans streptococci (Yamashita et al., 1993) which convert sucrose to water insoluble glucans and form biofilm wherein metabolic products of bacteria causes demineralization of tooth surfaces (Loesche, 1983). Three types of GTFs, GTF-I, GTF-SI and GTF-S are produced by *S. mutans* which are encoded by gtfB, gtfC and gtfD, respectively (Aoki et al., 1986; Hanada and Kuramitsu, 1988; Hanada and Kuramitsu, 1986). GTF-I in involved in the synthesis of water insoluble glucans in a primer independent manner; GTF-SI synthesizes both water soluble and insoluble glucans and GTF-S synthesizes water soluble glucans in primer dependent manner. In the case of S. sobrinus, glucan synthesis is mediated by four different types of GTFs designated as GTF-I, GTF-U, GTF-T and GTF-S which are encoded by the gtfI, gtfU, gtfT,and gtfS genes, respectively (Nanbu et al., 2000; Russel, 1994). GTF-I and GTF-F are responsible sucrose dependent colonization (Fukushima et al., 1981; 1982; Koga, 1986) whereas GTF-U and GTF-S mediate dental plaque formation. When these enzymes are mutated, the glucan synthesis is disrupted and caries formation is affected Yamashita et al., 1993). Those strains which cannot synthesise insoluble glucans cannot adhere to toth surface resulting in reduced cariogenicity (Yamashita et al., 1993; Munro et al., 1991). Thus, GTF is an important hallmark in the development of dental caries (Eto et al., 1999).

Certain phytochemicals exhibit anticaries activity by inhibiting the glucosyl transferases (Hada et al., 1989; Sakanaka et al., 1989; Otake et al., 1991). Tea catechins are found to inhibit GTF of *S. mutans* and *S. sobrinus* as reported by several workers. The ECCG fraction of tea polyphenols was found to be superior than other catechins in preventing caries (Otake et al., 1991; Hattori et al., 1990). The oolong tea catechins (Nakahara et al., 1993; Hameda et al., 1996), hop bract polyphenols of 36 - 40 kDa (Tagashira et al., 1997), gallotannins (Kakiuchi et al., 1986) and procyanidins from betel nuts (Hada et al., 1989) effectively prevent caries by inhibiting GTF. However, this mechanism slightly differs between different polyphenolic sources. *Andrographis paniculata* and *Cassia alata* could prevent caries by inhibiting the GTF and GBP whereas black tea and *Harrisonia perforate* inhibit adherence by inhibiting GTF activity (Limsong et al., 2004). The GTF inhibitors of Cacao (CEPWS) prevent glucan synthesis by inhibiting dextran dependent water insoluble glucan (WIG) synthesis by the GTF-I and 1,6- α- Dglucan synthesis by the GTF-T, and slightly stimulated highly branched 1,6aDglucan synthesis by the GTF-U. Similar GTF-I inhibiting activity was observed with ellagic acid (sawamura et al., 1992), maillard reaction products (Kobayashi et al., 1988), and mutastein (Hayashida et al., 1997).

The chemical fractions of Nidus vespae were found to inhibit cell associated and extracellular GTF of *S. mutans* (Xiao et al., 2007). Propolis extracts also follow the same mechanism in controlling dental caries. Green propolis prevents soluble glucan synthesis by inhibiting GTF. The flavanoids such as galangin, pinocembrin, kaempferol, sakuranetin and quercetin, and other phenolic compounds, e.g. cinnamic acid and its derivatives, present in the propolis extracts are responsible for the GTF inhibition (Koo et al., 2004; Lio et al., 1984; Park et al., 1997, 1998). The activity of flavanoids on sugar related enzymes like glyoxalase I (Lio et al., 1983), xanthine oxidase and a-glucosidase (Lio et al., 1984) makes them target compounds in anticaries treatment.

Tannins which are generally used as detoxifying agents were found to precipitate proteins and thereby inhibit key enzymes like aminoacylase, 3- alactosidase, and glucose isomerase (Paolino et al., 1980). Their ability to interact with proline rich salivary and bacterial enzymes makes them potential choices for dental caries treatment (Hogg and Embery, 1982; Wolinsky and Sote, 1984). These interactions mainly involve their ability to alter adhesion. Cell aggregation in a plaque is brought about by the interaction between GBP of *S, mutans* and glucans (McCabe and Hamelik, 1978; Russell, 1979). This mechanism is disrupted by several phytochemicals including plant lectins (Staat et al., 1978; Gibbons and Dankers, 1981; Wu-Yuan et al., 1978).

Inhibition of Acid Production

Another important virulent factor involved in caries formation is acid production from carbohydrates by mutans streptococci. Acid production is always related to energy producing processes. Less acid production is indicative of reduced growth. Several bioactive phytochemicals have an inhibitory role on acid production by cariogenic bacteria. The chemical fractions of Nidus vespae can reduce acid production even at low concentrations ml (Xiao et al., 2006). Extracts of cacao bean husk and Brazilian green propolis also show their anticariogenic activity by inhibiting acid production (Leitao et al., 2004).

The plant secondary metabolites, by virtue of their membrane permeability ability, inhibit cytosolic enzyme activities (Havsteen, 1983). The fatty acids produced by *Piper betle* L. act on glycolytic enzymes thereby reducing acid production which contributes to its antibacterial activity (Sheu and Freese, 1972;

Booth, 1985; Iwami et al., 1995). Polygonum cuspidatum root extract confers its bioactivity by inhibiting the metabolic enzymes resulting in reduced acid production (Havsteen, 1983). It also affects the acid tolerance of bacterial species (Song et al., 2006). Miswak exhibits its bioactivity by increasing plaque pH (Sofrata et al., 2006). The pH drop affected by the streptococcal acid production was alleviated by phytochemicals from Rheum undulatum. Thus, one of the important therapeutic approaches against dental caries is to disrupt the ability of mutans streptococci to produce acid and glucans on tooth surface.

Inhibition of Salivary and Bacterial Enzymes

Some phytochemicals exert their anticariogenic activity by inhibiting the salivary and bacterial amylases which are crucial in the development of caries. Several workers have reported that tea extracts inhibit the activity of salivary and bacterial amylases (Kashket and Paolino, 1988; Zhang and Kashket, 1998). However, black tea was found to be more active than green tea which is due the presence of high molecular weight polyphenols in black tea. The active principles responsible for this activity in black tea are simple catechins and theaavins which act upon salivary amylases. Some of the bioactive principles like GTF inhibitors, H-Tan (hydrolysable tannin: H-Tan) such as tannic acid and EGCg inhibited salivary R-amylase activity (Yanagida et al., 2000).

Other Mode of Inhibition

Polyphenols form complexes with proteins and polysaccharides present on bacterial cell wall which bind to cell surface components and inhibit extracellular enzymes (Haslam, 1996). They are also found to bind to metal ions thereby inhibiting the growth (Scalbert, 1991). The more the sites of metal chelation, greater the inhibition of growth (Smullen et al., 2007). Curcuma xanthorrhiza extract (Xan) exerts its activity by mediating morphological changes in *S. mutans* biofilm (Kim et al., 2008). The fatty acids and hydroxyl fatty acid ester components of *Piper betle* L. partition the lipids of bacterial cell membrane subsequently disrupting its structure and making it more permeable to nucleoid coagulating components of the extract (Nalina and Rahim, 2007). Similar activity is also observed with Magnolia bark extract (MBE) of *Magnolia officinalis* (Greenberg et al., 2007). Lipophilic antimicrobials are less effective against gram negative bacteria as they tightly bind to the outer cell wall and cannot enter the cells easily. Some polyphenols inhibit cell wall synthesis or activate those factors responsible for cell wall degradation. Some increase cell permeability while others act by interfering with protein and nucleic acid metabolism.

Tea polyphenols act by disrupting bacterial membranes (Ikigai et al., 1993). These compounds having an affinity for proteins, inhibit amylases and glucosyl transferases and bacterial adherence. The polyphenolic compounds from cacao beans such as epicatechins under go oxidative polymerization resulting in the formation of quinines. This further forms a water soluble GTF inhibitory compound by interacting with proteins and sugar (Ito et al., 2003). Same mechanism was also observed in oolong tea (Nakahara et al., 1993) and apple polyphenols (Yanagida et al., 2000). Polyphenols are also found to exert antimicrobial activity by inhibiting enzyme activity through sulfhydryl groups or non-specific interactions with proteins. The antimicrobial activity of alkaloids and terpenoids is due to their ability to intercalate with DNA and membrane disruption (Cowan, 1999). Few flavanoids and tannins, for example those from *Polygonum cuspidatum*, act by affecting the proton-translocating ATPase or FATPase, a membrane bound enzyme of mutans streptococci (Song et al., 2006). Powerdental, a potential anticariogenic agent inhibit glycolytic enzymes of *S. mutans* and also secretion of proinflammatory cytokine TNF – α (You et al., 2004).

Certain flavanoids for example, those from propolis which include quercetin, galangin, pinocembrin and caffeic acid, benzoic acid, cinnamic acid cause functional and structural changes in the cell wall or cell membrane (Marcucci, 1995; Cook and Samma, 1996; Mirzoeva et al., 1997). Their activity is mainly due to synergistic effects rather than individual components (Amoros et al., 1994; Bonhevi et al., 1994). This fact was supported by the fact that certain phytochemicals act by complex mechanisms involving formation of pseudomulticellular streptococci; disorganization of the cytoplasm, the cytoplasmatic membrane, and the cell wall; partial bacteriolysis; and inhibition of protein synthesis (Takaisikikuni and Schilcher, 1994). Few other antimicrobials act as calcium ion carriers thereby enhancing remineralisation (Cheng et al., 2008; Chu et al., 2006).

SUMMARY

Plant secondary metabolites have continued to fascinate researchers worldwide by their diverse biological properties and the use of plant products to dental caries is emerging as an interesting area in view of the growing concern over the prevalence of this oral disease. Hundreds of phytochemicals showing anticariogenic activities have been identified. These compounds interact with any one of the steps involved in caries development and help the patients to maintain a biologically compatible oral environment. This chapter shows strong evidences for the anticariogenic properties of plant products and advocates the incorporation of such compounds in oral health products like toothpastes, mouthwashes and dental floss.

ACKNOWLEDGMENT

This review work was supported by the Extra Mural Research Funding of AYUSH (Ayurveda, Yoga and Naturopathy, Unani, Siddha and Homoeopathy, [Grant number- Z. 28015/209 /2015-HPC]. Our sincere acknowledgments to Ms. Caroline Regina Delma Joseph Gilbert for her help during the chapter writing process.

REFERENCES

Addai, F.K., Nuamah, I.K. & Parkins, G. (2002). Brief chewing of Garcinia manii stick speedily reverses depression ofsaliva pH after a glucos rinse. *Med. Sci. Monitor, 8*(11), 746-750.

Afolayan, A. J., & Meyer, J. J. M. (1997). The antimicrobial activity of 3,5,7- trihydroxyflavone isolated from the shoots of Helichrysum aureonitens. *Journal of Ethnopharmacology, 57*(3), 177–181. doi:10.1016/S0378-8741(97)00065-2 PMID:9292410

Akpata, J. E., & Akimimisi, E. O. (1977). Antibacterial activity of extracts of some Nigerian chewing sticks. *Caries Research, 18*, 216–225.

Al-hebshi, N., Al-haroni, M., & Skaug, N. (2006). *In vitro* antimicrobial and resistance modifying activities of aqueous crude khat extracts against oral microorganisms. *Archives of Oral Biology*, *51*(3), 183–188. doi:10.1016/j.archoralbio.2005.08.001 PMID:16248981

Almas, K. (2001). The antimicrobial effects of seven different types of Asian chewing sticks. *Odonto-Stomatologie Tropicale*, *24*(96), 17–20. PMID:11887585

Almas & Al-Zeid. (2004). The immediate antimicrobial effect of a toothbrush and miswak on cariogenic bacteria: a clinical study. *J. Contemp. Dent. Pract.*, *5*(1), 105-114.

Alviano, W.S., Alviano, D.S., Diniz, C.G., Antoniolli, A.R., Alviano, C., Farias, L.M., Carvalho, M.A.R., Souza, M.M.G., & Bolognese, A.M. (2008). *In vitro* antioxidant potential of medicinal plant extracts and their activities against oral bacteria based on Brazilian folk medicine. *Arch. Oral Biol.*, *53*(6), 545-552.

Amadi, E. S., Oyeka, C. A., Onyeagba, R. A., Ogbogu, O. C., & Okoli, I. (2007). Antimicrobial screening of Breynia nivosus and Ageratum conyzoides against dental caries causing organisms. *The Journal of Biological Sciences*, *7*(2), 354–358. doi:10.3923/jbs.2007.354.358

Ambrosio, S. R., Furtado, N. A., de-Oliveira, D. C., da-Costa, F. B., Martins, C. H., de-Carvalho, T. C., ... Veneziani, R. C. (2008). Antimicrobial activity of kaurane diterpenes against oral pathogens. *Z Naturforsch C.*, *63*(5-6), 326–330. doi:10.1515/znc-2008-5-603 PMID:18669015

Amoros, M., Lurton, E., Boustie, J., Girre, L., Sauvager, F., & Cormier, M. (1994). Comparison of the anti-herpes simplex virus activities of propolis and 3-methyl-but-2-enyl caffeate. *Journal of Natural Products*, *57*(5), 644–647. doi:10.1021/np50107a013 PMID:8064297

Anand, T.D., Pothiraj, C., Gopinath, R.M. & Kayalvizhi, B. (2008). Effect of oil-pulling on dental caries causing bacteria. *African Journal of Microbiology Research, 2*, 63-66.

Aoki, H., Shiroza, T., Hayakawa, M., Sato, S., & Kuramitsu, H. K. (1986). Cloning of a Streptococcus mutans glucosyltransferase gene coding for insoluble glucan synthesis. *Infection and Immunity*, *53*, 587–594. PMID:3017865

Apisariyakul, A., Vanittana, N., & Buddasuki, D. (1995). Antifungal activity of turmeric oil extracted from Curcuma longa (Zingiberaceae). *Journal of Ethnopharmacology*, *49*(3), 163–169. doi:10.1016/0378-8741(95)01320-2 PMID:8824742

Avirutnant, W., & Pongpan, A. (1983). Antimicrobial activity of some Thai flowers and plants. *Mahidol Univ. Journal of Pharmaceutical Sciences*, *10*, 81–86.

Banskota, A. H., Tezuka, Y., Prasain, J. K., Matsushige, K., Saiki, I., & Kadota, S. (1998). Chemical constituents of Brazilian propolis and their cytotoxic activities. *Journal of Natural Products*, *61*(7), 896–900. doi:10.1021/np980028c PMID:9677271

Barea-Alvarez, M., Benito, M. T., Olano, A., Jimeno, M. L., & Moreno, F. J. (2014). Synthesis and Characterization of Isomaltulose-Derived Oligosaccharides Produced by Transglucosylation Reaction of Leuconostoc mesenteroides Dextransucrase. *Journal of Agricultural and Food Chemistry*, *62*(37), 9137–9144. doi:10.1021/jf5033735 PMID:25175804

Barroso, G. M. (1986). *Sistemática de Angiospermas do Brasil* (Vol. 3). Imprensa Universitária, Universidade Federal de Viçosa, Viçosa.

Beckers, H. J. (1988). Influence of xylitol on growth, establishment, and cariogenicity of *Streptococcus mutans* in dental plaque of rats. *Caries Research, 22*(3), 166–173. doi:10.1159/000261100 PMID:3163524

Biswas, K., Chattopadhyay, I., Ranajit, K. B., & Bandyopadhya, U. (2002). Biological activities and medical properties of neem (Azadirachta indica). *Current Science, 82*(11), 1336–1345.

Bonhevi, J. S., Coll, F. V., & Jord, R. E. (1994). The composition, active components and bacteriostatic activity of of Propolis in Dietetics. *Laboratori Agrari de la Generalitat de Catalunya, 08348 Cabrils (Barcelona), Spain. Journal of the American Oil Chemists' Society, 71*, 5.

Booth, I. R. (1985). Regulation of cytoplasmic pH in bacteria. *Microbiological Reviews, 49*, 329–378. PMID:3912654

Borris, R. P. (1996). Natural products research: Perspectives from a major pharmaceutical company. *Journal of Ethnopharmacology, 51*(1/3), 29–38. doi:10.1016/0378-8741(95)01347-4 PMID:9213624

Botelho, M. A., Nogueira, N. A. P., Bastos, G. M., Fonseca, S. G. C., Lemos, T. L. G., Matos, F. J. A., ... Brito, G. A. C. (2007). Antimicrobial activity of the essential oil from Lippia sidoides, carvacrol and thymol against oral pathogens. *Brazilian Journal of Medical and Biological Research, 40*(3), 349–356. doi:10.1590/S0100-879X2007000300010 PMID:17334532

Bowden, G. H., & Hamilton, I. R. (1998). Survival of Oral Bacteria. *Critical Reviews in Oral Biology and Medicine, 9*(1), 54–85. doi:10.1177/10454411980090010401 PMID:9488248

Bradshaw, D. J., & Marsh, P. D. (1998). Analysis of pH-driven disruption of oral microbial communities *in vitro. Caries Research, 32*(6), 456–462. doi:10.1159/000016487 PMID:9745120

Brighenti, F. L., Luppens, S. B., Delbem, A. C., Deng, D. M., Hoogenkamp, M. A., Gaetti-Jardim, E. Jr, ... ten Cate, J. M. (2008). Effect of Psidium cattleianum leaf extract on *Streptococcus mutans* viability, protein expression and acid production. *Caries Research, 42*(2), 148–154. doi:10.1159/000121439 PMID:18367836

Bruneton, J. (1995). *Pharmacognosy, Phytochemistry and Medicinal Plants*. Intercept. Ltd.

Cai, L., & Wu, C. D. (1996). Compounds from Syzygium aromaticum possessing growth inhibitory activity against oral pathogens. *Journal of Natural Products, 59*(10), 987–990. doi:10.1021/np960451q PMID:8904847

Carter, K., Landini, G., & Malmseys, A. D. (2001). Plaque removal characteristics of electric toothbrushes using an *in vitro* plaque model. *Journal of Clinical Periodontology, 28*(11), 1045–1049. doi:10.1034/j.1600-051X.2001.281109.x PMID:11686826

Ccahuana-Vasquez, R. A., Santos, S. S., Koga-Ito, C. Y., & Jorge, A. O. (2007). Antimicrobial activity of Uncaria tomentosa against oral human pathogens. *Brazilian Oral Research, 21*(1), 46–50. doi:10.1590/S1806-83242007000100008 PMID:17426895

Chan, W. R., Taylor, D. R., Willis, C. R., & Fehlhaber, H. W. (1968). The structure of neoandrographolide - a diterpene glucoside from andrographis paniculata nees. *Tetrahedron Letters*, *9*(46), 4803–4806. doi:10.1016/S0040-4039(00)75962-4

Cheng, L., Li, J., Hao, Y., & Zhou, X. (2008). Effect of compounds of Galla chinensis and their combined effects with fluoride on remineralization of initial enamel lesion *in vitro. Journal of Dentistry*, *36*(5), 369–373. doi:10.1016/j.jdent.2008.01.011 PMID:18308448

Cheng, L., Li, J., Hao, Y., & Zhou, X. (2008). Effect of compounds of Galla chinensis and their combined effects with fluoride on remineralization of initial enamel lesion *in vitro. Journal of Dentistry*, *36*(5), 369–373. doi:10.1016/j.jdent.2008.01.011 PMID:18308448

Chopra, R. N., Chopra, I. C., Handa, K. I., & Kapur, L. D. (1958). *Chopra's indegenous drugs of India* (2nd ed.). Calcutta: UN Dhur and Sons.

Chu, J., Li, J., Hao, Y., & Zhou, X. (2006). Effect of chemical compounds of *Galla chinensis* on enamel surface rehardening *in vitro. Chin J Stomatol*, *41*, 616–617. PMID:17129452

Cook, N., & Samma, S. (1996). Flavonoids-chemistry, metabolism, cardioprotective effects and dietary sources. *The Journal of Nutritional Biochemistry*, *7*(2), 66–76. doi:10.1016/0955-2863(95)00168-9

Cowan, M. M. (1999). Plant products as antimicrobial agents. *Clinical Microbiology Reviews*, *12*, 564–582. PMID:10515903

Daglia, M., Tarsi, T., Papetti, A., Grisoli, P., Dacarro, C., Pruzzo, C., & Gazzani, G. (2002). Antiadhesive Effect of Green and Roasted Coffee on Streptococcus mutans' Adhesive Properties on Saliva-Coated Hydroxyapatite Beads. *Journal of Agricultural and Food Chemistry*, *50*(5), 1225–1229. doi:10.1021/jf010958t PMID:11853508

Darout, I. A., Albandar, J. M., Skaug, N., & Ali, R. W. (2002). Salivary microbiota levels in relation to periodontal status, experience of caries and miswak use in Sudanese adults. *Journal of Clinical Periodontology*, *29*(5), 411–420. doi:10.1034/j.1600-051X.2002.290505.x PMID:12060423

Dhifi, W., Bellili, S., Jazi, S., Bahloul, N., & Mnif, W. (2016). Essential Oils' Chemical Characterization and Investigation of Some Biological Activities. A Critical Review. Medicines (Basel), 3(4).

Didry, N., Dubreuil, L., Trotin, F., & Pinkas, M. (1998). Antimicrobial activity of aerial parts of Drosera peltata Smith on oral bacteria. *Journal of Ethnopharmacology*, *60*(1), 91–96. doi:10.1016/S0378-8741(97)00129-3 PMID:9533437

Dixon, R. A., Dey, P. M., & Lamb, C. J. (1983). Phytoalexins: Enzymology and molecular biology. *Adv in Enzy and Rel A. Molecular Biology*, *53*, 1–36.

Duarte, S., Gregoire, S., Singh, A., Vorsa, N., Schaich, K., Bowen, W., & Koo, H. (2006). Inhibitory effects of cranberry polyphenols on formation and acidogenicity of *Streptococcus mutans* biofilms. *FEMS Microbiology Letters*, *257*(1), 50–56. doi:10.1111/j.1574-6968.2006.00147.x PMID:16553831

Dzink, J. L., & Socransky, S. S. (1985). Comparative in vitro activity of sanguinarine against oral microbial isolates. *Antimicrobial Agents and Chemotherapy*, *27*(4), 663–665. doi:10.1128/AAC.27.4.663 PMID:4004199

Ebel, J., & Grisebach, H. (1988). Defense strategies of soybean against the fungus Phytophthora megasperma f. sp. glycinea: A molecular analysis. *Trends in Biochemical Sciences*, *13*(1), 23–27. doi:10.1016/0968-0004(88)90014-X PMID:3072693

Edelstein, B. L., & Douglass, C. W. (1995). Dispelling the myth that 50 percent of U.S. schoolchildren have never had a cavity. *Public Health Reports (Washington, D.C.)*, *110*, 522–530. PMID:7480606

El-Said, F., Fadulu, S. O., Kuye, J. O., & Sofowora, E. A. (1971). Native cures in Nigeria: II The antimicrobial properties of the buffered extracts of chewing sticks. *Lloydia*, *34*, 172–174.

Eloff, J. N. (1998). Which extractant should be used for the screening and isolation of antimicrobial components from plants? *Journal of Ethnopharmacology*, *60*(1), 1–8. doi:10.1016/S0378-8741(97)00123-2 PMID:9533426

Elujoba, A. A., Odeleye, O. M., & Ogunyemi, C. M. (2005). Traditional Medical Development for medical and dental primary Health care Delivery System in Africa. *African Journal of Traditional, Complementary, and Alternative Medicines*, *2*, 46.

Eto, A., Saido, T. C., Fukushima, K., Tomioka, S., Imai, S., Nisizawa, T., & Hanada, N. (1999). Inhibitory effect of a self-derived peptide on glucosyltransferase of *Streptococcus mutans*. Possible novel anticaries measures. *The Journal of Biological Chemistry*, *274*(22), 15797–15802. doi:10.1074/jbc.274.22.15797 PMID:10336482

Fani, M. M., Kohanteb, J., & Dayaghi, M. (2007). Inhibitory activity of garlic Allium sativum extract on multi drug resistant *S.mutans*. *J. Indian Soc. Pedad. Prev. Dent*, *25*(4), 164–168. PMID:18007101

Fathilah, A. R., Rahim, Z. H., Othman, Y., & Yusoff, M. (2009). Effect of *Piper betle* and *Psidium guajava* Extracts on Dental Plaque Bacteria. *Pakistan Journal of Biological Sciences*, *12*(6), 518–521. doi:10.3923/pjbs.2009.518.521 PMID:19580002

Fessenden, R. J., & Fessenden, J. S. (1982). *Organic Chemistry* (2nd ed.). Boston: Willard Grant Press.

Fukushima, K., Motoda, R., Takada, K., & Ikeda, T. (1981). Resolution of Streptococcus mutans glucosyltransferase into two components essential to water-insoluble glucan synthesis. *FEBS Letters*, *128*(2), 213–216. doi:10.1016/0014-5793(81)80083-X PMID:6455304

Fukushima, K., Takada, K., Motoda, R., & Ikeda, T. (1982). Independence of water-insoluble glucan synthesis and adherence of Streptococcus mutans to smooth surfaces. *FEBS Letters*, *149*(2), 299–303. doi:10.1016/0014-5793(82)81121-6 PMID:6217990

Furiga, A., Lonvaud, A., & Badet, C. (2009). *In vitro* study of antioxidant capacity and antibacterial activity on oral anaerobes of a grape seed extract. *Food Chemistry*, *113*(4), 1037–1040. doi:10.1016/j.foodchem.2008.08.059

Gales, M. A., & Nguyen, T. M. (2000). Sorbitol compared with xylitol in prevention of dental caries. *The Annals of Pharmacotherapy*, *34*(1), 98–100. doi:10.1345/aph.19020 PMID:10669192

Ganguly, R., Mishra, P., & Sharma, A. (2001). Microbes and infection. *Indian Journal of Microbiology*, *41*, 211–213.

Gazi, M. I., Davies, T. J., Al-Bagieh, N., & Cox, S. W. (1992). The immediate and medium-term effects of Meswak on the composition of mixed saliva. *Journal of Clinical Periodontology*, *19*(2), 113–117. doi:10.1111/j.1600-051X.1992.tb00449.x PMID:1602035

Geissman, T. A. (1963). Flavonoid compounds, tannins, lignins and related compounds. In M. Florkin & E. H. Stotz (Eds.), *Pyrrole Pigments, Isoprenoid Compounds and Phenolic Plant Constituents* (p. 265). New York: Elsevier Press. doi:10.1016/B978-1-4831-9718-0.50018-7

Gibbons, R. J. (1984). Adherent interactions which may affect microbial ecology in the mouth. *Journal of Dental Research*, *63*(3), 378–385. doi:10.1177/00220345840630030401 PMID:6583240

Gibbons, R. J., & Dankers, I. (1981). Lectin-Like Constituents of Foods Which React with Components of Serum, Saliva, and *Streptococcus mutans*. *Applied and Environmental Microbiology*, *41*(4), 880–888. PMID:6786220

Gibbons, R. J., & Houte, J. V. (1975). Bacterial adherence in oral microbial ecology. *Annual Review of Microbiology*, *29*(1), 19–44. doi:10.1146/annurev.mi.29.100175.000315 PMID:1180512

Glass, R. L. (1982). The first international conference on the declining prevalence of dental caries. *Journal of Dental Research*, *61*(Spec Iss), 1304.

Greenberg, M., Urnezis, P., & Tian, M. (2007). Compressed mints and chewing gum containing magnolia bark extract are effective against bacteria responsible for oral malodor. *Journal of Agricultural and Food Chemistry*, *55*(23), 9465–9469. doi:10.1021/jf072122h PMID:17949053

Gregoire, S., Singh, A. P., Vorsa, N., & Koo, H. (2007). Influence of cranberry phenolics on glucan synthesis by glucosyltransferases and *Streptococcus mutans* acidogenicity. *Journal of Applied Microbiology*, *103*(5), 1960–1968. doi:10.1111/j.1365-2672.2007.03441.x PMID:17953606

Grenby, T. H., Phillips, A., & Mistry, M. (1989). Studies of the dental properties of lactitol compared with five other bulk sweeteners *in vitro*. *Caries Research*, *23*(5), 315–319. doi:10.1159/000261199 PMID:2766316

Hada, L. S., Kakiuchi, N., Hattori, M., & Namba, T. (1989). Identification of antibacterial principles against *Streptococcus mutans* inhibitory principles against glucosyltransferase from the seed of Areca catechu L. *Phytotherapy Research*, *3*(4), 140–144. doi:10.1002/ptr.2650030406

Hada, S., Kakiuchi, N., Hattori, M., & Namba, T. (1989). Identification of antibacterial principles against *Streptococcus mutans* and inhibitory principles against glucosyltransferase from the seed of Areca catechu L. *Phytotherapy Research*, *3*(4), 140–144. doi:10.1002/ptr.2650030406

Hamilton-Miller, J. M. T. (2001). Anti-cariogenic effects of tea (*Camellia sinensis*). *Journal of Medical Microbiology*, *50*(4), 299–302. doi:10.1099/0022-1317-50-4-299 PMID:11289514

Hanada, N., & Kuramitsu, H. K. (1988). Isolation and characterization of the Streptococcus mutans gtfC gene, coding for synthesis of both soluble and insoluble glucans. *Infection and Immunity*, *56*, 1999–2005. PMID:2969375

Hanada, N., & Kuramitsu, H. K. (1989). Isolation and characterization of the *Streptococcus mutans* gtfD gene, coding for primer-dependent soluble glucan synthesis. *Infection and Immunity*, *57*, 2079–2085. PMID:2543630

Haslam, E. (1996). Natural polyphenols (vegetable tannins) as drugs: Possible modes of action. *Journal of Natural Products*, *59*(2), 205–215. doi:10.1021/np960040+ PMID:8991956

Haslam, E. (1996). Natural polyphenols (vegetable tannins) as drugs: Possible modes of action. *Journal of Natural Products*, *59*(2), 205–215. doi:10.1021/np960040+ PMID:8991956

Hattori, M., Kusumoto, I. T., Namba, T., Ishigami, T., & Hara, Y. (1990). Effect of tea polyphenols on glucan synthesis by glucosyltransferase from *Streptococcus mutans*. *Chemical & Pharmaceutical Bulletin*, *38*(3), 717–720. doi:10.1248/cpb.38.717 PMID:2140716

Hauptmann, H., & Lacerda Nazario, L. (1950). Some constituents of leaves of *Cassia alata*. *Journal of the American Chemical Society*, *72*(4), 1492–1495. doi:10.1021/ja01160a019

Havsteen, B. (1983). Flavonoids, a class of natural products of high pharmacological potency. *Biochemical Pharmacology*, *32*(7), 1141–1148. doi:10.1016/0006-2952(83)90262-9 PMID:6342623

Havsteen, & Havsteen, B. (1983). Flavonoids, a class of natural products of high pharmacological potency. *Biochem. Pharmacol.*, *32*, 1141-1148.

Hayachibara, M. F., Koo, H., Rosalen, P. L., Duarte, S., Franco, E. M., Bowen, W. H., ... Cury, J. A. (2005). *In vitro* and in vivo effects of isolated fractions of Brazilian propolis on caries development. *Journal of Ethnopharmacology*, *101*(1-3), 110–115. doi:10.1016/j.jep.2005.04.001 PMID:15913934

Hayashida, O., Hasumi, K., & Endo, A. (1997). Chemical and Functional Properties of Mutastein, an Inhibitor of Insoluble Glucan Synthesis by *Streptococcus sobrinus*. *Bioscience, Biotechnology, and Biochemistry*, *61*(4), 588–591. doi:10.1271/bbb.61.588 PMID:9145515

Hayes, C. (2001). The effect of non-cariogenic sweeteners on the prevention of dental caries: A review of the evidence. *Journal of Dental Education*, *65*, 1106–1109. PMID:11699985

Hebbar, S. S., Harsha, V. H., Shripathi, V., & Hegde, G. R. (2004). Ethnomedicine of Dharwad district in Karnataka, India--plants used in oral health care. *Journal of Ethnopharmacology*, *94*(2-3), 261–266. doi:10.1016/j.jep.2004.04.021 PMID:15325728

Hebbar, S. S., Harsha, V. H., Shripathi, V., & Hegde, G. R. (2004). Ethnomedicineof Dharwad district in Karnataka India-plants used in oral health care. *Journal of Ethnopharmacology*, *94*(2-3), 261–266. doi:10.1016/j.jep.2004.04.021 PMID:15325728

Heisey, R. M. B., & Gorha, B. (1992). Antimicrobial effects of plant extracts on *Streptococcus mutans, Candida albicans, Trichophyton rubrum* and other micro-organisms. *Letters in Applied Microbiology*, *14*(4), 136–139. doi:10.1111/j.1472-765X.1992.tb00668.x

Hogg, S. D., & Embery, G. (1982). Blood-group reactive glycoprotein from human saliva interacts with lipoteichoic acid on the surface of *Streptococcus sanguis* cells. *Archives of Oral Biology*, *27*(3), 261–268. doi:10.1016/0003-9969(82)90060-7 PMID:6953942

Hogg, S. D., & Embery, G. (1982). Blood-group-reactive glycoprotein from human saliva interacts with lipoteichoic acid on the surface of *Streptococcus sanguis* cells. *Archives of Oral Biology*, *27*(3), 261–268. doi:10.1016/0003-9969(82)90060-7 PMID:6953942

Huang, Z. L., Mochizuki, T., Qu, W. M., Hong, Z. Y., Watanabe, T., Urade, Y., & Hayaishi, O. (2006). Altered sleep-wake characteristics and lack of arousal response to H3 receptor antagonist in histamine H1 receptor knockout mice. *Proceedings of the National Academy of Sciences of the United States of America*, *103*(12), 4687–4692. doi:10.1073/pnas.0600451103 PMID:16537376

Hwang, J. K., Chung, J. Y., Baek, N. I., & Park, J. H. (2004). Isopanduratin A from *Kaempferia pandurata* an active antibacterial agent against cariogenic *Streptococcus mutans. International Journal of Antimicrobial Agents*, *23*(4), 377–381. doi:10.1016/j.ijantimicag.2003.08.011 PMID:15081087

Hwang, J. K., Chung, J. Y., Baek, N. I., & Park, J. H. (2004). Isopanduratin A from Kaempferia pandurata as an active antibacterial agent against cariogenic Streptococcus mutans. *International Journal of Antimicrobial Agents*, *23*(4), 377–381. doi:10.1016/j.ijantimicag.2003.08.011 PMID:15081087

Iio, M., Uyeda, M., Iwanani, T., & Nakagawa, Y. (1984). Flavonoids as a possible preventive of dental caries. *Agricultural and Biological Chemistry*, *48*, 2143–2145.

Ikigai, H., Nakae, T., Hara, Y., & Shimamura, T. (1993). Bactericidal catechins damage the lipid bilayer. *Biochimica et Biophysica Acta*, *1147*(1), 132–136. doi:10.1016/0005-2736(93)90323-R PMID:8466924

Ikigai, H., Nakae, T., Hara, Y., & Shimamura, T. (1993). Bactericidal catechins damage the lipid bilayer. *Biochimica et Biophysica Acta*, *1147*(1), 132–136. doi:10.1016/0005-2736(93)90323-R PMID:8466924

Ito, K., Nakamura, Y., Tokunaga, T., Iijima, D., & Fukushima, K. (2003). Anti-cariogenic properties of a water-soluble extract from cacao. *Bioscience, Biotechnology, and Biochemistry*, *67*(12), 2567–2573. doi:10.1271/bbb.67.2567 PMID:14730134

Ito, M., Uyeda, M., Iwatani, T., & Nakagawa, Y. (1984). Flavonoids as a possible preventive of dental caries. *Agricultural and Biological Chemistry*, *48*, 2143–2145.

Iwami, Y., Schachtele, C. F., & Yamada, T. (1995). Effect of sucrose monolaurate on acid production, levels of glycolytic intermediates and enzyme activities of *Streptococcus mutans* NCTC 10449. *Journal of Dental Research*, *74*(9), 1613–1617. doi:10.1177/00220345950740091801 PMID:7560425

Iwu, M. M., Duncan, A. R., & Okunji, C. O. (1999). New Antimicrobials of plant origin. In Perspectives in New crops and New uses. ASHS Press.

Jagtap, A. G., & Karkera, S. G. (2000). Extract of Juglandaceae regia inhibits growth, in-vitro adherence, acid production and aggregation of Streptococcus mutans. *The Journal of Pharmacy and Pharmacology*, *52*(2), 235–242. doi:10.1211/0022357001773751 PMID:10714956

Jagtap, A. G., & Karkera, S. G. (2000). Extracts of *Juglandaceae regia* inhibits growth *in vitro* adherence, acid production and aggregation of *Streptococcus mutans. The Journal of Pharmacy and Pharmacology*, *52*(2), 235–242. doi:10.1211/0022357001773751 PMID:10714956

Jonathan, E. K., Anna, K. J., & Johannes, V. S. (2000). Zulu medicinal plants with antibacterial activity. *Journal of Ethnopharmacology*, *69*(3), 241–246. doi:10.1016/S0378-8741(99)00147-6 PMID:10722206

Kakiuchi, N., Hattori, M., Nishizawa, M., & Yamagishi, T. (1986). Studies on dental caries prevention by traditional medicines. VIII. Inhibitory effect of various tannins on glucan synthesis by glucosyltransferase from Streptococcus mutans. *Chemical & Pharmaceutical Bulletin*, *34*(2), 720–725. doi:10.1248/cpb.34.720 PMID:2939967

Kakiuchi, N., Hattori, M., Nishizawa, M., Yamagishi, T., Okuda, T., & Namba, T. (1986). Studies on dental caries prevention by traditional medicines. VIII. Inhibitory effect of various tannins on glucan synthesis by glucosyltransferase from Streptococcus mutans. *Chemical & Pharmaceutical Bulletin*, *34*(2), 720–725. doi:10.1248/cpb.34.720 PMID:2939967

Kashke, S., & Paolino, V. J. (1988). Inhibition of salivary amylase by water-soluble extracts of tea. *Archives of Oral Biology*, *33*(11), 845–846. doi:10.1016/0003-9969(88)90110-0 PMID:2476976

Kashket, S., Paolino, V. J., Lewis, D. A., & Van Houte, J. (1985). *In-vitro* inhibition of glucosyltransferase from the dental plaque bacterium *Streptococcus mutans* by common beverages and food extracts. *Archives of Oral Biology*, *30*(11-12), 822–826. doi:10.1016/0003-9969(85)90138-4 PMID:2938561

Katsura, H., Tsukiyama, R. I., Suzuki, A., & Kobayashi, M. (2001). *In vitro* antimicrobial activities of bakuchiol against oral microorganisms. *Antimicrobial Agents and Chemotherapy*, *45*(11), 3009–3013. doi:10.1128/AAC.45.11.3009-3013.2001 PMID:11600349

Keating, G., & O' Kennedy, R. (1997). Coumarin cytotoxicity activity on albino rats. John Wiley & Sons.

Kim, J. E., Kim, H. E., Hwang, J. K., Lee, H. J., Kwon, H. K., & Kim, B. I. (2008). Antibacterial characteristics of *Curcuma xanthorrhiza*extract on *Streptococcus mutans* biofilm. *Journal of Microbiology (Seoul, Korea)*, *46*(2), 228–232. doi:10.100712275-007-0167-7 PMID:18545974

Kim, J. H. (1997). Anti-bacterial action of onion (Allium cepa L.) extracts against oral pathogenic bacteria. *The Journal of Nihon University School of Dentistry*, *9*(3), 136–141. doi:10.2334/josnusd1959.39.136 PMID:9354029

King, R. M., & Robinson, H. (1987). *The Genera of the Eupatorieae (Asteraceae). Monographs in Systematic Botany*. St. Louis, MO: Missouri Botanical Garden.

Kobayashi, I. (1988). A study of the esthetic coloring of teeth by spectroradiometry-concerning the maxillary anterior teeth of the middle and old aged people. [Nippon Hotetsu Shikka Gakkai Zasshi]. *J Jpn Prosthodont Soc*, *32*(2), 439–448. doi:10.2186/jjps.32.439 PMID:3075715

Kocry, T. (1983). The use of chewing sticks in preventive oral hygiene. *Clinical Preventive Dentistry*, *5*, 11–14. PMID:6607809

Koga, T., Okahashi, N., Asakawa, H., & Hamada, S. (1985). Adherence of Streptococcus mutans to tooth surfaces. In S. Hamada, S. Michalek, H. Kiyono, L. Menaker, & J. R. McGhee (Eds.), Molecular Microbiology and Immunology of Streptococcus mutans (pp. 133–143). Academic Press.

Kohda, H., Kozai, K., Nagasaka, N., Miyake, Y., Suginaka, H., Hidaka, K., & Yamasaki, K. (1985). Prevention of dental caries by oriental Folk medicines- Active Principle of Zizyphi fructus for inhibition of insoluble glucan formation by carcenogenic bacterium *Streptococcus mutans. Planta Medica*, *52*(02), 119–120. doi:10.1055-2007-969095

Koo, H., Schobel, B., Scott-Anne, K., Watson, G., Bowen, W. H., Cury, J. A., ... Park, Y. K. (2005). Apigenin and *tt*-farnesol with fluoride effects on *S. mutans* biofilms and dental caries. *Journal of Dental Research, 84*(11), 1016–1020. doi:10.1177/154405910508401109 PMID:16246933

Koo, H., Schobel, B., Scott-Anne, K., Watson, G., Bowen, W. H., Cury, J. A., ... Park, Y. K. (2005). Apigenin and *tt*-farnesol with fluoride effects on *S. mutans* biofilms and dental caries. *Journal of Dental Research, 84*(11), 1016–1020. doi:10.1177/154405910508401109 PMID:16246933

Koo, H., Vacca-Smith, A. M., Bowen, W. H., Rosalen, P. L., Cury, J. A., & Park, Y. K. (2000). Effects of Apis mellifera propolis on the activities of streptococcal glucosyltransferases in solution and adsorbed onto saliva-coated hydroxyapatite. *Caries Research, 34*(5), 361–442. doi:10.1159/000016617 PMID:11014909

Kubo, I., Muroi, H., & Himejima, M. (1992). Antimicrobial activity of green tea flavor components and their combination effects. *Journal of Agricultural and Food Chemistry, 40*(2), 245–248. doi:10.1021/jf00014a015

Leclercq, R., Besbes, S., & Soussy, C. J. (1988). Resistance des streptocoques non groupables aux ß-lactamines. *Pathologie Biologie, 36*, 626–631. PMID:3054738

Lehner, T., Caldwell, J., & Smith, R. (1985). Local passive immunisation by monoclonal antibodies against streptococcal antigen I/II in the prevention of dental caries. *Infection and Immunity, 50*, 796–799. PMID:4066030

Lehner, T., Ma, J. K.-C., Grant, K., & Smith, R. (1986). Local passive immunisation by monoclonal antibodies against *Streptococcus mutans* antigen in the prevention of dental caries. In T. Lehner & G. Cimasoi (Eds.), *Borderland between Caries and Periodontal Disease* (pp. 356–357). Geneva: Medecine et Hygiene.

Leitão, D. P., Filho, A. A., Polizello, A. C., Bastos, J. K., & Spadaro, A. C. (2004). Comparative evaluation of in-vitro effects of brazilian green propolis and Baccharis dracunculifolia extracts on cariogenic factors of *Streptococcus mutans. Biological & Pharmaceutical Bulletin, 27*(11), 1834–1839. doi:10.1248/bpb.27.1834 PMID:15516733

Leita˜o, D. P. S., Da Silva Filho, A. A., Polizello, A. C. M., Bastos, J. K., & Spadaro, A. C. C. (2004). Comparative evaluation of in vitro effects of Brazilian green propolis and Baccharis dracunculifolia extracts on cariogenic factors of *Streptococcus mutans. Biological & Pharmaceutical Bulletin, 27*(11), 1834–1839. doi:10.1248/bpb.27.1834 PMID:15516733

Limsong, J., Benjavongkulchai, E., & Kuvatanasuchati, J. (2004). Inhibitory effect of some herbal extracts on adherence of *Streptococcus mutans. Journal of Ethnopharmacology, 92*(2-3), 281–289. doi:10.1016/j.jep.2004.03.008 PMID:15138013

Lio, M., Himeno, S., Miyauchi, K., Mikumo, K., & Ohta, N. (1983). Article. *Nippon Nogeikagaku Kaishi, 57*, 765. doi:10.1271/nogeikagaku1924.57.765

Lio, M., & Uyeda, M., lwanami, T., & Nakagawa, Y. (1984). Flavanoids as a possible preventive of dental caries. *Agricultural and Biological Chemistry, 48*, 2143–2145.

Loesche, W. J. (1986). Role of *Streptococcus mutans* in human dental decay. *Microbiological Reviews*, *50*, 353–380. PMID:3540569

Loesche, W. J. (1987). Role of *Streptococcus mutans* in human dental decay. *Microbiological Reviews*, *50*, 353–380. PMID:3540569

Loesche, W. J., & Grossman, N. S. (2001). Periodontal disease as a specific, albeit chronic, infection: Diagnosis and treatment. *Clinical Microbiology Reviews*, *14*(4), 727–752. doi:10.1128/CMR.14.4.727-752.2001 PMID:11585783

Ma, J. K., Hiatt, A., Hein, M., Vine, N. D., Wang, F., Stabila, P., ... Lehner, T. (1995). Generation and assembly of secretory antibodies in plants. *Science*, *268*(5211), 716–719. doi:10.1126cience.7732380 PMID:7732380

Ma, J. K., Hikmat, B. Y., Wycoff, K., Vine, N. D., Chargelegue, D., Yu, L., ... Lehner, T. (1998). Characterization of a recombinant plant monoclonal secretory antibody and preventive immunotherapy in humans. *Nature Medicine*, *4*(5), 601–606. doi:10.1038/nm0598-601 PMID:9585235

Ma, J.K., & Lehner, T. (1990). Prevention of colonization of Streptococcus mutans by topical application of monoclonal antibodies in human subjects. *Arch Oral Biol, 35*, 115S-122S.

Ma, J. K., Lehner, T., Stabila, P., Fux, C. I., & Hiatt, A. (1994). Assembly of monoclonal antibodies with IgG1 and IgA heavy chain domains in transgenic tobacco plants. *European Journal of Immunology*, *24*(1), 131–138. doi:10.1002/eji.1830240120 PMID:8020548

Mabogo, D. E. N. (1990). *The ethnobotany of the vhaVenda* (MSc Thesis). University of Pretoria, Pretoria, South Africa.

Mäkinen, K. K., Chen, C. Y., Mäkinen, P. L., Bennett, C. A., Isokangas, P. J., Isotupa, K. P., & Pape, H. R. Jr. (1996). Properties of whole saliva and dental plaque in relation to 40-month consumption of chewing gums containing xylitol, sorbitol of sucrose. *Caries Research*, *30*(3), 180–188. doi:10.1159/000262157 PMID:8860027

Mäkinen, K. K., Pemberton, D., Mäkinen, P. L., Chen, C. Y., Cole, J., Hujoel, P. P., ... Lambert, P. (1996). Polyol-combinant saliva stimulants and oral health in Veterans Affairs patients: An exploratory study. *Special Care in Dentistry*, *16*(3), 104–115. doi:10.1111/j.1754-4505.1996.tb00843.x PMID:9084323

Mandel, I. (1993). Dental caries: another extinct disease? In W. H. Bowen & L. A. Tabak (Eds.), *Cariology for the nineties* (pp. 1–10). Rochester, NY: University of Rochester Press.

Mann, A., Banso, A., & Clifford, L. C. (2008). Antifungal properties of Anogeissus leiocarpus and Terminalia avicennioides. *Tanzania Journal of Health Research*, *10*, 34–38. doi:10.4314/thrb.v10i1.14339 PMID:18680963

Marcucci, M. (1995). Propolis: Chemical composition, biological properties and therapeutic activity. *Apidologie*, *26*(2), 83–99. doi:10.1051/apido:19950202

Marcucci, M. C., Ferreres, F., Garcı'a-Viguera, C., Bankova, V. S., De Castro, S. L., Dantas, A. P., ... Paulino, N. (2001). Phenolic compounds from Brazilian propolis with pharmacological activities. *Journal of Ethnopharmacology*, *74*(2), 105–112. doi:10.1016/S0378-8741(00)00326-3 PMID:11167028

Marsh, P. D., & Bradsha, D. J. (1997). Physiological approaches to the control of oral biofilms. *Advances in Dental Research, 11*(1), 176–185. doi:10.1177/08959374970110010901 PMID:9524454

Martino, P. D., Gagniere, H., Berry, H., & Bret, L. (2002). Antimicrobial activity of extracts of some medicinal plants. Antimicrobial Agents. *Microbes and Infection, 4*, 613–620. PMID:12048030

Mathias, A. J., Somashekar, R. K., Sumithra, S., & Subramanya, S. (2000). An assessment of reservoirs of multi-resistant nosocomial pathogens in a secondary care hospital. *Antimicrob. Agents Indian J. Microbio, 40*, 183–190.

Matsumoto, M., Minami, T., Sasaki, H., Sobue, S., Hamada, S., & Ooshima, T. (1999). Inhibitory effects of oolong tea extract on caries-inducing properties of mutans streptococci. *Caries Research, 33*(6), 441–445. doi:10.1159/000016549 PMID:10529529

McCabe, M. M., & Hamelik, R. M. (1978). Multiple Forms of Dextran-binding Proteins from *Streptococcus mutans. Advances in Experimental Medicine and Biology, 107*, 749–759. doi:10.1007/978-1-4684-3369-2_84 PMID:742510

Medina, E., Brenes, M., Romero, C., García, A., & Castro, A. (2007). Main Antimicrobial Compounds in Table Olives. *Journal of Agricultural and Food Chemistry, 55*(24), 9817–9823. doi:10.1021/jf0719757 PMID:17970590

Mirzoeva, O., Grishanin, R., & Calder, P. C. (1997). Antimicrobial action of propolis and some of its components: The effects on growth, membrane potential and motility of bacteria. *Microbiological Research, 152*(3), 239–246. doi:10.1016/S0944-5013(97)80034-1 PMID:9352659

Mitsunaga, T., Abe, I., Kontani, M., Ono, H., & Tanaka, T. (1997). Inhibitory effects of bark proanthocyanidins on the activities of glucosyltransferases of *Streptococcus sobrinus. Journal of Wood Chemistry and Technology, 17*(3), 327–340. doi:10.1080/02773819708003134

Moerman, D. E. (1996). An analysis of the food plants and drug plants of Native North America. *Journal of Ethnopharmacology, 52*(1), 1–22. doi:10.1016/0378-8741(96)01393-1 PMID:8733114

Mtunzi, F. M., Ikechukwu, P. E., Tshifhiwa, M., Ezekiel, D., & Michael, J. K. (2017). Phytochemical Profiling, Antioxidant and Antibacterial Activities of Leaf Extracts from Rhus leptodictya. *International Journal of Pharmacognosy and Phytochemical Research, 9*(8), 1090–1099. doi:10.25258/phyto.v9i08.9616

Munro, C., Michalek, S. M., & Macrina, F. L. (1991). Cariogenicity of *Streptococcus mutans* V403 glucosyltransferase and fructosyltransferase mutants constructed by allelic exchange. *Infection and Immunity, 59*, 2316–2323. PMID:1828790

Murray, R. D. H., Mendez, J., & Brown, S. A. (1982). Coumarin activity in plants and bioorganism aspects. John Wiley.

Nakahara, K., Kawabata, S., Ono, H., Ogura, K., Tanaka, T., Ooshima, T., & Hamada, S. (1993). Inhibitory effect of oolong tea polyphenols on glycosyltransferases of mutans streptococci. *Applied and Environmental Microbiology, 59*, 968–973. PMID:8489234

Nakahara, K., Kawabata, S., Ono, H., Ogura, K., Tanaka, T., Ooshima, T., & Hamada, S. (1993). Inhibitory effect of oolong tea polyphenols on glucosyltransferases of mutans streptococci. *Applied and Environmental Microbiology, 59*(4), 968–973. PMID:8489234

Nakahara, K., Kawabata, S., Ono, H., Ogura, K., Tanaka, T., Ooshima, T., & Hamada, S. (1993). Inhibitory effect of oolong tea polyphenols on glycosyltransferases of mutans streptococci. *Applied and Environmental Microbiology, 59*, 968–973. PMID:8489234

Nalina, T., & Rahim, Z. H. A. (2007). The crude aqueous extract of Piper betle L. and its antibacterial effect towards *Streptococcus mutans. American Journal of Biochemistry and Biotechnology, 3*(1), 10–15. doi:10.3844/ajbbsp.2007.10.15

Namba, T., Morita, O., Huang, S. L., Goshima, K., Hattori, M., & Kakiuchi, N. (1988). Studies on cardioactive crude drugs. I. Effect of coumarins on cultured myocardial cells. *Planta Medica, 54*(04), 277–282. doi:10.1055-2006-962432 PMID:3222369

Namba, T., Tsunezuka, M., & Hattori, M. (1982). *Dental Caries Prevention by Traditional Chinese Medicines. Part II.* Potent Antibacterial Action of Magnoliae Cortex Extracts Against *Streptococcus Mutans. Planta Medica, 44*(2), 100–106. doi:10.1055-2007-971412 PMID:7071194

Nanbu, A., Hayakawa, M., Takada, K., Shinozaki, N., Abiko, Y., & Fukushima, K. (2000). Production, characterization, and application of monoclonal antibodies which distinguish four glucosyltransferases from Streptococcus sobrinus. *FEMS Immunology and Medical Microbiology, 27*(1), 9–15. doi:10.1111/j.1574-695X.2000.tb01405.x PMID:10617784

Nascimento, G. G. F., Locatelli, J., Freitas, P. C., & Silva, G. L. (2000). Antibacterial activity of plant extracts and phytochemicals on antibiotic-resistant bacteria. *Brazilian Journal of Microbiology, 31*(4), 247–256. doi:10.1590/S1517-83822000000400003

National Institutes of Health. (1987). *Oral health of United States adults. The national survey of oral health in U.S. employed adults and seniors, 1985-1986. National findings. NIH publication no. 87-2868.* Washington, DC: US Government Printing Office.

National Institutes of Health. (1989). *Oral health of United States children. The national survey of dental caries in U.S. school children, 1986-1987. National and regional findings. NIH publication no. 89-2247.* Washington, DC: US Government Printing Office.

Nostro, A., Cannatelli, M. A., Crisafi, G., Musolino, A. D., Procopio, F., & Alonzo, V. (2004). Modifications of hydrophobicity, in vitro adherence and cellular aggregation of Streptococcus mutans by Helichrysum italicum extract. *Letters in Applied Microbiology, 38*(5), 423–427. doi:10.1111/j.1472-765X.2004.01509.x PMID:15059215

Odebiyi, O. O., & Sofowora, A. (1979). Antimicrobial alkaloids from Nigeria chewing sticks. *Planta Medica, 36*, 204–204. doi:10.1055-0028-1097271 PMID:482432

Odusanya, S. A., Songonuga, O. O., & Folayan, J. O. (1979). Fluoriderion distribution in some African chewing sticks. *IRCS Medical Science [Microform], 7*, 58.

Onishi, M., Ozaki, F., Yoshino, F., & Murakami, Y. (1981). An experimental evidence of caries preventive activity of non- fluoride component in tea. *J. Dent. Health (Tokyo)*, *31*(2), 158–161. doi:10.5834/jdh.31.158 PMID:6948004

Ooshima, T., Izumitani, A., Sobue, S., Okahashi, N., & Hamada, S. (1983). Non-cariogenicity of the disaccharide palatinose in experimental dental caries of rats. *Infection and Immunity*, *39*(1), 43–49. PMID:6822422

Ooshima, T., Minami, T., Aono, W., Izumitani, A., Sobue, S., Fujiwara, T., ... Hamada, S. (1993). Oolong tea polyphenols inhibit experimental dental caries in SPF rats infected with *mutans streptococci*. *Caries Research*, *27*(2), 124–129. doi:10.1159/000261529 PMID:8319255

Ooshima, T., Minami, T., Aono, W., Izumitani, A., Sobue, S., Fujiwara, T., ... Hamada, S. (1993). Oolong tea polyphenols inhibit experimental dental caries in SPF rats infected with *Mutans streptococci*. *Caries Research*, *27*(2), 124–129. doi:10.1159/000261529 PMID:8319255

Ooshima, T., Minami, T., Aono, W., Tamura, Y., & Hamada, S. (1994). Reduction of dental plaque deposition in humans by Oolong tea extract. *Caries Research*, *28*(3), 146–149. doi:10.1159/000261636 PMID:8033186

Ooshima, T., Minami, T., Matsumoto, M., Fujiwara, T., Sobue, S., & Hamada, S. (1998). Comparison of the cariostatic effects between regimens to administer oolong tea polyphenols in SPF rats. *Caries Research*, *32*(1), 75–80. doi:10.1159/000016433 PMID:9438575

Ooshima, T., Osaka, Y., Sasaki, H., Osawa, K., Yasuda, H., Matsumura, M., ... Matsumoto, M. (2000). Caries inhibitory activity of cacao bean husk extract in in vitro and animal experiments. *Archives of Oral Biology*, *45*(8), 639–645. doi:10.1016/S0003-9969(00)00042-X PMID:10869475

Ooshima, T., Sobue, S., Hamada, S., & Kotani, S. (1981). Susceptibility of rats, hamsters, and mice to carious infection by Streptococcus mutans serotype c and d organisms. *Journal of Dental Research*, *60*(4), 855–859. doi:10.1177/00220345810600041701 PMID:6937525

Osawa, K., Yasuda, H., Maruyama, T., Morita, H., Takeya, K., & Itokawa, H. (1992). Isoflavanones from the heartwood of Swartzia polyphylla and their antibacterial activity against cariogenic bacteria. *Chemical & Pharmaceutical Bulletin*, *40*(11), 2970–2974. doi:10.1248/cpb.40.2970 PMID:1477911

Osman, A. M., Younes, M. E. G., & Sheta, A. E. (1974). Triterpenoids of the leaves of *Psidium guajava*. *Phytochemistry*, *13*(9), 2015–2016. doi:10.1016/0031-9422(74)85152-6

Otake, S., Makimura, M., Kuroki, T., Nishihara, Y., & Hirasawa, M. (1991). Anticaries effects of polyphenolic compounds from Japanese green tea. *Caries Research*, *25*(6), 438–443. doi:10.1159/000261407 PMID:1667297

Otake, S., Makimura, M., Kuroki, T., Nishihara, Y., & Hirasawa, M. (1991). Anticaries effects of polyphenolic compounds from Japanese green tea. *Caries Research*, *25*(6), 438–443. doi:10.1159/000261407 PMID:1667297

Pai, M. R., Acharya, L. D., & Udupa, N. (2004). Evaluation of antiplaque activity of Azadirachta indica leaf extract gel--a 6-week clinical study. *Journal of Ethnopharmacology*, *90*(1), 99–103. doi:10.1016/j.jep.2003.09.035 PMID:14698516

Paolino, C. J., Kashket, S., & Sparagna, C. A. (1980). Inhibition of dextran synthesis by tannic acid. *IADR Prog. Absrr.*, *58*(389).

Park, K. M., You, J. S., Lee, H. Y., Baek, N. I., Hwang, J. K., & Kuwanon, G. (2003). An antibacterial agent from the root bark of Morus alba against oral pathogens. *Journal of Ethnopharmacology*, *84*(2-3), 181–185. doi:10.1016/S0378-8741(02)00318-5 PMID:12648813

Park, Y. K., Ikegaki, M., Abreu, J. A. S., & Alcici, N. M. F. (1997). Comparison of the flavonoid aglycone contents of Apis mellifera propolis from various regions of Brazil. *Arquivos de Biologia e Tecnologia*, *40*(1), 97–106.

Park, Y. K., Koo, M. H., Abreu, J. A., Ikegaki, M., Cury, J. A., & Rosalen, P. L. (1998). Antimicrobial activity of propolis on oral microorganisms. *Current Microbiology*, *36*(1), 24–28. doi:10.1007002849900274 PMID:9405742

Pathak, D., Pathak, K., & Singla, A. (1991). Flavonoids as medical agents. *Fitoterapia*, *62*, 371–389.

Piller, N. B. A. (1975). Comparison of the effectiveness of some anti-inflammatory drugs on thermal oedema. *British Journal of Experimental Pathology*, *56*, 554–559. PMID:1222119

Pongpan, A., Chumsri, P., & Taworasate, T. (1982). The antimicrobial activity of some Thai Medicinal-Plants. *Mahidol Univ. J. Pharm. Sci*, *9*, 88–91.

Prabu, G. R., Gnanamani, A., & Sadulla, S. (2006). Guaijaverin a plant flavonoid as potential antiplaque agent against *Streptococcus mutans*. *Journal of Applied Microbiology*, *101*(2), 487–495. doi:10.1111/j.1365-2672.2006.02912.x PMID:16882158

Prashant, G., Chandu, G., Murulikrishna, K., & Shafiulla, M. (2007). The effect of mango and neem extract on four organisms causing dental caries: *Streptococcus mutans, Streptococcus salivavius, Streptococcus mitis,* and *Streptococcus sanguis*: An *in vitro* study. *Indian Journal of Dental Research*, *18*(4), 148–151. doi:10.4103/0970-9290.35822 PMID:17938488

Puupponen-Pimia, R., Nohynek, L., Hartmann-Schmidlin, S., Kahkonen, M., Heinonen, M., Maatta-Riihinen, K., & Oksman-Caldentey, K. M. (2005, April). (205). Berry phenolics selectively inhibit the growth of intestinal pathogens. *Journal of Applied Microbiology*, *98*(4), 991–1000. doi:10.1111/j.1365-2672.2005.02547.x PMID:15752346

Rahim, E. N. A. A., Ismail, A., Omar, M. N., Rahmat, U. N., & Ahmad, W. A. N. W. (2018). GC-MS Analysis of Phytochemical Compounds in Syzygium polyanthum Leaves Extracted using Ultrasound-Assisted Method. *Pharmacognosy Journal*, *10*(1), 110–119. doi:10.5530/pj.2018.1.20

Rao, K. (2000). *Materials for the database of medicinal plants*. Bangalore: Karnataka State Council for Science and Technology.

Robinson, C., Shore, R. C., Bonass, W. A., Brookes, S. J., Boteva, E., & Kirkham, J. (1998). Identification of human serum albumin in human caries lesions of enamel: The role of putative inhibitors of remineralisation. *Caries Research*, *32*(3), 193–199. doi:10.1159/000016452 PMID:9577984

Rotimi, V. O., Laughon, B. E., Bartlet, J. G., & Mesodomi, H. A. (1988). Activities of Nigeria chewing stick against Bacteroides gingivalis and Bacteroides melaninogenicus. *Nig. J. Microbiol.*, *9*, 13–16.

Rozen, R., Bachrach, G., Bronshteyn, M., Gedalia, I., & Steinberg, D. (2001). The role of fructans on dental biofilm formation by *Streptococcus sobrinus, Streptococcus mutans, Streptococcus gordonii* and *Actinomyces viscosus. FEMS Microbiology Letters*, *195*(2), 205–210. doi:10.1111/j.1574-6968.2001.tb10522.x PMID:11179653

Russell, M. W. (1979). Purification and properties of a protein surface antigen of *Streptococcus mutans. Microbios*, *25*, 7–18. PMID:393961

Russell, R. R. (1994). The application of molecular genetics to the microbiology of dental caries. *Caries Research*, *28*(2), 69–82. doi:10.1159/000261625 PMID:8156565

Saeki, Y., Ito, Y., Shibata, M., Sato, Y., Okuda, K., & Takazoe, I. (1989). Antimicrobial action of natural substances on oral bacteria. Bull.Tokyo Dent. Coll., 30(3), 129-135.

Sakanaka, S., Kim, M., Taniguchi, M., & Yamamoto, T. (1989). Antibacterial substances in Japanese green tea extract against *Streptococcus mutans*, a cariogenic bacterium. *Agricultural and Biological Chemistry*, *53*, 2307–2311.

Sakanaka, S., Kim, M., Taniguchi, M., & Yamato, T. (1989). Antibacterial substances in Japanese green tea extract against Streptococcus mutans, a cariogenic bacterium. *Agricultural and Biological Chemistry*, *53*, 2307–2311.

Sakanaka, S., Sato, T., Kim, M., & Yamamoto, T. (1990). Inhibitory effects of green tea polyphenols on glucan synthesis and cellular adherence of cariogenic *streptococci. Agricultural and Biological Chemistry*, *54*, 2925–2929.

Sakanaka, S., Shimura, N., Aizawa, M., Kim, M., & Yamamoto, T. (1992). Preventive effect of green tea polyphenols against dental caries in conventional rats. *Bioscience, Biotechnology, and Biochemistry*, *56*(4), 592–594. doi:10.1271/bbb.56.592 PMID:27280653

Santos, A. (2003). Evidence-based control of plaque and gingivitis. *Journal of Clinical Periodontology*, *30*(s5), 13–16. doi:10.1034/j.1600-051X.30.s5.5.x PMID:12787197

Sasaki, H., Matsumoto, M., Tanaka, T., Maeda, M., Nakai, M., Hamada, S., & Ooshima, T. (2004). Antibacterial activity of polyphenol components in Oolong tea extract against *Streptococcus mutans. Caries Research*, *38*(1), 2–8. doi:10.1159/000073913 PMID:14684970

Sato, M., Fujiwara, S., Tsuchiya, H., Fujii, T., Tinuma, M., Tosa, H., & Ohkawa, Y. (1996). Flavones with antibacterial activity against cariogenic bacteria. *Journal of Ethnopharmacology*, *54*(2-3), 171–176. doi:10.1016/S0378-8741(96)01464-X PMID:8953432

Sato, M., Tanaka, H., Fujiwara, S., Hirata, M., Yamaguchi, R., Etoh, H., & Tokuda, C. (2003). Antibacterial property of isoflavonoids isolated from Erythrina variegata against cariogenic oral bacteria. *Phytomedicine*, *10*(5), 427–433. doi:10.1078/0944-7113-00225 PMID:12834009

Sawamura, S., Tonosaki, Y., & Hamada, S. (1992). Inhibitory effects of ellagic acid on glucosyltransferases from mutans streptococci. *Bioscience, Biotechnology, and Biochemistry*, *56*(5), 766–768. doi:10.1271/bbb.56.766 PMID:27286205

Sawamura, S., Tonosaki, Y., & Hamada, S. (1992). Hamada S. Inhibitory effect of ellagic acid on glucosyltransferase from *mutans streptococci. Bioscience, Biotechnology, and Biochemistry*, *56*(5), 766–768. doi:10.1271/bbb.56.766 PMID:27286205

Scalbert, A. (1991). Antimicrobial properties of tannins. *Phytochemistry*, *30*(12), 3875–3883. doi:10.1016/0031-9422(91)83426-L

Schilling, K., & Doyle, R. J. (1995). Bacterial adhesion to hydroxylapatite. *Methods in Enzymology*, *253*, 536–542. doi:10.1016/S0076-6879(95)53045-2 PMID:7476417

Schilling, K. M., & Bowen, W. H. (1992). Glucans synthesized *in situ* in experimental salivary pellicle function as specific binding sites for *Streptococcus mutans. Infection and Immunity*, *60*, 284–295. PMID:1530843

Schmidt, H. (1988). *Phenol oxidase (E.I.14.18.1), a marker enzyme for defense cells. Progress in histochemistry and cytochemistry* (Vol. 17). New York, N.Y.: Gustav Fischer.

Shapiro, S., Meier, A., & Guggenheim, B. (1994). The antimicrobiol activity of essential oils and essential oil components toward oral bacteria. *Oral Microbiology and Immunology*, *9*(4), 202–208. doi:10.1111/j.1399-302X.1994.tb00059.x PMID:7478759

Sheu, C. W., & Freese, E. (1972). Effects of fatty acid on growth and envelope proteins of *Bacillus subtilis. Journal of Bacteriology*, *111*, 516–524. PMID:4626502

Smith, R., & Lehner, T. (1989). Characterisation of monoclonal antibodies binding to common protein epitopes on the cell surface of *Streptococcus mutans* and *Streptococcus sobrinus. Oral Microbiology and Immunology*, *4*(3), 153–158. doi:10.1111/j.1399-302X.1989.tb00243.x PMID:2639299

Smullen, J., Koutsou, G. A., Foster, H. A., Zumbe´, A., & Storey, D. M. (2007). The antibacterial activity of plant extracts containing polyphenols against *Streptococcus mutans. Caries Research*, *41*(5), 342–349. doi:10.1159/000104791 PMID:17713333

Sofowora, A. (1993). *Medicinal Plant and Traditional Medicine in Africa*. Ibadan, Nigeria: Spectrum Books.

Sofrata, A., Claesson, R., Lingstrom, P., & Gustafsson, A. (2008). Strong antibacterial effect of miswak against oral microorganisms associated with periodontitis and caries. *Journal of Periodontology*, *79*(8), 1474–1479. doi:10.1902/jop.2008.070506 PMID:18672998

Sofrata, A., Lingstrom, P., Baljoon, M., & Gustafsson, A. (2007). The effect of miswak extract on plaque pH. An *in vivo* study. *Caries Research*, *41*(6), 451–454. doi:10.1159/000107931 PMID:17823507

Sofrata, A., Lingstrom, P., Baljoon, M., & Gustafsson, A. (2007). The effect of miswak extract on plaque pH. An *in vivo* study. *Caries Research*, *41*(6), 451–454. doi:10.1159/000107931 PMID:17823507

Song, J. H., Kim, S. K., Chang, K. W., Han, S. K., Yi, H. K., & Jeon, J. G. (2006). *In vitro* inhibitory effects of Polygonum cuspidatum on bacterial viability and virulence factors of *Streptococcus mutans* and *Streptococcus sobrinus*. *Archives of Oral Biology*, *51*(12), 1131–1140. doi:10.1016/j.archoralbio.2006.06.011 PMID:16914113

Song, J. H., Kim, S. K., Chang, K. W., Han, S. K., Yi, H. K., & Jeon, J. G. (2006). *In vitro* inhibitory effects of Polygonum cuspidatum on bacterial viability and virulence factors of Streptococcus mutans and *Streptococcus sobrinus*. *Archives of Oral Biology*, *51*(12), 1131–1140. doi:10.1016/j.archoralbio.2006.06.011 PMID:16914113

Staat, R. H., Doyle, R. J., Langley, S. D., & Suddick, R. P. (1978). Modification of in vitro adherence of Streptococcus mutans by plant lectins. *Advances in Experimental Medicine and Biology*, *107*, 639–647. doi:10.1007/978-1-4684-3369-2_72 PMID:742505

Stary, F. (1996). *The Natural Guide to Medicinal Herbs and Plants*. Barnes & Noble Inc.

Steinberg, D., Feldman, M., Ofek, I., & Weiss, E. I. (2004). Effect of a high-molecular-weight component of cranberry on constituents of dental biofilm. *The Journal of Antimicrobial Chemotherapy*, *54*(1), 86–89. doi:10.1093/jac/dkh254 PMID:15163648

Steinberg, D., Feldman, M., Ofek, I., & Weiss, E. I. (2004). Effect of a high-molecular-weight component of cranberry on constituents of dental biofilm. *The Journal of Antimicrobial Chemotherapy*, *54*(1), 86–89. doi:10.1093/jac/dkh254 PMID:15163648

Steinberg, D., Feldman, M., Ofek, I., & Weiss, E. I. (2005). Cranberry high molecular weight constituents promote *Streptococcus sobrinus* desorption from artificial biofilm. *International Journal of Antimicrobial Agents*, *25*(3), 247–251. doi:10.1016/j.ijantimicag.2004.10.014 PMID:15737520

Stephen, K. (1993). Caries in young populations–worldwide. In W. H. Bowen & L. A. Tabak (Eds.), *Cariology for the nineties* (pp. 71–84). Rochester, NY: University of Rochester Press.

Stern, J. L., Hagerman, A. E., Steinberg, P. D., & Mason, P. K. (1996). Phlorotannin-protein interactions. *Journal of Chemical Ecology*, *22*(10), 1887–1899. doi:10.1007/BF02028510 PMID:24227114

Tagashira, M., Uchiyama, K., Yoshimura, T., Shirota, M., & Uemitsu, N. (1997). Inhibition by hop bract polyphenols of cellular adherence and water-insoluble glucan synthesis *of mutans streptococci*. *Bioscience, Biotechnology, and Biochemistry*, *61*(2), 332–335. doi:10.1271/bbb.61.332 PMID:9058972

Tagashira, M., Uchiyama, K., Yoshimura, T., Shirota, M., & Uemitsu, N. (1997). Inhibition by Hop Bract Polyphenols of Cellular Adherence and Water-insoluble Glucan Synthesis of *Mutans Streptococci*. *Bioscience, Biotechnology, and Biochemistry*, *61*(2), 332–335. doi:10.1271/bbb.61.332 PMID:9058972

Taiwo, O., Xu, H., & Lee, S. (1999). Antibacterial activities of extracts from Nigerian chewing sticks. *Phytotherapy Research*, *13*(8), 675–679. doi:10.1002/(SICI)1099-1573(199912)13:8<675::AID-PTR513>3.0.CO;2-X PMID:10594937

Takaisikikuni, N. B., & Schilcher, H. (1994). Electron-microscopic and micro calorimetric investigations of the possible mechanism of the antibacterial action of a defined propolis provenance. *Planta Medica, 60*(03), 222–227. doi:10.1055-2006-959463

Takarada, K., Kimizuka, R., Takahashi, N., Honma, K., Okuda, K., & Kato, T. A. (2004). Comparison of the antibacterial efficacies of essential oils against oral pathogens. *Oral Microbiology and Immunology, 19*(1), 61–64. doi:10.1046/j.0902-0055.2003.00111.x PMID:14678476

Taweechaisupapong, S., Wongkham, S., Chareonsuk, S., Suparee, S., Srilalai, P. S., & Chaiyarak, S. (2000). Selective activity of Streblus asper on *Mutans streptococci. Journal of Ethnopharmacology, 70*(1), 73–79. doi:10.1016/S0378-8741(99)00140-3 PMID:10720792

Tella, A. (1976). Analgesic and antimicrobial properties of *Vernonia amygdalina. British Journal of Clinical Pharmacology, 7*, 295–297.

Thastrup, O., Knudsen, J. B., Lemmich, J., & Winther, K. (1985). Inhibitions of human platelet aggregation by dihydropyrano-and dihydrofuranocoumarins, a new class of cAMP phosphodiesterase inhibitors. *Biochemical Pharmacology, 34*(12), 2137–2140. doi:10.1016/0006-2952(85)90407-1 PMID:2988567

Toukairin, T., Uchino, K., Iwamoto, M., Murakami, S., Tatebayashi, T., Ogawara, H., & Tonosaki, Y. (1991). New polyphenolic 5′-nucleotidase inhibitors isolated from the wine grape "Koshu" and their biological effects. *Chemical & Pharmaceutical Bulletin, 39*(6), 1480–1483. doi:10.1248/cpb.39.1480 PMID:1934168

Trongtokit, Y., Rongsriyam, Y., Komilamisra, N., Krisackphong, P., & Apiwathnasorn, C. (2004). Laboratory and field trial of developing medicinal local Thai plant products against four species of mosquito vectors. *The Southeast Asian Journal of Tropical Medicine and Public Health, 35*, 325–333. PMID:15691131

Tsuchiya, H., Sato, M., Iinuma, M., Yokoyama, J., Ohyama, M., Tanaka, T., ... Namikawa, I. (1994). Inhibition of the growth of cariogenic bacteria *in vitro* by plant flavanones. *Experientia, 50*(9), 846–849. doi:10.1007/BF01956469 PMID:7925853

Tsuchiya, H., Sato, M., Tinuma, M., Yokoyama, J., Ohyama, M., Tanaka, T., ... Namikawa, I. (1994). Inhibition of the growth of cariogenic bacteria *in vitro* by plant flavanones. *Experientia, 50*(9), 846–849. doi:10.1007/BF01956469 PMID:7925853

Usha, R., Sashidharan, S., & Palaniswamy, M. (2010). Antimicrobial Activity of a Rarely Known Species, Morinda citrifolia L. Ethnobotanical Leaflets, 2010(3).

Uzel, A., Sorkun, K., Oncag, O., Cogulu, D., Gencav, O., & Salih, B. (2005). Chemical compositions and antimicrobial activities of four different *Anatolian propolis* samples. *Microbial Res, 160*(2), 189–195. doi:10.1016/j.micres.2005.01.002 PMID:15881836

Vanka, A., Tandon, S., Rao, S. R., Udupa, N., & Ramkumar, P. (2001). The effect of indigenous Neem Azadirachta indica [correction of (Adirachta indica)] mouth wash on Streptococcus mutans and lactobacilli growth. *Indian Journal of Dental Research, 12*(3), 133–144. PMID:11808064

Vasconcelos, L. C., Sampaio, F. C., Sampaio, M. C., Pereira, M. S. V., Higino, J. S., & Peixoto, M. H. P. (2006). Minimum inhibitory concentration of adherence of Punica granatum Linn (pomegranate) gel against *S. mutans, S. mitis* and *C. albicans. Brazilian Dental Journal, 17*(3), 223–227. doi:10.1590/S0103-64402006000300009 PMID:17262129

Venditti, M., Baiocchi, P., Santini, C., Brandimarte, C., Serra, P., Gentile, G., ... Martino, P. (1989). Antimicrobial susceptibilities of Streptococcus species that cause septicemia in neutropenic patients. *Antimicrobial Agents and Chemotherapy, 33*(4), 580–582. doi:10.1128/AAC.33.4.580 PMID:2729950

Weintraub, J. A., Hilton, J. F., White, J. M., Hoover, C. I., Wycoff, K. L., Yu, L., ... Featherstone, J. D. B. (2005). Featherstone JD: Clinical trial of a plant-derived antibody on recolonization of *mutans streptococci. Caries Research, 39*(3), 241–250. doi:10.1159/000084805 PMID:15914988

Weiss, E. L., Lev-Dor, R., Sharon, N., & Ofek, I. (2002). Inhibitory effect of a high-molecular-weight constituent of cranberry on adhesion of oral bacteria. *Critical Reviews in Food Science and Nutrition, 42*(sup3), 285–292. doi:10.1080/10408390209351917 PMID:12058987

Westergren, G., & Olsson, J. (1983). Hydrophobicity and adherence of oral *Streptococci* after repeated subculture *in vitro. Infection and Immunity, 40*, 432–435. PMID:6832836

Whittaker, C. J., Klier, C. M., & Kolenbrander, P. E. (1996). Mechanisms of adhesion by oral bacteria. *Annual Review of Microbiology, 50*(1), 513–552. doi:10.1146/annurev.micro.50.1.513 PMID:8905090

Wolinsky, L. E., Mania, S., Nachnani, S., & Ling, S. (1996). The inhibiting effect of aqueous azadirachta indica (neem) extract upon bacterial properties influencing *in vitro* plaque formation. *Journal of Dental Research, 75*(2), 816–822. doi:10.1177/00220345960750021301 PMID:8655780

Wolinsky, L. E., & Sote, E. O. (1984). Isolation of natural plaque-inhibiting substances from Nigerian chewing sticks. *Caries Research, 18*(3), 216–225. doi:10.1159/000260768 PMID:6584212

Wolinsky, L. E., & Sote, E. O. (1984). Isolation of natural plaqueinhibiting substances from "Nigerian chewing sticks". *Caries Research, 18*(3), 216–225. doi:10.1159/000260768 PMID:6584212

Wongkham, S., Laupattarakasaem, P., Pienthaweechai, K., Areejitranusorn, P., Wongkham, C., & Techanitiswad, T. (2001). Antimicrobial activity of *Streblus asper* leaf extract. *Phytotherapy Research, 15*(2), 119–121. doi:10.1002/ptr.705 PMID:11268109

Wu, C. D., Darout, I. A., & Skaug, N. (2001). Chewing sticks: Timeless natural toothbrush for oral cleansing. *Journal of Periodontal Research, 36*(5), 275–28410. doi:10.1034/j.1600-0765.2001.360502.x PMID:11585114

Wu-Yuan, C. D., Green, L., & Birch, W. X. (1990). *In vitro* screening of Chinese medicinal toothpastes: Their effects on growth and plaque formation of *mutans streptococci. Caries Research, 24*(3), 198–202. doi:10.1159/000261265 PMID:2364405

Wu-Yuan, C. D., Tai, S., & Slade, H. D. (1978). Dextran/glucan binding by Streptococcus mutans: The role of molecular size and binding site in agglutination. *Advances in Experimental Medicine and Biology, 107*, 737–748. doi:10.1007/978-1-4684-3369-2_83 PMID:742509

Wunder, D., & Bowen, W. H. (1999). Action of agents on glucosyltransferases from *Streptococcus mutans* in solution and adsorbed to experimental pellicle. *Archives of Oral Biology*, *44*(3), 203–214. doi:10.1016/S0003-9969(98)00129-0 PMID:10217511

Xiao, J., Liu, Y., Zuo, Y. L., Li, J. Y., Ye, L., & Zhou, X. D. (2006). Effects of *Nidus Vespae* extract and chemical fractions on the growth and acidogenicity of oral microorganisms. *Archives of Oral Biology*, *51*(9), 804–813. doi:10.1016/j.archoralbio.2006.03.014 PMID:16723116

Xiao, J., Zhou, X. D., Feng, J., Hao, Y. Q., & Li, J. Y. (2007). Activity of *Nidus Vespae* extract and chemical fractions against *Streptococcus mutans* biofilms. *Letters in Applied Microbiology*, *45*(5), 547–552. doi:10.1111/j.1472-765X.2007.02230.x PMID:17916132

Xiao, Y., Liu, T. J., Huang, Z. W., Zhou, X. D., & Li, J. Y. (2004). The effect of natural medicine on adherence of *Streptococcus mutans* to salivary acquired pellicle. *Sichuan Da Xue Xue Baoxue Ban.*, *35*, 87–89.

Yamamoto, H., & Ogawa, T. (2002). Antimicrobial activity of perilla seed polyphenols against oral pathogenic bacteria. *Bioscience, Biotechnology, and Biochemistry*, *66*(4), 921–924. doi:10.1271/bbb.66.921 PMID:12036078

Yamanaka, A., Kimizuka, R., Kato, T., & Okuda, K. (2004). Inhibitory effects of cranberry juice on attachment of oral *streptococci* and biofilm formation. *Oral Microbiology and Immunology*, *19*(3), 150–154. doi:10.1111/j.0902-0055.2004.00130.x PMID:15107065

Yamanaka, A., Kimizuka, R., Kato, T., & Okuda, K. (2004). Inhibitory effects of cranberry juice on attachment of oral *streptococci* and biofilm formation. *Oral Microbiology and Immunology*, *19*(3), 150–154. doi:10.1111/j.0902-0055.2004.00130.x PMID:15107065

Yamashita, Y., Bowen, W. H., Burne, R. A., & Kuramitsu, H. K. (1993). Role of the Streptococcus mutans gtf genes in caries induction in the specific-pathogen-free rat model. *Infection and Immunity*, *61*, 3811–3817. PMID:8359902

Yanagida, A., Kanda, T., Tanabe, M., Matsudaira, F., & Cordeiro, J. G. O. (2000). Inhibitory effects of apple polyphenols and related compounds on cariogenic factors of *Mutans Streptococci. Journal of Agricultural and Food Chemistry*, *48*(11), 5666–5671. doi:10.1021/jf000363i PMID:11087536

Yanagida, A., Kanda, T., Tanabe, M., Matsudaira, F., & Oliveira, C. J. G. (2000). Inhibitory effects of apple polyphenols and related compounds on cariogenic factors of mutans streptococci. *Journal of Agricultural and Food Chemistry*, *48*(11), 5666–5671. doi:10.1021/jf000363i PMID:11087536

Yanagida, A., Kanda, T., Tanabe, M., Matsudaira, F., & Oliveira, C. J. G. (2000). Inhibitory effects of apple polyphenols and related compounds on cariogenic factors of *mutans streptococci. Journal of Agricultural and Food Chemistry*, *48*(11), 5666–5671. doi:10.1021/jf000363i PMID:11087536

Yatsuda, R., Rosalen, P. L., Cury, J. A., Murata, R. M., Rehder, V. L., Melo, L. V., & Koo, H. (2005). Effects of Mikania genus plants on growth and cell adherence of *mutans streptococci. Journal of Ethnopharmacology*, *97*(2), 183–189. doi:10.1016/j.jep.2004.09.042 PMID:15707750

Yatsuda, R., Rosalen, P. L., Cury, J. A., Murata, R. M., Rehder, V. L., Melo, L. V., & Koo, H. (2005). Effects of Mikania genus plants on growth and cell adherence of mutans streptococci. *Journal of Ethnopharmacology*, *97*(2), 183–189. doi:10.1016/j.jep.2004.09.042 PMID:15707750

You, Y. O., Shin, H. Y., Yu, H. H., Yoo, S. J., Kim, S. H., Kim, Y. K., ... Kim, H. M. (2004). Effect of Powerdental on caries-inducing properties of *Streptococcus mutans* and TNF-alpha secretion from HMC-1 cells. *Journal of Ethnopharmacology*, *92*(2-3), 331–335. doi:10.1016/j.jep.2004.03.023 PMID:15138020

Yu, H. H., Lee, J. S., Lee, K. H., Kim, K. Y., & You, Y. O. (2007). Saussurea lappa inhibits the growth, acid production, adhesion, and water-insoluble glucan synthesis of *Streptococcus mutans*. *Journal of Ethnopharmacology*, *111*(2), 413–417. doi:10.1016/j.jep.2006.12.008 PMID:17234374

Zhang, J., & Kashket, S. (1998). Inhibition of salivary amylase by black and green teas and their effects on the intraoral hydrolysis of starch. *Caries Research*, *32*(3), 233–238. doi:10.1159/000016458 PMID:9577990

APPENDIX

Table. 1. Plants containing antimicrobial activity against cariogenic organisms

Plant	Common name	Part of plant	Type of Extract	Compound	Organisms	Inhibitory mechanisms	Author
Accacia arabica	Sunt garad	Branch	Aqueous	?	*Streptococcus mutans*	Growth	Almas, 2001
Allium sativum	Garlic	Fruit	Aqueous	?	*S. mutans*	Growth	Fani *et al.*, 2008
Andrographis paniculata	Indian Echinacea	Whole plant	Ethanol	Polyphenols	*S. mutans*	Growth, Adherence; Glucan-binding lectin; Glucosyltransferase	Limsong *et al.*, 2004
Annona senegalensis	Wild custard-apple	Bark	Ethanol	Polyphenols	*S. mutans, Actinomyces naeslundii,*	Growth	Mabogo, 1990
Anogeissus leiocarpus	Axle-. wood tree	Root	Methanol	alkaloids, glycosides, phenols, steroids, tannins, ellagic acids, anthraquinones, saponins and flavonoids	*S. mutans, Streptococcus gordonii*	Growth	Taiwo *et al.*, 1999; Mann *et al.*, 2008
Aristolochia cymbifera Duch	Milhomem	Stem	Aqueous, ethanol	Polyphenols	*S. mutans, Lactobacillus casei, Fusobacterium nucleatum*	Growth	Alviano *et al.*, 2008
Arnica Montana	Leopard's bane	Flower	Ethanol	sesquiterpene lactones	*A. naeslundii*	Growth	Koo *et al.*, 2000
Artocarpus heterophyllus	Jackfruit	Wood	hexane/acetone, benzene/acetone	Flavonoids	*S. mutans, Actinomyces, Lactobacillus Other oral streptococci*	Growth	Sato *et al.*, 1996
Asarum sieboldii	Wild ginger	Whole plant	Ethanol, aqueous	alkaloids, phenolics, glycosides, peptides, terpenoids,	*S. mutans*	Acid production • adhesion, glucan synthesis,	Yu *et al.*, 2007
Aspilia foliacea	?	Leaves	Ethanol	kaurane diterpenes	*S. sobrinus, S. mutans, S. mitis, S. sanguinis, L. casei*	Growth	Ambrosio *et al.*, 2008
Azadirachta indica	Neem	Leaves	Aqueous, extract gel	?	*S. mutans, S. salivarius, S. sanguis, Lactobacilli*	Growth, aggregation, adhesion, glucan synthesis,	Prashant *et al.*, 2007; Pai *et al.*, 2004,
Baccharis dracunculifolia	Alecrim	Leaves	Dichlorometane	Polyphenols	*S. mutans*	Glucosyltransferase, acid production	Leitao *et al.*, 2004
Baphia nittida	Cam wood	Leaves	Aqueous	Flavonoids, tannins, protein, saponins, steroids	*S. mutans*	Growth	Amadi *et al.*, 2007
Caesalpinia pyramidalis Tul.	Catingueira	Leaves	Aqueous, ethanol	Polyphenols	*S. mutans, Lactobacillus casei, Fusobacterium nucleatum*	Growth	Alviano *et al.*, 2008
Calotropis gigantea,	Giant milkweed	Whole plant	Ash	?	?	Anti caries	Hebbar *et al.*, 2004
Cardamom Elattaria cardamomum	Cardamom	Seed	Hexane	Monoterpenoids, monoterpene acetate	*S. mutans*	Growth	Kubo *et al.*, 1992
Cassia alata L.	Candlebrush	Nees	Ethanol	Polyphenols	*S. mutans*	Growth, Adherence; Glucan-binding lectin; Glucosyltransferase	Limsong *et al.*, 2004
Cassia hirsuta, L.	Woolly senna	seeds	?	?	?	Anti caries	Hebbar *et al.* 2004
Cassia tora	TORA	seeds	?	?	?	Anti caries	Hebbar *et al.* 2004
Catha edulis	Khat	Leaves	Aqueous	?	*S. mutans, S. sobrinus, Lactobacillus acidophilus, L. fermentum*	Growth	Al-hebshi *et al.*, 2006

continued on following page

Table 1. Continued

Plant	Common name	Part of plant	Type of Extract	Compound	Organisms	Inhibitory mechanisms	Author
Celastrus scandens	American bittersweet	root bark	Methanol/ dichloromethane	Polyphenols	*S. mutans*	Growth	Heise *et al.*, 1992
Cocos nucifera Linn.	Coco cravo	Husk fiber	Aqueous, ethanol	Polyphenols	*S. mutans, Lactobacillus casei, Fusobacterium nucleatum*	Growth	Alviano *et al.*, 2008
Coffea arabica	Coffee	seeds	Aqueous	trigonelline and nicotinic and chlorogenic acids	*S. mutans*	Growth, Adherence	Daglia *et al.*, 2002
Coffea robusta	Coffee	seeds		trigonelline and nicotinic and chlorogenic acids	*S. mutans*	Growth, Adherence	Daglia *et al.*, 2002
Curcuma xanthorrhiza	Javanese turmeric	Root	Methanol	Polyphenols	*S. mutans*	Growth, biofilms, acidogenesis, morphological alteration	Kim *et al.*, 2008
Dicerocarym senecioides,	Boot protectors	Root	Ethanol	Polyphenols	*S. mutans, A. naeslundii, Actinobacillus actinomycetemcomitans*	Growth	Mabogo, 1990; More *et al.*, 2008
Digitaria sanguinalis	Crab grass	seeds	Methanol dichloromethane	Polyphenols	*S. mutans*	Growth	Heise *et al.*, 1992
Drosera peltata Smith	Insectivores plant	Whole plant	Petroleum ether, chloroform, diethylether, ethyl acetate ethanol	Polyphenols	*S. mutans, S. sobrinus, S. rattus, S. cricetus, S. sanguis, S. milleri, S. mitis S. constellatus, S. oralis, S. salivarius*	Growth	Didry *et al.*, 1997
Duroia hirsuta	?	Leaf	?	blackening agent	?	Dental caries	Elujoba *et al.*, 2005
Englerophytum magalismontanum	Transvaal milkplum	Bark	Ethanol	Polyphenols	*S. mutans, A. naeslundii,*	Growth	Mabogo, 1990; More *et al.*, 2008
Erythrina lysistemon	Coral Tree	Bark	Ethanol	Polyphenols	*S. mutans, A. naeslundii, A. actinomycetemcomitans*	Growth	Mabogo, 1990; More *et al.*, 2008
Euclea divinorum,	Magic guarri	Bark, leaves	Ethanol	Polyphenols	*S. mutans, A. naeslundii, A. actinomycetemcomitans*	Growth	Mabogo, 1990; More *et al.*, 2008
Euclea natalensis,	Chipambati	Leves	Ethanol	Polyphenols	*S. mutans, A. naeslundii, Privotella intermedia, A. actinomycetemcomitans*	Growth	Mabogo, 1990; More *et al.*, 2008
Galla chinensis	Chinese Gall	Leaves	Ethanol	alkaloids or glycosides	?	Remineralization, pH modification	Cheng *et al.*, 2008
Garcinia kola	Garcinia	Root	Methanol	?	*S. mutans, S. gordonii A. naeslundii*	Growth	Taiwo *et al.*, 1999
Ginkgo biloba	Ginkgo	Leaves	Methanol/ dichloromethane	Polyphenols	*S. mutans*	Growth	Heise *et al.*, 1992
Harrisonia perforata	Candle Stick	twigs	Ethanol	Polyphenols	*S. mutans*	Growth, Adherence; Glucan-binding lectin; Glucosyltransferase	Limsong *et al.*, 2004
Juniprrous virginiana	Eastern red cedar	Leaves	Methanol dichloromethane	Polyphenols	*S. mutans*	Growth	Heise *et al.*, 1992
Kaempferia pandurata	Krachai	Whole plant	Methanol	Isopanduratin A	*S. mutans, S. salivarius, S. sanguis S. sobrinus, L. acidophilus, L. casei, A. viscosus*	Growth, cell membrane damage	Hwang *et al.*, 2004
Labramia bojeri	Beach Apricot Tree	Seed	?	Protein	*S. mutans, S. sobrinus, S. mitis, S. sanguinis, S. oralis*	Adherence, biofilm	Oliveira *et al.*, 2007
Leucas aspera,	Common Leucas	Leaves	?	?	?	Anti caries	Hebbar *et al.*2004

continued on following page

Table 1. Continued

Plant	Common name	Part of plant	Type of Extract	Compound	Organisms	Inhibitory mechanisms	Author
Lippia sidoides	Pepper-rosmarin	Leaves	hydro-distillation	Essential oil	*S. mutans, S. sanguis, S. salivarius, S. mitis, C. albicans*	Growth	Botelho *et al.*, 2007
Magnolia officinalis	Magnolia	?	ethanol	magnolol,	*S. milleri*	GTF activity	Huang *et al.*, 2006
Mallotus japonicus	Wrapper Plant	flower	hexane/acetone, benzene/acetone	Flavonoids	*S. mutans, Actinomyces, Lactobacillus Other oral streptococci*	Growth	Sato *et al.*, 1996
Malus Pumila	Paradise apple	Fruit	Methyl acetate	phenolic acids, flavonols, dihydrochalcones, tannins	*S. mutans, S. sobrinus*	Growth, Adherence, glucosyltransferase	Yanagida, *et al.*, 2000
Manetta glandulosa,		Leaf	?	blackening agent	?	Dental caries	Elujoba *et al.*, 2005
Manettia divaricata,	Firecracker vine	Leaf	?	blackening agent	?	Dental caries	Elujoba *et al.*, 2005
Mangifera indica	Mango	Leaves	Aqueous	?	*S. mutans, S. salivarius,*	Growth	Prashant *et al.*, 2007
Melaphis chinensis	Chinese Gall	Leaves	Ethanol	gallotannins	*S. mutans, S. sobrinus,*	Growth, Adherence, Glucosyltransferase	Wu-Yuan *et al.*,1978; Wolinsky,*et al.*, 1996
Mikania glomerata	Sprengel	Arial parts	Ethanol, hexane ethyl acetate	Octadecen, cupressenic acid, coumarin, kaurenoic acid, diterpenic acid	*S. sobrinus, S. mutans, S. cricetus*	Growth, adherence	Yatsuda *et al.*, 2005
Morus alba	Mulberry	Bark	Ethanol	Kuwanon G;	*S. mutans*	Growth, morphological modification	Park *et al.*, 2003
Nauclea latifolia	Pin Cushion Tree	Stem	Aqueous	indole-quinolizidine alkaloids, glycoalkaloids, sapponins	*Streptococcus sp.,*	Dental caries	Iwu *et al.*, 1999
Neea parviflora	?	Leaf	?	blackening agent	?	Dental caries	Elujoba *et al.*, 2005
Parinari curatellifolia	Cork tree	Root	Ethanol	Polyphenols	*S. mutans, A. naeslundii, A. actinomycetemcomitans*	Growth	Mabogo, 1990; More *et al.*, 2008
Perilla frutescens	Perilla	Seed	Ethanol	Polyphenols	*S. mutans, S. sobrinus, S. salivarius, S. mitis, S. sanguinis, S. oralis,*	Growth	Yamamoto and Ogawa, 2002
Piper betle L	Tropical vine	leaves	Aqueous	Stearic acid, palmitic acid, hydroxybenzeneacetic acid, hydroxy esters	*S. mutans*	Growth	Nalina and Rahim, 2007
Polygonum cuspidatum	Japanese knotweed	Root	Methanol, hexane, ethyl acetate	alkaloids, sterol/ terpenes, tannins, flavonoids, carbohydrates	*S. mutans, S. sobrinus,*	Growth, biofilm, adherence, Glucosyltransferase, acid production	Song *et al.*, 2006, 2007
Psidium cattleianum	Cattley guava	Leaves	Aqueous	flavonoids (kaempferol, quercetin and cyanidin) tannin (ellagic acid)	*S. mutans*	Growth, acid production, metabolic protein	Brighenti *et al.*, 2008
Psidium guajava	Guava	Leaves	Ethanol	Flavanoids (quercetin-3-O-a-L-arabinopyranoside)	*S. mutans*	Growth, biofilm acid production, adherence, surface hydrophobicity	Prabu *et al.*, 2006; Limsong *et al.*, 2004
Psoralea corylifolia	Malay Tea	seed	Diethyl ether	phenolic isoprenoid (Bakuchiol)	*S. mutans, S. sobrinus, L. acidophilus*	Growth, adherence	Katsura *et al.*, 2001
Punica granatum	Pomegranate	Fruits	Carbopol, water and triethanolamine gel	tannin and polyphenolics	*S. mutans, S. sanguinis, C. albicans, S. mitis*	Growth, biofilm, adhesion, glucan synthesis, acid production	Kakiuchi *et al.* 1986; Vasconcelos *et al.*, 2006

continued on following page

Table 1. Continued

Plant	Common name	Part of plant	Type of Extract	Compound	Organisms	Inhibitory mechanisms	Author
Rheum undulatum	Garden rhubarb	Root	Methanol	Anthraquinones Cardiac glycosides, Coumarines, Phenolics, Sterol/ terpenes	*S. mutans, S. sobrinus,*	Growth, acid production, adhesion, glucan synthesis	Song *et al.*, 2006
Ruta graveolens	Common rue	Leaves	hexane/acetone, benzene/acetone	Flavonoids	*Mutans streptococci, Actinomyces, Lactobacillus Other oral streptococci*	Growth	Sato *et al.*, 1996
Salvadora persica	Salt Bush	Stem	Aqueous	Chewing stick	*S. mutans, Lactobacilli,*	Growth, pH modification	Almas, 2001; Sofrata *et al.*, 2008 ; Almas and Zeid, 2004; Sofrata *et al.*, 2007
Saussurea lappa	Costus	Root	ethanol	Phenolics, Steroids, terpenoids	*S. mutans*	Acid production; Adherence, Glucosyltransferase	Yu *et al.*, 2007
Sesamum indicum	Sesamum	Seed	?	Essential oil	*S. mutans, L. acidophilus*	Growth	Durai Anand *et al.*, 2008
Solanum panduriforme	Yellow Bitter-apple	Bark	Ethanol	Polyphenols	*S. mutans, A. naeslundii, A. actinomycetemcomitans*	Growth	Mabogo, 1990; More *et al.*, 2008
Sophora exigua	Toromiro Tree	Root	Chloroform	Ftavonoid (phytoalexin)	*S. mutans, S. sobrinus, S. salivarius, S. mitis, S. sanguinis, S. oralis, A. viscosus, S. rattus, S. cricetus, A. israefif, A. naeslundii L. casei*	Growth	Tsuchiya, *et al.*, 1994
Spondias mombin	Hogplum	Leaves	Aqueous	Flavonoids, tannins, protein, saponins, steroids	*S. mutans*	Growth	Amadi *et al.*, 2007
Streblus asper	Siamense Rough Bush	Leaves	Ethanol	Polyphenols	*S. mutans*	Growth, Adherence; Glucan-binding lectin; Glucosyltransferase	Limsong *et al.*, 2004 Taweechaisupapong *et al.*, 2000; Wongkham *et al.*, 2001
Syzygium aromaticum	Clove	Clove buds	Aqueous, ethanol	?	*S. mutans*	Adhesion, cell surface hydrophobicity, Glucosyltransferase	Rahim *et al.*, 2018
Talisia esculenta	Pitomba	Lectins	?	Lectins	*S. mutans, S. sobrinus, S. mitis, S. sanguinis, S. oralisya*	Adherence, biofilm	Oliveira *et al.*, 2007
Terminalia chebula	Myrobalan	Dried fruit	Aqueous	Polyphenols	*S. mutans*	Growth, Acid production, Adhesion, Glucosyltransferase	Jagtap and Karkera, 1999; Usha *et al.*, 2007;
Theobroma cacao	Cacao	Bean husk	Ethanol, aqueous	Polyphenols Protein, polysaccharide	*S. mutans, S. sobrinus,*	Glucosyltransferase, acid production, adherence, cell hydrophobicity,	Ooshima *et al.*, 2000a, b; Ito *et al.*, 2003
Uncaria tomentosa	Cat's claw	Leaves	?	alkaloids, triterpens, steroids, phenolic compound, glicosids, tanin and flavonoids	*S. mutans*	Growth	Ccahuana-Vasquez *et al.*, 2006
Vaccinium macrocarpon	cranberry	fruit	?	phenolic acids, flavonols and proanthocyanidins	*S. mutans*	Adhesion, Glucosyltransferase, acid production	Gregoire *et al.*, 2007
Vitis vinifera	Grape	seeds	?	polyphenols, mainly catechin, epicatechin	*S. mutans S. sobrinus, L. rhamnosus A. viscosus*	Growth, biofilm	Furiga *et al.*, 2009
Ziziphus joazeiro Mart.	Juazeiro	Innerbark	Aqueous, ethanol	Polyphenols	*S. mutans, Lactobacillus casei, Fusobacterium nucleatum*	Growth	Alviano *et al.*, 2008

Compilation of References

Abadi, M., Agarwal, A., Barham, P., Brevdo, E., Chen, Z., Citro, C., . . . Zheng, X. (2015). *TensorFlow: Large-Scale Machine Learning on Heterogeneous Systems*. Retrieved from https://www.tensorflow.org/

Abdel-Mottaleb, J. Z. M. (2004). Automatic Human identification based on Dental X-ray images. *SPIE Conference on Defense and Security-Biometric Technology for human Identification.*

Adatrao, S., & Mittal, M. (2016, October). An analysis of different image preprocessing techniques for determining the centroids of circular marks using Hough transform. In *Frontiers of Signal Processing (ICFSP), International Conference on* (pp. 110-115). IEEE. 10.1109/ICFSP.2016.7802966

Addai, F.K., Nuamah, I.K. & Parkins, G. (2002). Brief chewing of Garcinia manii stick speedily reverses depression ofsaliva pH after a glucos rinse. *Med. Sci. Monitor, 8*(11), 746-750.

Adel Kauzman, B. D. S., Pavone, M., Blanas, N., & Bradley, G. (2004). Pigmented lesions of the oral cavity: Review, differential diagnosis, and case presentations. *Journal - Canadian Dental Association, 70*(10), 682–683. PMID:15530266

Afolayan, A. J., & Meyer, J. J. M. (1997). The antimicrobial activity of 3,5,7- trihydroxyflavone isolated from the shoots of Helichrysum aureonitens. *Journal of Ethnopharmacology, 57*(3), 177–181. doi:10.1016/S0378-8741(97)00065-2 PMID:9292410

Ahmed, S. A., Taib, M. N., Khalid, N. E. A., Ahmad, R., & Taib, H. (2011). *Performance of compound enhancement algorithms on dental radiograph images*. WASET.

Akhoondali, H., Zoroofi, R. A., & Shirani, G. (2009). Rapid automatic segmentation and visualization of teeth in CT-scan data. *Journal of Applied Sciences (Faisalabad), 9*(11), 2031–2044. doi:10.3923/jas.2009.2031.2044

Akpata, J. E., & Akimimisi, E. O. (1977). Antibacterial activity of extracts of some Nigerian chewing sticks. *Caries Research, 18*, 216–225.

Alagumariappan, P., & Krishnamurthy, K. (2017). Analysis of Normal and Abnormal Electrogastrograms using Teager-Kaiser Energy. *Proceedings of the 2nd World Research Journals Congress*, 134-136.

Alagumariappan, P., & Krishnamurthy, K. (2018). An Approach Based on Information Theory for Selection of Systems for Efficient Recording of Electrogastrograms. In *Proceedings of the International Conference on Computing and Communication Systems* (pp. 463-471). Springer. 10.1007/978-981-10-6890-4_45

Alagumariappan, P., Krishnamurthy, K., Kandiah, S., & Ponnuswamy, M. J. (2017). Effect of electrode contact area on the information content of the recorded electrogastrograms: An analysis based on Rényi entropy and Teager-Kaiser Energy. *Polish Journal of Medical Physics and Engineering, 23*(2), 37–42. doi:10.1515/pjmpe-2017-0007

Alagumariappan, P., Rajagopal, A., & Krishnamurthy, K. (2016). Complexity Analysis on Normal and Abnormal Electrogastrograms Using Tsallis Entropy. In *3rd International Electronic and Flipped Conference on Entropy and Its Applications*. Multidisciplinary Digital Publishing Institute. 10.3390/ecea-3-A003

Alamri, H. M., Sadrameli, M., Alshalhoob, M. A., & Alshehri, M. A. (2012). Applications of CBCT in dental practice: A review of the literature. *General Dentistry*, *60*(5), 390–400. PMID:23032226

Alazab, M., Islam, M., & Venkatraman, S. (2009). Towards Automatic Image Segmentation Using Optimised Region Growing Technique. In AI 2009: Advances in Artificial Intelligence. AI 2009. Springer. doi:10.1007/978-3-642-10439-8_14

Alginahi, Y. (2010). Preprocessing techniques in character recognition. In *Character Recognition*. InTech. doi:10.5772/9776

Al-hebshi, N., Al-haroni, M., & Skaug, N. (2006). *In vitro* antimicrobial and resistance modifying activities of aqueous crude khat extracts against oral microorganisms. *Archives of Oral Biology*, *51*(3), 183–188. doi:10.1016/j.archoralbio.2005.08.001 PMID:16248981

Ali, M., Son, L. H., Khan, M., & Tung, N. T. (2017). Segmentation of Dental X-ray Images in Medical Imaging using Neutrosophic Orthogonal Matrices. *Expert Systems with Applications*. doi:10.1016/j.eswa.2017.09.027

Almas & Al-Zeid. (2004). The immediate antimicrobial effect of a toothbrush and miswak on cariogenic bacteria: a clinical study. *J. Contemp. Dent. Pract.*, *5*(1), 105-114.

Almas, K. (2001). The antimicrobial effects of seven different types of Asian chewing sticks. *Odonto-Stomatologie Tropicale*, *24*(96), 17–20. PMID:11887585

Alviano, W.S., Alviano, D.S., Diniz, C.G., Antoniolli, A.R., Alviano, C., Farias, L.M., Carvalho, M.A.R., Souza, M.M.G., & Bolognese, A.M. (2008). *In vitro* antioxidant potential of medicinal plant extracts and their activities against oral bacteria based on Brazilian folk medicine. *Arch. Oral Biol.*, *53*(6), 545-552.

Amadi, E. S., Oyeka, C. A., Onyeagba, R. A., Ogbogu, O. C., & Okoli, I. (2007). Antimicrobial screening of Breynia nivosus and Ageratum conyzoides against dental caries causing organisms. *The Journal of Biological Sciences*, *7*(2), 354–358. doi:10.3923/jbs.2007.354.358

Ambikapathy, B., & Krishnamurthy, K. (2018). Analysis of electromyograms recorded using invasive and noninvasive electrodes: A study based on entropy and Lyapunov exponents estimated using artificial neural networks. *Journal of Ambient Intelligence and Humanized Computing*, 1–9.

Ambrosio, S. R., Furtado, N. A., de-Oliveira, D. C., da-Costa, F. B., Martins, C. H., de-Carvalho, T. C., ... Veneziani, R. C. (2008). Antimicrobial activity of kaurane diterpenes against oral pathogens. *Z Naturforsch C.*, *63*(5-6), 326–330. doi:10.1515/znc-2008-5-603 PMID:18669015

Amer, Y. Y., & Aqel, M. J. (2015). An Efficient Segmentation Algorithm for Panoramic Dental Images. *Procedia Computer Science*, *65*, 718–725. doi:10.1016/j.procs.2015.09.016

Amoros, M., Lurton, E., Boustie, J., Girre, L., Sauvager, F., & Cormier, M. (1994). Comparison of the anti-herpes simplex virus activities of propolis and 3-methyl-but-2-enyl caffeate. *Journal of Natural Products*, *57*(5), 644–647. doi:10.1021/np50107a013 PMID:8064297

Anand, T.D., Pothiraj, C., Gopinath, R.M. & Kayalvizhi, B. (2008). Effect of oil-pulling on dental caries causing bacteria. *African Journal of Microbiology Research, 2*, 63-66.

Anandh, K. R., Sujatha, C. M., & Ramakrishnan, S. (2016). A Method to Differentiate Mild Cognitive Impairment and Alzheimer in MR Images using Eigen Value Descriptors. *Journal of Medical Systems*, *40*(25). PMID:26547845

Anandh, K. R., Sujatha, C. M., & Ramakrishnan, S. (2016). Laplace Beltrami eigen value based classification of normal and Alzheimer MR images using parametric and non-parametric classifiers. *Expert Systems with Applications*, *59*(C), 208–216.

Ana, P. A., Velloso, W. F. Jr, & Zezell, D. M. (2008). Three-dimensional finite element thermal analysis of dental tissues irradiated with Er, Cr: YSGG laser. *The Review of Scientific Instruments*, *79*(9), 093910. doi:10.1063/1.2953526 PMID:19044431

Anjna, E., & Kaur, E. R. (2017). Review of Image Segmentation Technique. *International Journal (Toronto, Ont.)*, *8*(4).

Annadurai, S. (2007). *Fundamentals of digital image processing*. Pearson Education India.

Anusavice, K. J., Shen, C., & Rawls, H. R. (2013). *Phillips' science of dental materials*. Elsevier Health Sciences.

Aoki, H., Shiroza, T., Hayakawa, M., Sato, S., & Kuramitsu, H. K. (1986). Cloning of a Streptococcus mutans glucosyltransferase gene coding for insoluble glucan synthesis. *Infection and Immunity*, *53*, 587–594. PMID:3017865

Apisariyakul, A., Vanittana, N., & Buddasuki, D. (1995). Antifungal activity of turmeric oil extracted from Curcuma longa (Zingiberaceae). *Journal of Ethnopharmacology*, *49*(3), 163–169. doi:10.1016/0378-8741(95)01320-2 PMID:8824742

Apparsamy, S., & Krishnamurthy, K. (2018). Design and Development of Capacitance Sensor Array for Analyzing the Non-Homogeneity in Biological Materials: A Smart Sensing Approach. In Expert System Techniques in Biomedical Science Practice (pp. 70-96). IGI Global. doi:10.4018/978-1-5225-5149-2.ch004

Arai, Y., Tammisalo, E., Iwai, K., Hashimoto, K., & Shinoda, K. (1999). Development of a compact computed tomographic apparatus for dental use. *Dento Maxillo Facial Radiology*, *28*(4), 245–248. doi:10.1038j.dmfr.4600448 PMID:10455389

Arenholt-Bindslev, D. (1992). Dental amalgam—environmental aspects. *Advances in Dental Research*, *6*(1), 125–130. doi:10.1177/08959374920060010501 PMID:1292452

Arik, Ö., Ibragimov, B., & Xing, L. (2017). Fully automated quantitative cephalometry using convolutional neural networks. *Journal of Medical Imaging (Bellingham, Wash.)*, *4*(1), 014501. doi:10.1117/1.JMI.4.1.014501 PMID:28097213

Arivazhagan, S., Ganesan, L., & Kumar, T. G. S. (2006). Texture classification using ridgelet transform. *Pattern Recognition Letters*, *27*(16), 1875–1883. doi:10.1016/j.patrec.2006.04.013

Arunachalam, K., Jacob, L. V., & Kamalanand, K. (2014). Design and analysis of finite element based sensors for diagnosis of liver disorders using biocompatible metals. *Technology and Health Care*, *22*(6), 867–875. PMID:25391529

Avirutnant, W., & Pongpan, A. (1983). Antimicrobial activity of some Thai flowers and plants. *Mahidol Univ. Journal of Pharmaceutical Sciences*, *10*, 81–86.

Babshet, M., Acharya, A. B., & Naikmasur, V. G. (2010). Age estimation in Indians from pulp/tooth area ratio of mandibular canines. *Forensic Sci. Int.*, *197*(1-3), 125e1–e4.

Backes, A. R., Gonçalves, W. N., Martinez, A. S., & Bruno, O. M. (2010). Texture analysis and classification using deterministic tourist walk. *Pattern Recognition*, *43*(3), 685–694. doi:10.1016/j.patcog.2009.07.017

Bagchi, P., & Joshi, N. (2012). Role of radiographic evaluation in treatment planning for dental implants: A review. *J Dent Allied Sci*, *1*(1), 21–25. doi:10.4103/2277-4696.159112

Banks, J. (2013). Adding value in additive manufacturing: Researchers in the United Kingdom and Europe look to 3D printing for customization. *IEEE Pulse*, *4*(6), 22–26. doi:10.1109/MPUL.2013.2279617 PMID:24233187

Banskota, A. H., Tezuka, Y., Prasain, J. K., Matsushige, K., Saiki, I., & Kadota, S. (1998). Chemical constituents of Brazilian propolis and their cytotoxic activities. *Journal of Natural Products*, *61*(7), 896–900. doi:10.1021/np980028c PMID:9677271

Banumathi, A., Vijayakumari, B., Geetha, A., Shanmugavadivu, N., & Raju, S. (2007). Performance analysis of various techniques applied in human identification using dental X-rays. *Journal of Medical Systems*, *31*(3), 210–218. doi:10.100710916-007-9057-0 PMID:17622024

Baratto Filho, F., Zaitter, S., Haragushiku, G. A., de Campos, E. A., Abuabara, A., & Correr, G. M. (2009). Analysis of the internal anatomy of maxillary first molars by using different methods. *Journal of Endodontics*, *35*(3), 337–342. doi:10.1016/j.joen.2008.11.022 PMID:19249591

Barboza, E., Marana, A. & Oliveira, D. (2012). A Multibiometric Approach in a Semi-Automatic Dental Recognition Using DIFT Technique and Dental Shape Features. *SIBGRAPI 2012 - Workshop of Theses and Dissertations*, 13-18.

Barea-Alvarez, M., Benito, M. T., Olano, A., Jimeno, M. L., & Moreno, F. J. (2014). Synthesis and Characterization of Isomaltulose-Derived Oligosaccharides Produced by Transglucosylation Reaction of Leuconostoc mesenteroides Dextransucrase. *Journal of Agricultural and Food Chemistry*, *62*(37), 9137–9144. doi:10.1021/jf5033735 PMID:25175804

Barker, T. M., Earwaker, W. J. S., & Lisle, D. A. (1994). Accuracy of stereolithographic models of human anatomy. *Journal of Medical Imaging and Radiation Oncology*, *38*(2), 106–111. PMID:8024501

Barroso, G. M. (1986). *Sistemática de Angiospermas do Brasil* (Vol. 3). Imprensa Universitária, Universidade Federal de Viçosa, Viçosa.

Bartlett, S. (2013). Printing organs on demand. *The Lancet. Respiratory Medicine*, *1*(9), 684. doi:10.1016/S2213-2600(13)70239-X PMID:24429271

Bashir, U., Gupta, M., & Ahuja, R. (2016). Implant systems. *International Journal of Applied Dental Sciences*, *2*(2), 35–41. PMID:26783652

Bastone, E. B., Freer, T. J., & McNamara, J. R. (2000). Epidemiology of dental trauma: A review of the literature. *Australian Dental Journal*, *45*(1), 2–9. doi:10.1111/j.1834-7819.2000.tb00234.x PMID:10846265

Beckers, H. J. (1988). Influence of xylitol on growth, establishment, and cariogenicity of *Streptococcus mutans* in dental plaque of rats. *Caries Research*, *22*(3), 166–173. doi:10.1159/000261100 PMID:3163524

Bergman, T. L., & Incropera, F. P. (2011). *Introduction to heat transfer*. John Wiley & Sons.

Bertassoni, L. E., Cecconi, M., Manoharan, V., Nikkhah, M., Hjortnaes, J., Cristino, A. L., ... Khademhosseini, A. (2014). Hydrogel bioprinted microchannel networks for vascularization of tissue engineering constructs. *Lab on a Chip*, *14*(13), 2202–2211. doi:10.1039/C4LC00030G PMID:24860845

Bhadauria, H. S., Singh, A., & Dewal, M. L. (2013). An integrated method for haemorrhage segmentation from brain CT imaging. *Computers & Electrical Engineering*, *39*(5), 1527–1536. doi:10.1016/j.compeleceng.2013.04.010

Bhandari, A. K., Kumar, A., & Singh, G. K. (2015). Modified artificial bee colony based computationally efficient multilevel thresholding for satellite image segmentation using Kapur's, Otsu and Tsallis functions. *Expert Systems with Applications*, *42*(3), 1573–1601. doi:10.1016/j.eswa.2014.09.049

Bhandari, A. K., Kumar, A., & Singh, G. K. (2015). Tsallis entropy based multilevel thresholding for colored satellite image segmentation using evolutionary algorithms. *Expert Systems with Applications*, *42*(22), 8707–8730. doi:10.1016/j.eswa.2015.07.025

Bharti, R., Wadhwani, K. K., Tikku, A. P., & Chandra, A. (2010). Dental amalgam: An update. *Journal of conservative dentistry. JCD*, *13*(4), 204. PMID:21217947

Bhaskar, S. N. (1968). Oral pathology in the dental office: Survey of 20, 575 biopsy specimens. *The Journal of the American Dental Association*, *76*(4), 761–766. doi:10.14219/jada.archive.1968.0119 PMID:5237768

Bhat, M., & Patil, T. (2014, May). Adaptive clip limit for contrast limited adaptive histogram equalization (CLAHE) of medical images using least mean square algorithm. In *Advanced Communication Control and Computing Technologies (ICACCCT), 2014 International Conference on* (pp. 1259-1263). IEEE.

Biswas, K., Chattopadhyay, I., Ranajit, K. B., & Bandyopadhya, U. (2002). Biological activities and medical properties of neem (Azadirachta indica). *Current Science*, *82*(11), 1336–1345.

Blum, I. R., Lynch, C. D., & Wilson, N. H. (2014). Factors influencing repair of dental restorations with resin composite. *Clinical, Cosmetic and Investigational Dentistry*, *6*, 81. doi:10.2147/CCIDE.S53461 PMID:25378952

Boehm, H. F., Lutz, M., Korner, M., Mutschler, W., Reiser, M., & Pfeifer, K.-J. (2009). Using radon transform of standard radiographs of the hip to differentiate between post-menopausal women with and without fracture of the proximal femur. *Osteoporosis International*, *20*(2), 323–333. doi:10.100700198-008-0663-6 PMID:18560746

Bonhevi, J. S., Coll, F. V., & Jord, R. E. (1994). The composition, active components and bacteriostatic activity of of Propolis in Dietetics. *Laboratori Agrari de la Generalitat de Catalunya, 08348 Cabrils (Barcelona), Spain. Journal of the American Oil Chemists' Society*, *71*, 5.

Booth, I. R. (1985). Regulation of cytoplasmic pH in bacteria. *Microbiological Reviews*, *49*, 329–378. PMID:3912654

Borchartt, T. B., Conci, A., Lima, R. C., Resmini, R., & Sanchez, A. (2012). Breast thermography from an image processing viewpoint: A survey. *Signal Processing*, *93*(10), 2785–2803. doi:10.1016/j.sigpro.2012.08.012

Bornstein, M., M, Scarfe, W. C., & Jacobs, R. (2014). Cone beam computed tomography in implant dentistry: A systematic review focusing on guidelines, indications, and radiation dose risks. *The International Journal of Oral and Maxillofacial Implants, 29*, 55-77.

Bornstein, M. M., Scarfe, W. C., Vaughn, V. M., & Jacobs, R. (2014). Cone beam computed tomography in implant dentistry: A systematic review focusing on guidelines, indications, and radiation dose risks. *International journal of oral & maxillofacial implants*, 29. PMID:24660190

Borris, R. P. (1996). Natural products research: Perspectives from a major pharmaceutical company. *Journal of Ethnopharmacology*, *51*(1/3), 29–38. doi:10.1016/0378-8741(95)01347-4 PMID:9213624

Bors, A. G., Kechagias, L., & Pitas, I. (2002). Binary morphological shape-based interpolation applied to 3-D tooth reconstruction. *IEEE Transactions on Medical Imaging*, *21*(2), 100–108. doi:10.1109/42.993129 PMID:11929098

Botelho, M. A., Nogueira, N. A. P., Bastos, G. M., Fonseca, S. G. C., Lemos, T. L. G., Matos, F. J. A., ... Brito, G. A. C. (2007). Antimicrobial activity of the essential oil from Lippia sidoides, carvacrol and thymol against oral pathogens. *Brazilian Journal of Medical and Biological Research*, *40*(3), 349–356. doi:10.1590/S0100-879X2007000300010 PMID:17334532

Bowden, G. H., & Hamilton, I. R. (1998). Survival of Oral Bacteria. *Critical Reviews in Oral Biology and Medicine*, *9*(1), 54–85. doi:10.1177/10454411980090010401 PMID:9488248

Braden, M., & Clarke, R. L. (1984). Water absorption characteristics of dental microfine composite filling materials: I. Proprietary materials. *Biomaterials*, *5*(6), 369–372. doi:10.1016/0142-9612(84)90038-3 PMID:6525397

Bradshaw, D. J., & Marsh, P. D. (1998). Analysis of pH-driven disruption of oral microbial communities *in vitro*. *Caries Research*, *32*(6), 456–462. doi:10.1159/000016487 PMID:9745120

Brighenti, F. L., Luppens, S. B., Delbem, A. C., Deng, D. M., Hoogenkamp, M. A., Gaetti-Jardim, E. Jr, ... ten Cate, J. M. (2008). Effect of Psidium cattleianum leaf extract on *Streptococcus mutans* viability, protein expression and acid production. *Caries Research*, *42*(2), 148–154. doi:10.1159/000121439 PMID:18367836

Bro-Nielsen, M., Larsen, P., & Kreiborg, S. (1996). Virtual teeth: a 3D method for editing and visualizing small structures in CT scans. In Proc. Computer Assisted Radiology (CAR'96) (pp. 921-924). Academic Press.

Brown, W. S., Dewey, W. A., & Jacobs, H. R. (1970). Thermal properties of teeth. *Journal of Dental Research*, *49*(4), 752–755. doi:10.1177/00220345700490040701 PMID:5269374

Bruneton, J. (1995). *Pharmacognosy, Phytochemistry and Medicinal Plants*. Intercept. Ltd.

Brunthaler, A., König, F., Lucas, T., Sperr, W., & Schedle, A. (2003). Longevity of direct resin composite restorations in posterior teeth: A review. *Clinical Oral Investigations*, *7*(2), 63–70. doi:10.100700784-003-0206-7 PMID:12768463

Buchaillard, S. I., Ong, S. H., Payan, Y., & Foong, K. (2007). 3D statistical models for tooth surface reconstruction. *Computers in Biology and Medicine*, *37*(10), 1461–1471. doi:10.1016/j.compbiomed.2007.01.003 PMID:17336957

Bullmore, E., Fadili, J., Maxim, V., Sendur, L., Whitcher, B., Suckling, J., ... Breakspear, M. (2004). Wavelets and functional magnetic resonance imaging of the human brain. *NeuroImage*, *23*, 234–249. doi:10.1016/j.neuroimage.2004.07.012 PMID:15501094

Cai, L., & Wu, C. D. (1996). Compounds from Syzygium aromaticum possessing growth inhibitory activity against oral pathogens. *Journal of Natural Products*, *59*(10), 987–990. doi:10.1021/np960451q PMID:8904847

Campbell, D. J. (1995). A brief history of dental radiography. *The New Zealand Dental Journal*, *91*(406), 127–133. PMID:8602286

Candes, E. J., & Donoho, D. L. (1999). Ridgelets: A key to higher-dimensional Intermittency. *Philosophical Transactions of Royal Society London*, *357*(1760), 2495-2509.

Candes, E., Demanet, L., Donoho, D., & Ying, L. (2006). Fast discrete curvelettransforms. *Multiscale Modeling & Simulation*, *5*(3), 861–899. doi:10.1137/05064182X

Capitelli, F., Colao, F., Provenzano, M. R., Fantoni, R., Brunetti, G., & Senesi, N. (2002). Determination of heavy metals in soils by laser induced breakdown spectroscopy. *Geoderma*, *106*(1-2), 45–62. doi:10.1016/S0016-7061(01)00115-X

Carter, C. N., Pusateri, R. J., Chen, D., Ahmed, A. H., & Farag, A. A. (2010, September). Shape from shading for hybrid surfaces as applied to tooth reconstruction. In *Image Processing (ICIP), 2010 17th IEEE International Conference on* (pp. 4049-4052). IEEE. 10.1109/ICIP.2010.5653040

Carter, K., Landini, G., & Malmseys, A. D. (2001). Plaque removal characteristics of electric toothbrushes using an *in vitro* plaque model. *Journal of Clinical Periodontology*, *28*(11), 1045–1049. doi:10.1034/j.1600-051X.2001.281109.x PMID:11686826

Caselles, V., Catte, F., Coll, T., & Dibos, F. (1993). A geometric model for active contours in image processing. *Numerische Mathematik*, *66*, 1–31. doi:10.1007/BF01385685

Caselles, V., Ron, K., & Sapiro, G. (1997). Geodesic active contours. *International Journal of Computer Vision*, *22*(1), 61–79. doi:10.1023/A:1007979827043

Cavalcanti, M., Ruprecht, A., Johnson, W., Thomas, S., Jakobsen, J., & Paulo, S. (1999). Oral and maxillofacial radiology. *Oral Surgery, Oral Medicine, and Oral Pathology*, *88*, 353–357. doi:10.1016/S1079-2104(99)70042-9

Ccahuana-Vasquez, R. A., Santos, S. S., Koga-Ito, C. Y., & Jorge, A. O. (2007). Antimicrobial activity of Uncaria tomentosa against oral human pathogens. *Brazilian Oral Research*, *21*(1), 46–50. doi:10.1590/S1806-83242007000100008 PMID:17426895

Chaabene, M., Ali, R. B., Ejbali, R., & Zaied, M. (2017). Hybrid approach for detection of dental caries based on the methods FCM and level sets. *SPIE 10341, Ninth International Conference on Machine Vision*. doi: 10.1117/12.2268437

Chae, M. P., Rozen, W. M., McMenamin, P. G., Findlay, M. W., Spychal, R. T., & Hunter-Smith, D. J. (2015). Emerging applications of bedside 3D printing in plastic surgery. *Frontiers in Surgery*, *2*, 25.

Chalmers, J. M. (2008). Minimal intervention dentistry: A new focus for dental hygiene. *Dentistry Today*, *27*(4), 132–134. PMID:18497207

Chan, W. R., Taylor, D. R., Willis, C. R., & Fehlhaber, H. W. (1968). The structure of neoandrographolide - a diterpene glucoside from andrographis paniculata nees. *Tetrahedron Letters*, *9*(46), 4803–4806. doi:10.1016/S0040-4039(00)75962-4

Chappard, C., Brunet-Imbault, B., Lemineur, G., Giraudeau, B., Basillais, A., Harba, R., & Claude, L. B. (2005). Anisotropy changes in post-menopausal osteoporosis: Characterization by a new index applied to trabecular bone radiographic images. *Osteoporosis International*, *16*(10), 1193–1202. doi:10.100700198-004-1829-5 PMID:15685395

Chappard, D., Baslé, M.-F., Legrand, E., & Audran, M. (2008). Trabecular bone microarchitecture: A review. *Morphologie*, *92*(299), 162–170. doi:10.1016/j.morpho.2008.10.003 PMID:19019718

Chattopadhyay, C., Kim, D. W., Gombos, D. S., Oba, J., Qin, Y., Williams, M. D., & Patel, S. P. (2016). Uveal melanoma: From diagnosis to treatment and the science in between. *Cancer*, *122*(15), 2299–2312. doi:10.1002/cncr.29727 PMID:26991400

Chattopadhyay, S., Davis, R. M., Menezes, D. D., Singh, G., Acharya, R. U., & Tamura, T. (2012). Application of Bayesian classifier for the diagnosis of dental pain. *Journal of Medical Systems*, *36*(3), 1425–1439. doi:10.100710916-010-9604-y PMID:20945154

Chaubey, A. K. (2016). Comparison of The Local and Global Thresholding Methods in Image Segmentation. *World Journal of Research and Review*, *2*(1), 1-4.

Chauhan, A. S., Silakari, S., & Dixit, M. (2014, April). Image segmentation methods: A survey approach. In *Communication Systems and Network Technologies (CSNT), 2014 Fourth International Conference on* (pp. 929-933). IEEE. 10.1109/CSNT.2014.191

Chen, E., & Abbott, P. V. (2009). Dental pulp testing: A review. *International Journal of Dentistry*. . doi:10.1155/2009/365785

Chen, H., & Jain, A. K. (2004). Tooth contour extraction for matching dental radiographs. Proc. of International Conference on Pattern Recognition, 3, 522-525.

Chen, H., & Jain, A. K. (2005, January). Dental biometrics: Alignment and matching of dental radiographs. In Application of Computer Vision, 2005. WACV/MOTIONS'05 Volume 1. Seventh IEEE Workshops on (Vol. 1, pp. 316-321). IEEE.

Chen, E., & Abbott, P. V. (2011). Evaluation of accuracy, reliability, and repeatability of five dental pulp tests. *Journal of Endodontics*, *37*(12), 1619–1623. doi:10.1016/j.joen.2011.07.004 PMID:22099893

Chen, G. Y., & Kegl, B. (2010). Invariant pattern recognition using contourlets and AdaBoost. *Pattern Recognition*, *43*(3), 579–583. doi:10.1016/j.patcog.2009.08.020

Cheng, L., Li, J., Hao, Y., & Zhou, X. (2008). Effect of compounds of Galla chinensis and their combined effects with fluoride on remineralization of initial enamel lesion *in vitro*. *Journal of Dentistry*, *36*(5), 369–373. doi:10.1016/j.jdent.2008.01.011 PMID:18308448

Cheng, R., Yang, H., Shao, M. Y., Hu, T., & Zhou, X. D. (2009). Dental erosion and severe tooth decay related to soft drinks: A case report and literature review. *Journal of Zhejiang University. Science. B.*, *10*(5), 395–399. doi:10.1631/jzus.B0820245 PMID:19434767

Chen, H., & Jain, A. K. (2005, January). Dental biometrics: Alignment and matching of dental radiographs. *IEEE Transactions on Pattern Analysis and Machine Intelligence*, *27*(8), 1319–1326. doi:10.1109/TPAMI.2005.157 PMID:16119269

Chen, Y., Zhang, J., & Yang, J. (2012). An anisotropic images segmentation and bias correction method. *Magnetic Resonance Imaging*, *30*(1), 85–95. doi:10.1016/j.mri.2011.09.003 PMID:22055751

Chi, Z., Yan, H., & Pham, T. (1996). *Fuzzy algorithms: with applications to image processing and pattern recognition* (Vol. 10). World Scientific. doi:10.1142/9789812830111

Choksi, S. K., Brady, J. M., Dang, D. H., & Rao, M. S. (1994). Detecting approximal dental caries with transillumination: a clinical evaluation. *Journal of the American Dental Association, 125*(8), 1098-1102.

Chollet, F. (2015). *keras*. GitHub. Retrieved from https://github.com/fchollet/keras

Chopra, R. N., Chopra, I. C., Handa, K. I., & Kapur, L. D. (1958). *Chopra's indegenous drugs of India* (2nd ed.). Calcutta: UN Dhur and Sons.

Chu, J., Li, J., Hao, Y., & Zhou, X. (2006). Effect of chemical compounds of *Galla chinensis* on enamel surface rehardening *in vitro*. *Chin J Stomatol*, *41*, 616–617. PMID:17129452

Chunming, L., Chenyang, X., Changfeng, G., & Martin, D. F. (2005) Level set evolution without re-initialization: A new variational formulation. *Proceedings of the IEEE international conference on computer vision and pattern recognition, 1*, 430-436. 10.1109/CVPR.2005.213

Cleghorn, B. M., Christie, W. H., & Dong, C. C. S. (2008). Anomalous mandibular premolars: A mandibular first premolar with three roots and a mandibular second premolar with a C-shaped canal system. *International Endodontic Journal*, *41*(11), 1005–1014. doi:10.1111/j.1365-2591.2008.01451.x PMID:19133090

Cohen, A., Laviv, A., Berman, P., Nashef, R., & Abu-Tair, J. (2009). Mandibular reconstruction using stereolithographic 3-dimensional printing modeling technology. *Oral Surgery, Oral Medicine, Oral Pathology, Oral Radiology, and Endodontics*, *108*(5), 661–666. doi:10.1016/j.tripleo.2009.05.023 PMID:19716728

Cook, N., & Samma, S. (1996). Flavonoids-chemistry, metabolism, cardioprotective effects and dietary sources. *The Journal of Nutritional Biochemistry*, *7*(2), 66–76. doi:10.1016/0955-2863(95)00168-9

Côrtes, D. F., Ekstrand, K. R., Elias-Boneta, A. R., & Ellwood, R. P. (2000). An in vitro Comparison of the Ability of Fibre–Optic Transillumination, Visual Inspection and Radiographs to Detect Occlusal Caries and Evaluate Lesion Depth. *Caries Research*, *34*(6), 443–447. doi:10.1159/000016621 PMID:11093016

Cowan, M. M. (1999). Plant products as antimicrobial agents. *Clinical Microbiology Reviews*, *12*, 564–582. PMID:10515903

Craig, R. G., & Powers, J. M. (Eds.). (1989). Restorative dental materials. Academic Press.

Cremers, D. (2003) A multiphase levelset framework for variational motion segmentation. Proc. In Scale Space Meth. Comput. Vis., 599–614.

Croll, T. P., & Nicholson, J. W. (2002). Glass ionomer cements in pediatric dentistry: Review of the literature. *Pediatric Dentistry*, *24*(5), 423–429. PMID:12412956

Cui, X., Boland, T., & D'Lima, D., & Lotz, M. (2012). Thermal inkjet printing in tissue engineering and regenerative medicine. *Recent Patents on Drug Delivery & Formulation*, *6*(2), 149–155. doi:10.2174/187221112800672949 PMID:22436025

Cui, Z., & Zhang, G. (2010). A novel medical image dynamic fuzzy classification model based on ridgelet transform. *Journal of Software*, *5*(5), 458–465.

Cummings, K. M., Rolf, J. C., Rosenflanz, A. Z., Rusin, R. P., & Swanson, J. E. (2006). *U.S. Patent No. 6,984,261*. Washington, DC: U.S. Patent and Trademark Office.

Cunningham, L. L. Jr, Madsen, M. J., & Peterson, G. (2005). Stereolithographic modeling technology applied to tumor resection. *Journal of Oral and Maxillofacial Surgery*, *63*(6), 873–878. doi:10.1016/j.joms.2005.02.027 PMID:15944992

Daglia, M., Tarsi, T., Papetti, A., Grisoli, P., Dacarro, C., Pruzzo, C., & Gazzani, G. (2002). Antiadhesive Effect of Green and Roasted Coffee on Streptococcus mutans' Adhesive Properties on Saliva-Coated Hydroxyapatite Beads. *Journal of Agricultural and Food Chemistry*, *50*(5), 1225–1229. doi:10.1021/jf010958t PMID:11853508

Darout, I. A., Albandar, J. M., Skaug, N., & Ali, R. W. (2002). Salivary microbiota levels in relation to periodontal status, experience of caries and miswak use in Sudanese adults. *Journal of Clinical Periodontology*, *29*(5), 411–420. doi:10.1034/j.1600-051X.2002.290505.x PMID:12060423

Das, A. (2015). *Guide to signals and patterns in image processing: foundations, methods and applications*. Springer. doi:10.1007/978-3-319-14172-5

Davies, G. M., Worthington, H. V., Clarkson, J. E., Thomas, P., & Davies, R. M. (2001). The use of fibre-optic transillumination in general dental practice. *British Dental Journal*, *191*(3), 145–147. doi:10.1038j.bdj.4801123 PMID:11523886

Davis, A. B. (1981). *Medicine and its technology: an introduction to the history of medical instrumentation*. Academic Press.

Davis, S., Gluskin, A. H., Livingood, P. M., & Chambers, D. W. (2010). Analysis of temperature rise and the use of coolants in the dissipation of ultrasonic heat buildup during post removal. *Journal of Endodontics*, *36*(11), 1892–1896. doi:10.1016/j.joen.2010.08.027 PMID:20951308

De Albuquerque, M. P., Esquef, I. A., & Mello, A. G. (2004). Image thresholding using Tsallis entropy. *Pattern Recognition Letters*, *25*(9), 1059–1065. doi:10.1016/j.patrec.2004.03.003

Défossez, H., Hall, R. M., Walker, P. G., Wroblewski, B. M., Siney, P. D., & Purbach, B. (2003). Determination of the trabecular bone direction from digitised radiographs. *Medical Engineering & Physics Journal*, *25*(9), 719-729.

De Luca, A., & Termini, S. (1972). A definition of a nonprobabilistic entropy in the setting of fuzzy sets theory. *Information and Control*, *20*(4), 301–312. doi:10.1016/S0019-9958(72)90199-4

De Vos, W., Casselman, J., & Swennen, G. R. J. (2009). Cone-beam computerized tomography (CBCT) imaging of the oral and maxillofacial region: A systematic review of the literature. *International Journal of Oral and Maxillofacial Surgery*, *38*(6), 609–625. doi:10.1016/j.ijom.2009.02.028 PMID:19464146

DelBalso, A. M., Greiner, F. G., & Licata, M. (1994). Role of diagnostic imaging in evaluation of the dental implant patient. *Radiographics*, *14*(4), 699–719. doi:10.1148/radiographics.14.4.7938761 PMID:7938761

Deng, J., Dong, W., Socher, R., Li, L. J., Li, K., & Fei-Fei, L. (2009). ImageNet: A large-scale hierarchical image database. *The Conference on Computer Vision and Pattern Recognition (CVPR)*, 248-255.

Denry, I., & Holloway, J. A. (2010). Ceramics for dental applications: A review. *Materials (Basel)*, *3*(1), 351–368. doi:10.3390/ma3010351

Dental Radiographs (X-Rays). (n.d.). Retrieved from California Dental Association: http://www.cda.org

Dérand, P., & Hirsch, J. M. (2009). Virtual bending of mandibular reconstruction plates using a computer-aided design. *Journal of Oral and Maxillofacial Surgery*, *67*(8), 1640–1643. doi:10.1016/j.joms.2009.03.039 PMID:19615575

Deserno, T. M. (2010). Fundamentals of biomedical image processing. In *Biomedical Image Processing* (pp. 1–51). Berlin: Springer. doi:10.1007/978-3-642-15816-2_1

Dettori, L., & Lindsay, S. (2007). A comparison of wavelet, ridgelet, and curvelet-based texture classification algorithms in computed tomography. *Computers in Biology and Medicine*, *37*(4), 486–498. doi:10.1016/j.compbiomed.2006.08.002 PMID:17054933

Dhawan, A. P. (2011). *Medical Image Analysis* (2nd ed.). Wiley-IEEE Press. doi:10.1002/9780470918548

Dhifi, W., Bellili, S., Jazi, S., Bahloul, N., & Mnif, W. (2016). Essential Oils' Chemical Characterization and Investigation of Some Biological Activities. A Critical Review. Medicines (Basel), 3(4).

Didry, N., Dubreuil, L., Trotin, F., & Pinkas, M. (1998). Antimicrobial activity of aerial parts of Drosera peltata Smith on oral bacteria. *Journal of Ethnopharmacology*, *60*(1), 91–96. doi:10.1016/S0378-8741(97)00129-3 PMID:9533437

Dighe, S. C., & Shriram, R. (2012, November). Dental biometrics for human identification based on dental work and image properties in Periapical radiographs. In TENCON 2012-2012 IEEE Region 10 Conference (pp. 1-6). IEEE. doi:10.1109/TENCON.2012.6412216

Digital Dental Periapical X-Ray Database for Caries Screening. (n.d.). Vahab – LabArchives, Your Electronic Lab Notebook, Biomed Central Edition.

Ding, K., Xiao, L., & Weng, G. (2017). Active contours driven by region-scalable fitting and opti-mized Laplacian of Gaussian energy for image segmentation. *Signal Processing*, *134*, 224–233. doi:10.1016/j.sigpro.2016.12.021

Dixon, R. A., Dey, P. M., & Lamb, C. J. (1983). Phytoalexins: Enzymology and molecular biology. *Adv in Enzy and Rel A. Molecular Biology*, *53*, 1–36.

Do, M. N., & Vetterli, M. (2003). The finite ridgelet transform for image representation. *IEEE Transactions on Image Processing*, *12*(1), 16–28. doi:10.1109/TIP.2002.806252 PMID:18237876

Do, M. N., & Vetterli, M. (2005). The contourlet transform: An efficient directional multiresolution image representation. *IEEE Transactions on Image Processing*, *14*(12), 2091–2106. doi:10.1109/TIP.2005.859376 PMID:16370462

Dong-ri, S., & Fu-yuan, G. (2010, April). The segmentation algorithm of dental CT images based on fuzzy maximum entropy and region growing. In *Bioinformatics and Biomedical Technology (ICBBT), 2010 International Conference on* (pp. 74-78). IEEE. 10.1109/ICBBT.2010.5479006

Dougherty, G. (2009). *Digital Image Processing for Medical Applications*. Cambridge University Press.

Douglas, R. D. (2000). Color stability of new-generation indirect resins for prosthodontic application. *The Journal of Prosthetic Dentistry*, *83*(2), 166–170. doi:10.1016/S0022-3913(00)80008-6 PMID:10668028

Duarte, S., Gregoire, S., Singh, A., Vorsa, N., Schaich, K., Bowen, W., & Koo, H. (2006). Inhibitory effects of cranberry polyphenols on formation and acidogenicity of *Streptococcus mutans* biofilms. *FEMS Microbiology Letters*, *257*(1), 50–56. doi:10.1111/j.1574-6968.2006.00147.x PMID:16553831

Duda, R. O., Hart, P. E., & Stork, D. G. (2001). *Pattern Classification*. New York-John Wiley & Sons.

DXR-database. (n.d.). Retrieved from https://mynotebook.labarchives.com/share/Vahab/MjAuOHw4NTc2Mi8xNi9Uc-mVlTm9kZS83NzM5OTk2MDZ8NTIuOA==

Dzink, J. L., & Socransky, S. S. (1985). Comparative in vitro activity of sanguinarine against oral microbial isolates. *Antimicrobial Agents and Chemotherapy*, *27*(4), 663–665. doi:10.1128/AAC.27.4.663 PMID:4004199

Ebel, J., & Grisebach, H. (1988). Defense strategies of soybean against the fungus Phytophthora megasperma f. sp. glycinea: A molecular analysis. *Trends in Biochemical Sciences*, *13*(1), 23–27. doi:10.1016/0968-0004(88)90014-X PMID:3072693

Eckardt, A., & Swennen, G. R. (2005). Virtual planning of composite mandibular reconstruction with free fibula bone graft. *The Journal of Craniofacial Surgery*, *16*(6), 1137–1140. doi:10.1097/01.scs.0000186306.32042.96 PMID:16327572

Edelstein, B. L., & Douglass, C. W. (1995). Dispelling the myth that 50 percent of U.S. schoolchildren have never had a cavity. *Public Health Reports (Washington, D.C.)*, *110*, 522–530. PMID:7480606

El Sherbini, A. M., El Sherbini, T. M., Hegazy, H., Cristoforetti, G., Legnaioli, S., Palleschi, V., ... Tognoni, E. (2005). Evaluation of self-absorption coefficients of aluminum emission lines in laser-induced breakdown spectroscopy measurements. *Spectrochimica Acta. Part B, Atomic Spectroscopy*, *60*(12), 1573–1579. doi:10.1016/j.sab.2005.10.011

Elazab, A., Wang, C., Jia, F., Wu, J., Li, G., & Hu, Q. (2015). Segmentation of Brain Tissues from Magnetic Resonance Images Using Adaptively regularized Kernel-Based Fuzzy *C*-Means Clustering. *Computational and Mathematical Methods in Medicine*, *485495*, 1–12. doi:10.1155/2015/485495 PMID:26793269

Ellwood, R. P., Goma, J., & Pretty, I. A. (2012). Caries clinical trial methods for the assessment of oral care products in the 21st century. *Advances in Dental Research*, *24*(2), 32–35. doi:10.1177/0022034512449464 PMID:22899676

Eloff, J. N. (1998). Which extractant should be used for the screening and isolation of antimicrobial components from plants? *Journal of Ethnopharmacology*, *60*(1), 1–8. doi:10.1016/S0378-8741(97)00123-2 PMID:9533426

El-Said, F., Fadulu, S. O., Kuye, J. O., & Sofowora, E. A. (1971). Native cures in Nigeria: II The antimicrobial properties of the buffered extracts of chewing sticks. *Lloydia*, *34*, 172–174.

El-Sherbini, T. M. (2016). Impact of Physics on Medical Sciences and Applications: Lasers and Nanotechnology. *Journal of Medical Physics and Applied Sciences*, *1*(1).

Elujoba, A. A., Odeleye, O. M., & Ogunyemi, C. M. (2005). Traditional Medical Development for medical and dental primary Health care Delivery System in Africa. *African Journal of Traditional, Complementary, and Alternative Medicines*, *2*, 46.

Eskicioglu, A. M., & Fisher, P. S. (1995). Image quality measures and their performance. *IEEE Transactions on Communications*, *43*(12), 2959–2965. doi:10.1109/26.477498

Estrela, C., Bueno, M. R., Azevedo, B. C., Azevedo, J. R., & Pécora, J. D. (2008). A new periapical index based on cone beam computed tomography. *Journal of Endodontics*, *34*(11), 1325–1331. doi:10.1016/j.joen.2008.08.013 PMID:18928840

Estrela, C., Bueno, M. R., Leles, C. R., Azevedo, B., & Azevedo, J. R. (2008). Accuracy of cone beam computed tomography and panoramic and periapical radiography for detection of apical periodontitis. *Journal of Endodontics*, *34*(3), 273–279. doi:10.1016/j.joen.2007.11.023 PMID:18291274

Estrela, C., Bueno, M. R., Sousa-Neto, M. D., & Pécora, J. D. (2008). Method for determination of root curvature radius using cone-beam computed tomography images. *Brazilian Dental Journal*, *19*(2), 114–118. doi:10.1590/S0103-64402008000200005 PMID:18568224

Eto, A., Saido, T. C., Fukushima, K., Tomioka, S., Imai, S., Nisizawa, T., & Hanada, N. (1999). Inhibitory effect of a self-derived peptide on glucosyltransferase of *Streptococcus mutans*. Possible novel anticaries measures. *The Journal of Biological Chemistry*, *274*(22), 15797–15802. doi:10.1074/jbc.274.22.15797 PMID:10336482

Evans, L. (1998). *Partial Differential Equations*. Providence, RI: American Mathematical Society.

Evens, R. G. (1995). Roentgen retrospective: One hundred years of a revolutionary technology. *Journal of the American Medical Association*, *274*(11), 912–916. doi:10.1001/jama.1995.03530110074039 PMID:7674507

Fahmy, G., Nassar, D., Haj-Said, E., Chen, H., Nomir, O., & Zhou, J. (2004). Towards an automated dental identification system (ADIS). *International conference on biometric authentication*, *3*, 522–525. 10.1007/978-3-540-25948-0_107

Fani, M. M., Kohanteb, J., & Dayaghi, M. (2007). Inhibitory activity of garlic Allium sativum extract on multi drug resistant *S.mutans*. *J. Indian Soc. Pedad. Prev. Dent*, *25*(4), 164–168. PMID:18007101

Fathilah, A. R., Rahim, Z. H., Othman, Y., & Yusoff, M. (2009). Effect of *Piper betle* and *Psidium guajava* Extracts on Dental Plaque Bacteria. *Pakistan Journal of Biological Sciences*, *12*(6), 518–521. doi:10.3923/pjbs.2009.518.521 PMID:19580002

Featherstone, J. D. (1999). Prevention and reversal of dental caries: Role of low level fluoride. *Community Dentistry and Oral Epidemiology*, *27*(1), 31–40. doi:10.1111/j.1600-0528.1999.tb01989.x PMID:10086924

Feng, C., Zhao, D., & Huang, M. (2017). Image segmentation and bias correction using local inhomogeneous iNtensity clustering (LINC): A region-based level set method. *Neurocomputing*, *219*(C), 107–129. doi:10.1016/j.neucom.2016.09.008

Fennis-Ie, Y. L., Verdonschot, E. H., & Van't Hof, M. A. (1998). Performance of some diagnostic systems in the prediction of occlusal caries in permanent molars in 6-and 11-year-old children. *Journal of Dentistry*, *26*(5-6), 403–408. doi:10.1016/S0300-5712(97)00060-2 PMID:9699429

Ferracane, J. L. (1995). Current trends in dental composites. *Critical Reviews in Oral Biology and Medicine*, *6*(4), 302–318. doi:10.1177/10454411950060040301 PMID:8664421

Ferracane, J. L. (2011). Resin composite—state of the art. *Dental Materials*, *27*(1), 29–38. doi:10.1016/j.dental.2010.10.020 PMID:21093034

Fessenden, R. J., & Fessenden, J. S. (1982). *Organic Chemistry* (2nd ed.). Boston: Willard Grant Press.

Figueiredo, F. E., Martins-Filho, P. R., & Faria-E-Silva, A. L. (2015). Do metal post-retained restorations result in more root fractures than fiber post-retained restorations? A systematic review and meta-analysis. *Journal of Endodontics*, *41*(3), 309–316. doi:10.1016/j.joen.2014.10.006 PMID:25459568

Firdousi, R., & Parveen, S. (2014). Local Thresholding Techniques in Image Binarization. *International Journal Of Engineering And Computer Science*, *3*(3), 4062–4065.

Forrai, J. (2005). *Culture history of dentistry*. Academic Press.

Frederiksen, N. L. (1995). Diagnostic imaging in dental implantology. *Oral Surgery, Oral Medicine, Oral Pathology, Oral Radiology, and Endodontology*, *80*(5), 540–554. doi:10.1016/S1079-2104(05)80153-2 PMID:8556464

Fredo, A. R. J., Kavitha, G., & Ramakrishnan, S. (2014). Segmentation and morphometric analysis of sub-cortical regions in autistic MR brain images using fuzzy Gaussian model based distance regularized multi-phase level set. *International Journal of Biomedical Engineering and Technology*, *15*(3), 211–223. doi:10.1504/IJBET.2014.064647

Fredo, A. R. J., Kavitha, G., & Ramakrishnan, S. (2015). Segmentation and analysis of corpus callosum in autistic MR brain images using reaction diffusion level sets. *Journal of Medical Imaging and Health Informatics*, *5*(4), 737–741. doi:10.1166/jmihi.2015.1442

Fukushima, K., & Miyake, S. (1982). Neocognitron: A new algorithm for pattern recognition tolerant of deformations and shifts in position. *Pattern Recognition*, *15*(6), 455–469. doi:10.1016/0031-3203(82)90024-3

Fukushima, K., Motoda, R., Takada, K., & Ikeda, T. (1981). Resolution of Streptococcus mutans glucosyltransferase into two components essential to water-insoluble glucan synthesis. *FEBS Letters*, *128*(2), 213–216. doi:10.1016/0014-5793(81)80083-X PMID:6455304

Fukushima, K., Takada, K., Motoda, R., & Ikeda, T. (1982). Independence of water-insoluble glucan synthesis and adherence of Streptococcus mutans to smooth surfaces. *FEBS Letters*, *149*(2), 299–303. doi:10.1016/0014-5793(82)81121-6 PMID:6217990

Fullerton, J. N., Frodsham, G. C., & Day, R. M. (2014). 3D printing for the many, not the few. *Nature Biotechnology*, *32*(11), 1086–1087. doi:10.1038/nbt.3056 PMID:25380438

Furiga, A., Lonvaud, A., & Badet, C. (2009). *In vitro* study of antioxidant capacity and antibacterial activity on oral anaerobes of a grape seed extract. *Food Chemistry*, *113*(4), 1037–1040. doi:10.1016/j.foodchem.2008.08.059

Gahleitner, A., Watzek, G., & Imhof, H. (2003). Dental CT: Imaging technique, anatomy, and pathologic conditions of the jaws. *European Radiology*, *13*(2), 366–376. PMID:12599003

Gales, M. A., & Nguyen, T. M. (2000). Sorbitol compared with xylitol in prevention of dental caries. *The Annals of Pharmacotherapy*, *34*(1), 98–100. doi:10.1345/aph.19020 PMID:10669192

Ganguly, R., Mishra, P., & Sharma, A. (2001). Microbes and infection. *Indian Journal of Microbiology*, *41*, 211–213.

Gao, H., & Chae, O. (2010). Individual tooth segmentation from CT images using level set method with shape and intensity prior. *Pattern Recognition*, *43*(7), 2406–2417. doi:10.1016/j.patcog.2010.01.010

Gao, J., Xu, W., & Geng, J. (2006). 3D shape reconstruction of teeth by shadow speckle correlation method. *Optics and Lasers in Engineering*, *44*(5), 455–465. doi:10.1016/j.optlaseng.2005.04.013

Garg, N. (2013). Binarization Techniques used for grey scale images. *International Journal of Computers and Applications*, *71*(1).

Gayathri, V., Menon, H. P., & Viswa, A. (2014). *Challenges in Edge Extraction of Dental X-Ray Images Using Image Processing Algorithms–A Review*. Academic Press.

Gayathri, V., & Menon, H. P. (2014). Challenges in edge extraction of dental x-ray images using image processing algorithms - a review. *Int. J. Comput. Sci. Inf. Technol.*, *5*, 5355–5358.

Gazi, M. I., Davies, T. J., Al-Bagieh, N., & Cox, S. W. (1992). The immediate and medium-term effects of Meswak on the composition of mixed saliva. *Journal of Clinical Periodontology*, *19*(2), 113–117. doi:10.1111/j.1600-051X.1992.tb00449.x PMID:1602035

Geissman, T. A. (1963). Flavonoid compounds, tannins, lignins and related compounds. In M. Florkin & E. H. Stotz (Eds.), *Pyrrole Pigments, Isoprenoid Compounds and Phenolic Plant Constituents* (p. 265). New York: Elsevier Press. doi:10.1016/B978-1-4831-9718-0.50018-7

Gepreel, M. A. H., & Niinomi, M. (2013). Biocompatibility of Ti-alloys for long-term implantation. *Journal of the Mechanical Behavior of Biomedical Materials*, *20*, 407–415. doi:10.1016/j.jmbbm.2012.11.014 PMID:23507261

Gibbons, R. J. (1984). Adherent interactions which may affect microbial ecology in the mouth. *Journal of Dental Research*, *63*(3), 378–385. doi:10.1177/00220345840630030401 PMID:6583240

Gibbons, R. J., & Dankers, I. (1981). Lectin-Like Constituents of Foods Which React with Components of Serum, Saliva, and *Streptococcus mutans*. *Applied and Environmental Microbiology*, *41*(4), 880–888. PMID:6786220

Gibbons, R. J., & Houte, J. V. (1975). Bacterial adherence in oral microbial ecology. *Annual Review of Microbiology*, *29*(1), 19–44. doi:10.1146/annurev.mi.29.100175.000315 PMID:1180512

Girshick, R. (2015). Fast R-CNN. *IEEE International Conference on Computer Vision (ICCV)*, 1440-1448.

Girshick, R., Donahue, J., Darrell, T., & Malik, J. (2015). Region-Based Convolutional Networks for Accurate Object Detection and Segmentation. *IEEE Transactions on Pattern Analysis and Machine Intelligence*, *38*(1), 142–158. doi:10.1109/TPAMI.2015.2437384 PMID:26656583

Glass, R. L. (1982). The first international conference on the declining prevalence of dental caries. *Journal of Dental Research*, *61*(Spec Iss), 1304.

Goaz, P. W., & White, S. C. (1982). *Oral Radiology-Principles and Interpretation*. St. Louis, MO: The C.V. Mosby Company.

Gonzalez, R. C., & Woods, R. E. (2001). Digital Image Processing. Upper Saddle River, NJ: Prentice Hall.

Gonzalez, R. C., & Woods, R. E. (2002). *Digital image processing*. Academic Press.

Gonzalez, Woods, & Eddins. (2010). Digital Image Processing Using MATLAB (2nd ed.). Tata McGraw Hill Education Private Limited.

Gonzalez, R. C., & Woods, R. E. (2006). *Digital Image Processing* (3rd ed.). Prentice-Hall Inc.

Goodfellow, I., Bengio, Y., & Courville, A. (2016). *Deep Learning*. MIT Press.

Gormez, O., & Yilmaz, H. H. (2009). Image post-processing in dental practice. *European Journal of Dentistry*, *3*(4), 343. PMID:19826609

Grabisch, M., Nguyen, H. T., & Walker, E. A. (2013). Fundamentals of Uncertainty Calculi with Applications to Fuzzy Inference. Springer Science & Business Media.

Graham Leedham, C. Y., Takru, K., Tan, J. H. N., & Mian, L. (2003, August). Comparison of some thresholding algorithms for text/background segmentation in difficult document images. In *Proceedings of the seventh international conference on document analysis and recognition* (*Vol. 2*, pp. 859-864). Academic Press. 10.1109/ICDAR.2003.1227784

Greenberg, M., Urnezis, P., & Tian, M. (2007). Compressed mints and chewing gum containing magnolia bark extract are effective against bacteria responsible for oral malodor. *Journal of Agricultural and Food Chemistry*, *55*(23), 9465–9469. doi:10.1021/jf072122h PMID:17949053

Gregoire, S., Singh, A. P., Vorsa, N., & Koo, H. (2007). Influence of cranberry phenolics on glucan synthesis by glucosyltransferases and *Streptococcus mutans* acidogenicity. *Journal of Applied Microbiology, 103*(5), 1960–1968. doi:10.1111/j.1365-2672.2007.03441.x PMID:17953606

Gregory, J. S., Stewart, A., Undrill, P. E., Reid, D. M., & Aspden, R. M. (2004). Identification of hip fracture patients from radiographs using Fourier analysis of the trabecular structure: A cross-sectional study. *BMC Medical Imaging, 4*(4), 1–11. PMID:15469614

Grenby, T. H., Phillips, A., & Mistry, M. (1989). Studies of the dental properties of lactitol compared with five other bulk sweeteners *in vitro*. *Caries Research, 23*(5), 315–319. doi:10.1159/000261199 PMID:2766316

Grenier, T., Revol-Muller, C., Costes, N., Janier, M., & Gimenez, G. (2005). Automated seeds location for whole body NaF PET segmentation. *IEEE Transactions on Nuclear Science, 52*(5), 1401–1405. doi:10.1109/TNS.2005.859354

Gross, B. C., Erkal, J. L., Lockwood, S. Y., Chen, C., & Spence, D. M. (2014). *Evaluation of 3D printing and its potential impact on biotechnology and the chemical sciences*. Academic Press.

Gulsahi, A. (2011). Bone quality assessment for dental implants. In *Implant Dentistry* (pp. 437–452). Ankara, Turkey: The Most Promising Discipline of Dentistry.

Gupta, S., Patil, N., Solanki, J., Singh, R., & Laller, S. (2014). Oral implant imaging: A review. *The Malaysian Journal of Medical Sciences, 22*(3), 7–17. PMID:26715891

Gustafson, G., & Malmö, D. O. (1950). Age determination on teeth. *The Journal of the American Dental Association, 41*(1), 45–54. doi:10.14219/jada.archive.1950.0132 PMID:15428197

Hada, L. S., Kakiuchi, N., Hattori, M., & Namba, T. (1989). Identification of antibacterial principles against *Streptococcus mutans* inhibitory principles against glucosyltransferase from the seed of Areca catechu L. *Phytotherapy Research, 3*(4), 140–144. doi:10.1002/ptr.2650030406

Haddad, E. E., Lauritano, D., Candotto, V., &Carinci, F. (2014). Guided bone regeneration is a reliable technique in implant dentistry: An overview and a case report. *OA Dentistry, 2*(1).

Haggard, P., & de Boer, L. (2014). Oral somatosensory awareness. *Neuroscience and Biobehavioral Reviews, 47*, 469–484. doi:10.1016/j.neubiorev.2014.09.015 PMID:25284337

Halim, A. (2008). *Human Anatomy: Volume 3. Head, Neck and Brain*. New Delhi, India: I. K. International Publisher.

Hall, E. L., Kruger, R. P., Dwyer, S. J., Hall, D. L., Mclaren, R. W., & Lodwick, G. S. (1971). A survey of preprocessing and feature extraction techniques for radiographic images. *IEEE Transactions on Computers, 100*(9), 1032–1044. doi:10.1109/T-C.1971.223399

Hamilton-Miller, J. M. T. (2001). Anti-cariogenic effects of tea (*Camellia sinensis*). *Journal of Medical Microbiology, 50*(4), 299–302. doi:10.1099/0022-1317-50-4-299 PMID:11289514

Hanada, N., & Kuramitsu, H. K. (1988). Isolation and characterization of the Streptococcus mutans gtfC gene, coding for synthesis of both soluble and insoluble glucans. *Infection and Immunity, 56*, 1999–2005. PMID:2969375

Hanada, N., & Kuramitsu, H. K. (1989). Isolation and characterization of the *Streptococcus mutans* gtfD gene, coding for primer-dependent soluble glucan synthesis. *Infection and Immunity, 57*, 2079–2085. PMID:2543630

Hannig, M., & Bott, B. (1999). In-vitro pulp chamber temperature rise during composite resin polymerization with various light-curing sources. *Dental Materials, 15*(4), 275–281. doi:10.1016/S0109-5641(99)00047-0 PMID:10551096

Hanson, M., & Pleva, J. (1991). The dental amalgam issue. A review. *Experientia*, *47*(1), 9–22. doi:10.1007/BF02041243 PMID:1999251

Haralick, R. M., Shanmugam, K., & Dinstein, I. (1973). Textural features for image classification. *IEEE Transactions on Systems, Man, and Cybernetics*, *SMC-3*(6), 610–621. doi:10.1109/TSMC.1973.4309314

Harandi, A. A., Pourghassem, H., & Mahmoodian, H. (2011, December). Upper and lower jaw segmentation in dental X-ray image using modified active contour. In *Intelligent Computation and Bio-Medical Instrumentation (ICBMI), 2011 International Conference on* (pp. 124-127). IEEE. 10.1109/ICBMI.2011.88

Harrell, W. E. Jr. (2007). Three-dimensional diagnosis and treatment planning: The use of 3D facial imaging and 3D cone beam CT in orthodontics and dentistry. *Australian Dental Practice*, *18*, 102–113.

Haslam, E. (1996). Natural polyphenols (vegetable tannins) as drugs: Possible modes of action. *Journal of Natural Products*, *59*(2), 205–215. doi:10.1021/np960040+ PMID:8991956

Hassanpour, H., Samadiani, N., & Salehi, S. M. (2015). Using morphological transforms to enhance the contrast of medical images. *The Egyptian Journal of Radiology and Nuclear Medicine*, *46*(2), 481–489. doi:10.1016/j.ejrnm.2015.01.004

Hattori, M., Kusumoto, I. T., Namba, T., Ishigami, T., & Hara, Y. (1990). Effect of tea polyphenols on glucan synthesis by glucosyltransferase from *Streptococcus mutans*. *Chemical & Pharmaceutical Bulletin*, *38*(3), 717–720. doi:10.1248/cpb.38.717 PMID:2140716

Hauptmann, H., & Lacerda Nazario, L. (1950). Some constituents of leaves of *Cassia alata*. *Journal of the American Chemical Society*, *72*(4), 1492–1495. doi:10.1021/ja01160a019

Havsteen, & Havsteen, B. (1983). Flavonoids, a class of natural products of high pharmacological potency. *Biochem. Pharmacol.*, *32*, 1141-1148.

Havsteen, B. (1983). Flavonoids, a class of natural products of high pharmacological potency. *Biochemical Pharmacology*, *32*(7), 1141–1148. doi:10.1016/0006-2952(83)90262-9 PMID:6342623

Hayachibara, M. F., Koo, H., Rosalen, P. L., Duarte, S., Franco, E. M., Bowen, W. H., ... Cury, J. A. (2005). *In vitro* and in vivo effects of isolated fractions of Brazilian propolis on caries development. *Journal of Ethnopharmacology*, *101*(1-3), 110–115. doi:10.1016/j.jep.2005.04.001 PMID:15913934

Hayashida, O., Hasumi, K., & Endo, A. (1997). Chemical and Functional Properties of Mutastein, an Inhibitor of Insoluble Glucan Synthesis by *Streptococcus sobrinus*. *Bioscience, Biotechnology, and Biochemistry*, *61*(4), 588–591. doi:10.1271/bbb.61.588 PMID:9145515

Hayes, C. (2001). The effect of non-cariogenic sweeteners on the prevention of dental caries: A review of the evidence. *Journal of Dental Education*, *65*, 1106–1109. PMID:11699985

Hebbar, S. S., Harsha, V. H., Shripathi, V., & Hegde, G. R. (2004). Ethnomedicine of Dharwad district in Karnataka, India--plants used in oral health care. *Journal of Ethnopharmacology*, *94*(2-3), 261–266. doi:10.1016/j.jep.2004.04.021 PMID:15325728

Hechler, S. L. (2008). Cone-beam CT: Applications in orthodontics. *Dental Clinics*, *52*(4), 809–823. doi:10.1016/j.cden.2008.05.001 PMID:18805230

Heisey, R. M. B., & Gorha, B. (1992). Antimicrobial effects of plant extracts on *Streptococcus mutans, Candida albicans, Trichophyton rubrum* and other micro-organisms. *Letters in Applied Microbiology*, *14*(4), 136–139. doi:10.1111/j.1472-765X.1992.tb00668.x

He, K., Gkioxari, G., Dollár, P., & Girshick, R. (2017). Mask R-CNN. *IEEE International Conference on Computer Vision (ICCV)*, 2980-2988.

He, K., Zhang, X., Ren, S., & Sun, J. (2016). Deep Residual Learning for Image Recognition. *2016 IEEE Conference on Computer Vision and Pattern Recognition (CVPR)*, 770-778. 10.1109/CVPR.2016.90

Hemanth, D. J., & Anitha, J. (2012). Image Pre-processing and Feature Extraction Techniques for Magnetic Resonance Brain Image Analysis. In *Computer Applications for Communication, Networking, and Digital Contents* (pp. 349–356). Berlin: Springer. doi:10.1007/978-3-642-35594-3_47

Hewitt, G. F., Shires, G. L., & Bott, T. R. (1994). *Process heat transfer* (Vol. 113). Boca Raton, FL: CRC Press.

Hintze, H., Wenzel, A., Danielsen, B., & Nyvad, B. (1998). Reliability of visual examination, fibre-optic transillumination, and bite-wing radiography, and reproducibility of direct visual examination following tooth separation for the identification of cavitated carious lesions in contacting approximal surfaces. *Caries Research*, *32*(3), 204–209. doi:10.1159/000016454 PMID:9577986

Hirogaki, Y., Sohmura, T., Satoh, H., Takahashi, J., & Takada, K. (2001). Complete 3-D reconstruction of dental cast shape using perceptual grouping. *IEEE Transactions on Medical Imaging*, *20*(10), 1093–1101. doi:10.1109/42.959306 PMID:11686444

Hogg, S. D., & Embery, G. (1982). Blood-group reactive glycoprotein from human saliva interacts with lipoteichoic acid on the surface of *Streptococcus sanguis* cells. *Archives of Oral Biology*, *27*(3), 261–268. doi:10.1016/0003-9969(82)90060-7 PMID:6953942

Ho, J. T., Wu, J., Huang, H. L., Chen, M. Y. C., Fuh, L. J., & Hsu, J. T. (2013). Trabecular bone structural parameters evaluated using dental cone-beam computed tomography: Cellular synthetic bones. *Biomedical Engineering Online*, *12*(1), 115. doi:10.1186/1475-925X-12-115 PMID:24207062

Holambe, S., & Kumbhar, P. (2016). Comparison between Otsu's Image Thresholding Technique and Iterative Triclass. *International Journal of Computer Trends and Technology*, *33*(2), 80–82. doi:10.14445/22312803/IJCTT-V33P117

Hosang, J. (2016). *Faster RCNN TF*. GitHub. Retrieved from https://github.com/smallcorgi/Faster-RCNN_TF

Hosntalab, M., Zoroofi, R. A., Tehrani-Fard, A. A., & Shirani, G. (2010). Classification and numbering of teeth in multi-slice CT images using wavelet-Fourier descriptor. *International Journal of Computer Assisted Radiology and Surgery*, *5*(3), 237–249. doi:10.100711548-009-0389-8 PMID:20033505

Hoy, M. B. (2013). 3D printing: Making things at the library. *Medical Reference Services Quarterly*, *32*(1), 93–99. doi:10.1080/02763869.2013.749139 PMID:23394423

Huang, J., Rathod, V., Sun, C., Zhu, M., Korattikara, A., Fathi, A., ... Murphy, K. (2017, 7). Speed/Accuracy Trade-Offs for Modern Convolutional Object Detectors. *The IEEE Conference on Computer Vision and Pattern Recognition (CVPR)*. 10.1109/CVPR.2017.351

Huang, Z. L., Mochizuki, T., Qu, W. M., Hong, Z. Y., Watanabe, T., Urade, Y., & Hayaishi, O. (2006). Altered sleep-wake characteristics and lack of arousal response to H3 receptor antagonist in histamine H1 receptor knockout mice. *Proceedings of the National Academy of Sciences of the United States of America*, *103*(12), 4687–4692. doi:10.1073/pnas.0600451103 PMID:16537376

Hubel, D. H., & Wiesel, T. N. (1959). Receptive Fields of Single Neurons in the Cat's Striate Cortex. *The Journal of Physiology*, *148*(3), 574–591. doi:10.1113/jphysiol.1959.sp006308 PMID:14403679

Hubel, D. H., & Wiesel, T. N. (1962). Receptive fields, binocular interaction and functional architecture in the cat's visual cortex. *The Journal of Physiology*, *160*(1), 106–154. doi:10.1113/jphysiol.1962.sp006837 PMID:14449617

Huber, M. B., Carballido, G. J., Fritscher, K., Schubert, R., Haenni, M., Hengg, C., ... Link, T. M. (2009). Development and testing of texture discriminators for the analysis of trabecular bone in proximal femur radiographs. *Medical Physics*, *36*(11), 5089–5098. doi:10.1118/1.3215535 PMID:19994519

Huh, J., Nam, H., Kim, J., Park, J., Shin, S., & Lee, R. (2015). Studies of automatic dental cavity detection system as an auxiliary tool for diagnosis of dental caries in digital x-ray image. *Progress in Medical Physics*, *25*(1), 52–58. doi:10.14316/pmp.2015.26.1.52

Hwang, J. K., Chung, J. Y., Baek, N. I., & Park, J. H. (2004). Isopanduratin A from *Kaempferia pandurata* an active antibacterial agent against cariogenic *Streptococcus mutans*. *International Journal of Antimicrobial Agents*, *23*(4), 377–381. doi:10.1016/j.ijantimicag.2003.08.011 PMID:15081087

Ie, Y. L., Verdonschot, E. H., Schaeken, M. J. M., & Van't Hof, M. A. (1995). Electrical conductance of fissure enamel in recently erupted molar teeth as related to caries status. *Caries Research*, *29*(2), 94–99. doi:10.1159/000262048 PMID:7728835

Iio, M., Uyeda, M., Iwanani, T., & Nakagawa, Y. (1984). Flavonoids as a possible preventive of dental caries. *Agricultural and Biological Chemistry*, *48*, 2143–2145.

Ikigai, H., Nakae, T., Hara, Y., & Shimamura, T. (1993). Bactericidal catechins damage the lipid bilayer. *Biochimica et Biophysica Acta*, *1147*(1), 132–136. doi:10.1016/0005-2736(93)90323-R PMID:8466924

Imangaliyev, S., van der Veen, M., Volgenant, C., Keijser, B., Crielaard, W., & Levin, E. (2016). Deep Learning for Classification of Dental Plaque Images. In P. M. Pardalos, P. Conca, G. Giuffrida, & G. Nicosia (Eds.), Lecture Notes in Computer Science: Vol. 10122. Machine Learning, Optimization, and Big Data. MOD 2016 (pp. 407–410). Berlin: Springer. doi:10.1007/978-3-319-51469-7_34

ISIB. (2015). *Dataset*. Retrieved from http://www-o.ntust.edu.tw/~cweiwang/ISBI2015/challenge2/

Ismail, A. I., Sohn, W., Tellez, M., Amaya, A., Sen, A., Hasson, H., & Pitts, N. B. (2007). The International Caries Detection and Assessment System (ICDAS): An integrated system for measuring dental caries. *Community Dentistry and Oral Epidemiology*, *35*(3), 170–178. doi:10.1111/j.1600-0528.2007.00347.x PMID:17518963

ISO. (2016). *ISO 3950:2016 Dentistry--Designation system for teeth and areas of the oral cavity.*

Ito, K., Nakamura, Y., Tokunaga, T., Iijima, D., & Fukushima, K. (2003). Anti-cariogenic properties of a water-soluble extract from cacao. *Bioscience, Biotechnology, and Biochemistry*, *67*(12), 2567–2573. doi:10.1271/bbb.67.2567 PMID:14730134

Iwami, Y., Schachtele, C. F., & Yamada, T. (1995). Effect of sucrose monolaurate on acid production, levels of glycolytic intermediates and enzyme activities of *Streptococcus mutans* NCTC 10449. *Journal of Dental Research*, *74*(9), 1613–1617. doi:10.1177/00220345950740091801 PMID:7560425

Iwu, M. M., Duncan, A. R., & Okunji, C. O. (1999). New Antimicrobials of plant origin. In Perspectives in New crops and New uses. ASHS Press.

Izzaty, N., & Hisham, N. (2015). *Image segmentation using adaptive thresholding*. Academic Press.

Jac Fredo, A. R., Kavitha, G., & Ramakrishnan, S. (2014). Segmentation and analysis of brain sub-cortical regions using regularized multi-phase level set in autistic MR Images. *International Journal of Imaging Systems and Technology*, *24*(3), 256–262. doi:10.1002/ima.22101

Jac Fredo, A. R., Kavitha, G., & Ramakrishnan, S. (2015). Automated segmentation and analysis of corpus callosum in autistic MR images using fuzzy c-means based level set. *Journal of Medical and Biological Engineering, 35*(3), 331–337. doi:10.100740846-015-0047-2

Jafarzadeh, H., & Abbott, P. V. (2010). Review of pulp sensibility tests. Part II: Electric pulp tests and test cavities. *International Endodontic Journal, 43*(11), 945–958. doi:10.1111/j.1365-2591.2010.01760.x PMID:20726917

Jagtap, A. G., & Karkera, S. G. (2000). Extract of Juglandaceae regia inhibits growth, in-vitro adherence, acid production and aggregation of Streptococcus mutans. *The Journal of Pharmacy and Pharmacology, 52*(2), 235–242. doi:10.1211/0022357001773751 PMID:10714956

Jain, A. K., & Chen, H. (2004). Matching of dental X-ray images for human identification. *Pattern Recognition, 37*(7), 1519–1532. doi:10.1016/j.patcog.2003.12.016

Jain, A. K., Chen, H., & Minut, S. (2003). Dental biometrics: human identi'cation using dental radiographs. *Proc. Fourth International Conference on AVBPA*, 429-437.

Jain, A. K., Chen, H., & Minut, S. (2003, June). Dental biometrics: human identification using dental radiographs. In *International Conference on Audio-and Video-Based Biometric Person Authentication* (pp. 429-437). Springer. 10.1007/3-540-44887-X_51

Jain, A., Bolle, R., & Pankanti, S. (1999). *Biometrics-Personal Identi'cation in Networked Society*. Dordrecht: Kluwer Academic Publishers.

Jain, K. R., & Chauhan, N. C. (2017). An Automated Level Set Segmentation Approach for Lesion Detection in Dental Radiograph for Endodontic Treatment. *International Journal of Computers and Applications, 172*(5), 17–24. doi:10.5120/ijca2017914092

Jakubinek, M. B., O'Neill, C., Felix, C., Price, R. B., & White, M. A. (2008). Temperature excursions at the pulp–dentin junction during the curing of light-activated dental restorations. *Dental Materials, 24*(11), 1468–1476. doi:10.1016/j.dental.2008.03.012 PMID:18448161

Jandt, K. D., & Sigusch, B. W. (2009). Future perspectives of resin-based dental materials. *Dental Materials, 25*(8), 1001–1006. doi:10.1016/j.dental.2009.02.009 PMID:19332352

Jayadevappa, B. S., Kodhandarama, G. S., Santosh, S. V., & Rashid, W. T. (2010). Imaging of dental implants. *Journal of Oral Health Research.*

Jayadevappa, D., Srinivas Kumar, S., & Murty, D. S. (2011). Medical Image Segmentation Algorithms using Deformable Models: A Review. *IETE Technical Review, 28*(3), 248–251. doi:10.4103/0256-4602.81244

Jespersen, J. J., Hellstein, J., Williamson, A., Johnson, W. T., & Qian, F. (2014). Evaluation of dental pulp sensibility tests in a clinical setting. *Journal of Endodontics, 40*(3), 351–354. doi:10.1016/j.joen.2013.11.009 PMID:24565651

Ji, Z., Xia, Y., Sun, Q., Xia, D., & Feng, D. D. (2012). Local Gaussian distribution fitting based FCM algorithm for brain MR image segmentation. In Intelligent Science and Intelligent Data Engineering, 7202, 318-325. doi:10.1007/978-3-642-31919-8_41

Jiang, X., Zhang, R., & Nie, S. (2012). Image segmentation based on level set method. *Physics Procedia, 33*, 840–845. doi:10.1016/j.phpro.2012.05.143

Johnson, B. R. (1999). Considerations in the selection of a root-end filling material. *Oral Surgery, Oral Medicine, Oral Pathology, Oral Radiology, and Endodontics, 87*(4), 398–404. doi:10.1016/S1079-2104(99)70237-4 PMID:10225620

Johnson, W. T., & Guttmann, J. L. (2007). Obturation of cleaned and shaped root canal system. In S. Cohen & K. Hargreaves (Eds.), *Pathways of the pulp* (9th ed.). Philadelphia, PA: Elsevier.

Jonathan, E. K., Anna, K. J., & Johannes, V. S. (2000). Zulu medicinal plants with antibacterial activity. *Journal of Ethnopharmacology*, *69*(3), 241–246. doi:10.1016/S0378-8741(99)00147-6 PMID:10722206

Juergens, P., Krol, Z., Zeilhofer, H. F., Beinemann, J., Schicho, K., Ewers, R., & Klug, C. (2009). Computer simulation and rapid prototyping for the reconstruction of the mandible. *Journal of Oral and Maxillofacial Surgery*, *67*(10), 2167–2170. doi:10.1016/j.joms.2009.04.104 PMID:19761910

Jumb, V., Sohani, M., & Shrivas, A. (2014). Color Image Segmentation Using K-Means Clustering and Otsu's Adaptive Thresholding. *International Journal of Innovative Technology and Exploring Engineering*, *3*(9), 72–76.

Jung, A. (2015). *Image augmentation for machine learning experiments*. GitHub. Retrieved from https://github.com/aleju/imgaug

Kakiuchi, N., Hattori, M., Nishizawa, M., & Yamagishi, T. (1986). Studies on dental caries prevention by traditional medicines. VIII. Inhibitory effect of various tannins on glucan synthesis by glucosyltransferase from Streptococcus mutans. *Chemical & Pharmaceutical Bulletin*, *34*(2), 720–725. doi:10.1248/cpb.34.720 PMID:2939967

Kalra, D., Jain, G., Deoghare, A., & Lambade, P. (2010). Role of imaging in dental implants. *Journal of Indian Academy of Oral Medicine and Radiology*, *22*(1), 34–38. doi:10.5005/jp-journals-10011-1007

Kamalanand, K., & Jawahar, P. M. (2012). Coupled jumping frogs/particle swarm optimization for estimating the parameters of three dimensional HIV model. *BMC Infectious Diseases*, *12*(1Suppl 1), 82. doi:10.1186/1471-2334-12-S1-P82 PMID:22471518

Kamalanand, K., & Jawahar, P. M. (2013a). Particle swarm optimization based estimation of HIV-1 viral load in resource limited settings. *African Journal of Microbiological Research*, *7*(20), 2297–2304. doi:10.5897/AJMR12.1924

Kamalanand, K., & Jawahar, P. M. (2013b). Resource limited estimation of HIV-1 viral load from CD4 cell count using coupled bacterial foraging/jumping frogs algorithm. *The World Allergy Organization Journal*, *6*(S1), 189. doi:10.1186/1939-4551-6-S1-P189

Kamalanand, K., & Jawahar, P. M. (2014a). A graphical user interface for resource limited estimation of HIV-1 viral load using swarm intelligence techniques. *Journal of Bioinformatics and Intelligent Control*, *3*(2), 110–116. doi:10.1166/jbic.2014.1073

Kamalanand, K., & Jawahar, P. M. (2014b). Hybrid BFPSO algorithm based estimation of optimal drug dosage for antiretroviral therapy in HIV-1 infected patients. *BMC Infectious Diseases*, *14*(S3), E14. doi:10.1186/1471-2334-14-S3-E14

Kamalanand, K., & Mannar Jawahar, P. (2015). Comparison of Swarm Intelligence Techniques for Estimation of HIV-1 Viral Load. *IETE Technical Review*, *32*(3), 188–195. doi:10.1080/02564602.2014.1000981

Kamalanand, K., & Mannar Jawahar, P. (2016). Comparison of particle swarm and bacterial foraging optimization algorithms for therapy planning in HIV/AIDS patients. *International Journal of Biomathematics*, *9*(02), 1650024. doi:10.1142/S1793524516500248

Kamalanand, K., & Ramakrishnan, S. (2015). Effect of gadolinium concentration on segmentation of vasculature in cardiopulmonary magnetic resonance angiograms. *Journal of Medical Imaging and Health Informatics*, *5*(1), 147–151. doi:10.1166/jmihi.2015.1370

Kamalanand, K., Sridhar, B. T. N., Rajeshwari, P. M., & Ramakrishnan, S. (2010). Correlation of Dielectric Permittivity with Mechanical Properties in Soft Tissue-Mimicking Polyacrylamide Phantoms. *Journal of Mechanics in Medicine and Biology*, *10*(02), 353–360. doi:10.1142/S0219519410003411

Kamalanand, K., & Srinivasan, S. (2011a). Modeling of Normal and Atherosclerotic Blood Vessels using Finite Element Methods and Artificial Neural Networks. *World Academy of Science, Engineering and Technology*, *60*, 1314.

Kamalanand, K., & Srinivasan, S. (2011b). Modelling and analysis of normal and atherosclerotic blood vessel mechanics using 3D finite element models. *ICTACT Journal on Soft Computing: Special Issue on Fuzzy in Industrial and Process Automation*, *2*(1), 261–264. doi:10.21917/ijsc.2011.0040

Kamalanand, K., Srinivasan, S., & Ramakrishnan, S. (2011). Analysis of Normal and Atherosclerotic Blood Vessels Using 2D Finite Element Models. In *5th Kuala Lumpur International Conference on Biomedical Engineering 2011* (pp. 411-414). Springer. 10.1007/978-3-642-21729-6_105

Kandlikar, S., Garimella, S., Li, D., Colin, S., & King, M. R. (2005). *Heat transfer and fluid flow in minichannels and microchannels*. Elsevier.

Kang, J., & Ji, Z. (2010) Dental Plaque Quantification using Mean-shift-based Image Segmentation. *IEEE International Symposium on Computer, Communication, Control and Automation*. 10.1109/3CA.2010.5533758

Kapila, R., Dhaliwal, A., & Singh, N. (2014). History of X-rays in dentistry. *Annals of Dental Research*, *2*(1), 22–25.

Kapur, J. N., Sahoo, P. K., & Wong, A. K. (1985). A new method for gray-level picture thresholding using the entropy of the histogram. *Computer Vision Graphics and Image Processing*, *29*(3), 273–285. doi:10.1016/0734-189X(85)90125-2

Kashke, S., & Paolino, V. J. (1988). Inhibition of salivary amylase by water-soluble extracts of tea. *Archives of Oral Biology*, *33*(11), 845–846. doi:10.1016/0003-9969(88)90110-0 PMID:2476976

Kashket, S., Paolino, V. J., Lewis, D. A., & Van Houte, J. (1985). *In-vitro* inhibition of glucosyltransferase from the dental plaque bacterium *Streptococcus mutans* by common beverages and food extracts. *Archives of Oral Biology*, *30*(11-12), 822–826. doi:10.1016/0003-9969(85)90138-4 PMID:2938561

Kass, M., Witkin, A., & Terzopoulos, D. (1988). Snakes: Active contour models. *International Journal of Computer Vision*, *1*(4), 321–331. doi:10.1007/BF00133570

Kassner, A., & Thornhill, R. E. (2010). Texture analysis: A review of neurologic MR imaging applications. *AJNR. American Journal of Neuroradiology*, *31*(5), 809–816. doi:10.3174/ajnr.A2061 PMID:20395383

Katsigiannis, S., Keramidas, E. G., & Maroulis, G. (2010). Contourlet transform for texture representation of ultrasound thyroid images. *IFIP Advances in Information and Communication Technology*, 138-145.

Katsura, H., Tsukiyama, R. I., Suzuki, A., & Kobayashi, M. (2001). *In vitro* antimicrobial activities of bakuchiol against oral microorganisms. *Antimicrobial Agents and Chemotherapy*, *45*(11), 3009–3013. doi:10.1128/AAC.45.11.3009-3013.2001 PMID:11600349

Kaur, N., & Kaur, R. (2011). A review on various methods of image thresholding. *International Journal on Computer Science and Engineering*, *3*(10), 3441.

Kavitha, M. S., Asano, A., Taguchi, A., Kurita, T., & Sanada, M. (2012). Diagnosis of osteoporosis from dental panoramic radiographs using the support vector machine method in a computer-aided system. *BMC Medical Imaging*, *12*(1), 2–11. doi:10.1186/1471-2342-12-1 PMID:22248480

Kayalvizhi, M., & Kavitha, G., Sujatha, C. M., & Ramakrishnan, S. (2015). Study of Alzheimer's Disease progression in MR brain images based on segmentation and analysis of ventricles using Modified DRLSE Method and Minkowski Functionals. *Biomedical Sciences Instrumentation, 51*, 332–340. PMID:25996736

Kayalvizhi, M., Kavitha, G., & Sujatha, C. M. (2013). Analysis of ventricle regions in Alzheimer's brain MR images using level set based methods. *International Journal of Biomedical Engineering and Technology, 12*(3), 300–319. doi:10.1504/IJBET.2013.057266

Keating, G., & O' Kennedy, R. (1997). Coumarin cytotoxicity activity on albino rats. John Wiley & Sons.

Keshtkar, F., & Gueaieb, W. (2006). Segmentation of dental radiographs using a swarm intelligence approach. *IEEE Canadian Conference on Electrical and Computer Engineering*, 328-331. 10.1109/CCECE.2006.277656

Khaled, S. A., Burley, J. C., Alexander, M. R., & Roberts, C. J. (2014). Desktop 3D printing of controlled release pharmaceutical bilayer tablets. *International Journal of Pharmaceutics, 461*(1-2), 105–111. doi:10.1016/j.ijpharm.2013.11.021 PMID:24280018

Khan, M. W. (2014). A survey: Image segmentation techniques. *International Journal of Future Computer and Communication, 3*(2), 89–93. doi:10.7763/IJFCC.2014.V3.274

Khurshid, K., Siddiqi, I., Faure, C., & Vincent, N. (2009). Comparison of Niblack inspired binarization methods for ancient documents. *DRR, 7247*, 1–10.

Kim, D.-G. (2014). Can dental cone beam computed tomography assess bone mineral density? *Journal of Bone Metabolism, 21*(2), 117–126. doi:10.11005/jbm.2014.21.2.117 PMID:25006568

Kim, H. E., & Kim, B. I. (2015). An in vitro comparison of quantitative light-induced fluorescence-digital and spectrophotometer on monitoring artificial white spot lesions. *Photodiagnosis and Photodynamic Therapy, 12*(3), 378–384. doi:10.1016/j.pdpdt.2015.06.006 PMID:26117198

Kim, J. E., Kim, H. E., Hwang, J. K., Lee, H. J., Kwon, H. K., & Kim, B. I. (2008). Antibacterial characteristics of *Curcuma xanthorrhiza*extract on *Streptococcus mutans* biofilm. *Journal of Microbiology (Seoul, Korea), 46*(2), 228–232. doi:10.100712275-007-0167-7 PMID:18545974

Kim, J. E., Shin, J. M., Oh, S. O., Yi, W. J., Heo, M. S., Lee, S. S., ... Huh, K. H. (2013). The three-dimensional micro structure of trabecular bone: Analysis of site-specific variation in human jaw bone. *Imaging Science in Dentistry, 43*(4), 227–233. doi:10.5624/isd.2013.43.4.227 PMID:24380061

Kim, J. H. (1997). Anti-bacterial action of onion (Allium cepa L.) extracts against oral pathogenic bacteria. *The Journal of Nihon University School of Dentistry, 9*(3), 136–141. doi:10.2334/josnusd1959.39.136 PMID:9354029

King, R. M., & Robinson, H. (1987). *The Genera of the Eupatorieae (Asteraceae). Monographs in Systematic Botany.* St. Louis, MO: Missouri Botanical Garden.

Kirsch, J., Tchorz, J., Hellwig, E., Tauböck, T. T., Attin, T., & Hannig, C. (2016). Decision criteria for replacement of fillings: A retrospective study. *Clinical and Experimental Dental Research, 2*(2), 121–128. doi:10.1002/cre2.30 PMID:29744158

Klein, G. T., Lu, Y., & Wang, M. Y. (2013). 3D printing and neurosurgery—ready for prime time? *World Neurosurgery, 80*(3), 233–235. doi:10.1016/j.wneu.2013.07.009 PMID:23871811

Kobayashi, I. (1988). A study of the esthetic coloring of teeth by spectroradiometry-concerning the maxillary anterior teeth of the middle and old aged people. [Nippon Hotetsu Shikka Gakkai Zasshi]. *J Jpn Prosthodont Soc, 32*(2), 439–448. doi:10.2186/jjps.32.439 PMID:3075715

Kocry, T. (1983). The use of chewing sticks in preventive oral hygiene. *Clinical Preventive Dentistry*, *5*, 11–14. PMID:6607809

Koga, T., Okahashi, N., Asakawa, H., & Hamada, S. (1985). Adherence of Streptococcus mutans to tooth surfaces. In S. Hamada, S. Michalek, H. Kiyono, L. Menaker, & J. R. McGhee (Eds.), Molecular Microbiology and Immunology of Streptococcus mutans (pp. 133–143). Academic Press.

Kohda, H., Kozai, K., Nagasaka, N., Miyake, Y., Suginaka, H., Hidaka, K., & Yamasaki, K. (1985). Prevention of dental caries by oriental Folk medicines- Active Principle of Zizyphi fructus for inhibition of insoluble glucan formation by carcenogenic bacterium *Streptococcus mutans. Planta Medica*, *52*(02), 119–120. doi:10.1055-2007-969095

Kom, G., Tiedeu, A., & Kom, M. (2007). Automated detection of masses in mammograms by local adaptive thresholding. *Computers in Biology and Medicine*, *37*(1), 37–48. doi:10.1016/j.compbiomed.2005.12.004 PMID:16487954

Kondo, T., Ong, S. H., & Foong, K. W. (2004). Tooth segmentation of dental study models using range images. *IEEE Transactions on Medical Imaging*, *23*(3), 350–362. doi:10.1109/TMI.2004.824235 PMID:15027528

Koo, H., Schobel, B., Scott-Anne, K., Watson, G., Bowen, W. H., Cury, J. A., ... Park, Y. K. (2005). Apigenin and *tt*-farnesol with fluoride effects on *S. mutans* biofilms and dental caries. *Journal of Dental Research*, *84*(11), 1016–1020. doi:10.1177/154405910508401109 PMID:16246933

Koo, H., Vacca-Smith, A. M., Bowen, W. H., Rosalen, P. L., Cury, J. A., & Park, Y. K. (2000). Effects of Apis mellifera propolis on the activities of streptococcal glucosyltransferases in solution and adsorbed onto saliva-coated hydroxyapatite. *Caries Research*, *34*(5), 361–442. doi:10.1159/000016617 PMID:11014909

Kopelman, A., & Taub, E. (2005). *U.S. Patent No. 6,845,175*. Washington, DC: U.S. Patent and Trademark Office.

Krishnamurthy, K. (2016). Parameter Estimation of Nonlinear Biomedical Systems Using Extended Kalman Filter Algorithm: Development of Patient Specific Models. *Computational Tools and Techniques for Biomedical Signal Processing*, 76.

Krishnamurthy, K., & Jacob, L. (2014). Finite element based design of a capacitive sensor for diagnosis of cirrhotic and malignant liver. *Journal of Clinical and Experimental Hepatology*, *4*, S65. doi:10.1016/j.jceh.2014.02.126

Krishnamurthy, K., Sridhar, B. T. N., Rajeshwari, P. M., & Swaminathan, R. (2009). Correlation of Electrical Impedance with Mechanical Properties in Models of Tissue Mimicking Phantoms. In *13th International Conference on Biomedical Engineering* (pp. 1708-1711). Springer. 10.1007/978-3-540-92841-6_424

Krizhevsky, A., Sutskever, I., & Hinton, G. E. (2012). ImageNet classification with deep convolutional neural networks. *Annual Conference on Neural Information Processing Systems (NIPS)*, 2, 1097-1105.

Krol, Z., Chlebiej, M., Zerfass, P., Sader, R., Zeilhofer, H. R., Mikołajczak, P., & Keeve, E. (2002). Surgery planning tools for the osseous grafting treatment. *Biomedizinische Technik/Biomedical Engineering, 47*(s1a), 97-100.

Kubo, I., Muroi, H., & Himejima, M. (1992). Antimicrobial activity of green tea flavor components and their combination effects. *Journal of Agricultural and Food Chemistry*, *40*(2), 245–248. doi:10.1021/jf00014a015

Kumar, S. S., Moni, R. S., & Rajeesh, J. (2011). Automatic segmentation of liver and tumor for CAD of liver. *Journal of Advances in Information Technology, 2*(1), 63-70.

Kumar, V., & Satheesh, K. (2013). Applications of cone beam computed tomography (CBCT) in implant treatment planning. *JSM, 1*(1008).

Kumar, V. S., Anantharaj, U. J., Sakthioli, M., Vinnakota, R., & Krishnamurthy, K. (2016). Mathematical modelling of the effects of prebiotic concentration on lactobacillus casei growth. *International Journal of Infectious Diseases*, *45*, 205. doi:10.1016/j.ijid.2016.02.470

Kuo, H. C., Maryellen, L. G., Ingrid, R., Boone, J. M., Lindfors, K. K., Yang, K., & Edwards, A. (2014). Level Set Segmentation of Breast Masses in Contrast-Enhanced Dedicated Breast CT and Evaluation of Stopping Criteria. *Journal of Digital Imaging*, *27*(2), 1–11. doi:10.100710278-013-9652-1 PMID:24162667

Kvaal, S. I., Kolltveit, K. M., Thomsen, I. O., & Solheim, T. (1995). Age estimation of adults from dental radiographs. *Forensic Science International*, *74*(3), 175–185. doi:10.1016/0379-0738(95)01760-G PMID:7557754

Lai & Lin. (2006) Effective Segmentation for Dental X-Ray Images Using Texture-Based Fuzzy Inference System. *LNCS*, *5259*, 936–947.

Lai, Y. H., & Lin, P. L. (2008). *Effective Segmentation for Dental X-ray Images Using Texture-based Fuzzy Inference System. In Advanced Concepts for IntelligentVision Systems* (pp. 936–947). Berlin: Springer.

Lai, Y. H., & Lin, P. L. (2008, October). Effective segmentation for dental X-ray images using texture-based fuzzy inference system. In *International Conference on Advanced Concepts for Intelligent Vision Systems* (pp. 936-947). Springer. 10.1007/978-3-540-88458-3_85

Lakhani, K., Minocha, B., & Gugnani, N. (2016). Analyzing edge detection techniques for feature extraction in dental radiographs. *Perspectives on Science*, *8*, 395–398. doi:10.1016/j.pisc.2016.04.087

Lakshmi, V. S., Tebby, S. G., Shriranjani, D., & Rajinikanth, V. (2016). Chaotic cuckoo search and Kapur/Tsallis approach in segmentation of T. cruzi from blood smear images. *International Journal of Computer Science and Information Security*, *14*, 51–56.

Latha, M., & Kavitha, G. (2017). Segmentation and analysis of ventricles in Schizophrenic MR brain images using optimal region based energy minimization framework. *IEEE Proc. 4th International Conference on Signal Processing, Communications and Networking (ICSCN -2017)*. 10.1109/ICSCN.2017.8085735

Lauer, H. C., Kraft, E., Rothlauf, W., & Zwingers, T. (1990). Effects of the temperature of cooling water during high-speed and ultrahigh-speed tooth preparation. *The Journal of Prosthetic Dentistry*, *63*(4), 407–414. doi:10.1016/0022-3913(90)90228-5 PMID:2184223

Laura, F., Rubio, J. L., Ledesma-Carbayo, M. J., Pascau, J., Tellado, J. M., & Ramón, E. (2009) 3D liver segmentation in preoperative CT images using a level sets active surface method. *Proceedings of the 31st Annual International Conference of the IEEE Engineering in Medicine and Biology Society*, 3625-3628.

Leclercq, R., Besbes, S., & Soussy, C. J. (1988). Resistance des streptocoques non groupables aux ß-lactamines. *Pathologie Biologie*, *36*, 626–631. PMID:3054738

LeCun, Y., Bengio, Y., & Hinton, G. (2015). Deep learning. *Nature*, *521*(7553), 436–444. doi:10.1038/nature14539 PMID:26017442

LeCun, Y., Bottou, L., Bengio, Y., & Haffner, P. (1998). Gradient-based learning applied to document recognition. *Proceedings of the IEEE*, *86*(11), 2278–2323. doi:10.1109/5.726791

Lee, H., Park, M., & Kim, J. (2017). Cephalometric Landmark Detection in Dental X-ray Images Using Convolutional Neural Networks. In S. G. Armato & N. A. Petrick (Eds.), SPIE Medical Imaging, 10134 (pp. 1–6). Academic Press.

Lee, J. S. (1980). Digital image enhancement and noise filtering by use of local statistics. *IEEE Transactions on Pattern Analysis and Machine Intelligence*, *PAMI-2*(2), 165–168. doi:10.1109/TPAMI.1980.4766994 PMID:21868887

Lee, J.-G., Jun, S., Cho, Y.-W., Lee, H., Bae Kim, G., Beom Seo, J., & Kim, N. (2017). 7). Deep Learning in Medical Imaging: General Overview. *Korean Journal of Radiology*, *18*(4), 570. doi:10.3348/kjr.2017.18.4.570 PMID:28670152

Lehner, T., Caldwell, J., & Smith, R. (1985). Local passive immunisation by monoclonal antibodies against streptococcal antigen I/II in the prevention of dental caries. *Infection and Immunity*, *50*, 796–799. PMID:4066030

Lehner, T., Ma, J. K.-C., Grant, K., & Smith, R. (1986). Local passive immunisation by monoclonal antibodies against *Streptococcus mutans* antigen in the prevention of dental caries. In T. Lehner & G. Cimasoi (Eds.), *Borderland between Caries and Periodontal Disease* (pp. 356–357). Geneva: Medecine et Hygiene.

Leitão, D. P., Filho, A. A., Polizello, A. C., Bastos, J. K., & Spadaro, A. C. (2004). Comparative evaluation of in-vitro effects of brazilian green propolis and Baccharis dracunculifolia extracts on cariogenic factors of *Streptococcus mutans. Biological & Pharmaceutical Bulletin*, *27*(11), 1834–1839. doi:10.1248/bpb.27.1834 PMID:15516733

Levinson, M. H. (1999). Medicine's 10 Greatest Discoveries. *et Cetera, 56*(2), 232.

Li, C., Huang, R., Ding, Z., Gatenby, C., Metaxas, D., & Gore, J. (2008). A variational level set approach to segmentation and bias correction of images with intensity inhomogeneity. *Medical image computing and computer-assisted intervention, 11*(Pt 2), 1083-1091.

Li, S., Fevens, T., Krzyz˙ak, A., & Li, S. (2005). Level set segmentation for computer aided dental X-ray analysis. SPIE conference on medical imaging, 5747, 580-589. doi:10.1117/12.595537

Liang, S., Sa, Y., Jiang, T., Ma, X., Xing, W., Wang, Z., & Wang, Y. (2013). In vitro evaluation of halogen light-activated vs chemically activated in-office bleaching systems. *Acta Odontologica Scandinavica*, *71*(5), 1149–1155. doi:10.3109/00016357.2012.757355 PMID:23294115

Liang, S., Sa, Y., Sun, L., Ma, X., Wang, Z., Xing, W., ... Wang, Y. (2012). Effect of halogen light irradiation on hydrogen peroxide bleaching: An in vitro study. *Australian Dental Journal*, *57*(3), 277–283. doi:10.1111/j.1834-7819.2012.01702.x PMID:22924349

Li, C., Gore, J. C., & Davatzikos, C. (2014). Multiplicative intrinsic component optimization (MICO) for MRI bias field estimation and tissue segmentation. *Magnetic Resonance Imaging*, *32*(7), 913–923. doi:10.1016/j.mri.2014.03.010 PMID:24928302

Li, C., Huang, R., Ding, Z., Gatenby, J. C., Metaxas, D. N., & Gore, J. C. (2011). A level set method for image segmentation in the presence of intensity in homogeneities with application to MRI. *IEEE Transactions on Image Processing*, *20*(7), 2007–2016. doi:10.1109/TIP.2011.2146190 PMID:21518662

Li, C., Xu, C., Guiand, C., & Fox, M. D. (2005). Level set evolution without re-initialization: a new variational formulation. *IEEE Computer Society Conference on Computer Vision and Pattern Recognition*, *1*, 430-436.

Li, C., Xu, C., Gui, C., & Fox, M. D. (2010). Distance Regularized Level Set Evolution and Its Application to Image Segmentation. *IEEE Transactions on Image Processing*, *19*(12), 3243–3254. doi:10.1109/TIP.2010.2069690 PMID:20801742

Lienhard, J. H. (2013). *A heat transfer textbook*. Courier Corporation.

Limsong, J., Benjavongkulchai, E., & Kuvatanasuchati, J. (2004). Inhibitory effect of some herbal extracts on adherence of *Streptococcus mutans. Journal of Ethnopharmacology*, *92*(2-3), 281–289. doi:10.1016/j.jep.2004.03.008 PMID:15138013

Lin, M., Xu, F., Lu, T. J., &Bai, B. F. (2010). A review of heat transfer in human tooth—experimental characterization and mathematical modeling. *Dental Materials, 26*(6), 501-513.

Lin, P. L., Huang, P. Y., & Huang, P. W. (2012). An automatic lesiondetection method for dental X-ray images by segmentationusing variational level set. *Proc. 2012 International Conference on Machine Learning and Cybernetics*, 1821-1826.

Lindner, C., Bromiley, P. A., Ionita, M. C., & Cootes, T. F. (2015). Robust and Accurate Shape Model Matching Using Random Forest Regression-Voting. *IEEE Transactions on Pattern Analysis and Machine Intelligence*, *37*(9), 1862–1874. doi:10.1109/TPAMI.2014.2382106 PMID:26353132

Lingeshwar, D., Dhanasekar, B., & Aparna, I. N. (2010). Diagnostic imaging in implant dentistry. *International Journal of Oral Implantology & Clinical Research*, *1*(3), 147–153.

Lin, M., Liu, S., Niu, L., Xu, F., & Lu, T. J. (2011). Analysis of thermal-induced dentinal fluid flow and its implications in dental thermal pain. *Archives of Oral Biology*, *56*(9), 846–854. doi:10.1016/j.archoralbio.2011.02.011 PMID:21411060

Lin, P. L., Huang, P. W., Cho, Y. S., & Kuo, C. H. (2013). An automatic and effective tooth isolation method for dental radiographs. *Opto-Electronics Review*, *21*(1), 126–136. doi:10.247811772-012-0051-9

Lin, P. L., Huang, P. Y., Huang, P. W., Hsu, H. C., & Chen, C. C. (2014). Teeth segmentation of dental periapical radiographs based on local singularity analysis. *Computer Methods and Programs in Biomedicine*, *113*(2), 433–445. doi:10.1016/j.cmpb.2013.10.015 PMID:24252317

Lin, P. L., & Lai, Y. H. (2008). Effective segmentation for dental X-rayimages using texture-based fuzzy inference system. *Advanced Concepts for Intelligent Visions System*, *5259*, 936–947.

Lin, P. L., Lai, Y. H., & Huang, P. W. (2010). An effective classification and numbering system for dental bitewing radiographs using teeth region and contour information. *Pattern Recognition*, *43*(4), 1380–1392. doi:10.1016/j.patcog.2009.10.005

Lin, P. L., Lai, Y. H., & Huang, P. W. (2012). Dental biometrics: Human identification based on teeth and dental works in bitewing radiographs. *Pattern Recognition*, *45*(3), 934–946. doi:10.1016/j.patcog.2011.08.027

Linsuwanont, P., Palamara, J. E. A., & Messer, H. H. (2007). An investigation of thermal stimulation in intact teeth. *Archives of Oral Biology*, *52*(3), 218–227. doi:10.1016/j.archoralbio.2006.10.009 PMID:17109811

Linsuwanont, P., Versluis, A., Palamara, J. E., & Messer, H. H. (2008). Thermal stimulation causes tooth deformation: A possible alternative to the hydrodynamic theory? *Archives of Oral Biology*, *53*(3), 261–272. doi:10.1016/j.archoralbio.2007.10.006 PMID:18037388

Linwei, L., Guangwei, M., He, G., Dong, Z., & Weimin, Z. (2012). A new method for the measurement and analysis of three-dimensional morphological parameters of proximal male femur. *Biomedical Research*, *23*(2), 219–226.

Lio, M., Himeno, S., Miyauchi, K., Mikumo, K., & Ohta, N. (1983). Article. *Nippon Nogeikagaku Kaishi*, *57*, 765. doi:10.1271/nogeikagaku1924.57.765

Lio, M., & Uyeda, M., lwanami, T., & Nakagawa, Y. (1984). Flavanoids as a possible preventive of dental caries. *Agricultural and Biological Chemistry*, *48*, 2143–2145.

Lipski, M., Woźniak, K., Lichota, D., & Nowicka, A. (2011). Root surface temperature rise of mandibular first molar during root canal filling with high-temperature thermoplasticized Gutta-Percha in the dog. *Polish Journal of Veterinary Sciences*, *14*(4), 591–595. doi:10.2478/v10181-011-0088-6 PMID:22439330

Lipson, H. (2013). New world of 3-D printing offers" completely new ways of thinking": Q&A with author, engineer, and 3-D printing expert Hod Lipson. *IEEE Pulse*, *4*(6), 12–14. doi:10.1109/MPUL.2013.2279615 PMID:24215725

Lira, P., Giraldi, G., Neves, L. & Feijoo, R. (2014). Dental R-Ray Image Segmentation Using Texture Recognition. *Latin America Transactions*, *12*(4).

Li, S., Fevens, T., Krzyzak, A., Jin, C., & Li, S. (2007). Semi-automatic computer aided lesion detection in dental X-rays using variational level set. *Pattern Recognition*, *40*(10), 2861–2873. doi:10.1016/j.patcog.2007.01.012

Li, S., Fevens, T., Krzyzak, A., & Li, S. (2006). An automatic variational level set segmentation framework for computer aided dental X-rays analysis in clinical environments. *Computerized Medical Imaging and Graphics*, *30*(2), 65–74. doi:10.1016/j.compmedimag.2005.10.007 PMID:16500077

Litjens, G., Kooi, T., Ehteshami Bejnordi, B., Setio, A., Ciompi, F., Ghafoorian, M., & ... Sánchez, C. (2017). A Survey on Deep Learning in Medical Image Analysis. *Medical Image Analysis*, 42. PMID:28778026

Liu, W., Anguelov, D., Erhan, D., Szegedy, C., Reed, S., & Fu, C.-Y. (2016). SSD: Single Shot MultiBox Detector. In B. Leibe, J. Matas, & N. Sebe (Eds.), Computer Vision -- ECCV 2016 (pp. 21-37). Academic Press.

Liu, J., Wang, D., Lu, L., Wei, Z., Kim, L., Turkbey, E. B., ... Summers, R. M. (2017). Detection and diagnosis of colitis on computed tomography using deep convolutional neural networks. *Medical Physics*, *44*(9), 4630–4642. doi:10.1002/mp.12399 PMID:28594460

Li, X., Abaza, A., Nassar, D. E., & Ammar, H. (2006). Fast and accurate segmentation of dental x-ray records. *Lecture Notes in Computer Science*, *3832*, 688–696. doi:10.1007/11608288_92

Li, Y., Yang, C., Zhao, H., Qu, S., Li, X., & Li, Y. (2014). New developments of Ti-based alloys for biomedical applications. *Materials (Basel)*, *7*(3), 1709–1800. doi:10.3390/ma7031709 PMID:28788539

Loesche, W. J. (1986). Role of *Streptococcus mutans* in human dental decay. *Microbiological Reviews*, *50*, 353–380. PMID:3540569

Loesche, W. J., & Grossman, N. S. (2001). Periodontal disease as a specific, albeit chronic, infection: Diagnosis and treatment. *Clinical Microbiology Reviews*, *14*(4), 727–752. doi:10.1128/CMR.14.4.727-752.2001 PMID:11585783

Lofthag-Hansen, S., Huumonen, S., Gröndahl, K., & Gröndahl, H. G. (2007). Limited cone-beam CT and intraoral radiography for the diagnosis of periapical pathology. *Oral Surgery, Oral Medicine, Oral Pathology, Oral Radiology, and Endodontics*, *103*(1), 114–119. doi:10.1016/j.tripleo.2006.01.001 PMID:17178504

Lohbauer, U. (2009). Dental glass ionomer cements as permanent filling materials? – Properties, limitations and future trends. *Materials (Basel)*, *3*(1), 76–96. doi:10.3390/ma3010076

Lpez-Lpez, J. (2012). Computer-aided system for morphometric mandibular index computation (using dental panoramic radiographs). *Med. Oral Patol. Oral*, *17*, e624–e632. doi:10.4317/medoral.17637 PMID:22322489

Lu, H., Kot, A. C., & Shi, Y. Q. (2004). Distance-reciprocal distortion measure for binary document images. *IEEE Signal Processing Letters*, *11*(2), 228–231. doi:10.1109/LSP.2003.821748

Luiz Mialhe, F., Carlos Pereira, A., Pardi, V., & de Castro Meneghim, M. (2004). Comparison of three methods for detection of carious lesions in proximal surfaces versus direct visual examination after tooth separation. *The Journal of Clinical Pediatric Dentistry*, *28*(1), 59–62. doi:10.17796/jcpd.28.1.g121387868676514 PMID:14604144

Ma, J.K., & Lehner, T. (1990). Prevention of colonization of Streptococcus mutans by topical application of monoclonal antibodies in human subjects. *Arch Oral Biol, 35*, 115S-122S.

Mabogo, D. E. N. (1990). *The ethnobotany of the vhaVenda* (MSc Thesis). University of Pretoria, Pretoria, South Africa.

Machado, D. A., Giraldi, G., Novotny, A. A., Marques, R. S., & Conci, A. (2013) Topological Derivative Applied to Automatic Segmentation of Frontal Breast Thermograms. *IX Workshop de Visão Computacional.*

Mahoor, M. H., & Abdel-Mottaleb, M. (2005). Classification and numbering of teeth in dental bitewing images. *Pattern Recognition, 38*(4), 577–586. doi:10.1016/j.patcog.2004.08.012

Maitra, M., & Chatterjee, A. (2006). A slantlet transform based intelligent system for magnetic resonance brain image classification. *Biomedical Signal Processing and Control, 1*(4), 299–306. doi:10.1016/j.bspc.2006.12.001

Ma, J. K., Hiatt, A., Hein, M., Vine, N. D., Wang, F., Stabila, P., ... Lehner, T. (1995). Generation and assembly of secretory antibodies in plants. *Science, 268*(5211), 716–719. doi:10.1126cience.7732380 PMID:7732380

Ma, J. K., Hikmat, B. Y., Wycoff, K., Vine, N. D., Chargelegue, D., Yu, L., ... Lehner, T. (1998). Characterization of a recombinant plant monoclonal secretory antibody and preventive immunotherapy in humans. *Nature Medicine, 4*(5), 601–606. doi:10.1038/nm0598-601 PMID:9585235

Ma, J. K., Lehner, T., Stabila, P., Fux, C. I., & Hiatt, A. (1994). Assembly of monoclonal antibodies with IgG1 and IgA heavy chain domains in transgenic tobacco plants. *European Journal of Immunology, 24*(1), 131–138. doi:10.1002/eji.1830240120 PMID:8020548

Mäkinen, K. K., Chen, C. Y., Mäkinen, P. L., Bennett, C. A., Isokangas, P. J., Isotupa, K. P., & Pape, H. R. Jr. (1996). Properties of whole saliva and dental plaque in relation to 40-month consumption of chewing gums containing xylitol, sorbitol of sucrose. *Caries Research, 30*(3), 180–188. doi:10.1159/000262157 PMID:8860027

Mäkinen, K. K., Pemberton, D., Mäkinen, P. L., Chen, C. Y., Cole, J., Hujoel, P. P., ... Lambert, P. (1996). Polyol-combinant saliva stimulants and oral health in Veterans Affairs patients: An exploratory study. *Special Care in Dentistry, 16*(3), 104–115. doi:10.1111/j.1754-4505.1996.tb00843.x PMID:9084323

Malladi, R., Sethian, J. A., & Vemuri, B. C. (1995). Shape modeling with front propagation: A level set approach. *IEEE Transactions on Pattern Analysis and Machine Intelligence, 17*(2), 158–175. doi:10.1109/34.368173

Mandel, I. (1993). Dental caries: another extinct disease? In W. H. Bowen & L. A. Tabak (Eds.), *Cariology for the nineties* (pp. 1–10). Rochester, NY: University of Rochester Press.

Manic, K. S., Priya, R. K., & Rajinikanth, V. (2016). Image multithresholding based on kapur/tsallis entropy and firefly algorithm. *Indian Journal of Science and Technology, 9*(12). doi:10.17485/ijst/2016/v9i12/89949

Manickavasagam, K., Sutha, S., & Kamalanand, K. (2014). An automated system based on 2 d empirical mode decomposition and k-means clustering for classification of Plasmodium species in thin blood smear images. *BMC Infectious Diseases, 14*(Suppl 3), 13. doi:10.1186/1471-2334-14-S3-P13 PMID:24405683

Manickavasagam, K., Sutha, S., & Kamalanand, K. (2014). Development of systems for classification of different plasmodium species in thin blood smear microscopic images. *Journal of Advanced Microscopy Research, 9*(2), 86–92. doi:10.1166/jamr.2014.1194

Manisundar, N., Saravanakumar, B., Hemalatha, V. T., Manigandan, T., & Amudhan, A. (2014). Implant imaging-A literature review. *Biosciences Biotechnology Research Asia, 11*(1), 179–187. doi:10.13005/bbra/1251

Mann, A., Banso, A., & Clifford, L. C. (2008). Antifungal properties of Anogeissus leiocarpus and Terminalia avicennioides. *Tanzania Journal of Health Research, 10*, 34–38. doi:10.4314/thrb.v10i1.14339 PMID:18680963

Marcucci, M. (1995). Propolis: Chemical composition, biological properties and therapeutic activity. *Apidologie, 26*(2), 83–99. doi:10.1051/apido:19950202

Marcucci, M. C., Ferreres, F., Garcı'a-Viguera, C., Bankova, V. S., De Castro, S. L., Dantas, A. P., ... Paulino, N. (2001). Phenolic compounds from Brazilian propolis with pharmacological activities. *Journal of Ethnopharmacology*, *74*(2), 105–112. doi:10.1016/S0378-8741(00)00326-3 PMID:11167028

Marsh, P. D., & Bradsha, D. J. (1997). Physiological approaches to the control of oral biofilms. *Advances in Dental Research*, *11*(1), 176–185. doi:10.1177/08959374970110010901 PMID:9524454

Marthaler, T. M. (2004). Changes in dental caries 1953–2003. *Caries Research*, *38*(3), 173–181. doi:10.1159/000077752 PMID:15153686

Martınez, S., Nicolás, F., Pennini, F., & Plastino, A. (2000). Tsallis' entropy maximization procedure revisited. *Physica A*, *286*(3), 489–502. doi:10.1016/S0378-4371(00)00359-9

Martino, P. D., Gagniere, H., Berry, H., & Bret, L. (2002). Antimicrobial activity of extracts of some medicinal plants. Antimicrobial Agents. *Microbes and Infection*, *4*, 613–620. PMID:12048030

Matherne, R. P., Angelopoulos, C., Kulild, J. C., & Tira, D. (2008). Use of cone-beam computed tomography to identify root canal systems in vitro. *Journal of Endodontics*, *34*(1), 87–89. doi:10.1016/j.joen.2007.10.016 PMID:18155501

Mathias, A. J., Somashekar, R. K., Sumithra, S., & Subramanya, S. (2000). An assessment of reservoirs of multi-resistant nosocomial pathogens in a secondary care hospital. *Antimicrob. Agents Indian J. Microbio*, *40*, 183–190.

Matsumoto, M., Minami, T., Sasaki, H., Sobue, S., Hamada, S., & Ooshima, T. (1999). Inhibitory effects of oolong tea extract on caries-inducing properties of mutans streptococci. *Caries Research*, *33*(6), 441–445. doi:10.1159/000016549 PMID:10529529

McCabe, M. M., & Hamelik, R. M. (1978). Multiple Forms of Dextran-binding Proteins from *Streptococcus mutans*. *Advances in Experimental Medicine and Biology*, *107*, 749–759. doi:10.1007/978-1-4684-3369-2_84 PMID:742510

McMichen, F. R. S., Pearson, G., Rahbaran, S., & Gulabilava, K. (2003). A comparative study of selected physical properties of five root-canal sealers. *Journal of Int Endod*, *36*(9), 629–635. doi:10.1046/j.1365-2591.2003.00701.x PMID:12950578

Medina, E., Brenes, M., Romero, C., García, A., & Castro, A. (2007). Main Antimicrobial Compounds in Table Olives. *Journal of Agricultural and Food Chemistry*, *55*(24), 9817–9823. doi:10.1021/jf0719757 PMID:17970590

Mehra, P., Miner, J., D'Innocenzo, R., & Nadershah, M. (2011). Use of 3-d stereolithographic models in oral and maxillofacial surgery. *Journal of Maxillofacial and Oral Surgery*, *10*(1), 6–13. doi:10.100712663-011-0183-3 PMID:22379314

Mertz, L. (2013). Dream it, design it, print it in 3-D: What can 3-D printing do for you? *IEEE Pulse*, *4*(6), 15–21. doi:10.1109/MPUL.2013.2279616 PMID:24233186

Miki, Y., Muramatsu, C., Hayashi, T., Zhou, X., Hara, T., Katsumata, A., & Fujita, H. (2017). Classification of teeth in cone-beam CT using deep convolutional neural network. *Computers in Biology and Medicine*, *80*, 24–29. doi:10.1016/j.compbiomed.2016.11.003 PMID:27889430

Mirebeau, J. M., Fehrenbach, J., Risser, L., & Tobji, S. (2015). *Anisotropic diffusion in ITK*. arXiv preprint arXiv:1503.00992

Mirzoeva, O., Grishanin, R., & Calder, P. C. (1997). Antimicrobial action of propolis and some of its components: The effects on growth, membrane potential and motility of bacteria. *Microbiological Research*, *152*(3), 239–246. doi:10.1016/S0944-5013(97)80034-1 PMID:9352659

Mital, P., Mehta, N., Saini, A., Raisingani, D., & Sharma, M. (2014). Recent advances in detection and diagnosis of dental caries. *Journal of Evolution of Medical and Dental Sciences*, *3*(1), 177–192. doi:10.14260/jemds/1807

Mitiche, A., & Ayed, I. B. (2011). *Variational and Level Set Methods in Image Segmentation. In Springer Topics in Signal Processing*. Springer-Verlag Berlin Heidelberg. doi:10.1007/978-3-642-15352-5

Mitra, S. (2005). Anatomy: Part 3. In A. Tapadar (Ed.), *Anatomy: osteology, important landmarks and surface markings* (pp. 43–53). Kolkata, India: Academic Publishers.

Mitsunaga, T., Abe, I., Kontani, M., Ono, H., & Tanaka, T. (1997). Inhibitory effects of bark proanthocyanidins on the activities of glucosyltransferases of *Streptococcus sobrinus*. *Journal of Wood Chemistry and Technology*, *17*(3), 327–340. doi:10.1080/02773819708003134

Miziolek, A. W., Palleschi, V., & Schechter, I. (Eds.). (2006). *Laser induced breakdown spectroscopy*. Cambridge University Press.

Moerman, D. E. (1996). An analysis of the food plants and drug plants of Native North America. *Journal of Ethnopharmacology*, *52*(1), 1–22. doi:10.1016/0378-8741(96)01393-1 PMID:8733114

Mohamed, S. E., Christensen, L. V., & Harrison, J. D. (1983). Tooth contact patterns and contractile activity of the elevator jaw muscles during mastication of two different types of food. *Journal of Oral Rehabilitation*, *10*(1), 87–95. doi:10.1111/j.1365-2842.1983.tb00103.x PMID:6572242

Mokji, M. M., & AbuBakar, S. A. R. (2007). Gray level co-occurrence matrix computation based on Haar wavelet. Proceeding of Computer Graphics Imaging and Visualisation, 273-279. doi:10.1109/CGIV.2007.45

Mol, A., & van der Stelt, P. F. (1991). Application of digital image analysis in dental radiography for the description of periapical bone lesions: A preliminary study. *IEEE Transactions on Biomedical Engineering*, *38*(4), 357–359. doi:10.1109/10.133231 PMID:1855798

Moorthy, A., Hogg, C. H., Dowling, A. H., Grufferty, B. F., Benetti, A. R., & Fleming, G. J. P. (2012). Cuspal deflection and microleakage in premolar teeth restored with bulk-fill flowable resin-based composite base materials. *Journal of Dentistry*, *40*(6), 500–505. doi:10.1016/j.jdent.2012.02.015 PMID:22390980

Mtunzi, F. M., Ikechukwu, P. E., Tshifhiwa, M., Ezekiel, D., & Michael, J. K. (2017). Phytochemical Profiling, Antioxidant and Antibacterial Activities of Leaf Extracts from Rhus leptodictya. *International Journal of Pharmacognosy and Phytochemical Research*, *9*(8), 1090–1099. doi:10.25258/phyto.v9i08.9616

Munro, C., Michalek, S. M., & Macrina, F. L. (1991). Cariogenicity of *Streptococcus mutans* V403 glucosyltransferase and fructosyltransferase mutants constructed by allelic exchange. *Infection and Immunity*, *59*, 2316–2323. PMID:1828790

Mupparapu, M., & Singer, S. R. (2004). Implant imaging for the dentist. *Journal - Canadian Dental Association*, *70*(1), 32–32. PMID:14709253

Murakami, S., Tan, J. K., Kim, H., Ishikawa, S., & Aoki, T. (2012). Development of a quantitative method for the detection of periarticular osteoporosis using density features within ROIs from computed radiography images of the hand. In Technical Advancements in Biomedicine for Healthcare Applications (pp. 55-67). IGI Global.

Murray, R. D. H., Mendez, J., & Brown, S. A. (1982). Coumarin activity in plants and bioorganism aspects. John Wiley.

Nageswararao, A. V., Srinivasan, S., Babu, S., & Priya, E. (2016, January). Multifractal Analysis Based Segmentation of Short Axis CMRI Images. In *2nd International Conference on Biomedical Systems, Signals and Images*. Department of Applied Mechanics, IIT Madras.

Naidu, M. S. R., Kumar, P. R., & Chiranjeevi, K. (2017). *Shannon and fuzzy entropy based evolutionary image thresholding for image segmentation*. Alexandria Engineering Journal.

Nakahara, K., Kawabata, S., Ono, H., Ogura, K., Tanaka, T., Ooshima, T., & Hamada, S. (1993). Inhibitory effect of oolong tea polyphenols on glucosyltransferases of mutans streptococci. *Applied and Environmental Microbiology*, *59*(4), 968–973. PMID:8489234

Nakahara, K., Kawabata, S., Ono, H., Ogura, K., Tanaka, T., Ooshima, T., & Hamada, S. (1993). Inhibitory effect of oolong tea polyphenols on glycosyltransferases of mutans streptococci. *Applied and Environmental Microbiology*, *59*, 968–973. PMID:8489234

Nakamoto, T., Taguchi, A., Ohtsuka, M., Suei, Y., Fujita, M., Tsuda, M., ... Tanimoto, K. (2008). A computer-aided diagnosis system to screen for osteoporosis using dental panoramic radiographs. *Dento Maxillo Facial Radiology*, *37*(5), 274–281. doi:10.1259/dmfr/68621207 PMID:18606749

Nalina, T., & Rahim, Z. H. A. (2007). The crude aqueous extract of Piper betle L. and its antibacterial effect towards *Streptococcus mutans*. *American Journal of Biochemistry and Biotechnology*, *3*(1), 10–15. doi:10.3844/ajbbsp.2007.10.15

Namba, T., Morita, O., Huang, S. L., Goshima, K., Hattori, M., & Kakiuchi, N. (1988). Studies on cardioactive crude drugs. I. Effect of coumarins on cultured myocardial cells. *Planta Medica*, *54*(04), 277–282. doi:10.1055-2006-962432 PMID:3222369

Namba, T., Tsunezuka, M., & Hattori, M. (1982). *Dental Caries Prevention by Traditional Chinese Medicines. Part II.* Potent Antibacterial Action of Magnoliae Cortex Extracts Against *Streptococcus Mutans*. *Planta Medica*, *44*(2), 100–106. doi:10.1055-2007-971412 PMID:7071194

Nanbu, A., Hayakawa, M., Takada, K., Shinozaki, N., Abiko, Y., & Fukushima, K. (2000). Production, characterization, and application of monoclonal antibodies which distinguish four glucosyltransferases from Streptococcus sobrinus. *FEMS Immunology and Medical Microbiology*, *27*(1), 9–15. doi:10.1111/j.1574-695X.2000.tb01405.x PMID:10617784

Nance, R., Tyndall, D., Levin, L. G., & Trope, M. (2000). Identification of root canals in molars by tuned-aperture computed tomography. *International Endodontic Journal*, *33*(4), 392–396. doi:10.1046/j.1365-2591.2000.00330.x PMID:11307216

Nascimento, G. G. F., Locatelli, J., Freitas, P. C., & Silva, G. L. (2000). Antibacterial activity of plant extracts and phytochemicals on antibiotic-resistant bacteria. *Brazilian Journal of Microbiology*, *31*(4), 247–256. doi:10.1590/S1517-83822000000400003

National Institutes of Health. (1987). *Oral health of United States adults. The national survey of oral health in U.S. employed adults and seniors, 1985-1986. National findings. NIH publication no. 87-2868*. Washington, DC: US Government Printing Office.

National Institutes of Health. (1989). *Oral health of United States children. The national survey of dental caries in U.S. school children, 1986-1987. National and regional findings. NIH publication no. 89-2247*. Washington, DC: US Government Printing Office.

Nayar, S., Bhuminathan, S., & Bhat, W. M. (2015). Rapid prototyping and stereolithography in dentistry. *Journal of Pharmacy & Bioallied Sciences*, *7*(5Suppl 1), S216. doi:10.4103/0975-7406.155913 PMID:26015715

Nelson, S. J., & Ash, M. M. (2010). *Wheeler's Dental Anatomy, Physiology, and Occlusion* (9th ed.). Saunders Elsevier.

Nevriyanto, A., & Purnamasari, D. (2017, August). Image enhancement using the image sharpening, contrast enhancement, and Standard Median Filter (Noise Removal) with pixel-based and human visual system-based measurements. In *Electrical Engineering and Computer Science (ICECOS), 2017 International Conference on* (pp. 114-119). IEEE.

Newbrun, E. (1992). Preventing dental caries: current and prospective strategies. *Journal of the American Dental Association, 123*(5), 68-73.

Nomir, O., & Abdel-Mottaleb, M. (2005). A system for human identification from X-ray dental radiographs. *Pattern Recognition*, *38*(8), 1295–1305. doi:10.1016/j.patcog.2004.12.010

Nomir, O., & Abdel-Mottaleb, M. (2007). Human Identification from Dental X-Ray Images Based on the Shape and Appearance of the Teeth. *IEEE Transactions on Information Forensics and Security*, *2*(2), 188–197. doi:10.1109/TIFS.2007.897245

Nostro, A., Cannatelli, M. A., Crisafi, G., Musolino, A. D., Procopio, F., & Alonzo, V. (2004). Modifications of hydrophobicity, in vitro adherence and cellular aggregation of Streptococcus mutans by Helichrysum italicum extract. *Letters in Applied Microbiology*, *38*(5), 423–427. doi:10.1111/j.1472-765X.2004.01509.x PMID:15059215

Ntirogiannis, K., Gatos, B., & Pratikakis, I. (2014). A combined approach for the binarization of handwritten document images. *Pattern Recognition Letters*, *35*, 3–15. doi:10.1016/j.patrec.2012.09.026

Nurtanio, I., Purnama, I. K. E., Hariadi, M., & Purnomo, M. H. (2011). Cyst and tumor lesion segmentation on dental panoramic images using active contour models. *IPTEK J. Technol. Sci.*, *22*(3).

Odaira, C., Itoh, S., & Ishibashi, K. (2011). Clinical evaluation of a dental color analysis system: The Crystaleye Spectrophotometer. *Journal of Prosthodontic Research*, *55*(4), 199–205. doi:10.1016/j.jpor.2010.12.005 PMID:21296639

Odebiyi, O. O., & Sofowora, A. (1979). Antimicrobial alkaloids from Nigeria chewing sticks. *Planta Medica*, *36*, 204–204. doi:10.1055-0028-1097271 PMID:482432

Odusanya, S. A., Songonuga, O. O., & Folayan, J. O. (1979). Fluoriderion distribution in some African chewing sticks. *IRCS Medical Science [Microform]*, *7*, 58.

Oliveira, J., & Proença, H. (2011). Caries Detection in Panoramic Dental X-ray Images. *Computational Methods in Applied Sciences*, *19*, 175–190. doi:10.1007/978-94-007-0011-6_10

Oliveira, J., & Proenc, H. (2011). Caries Detection in Panoramic Dental X-ray Images. *Computational Vision and Medical Image Processing: Recent Trends, Computational Methods in Applied Sciences* 19. *Springer Science Business Media BV*, *10*. doi:10.1007/978-94-007-0011-6

Omachi, S., Saito, K., Aso, H., Kasahara, S., Yamada, S., & Kimura, K. (2007). Tooth shape reconstruction from CT images using spline curves. *Proceedings of the International Conference on Wavelet Analysis and Pattern Recognition*, 393–6.

Omachi, S., Saito, K., Aso, H., Kasahara, S., Yamada, S., & Kimura, K. (2008). Semi-automatic reconstruction of tooth shape from CT images by contour propagation. [in Japanese.]. *IEICE Transactions on Information and Systems*, *J91-D*, 2426–2429.

Onishi, M., Ozaki, F., Yoshino, F., & Murakami, Y. (1981). An experimental evidence of caries preventive activity of non- fluoride component in tea. *J. Dent. Health (Tokyo)*, *31*(2), 158–161. doi:10.5834/jdh.31.158 PMID:6948004

Onisor, I., Asmussen, E., & Krejci, I. (2011). Temperature rise during photo-polymerization for onlay luting. *American Journal of Dentistry*, *24*(4), 250. PMID:22016921

Ooshima, T., Izumitani, A., Sobue, S., Okahashi, N., & Hamada, S. (1983). Non-cariogenicity of the disaccharide palatinose in experimental dental caries of rats. *Infection and Immunity*, *39*(1), 43–49. PMID:6822422

Ooshima, T., Minami, T., Aono, W., Izumitani, A., Sobue, S., Fujiwara, T., ... Hamada, S. (1993). Oolong tea polyphenols inhibit experimental dental caries in SPF rats infected with *mutans streptococci*. *Caries Research*, *27*(2), 124–129. doi:10.1159/000261529 PMID:8319255

Ooshima, T., Minami, T., Aono, W., Tamura, Y., & Hamada, S. (1994). Reduction of dental plaque deposition in humans by Oolong tea extract. *Caries Research, 28*(3), 146–149. doi:10.1159/000261636 PMID:8033186

Ooshima, T., Minami, T., Matsumoto, M., Fujiwara, T., Sobue, S., & Hamada, S. (1998). Comparison of the cariostatic effects between regimens to administer oolong tea polyphenols in SPF rats. *Caries Research, 32*(1), 75–80. doi:10.1159/000016433 PMID:9438575

Ooshima, T., Osaka, Y., Sasaki, H., Osawa, K., Yasuda, H., Matsumura, M., ... Matsumoto, M. (2000). Caries inhibitory activity of cacao bean husk extract in in vitro and animal experiments. *Archives of Oral Biology, 45*(8), 639–645. doi:10.1016/S0003-9969(00)00042-X PMID:10869475

Ooshima, T., Sobue, S., Hamada, S., & Kotani, S. (1981). Susceptibility of rats, hamsters, and mice to carious infection by Streptococcus mutans serotype c and d organisms. *Journal of Dental Research, 60*(4), 855–859. doi:10.1177/00220345810600041701 PMID:6937525

Osawa, K., Yasuda, H., Maruyama, T., Morita, H., Takeya, K., & Itokawa, H. (1992). Isoflavanones from the heartwood of Swartzia polyphylla and their antibacterial activity against cariogenic bacteria. *Chemical & Pharmaceutical Bulletin, 40*(11), 2970–2974. doi:10.1248/cpb.40.2970 PMID:1477911

Osher, S., & Sethian, J. A. (1988). Fronts Propagating with Curvature Dependent speed: Algorithms Based on Hamilton-Jacobi Formulation. *Journal of Computational Physics, 79*(1), 12–49. doi:10.1016/0021-9991(88)90002-2

Oskui, I. Z., Ashtiani, M. N., Hashemi, A., & Jafarzadeh, H. (2013). Thermal analysis of the intact mandibular premolar: A finite element analysis. *International Endodontic Journal, 46*(9), 841–846. doi:10.1111/iej.12069 PMID:23480124

Osman, A. M., Younes, M. E. G., & Sheta, A. E. (1974). Triterpenoids of the leaves of *Psidium guajava. Phytochemistry, 13*(9), 2015–2016. doi:10.1016/0031-9422(74)85152-6

Otake, S., Makimura, M., Kuroki, T., Nishihara, Y., & Hirasawa, M. (1991). Anticaries effects of polyphenolic compounds from Japanese green tea. *Caries Research, 25*(6), 438–443. doi:10.1159/000261407 PMID:1667297

Otsu, N. (1979). A threshold selection method from gray-level histograms. *IEEE Transactions on Systems, Man, and Cybernetics, 9*(1), 62–66. doi:10.1109/TSMC.1979.4310076

Ozbolat, I. T., & Yu, Y. (2013). Bioprinting toward organ fabrication: Challenges and future trends. *IEEE Transactions on Biomedical Engineering, 60*(3), 691–699. doi:10.1109/TBME.2013.2243912 PMID:23372076

Özen, T., Kamburoğlu, K., Cebeci, A. R. I., Yüksel, S. P., & Paksoy, C. S. (2009). Interpretation of chemically created periapical lesions using 2 different dental cone-beam computerized tomography units, an intraoral digital sensor, and conventional film. *Oral Surgery, Oral Medicine, Oral Pathology, Oral Radiology, and Endodontics, 107*(3), 426–432. doi:10.1016/j.tripleo.2008.08.017 PMID:18996725

Pai, M. R., Acharya, L. D., & Udupa, N. (2004). Evaluation of antiplaque activity of Azadirachta indica leaf extract gel--a 6-week clinical study. *Journal of Ethnopharmacology, 90*(1), 99–103. doi:10.1016/j.jep.2003.09.035 PMID:14698516

Palani, T. K., Parvathavarthini, B., & Chitra, K. (2016). Segmentation of brain regions by integrating meta heuristic multilevel threshold with Markov random field. *Current Medical Imaging Reviews, 12*(1), 4–12. doi:10.2174/1573394711666150827203434

Pal, N. R., & Pal, S. K. (1993). A review on image segmentation techniques. *Pattern Recognition, 26*(9), 1277–1294. doi:10.1016/0031-3203(93)90135-J

Palomera-P'erez, M. A., Martinez-Perez, M. E., Benitez-P'erez, H., & Ortega-Arjona, J. L. (2010). Parallel multiscale feature extraction and region growing: Application in retinal blood vessel detection. *IEEE Transactions on Information Technology in Biomedicine*, *14*(2), 500–506. doi:10.1109/TITB.2009.2036604 PMID:20007040

Panda, S. P. (2016, March). Image contrast enhancement in spatial domain using fuzzy logic based interpolation method. In *Electrical, Electronics and Computer Science (SCEECS), 2016 IEEE Students' Conference on* (pp. 1-4). IEEE. 10.1109/SCEECS.2016.7509315

Pandey, P., Bhan, A., Dutta, M. K., & Travieso, C. M. (2017). Automatic Image Processing Based Dental Image Analysis Using Automatic Gaussian Fitting Energy and Level Sets. *International Conference and Workshop on Bioinspired Intelligence*. 10.1109/IWOBI.2017.7985529

Pandey, V., & Gupta, V. (2014). MRI image segmentation using shannon and non shannon entropy measures. *International Journal of Application or Innovation in Engineering & Management*, *3*(7), 41–46.

Paolino, C. J., Kashket, S., & Sparagna, C. A. (1980). Inhibition of dextran synthesis by tannic acid. *IADR Prog. Absrr.*, *58*(389).

Park, K. M., You, J. S., Lee, H. Y., Baek, N. I., Hwang, J. K., & Kuwanon, G. (2003). An antibacterial agent from the root bark of Morus alba against oral pathogens. *Journal of Ethnopharmacology*, *84*(2-3), 181–185. doi:10.1016/S0378-8741(02)00318-5 PMID:12648813

Park, Y. K., Ikegaki, M., Abreu, J. A. S., & Alcici, N. M. F. (1997). Comparison of the flavonoid aglycone contents of Apis mellifera propolis from various regions of Brazil. *Arquivos de Biologia e Tecnologia*, *40*(1), 97–106.

Park, Y. K., Koo, M. H., Abreu, J. A., Ikegaki, M., Cury, J. A., & Rosalen, P. L. (1998). Antimicrobial activity of propolis on oral microorganisms. *Current Microbiology*, *36*(1), 24–28. doi:10.1007002849900274 PMID:9405742

Pastor, M. (1995). *The digestive system*. Academic Press.

Patanachai, N., Covavisaruch, N., & Sinthanayothin, C. (2010). Wavelet transformation for dental X-ray radiographs segmentation technique. *8th International Conference on ICT and Knowledge Engineering*, 103-106.

Patel, S., & Kumar, N. A. (2009). Implant radiology. *Journal of the Indian Dental Association*, 3.

Pathak, D., Pathak, K., & Singla, A. (1991). Flavonoids as medical agents. *Fitoterapia*, *62*, 371–389.

Pauwels, R., Araki, K., Siewerdsen, J. H., & Thongvigitmanee, S. S. (2015). Technical aspects of dental CBCT: State of the art. *Dento Maxillo Facial Radiology*, *44*(1). doi:10.1259/dmfr.20140224 PMID:25263643

Peers, A., Hill, F. J., Mitropoulos, C. M., & Holloway, P. J. (1993). Validity and reproducibility of clinical examination, fibre-optic transillumination, and bite-wing radiology for the diagnosis of small approximal carious lesions: An in vitro study. *Caries Research*, *27*(4), 307–311. doi:10.1159/000261556 PMID:8402807

Perona, P., Shiota, T., & Malik, J. (1994) Anisotropic diffusion. In Geometry-driven diffusion in computer vision. Springer.

Perona, P., & Malik, J. (1990). Scale-space and edge detection using anisotropic diffusion. *IEEE Transactions on Pattern Analysis and Machine Intelligence*, *12*(7), 629–639. doi:10.1109/34.56205

Peutzfeldt, A. (1997). Resin composites in dentistry: The monomer systems. *European Journal of Oral Sciences*, *105*(2), 97–116. doi:10.1111/j.1600-0722.1997.tb00188.x PMID:9151062

Piller, N. B. A. (1975). Comparison of the effectiveness of some anti-inflammatory drugs on thermal oedema. *British Journal of Experimental Pathology*, *56*, 554–559. PMID:1222119

Pine, C. M. (1996). Fiber-optic transillumination (FOTI) in caries diagnosis. *Early Detection of Dental Caries*, *I*, 51–65.

Pineda, F. (1973). Roentgenographic investigation of the mesiobuccal root of the maxillary first molar. *Oral Surgery, Oral Medicine, and Oral Pathology*, *36*(2), 253–260.

Pisulkar, S. G., Pakhan, A. J., Godbole, S. R., & Dahane, T. M. (2016). Correlation of bone mineral density on primary implant stability in elderly edentulous patients: A literature review. *Scholars Academic Journals of Biosciences*, *4*(10A), 822–826.

Pongpan, A., Chumsri, P., & Taworasate, T. (1982). The antimicrobial activity of some Thai MedicinalPlants. *Mahidol Univ. J. Pharm. Sci*, *9*, 88–91.

Prabha, S., Anandh, K. R., Sujatha, C. M., & Ramakrishnan, S. (2014). Total variation based edge enhancement for level set segmentation and asymmetry analysis in breast thermograms. *IEEE Engineering in Medicine and Biology Society*, *36*, 6438–6441. PMID:25571470

Prabha, S., Suganthi, S. S., & Sujatha, C. M. (2015). An Approach to Analyze the Breast Tissues in Infrared Images Using Nonlinear Adaptive Level sets and Riesz Transform Features. *Technology and Health Care*, *23*(4), 429–442. doi:10.3233/THC-150915 PMID:26409908

Prabha, S., & Sujatha, C. M. (2014). Analysis of Breast Thermograms using Coherence Enhanced Diffusion Filter and Radon Transform. *Journal of the Instrument Society of India*, *44*(4), 271–274.

Prabha, S., Sujatha, C. M., & Ramakrishnan, S. (2015). Robust Anisotropic Diffusion Based Edge Enhancement for Level Set Segmentation and Asymmetry Analysis Of Breast Thermogramsusing Zernike Moments. *Journal of Biomedical Science Instrumentation*, *51*, 341–348. PMID:25996737

Prabu, G. R., Gnanamani, A., & Sadulla, S. (2006). Guaijaverin a plant flavonoid as potential antiplaque agent against *Streptococcus mutans*. *Journal of Applied Microbiology*, *101*(2), 487–495. doi:10.1111/j.1365-2672.2006.02912.x PMID:16882158

Prasad, M. S., Krishna, V. R., & Reddy, L. S. S. (2013). Investigations on Entropy Based Threshold Methods. *Asian Journal of Computer Science & Information Technology*, *1*(5).

Prashant, G., Chandu, G., Murulikrishna, K., & Shafiulla, M. (2007). The effect of mango and neem extract on four organisms causing dental caries: *Streptococcus mutans, Streptococcus salivavius, Streptococcus mitis,* and *Streptococcus sanguis*: An *in vitro* study. *Indian Journal of Dental Research*, *18*(4), 148–151. doi:10.4103/0970-9290.35822 PMID:17938488

Pratt, W. K. (2007). *Digital image processing* (4th ed.). Los Altos, CA: Wiley-Interscience. doi:10.1002/0470097434

Preiskorn, M., Zmuda, S., Trykowski, J., Panas, A., & Preiskorn, M. (2003). In vitro investigations of the heat transfer phenomena in human tooth. *Acta of Bioengineering and Biomechanics*, *5*(2), 23–36.

Pretty, I. A., & Sweet, D. (2001). A look at forensic dentistry-Part 1: The role of teeth in the determination of human identity. *British Dental Journal*, *190*(7), 359–366. PMID:11338039

Pretty, J. A., & Maupomé, G. (2004). A closer look at diagnosis in clinical dental practice:part 1. Reliability, validity, specificity and sensitivity of diagnostic procedures. *Journal - Canadian Dental Association*, *70*(4), 251–256. PMID:15120020

Priya, E., Srinivasan, S., & Ramakrishnan, S. (2014). Retrospective Non-Uniform Illumination Correction Techniques in Images of Tuberculosis. *Microscopy and Microanalysis*, *20*(5), 1382–1391. doi:10.1017/S1431927614012896 PMID:25115957

Pulkkinen, P., Partanen, J., Jalovaara, P., & Jamsa, T. (2004). Combination of bone mineral density and upper femur geometry improves the prediction of hip fracture. *Osteoporosis International, 15*(4), 274–280. doi:10.100700198-003-1556-3 PMID:14760516

Pushparaj, V., Gurunathan, U., & Arumugam, B. (2013). An effective dental shape extraction algorithm using contour information and matching by mahalanobis distance. *Journal of Digital Imaging, 26*(2), 259–268. doi:10.100710278-012-9492-4 PMID:22695751

Puupponen-Pimia, R., Nohynek, L., Hartmann-Schmidlin, S., Kahkonen, M., Heinonen, M., Maatta-Riihinen, K., & Oksman-Caldentey, K. M. (2005, April). (205). Berry phenolics selectively inhibit the growth of intestinal pathogens. *Journal of Applied Microbiology, 98*(4), 991–1000. doi:10.1111/j.1365-2672.2005.02547.x PMID:15752346

Rad, A. E., Rahim, M. S. M., Rehman, A., & Saba, T. (2016). Digital dental X-ray database for caries screening. *3D Research, 7*(2), 18.

Rad, A. E., Rahinm, M. S. M., Rehman, A., & Saba, T. (2016). Digital Dental X-ray Database for Caries Screening. *3D Res, 7*, 18.

Rad, A.E., Rahim, M.S.M., Rehman, A. & Saba, T. (2016). Digital dental x-ray database for caries screening. *3D Research, 7*(2), 1-5. . doi:10.100713319-016-0096-5

Rad, A. E., Mohd Rahim, M. S., Rehman, A., Altameem, A., & Saba, T. (2013). Evaluation of current dental radiographs segmentation approaches in computer-aided applications. *IETE Technical Review, 30*(3), 210–222. doi:10.4103/0256-4602.113498

Rad, A. E., Rahim, A. S. M., Kolivand, H., & Amin, I. B. M. (2017). Morphological region-based initial contour algorithm for level set methods in image segmentation. *Multimedia Tools and Applications, 76*(2), 2185–2201. doi:10.100711042-015-3196-y

Rad, A. E., Rahim, M. S. M., & Norouzi, A. (2013). Digital dental x-ray image segmentation and feature extraction. *Indonesian Journal of Electrical Engineering and Computer Science, 11*(6), 3109–3114.

Rad, A. E., Rahim, M. S. M., & Norouzi, A. (2014). Level set and morphological operation techniques in application of dental image segmentation. *Int. Sch. Sci. Res.Innov., 8*(4), 177–180.

Rahim, E. N. A. A., Ismail, A., Omar, M. N., Rahmat, U. N., & Ahmad, W. A. N. W. (2018). GC-MS Analysis of Phytochemical Compounds in Syzygium polyanthum Leaves Extracted using Ultrasound-Assisted Method. *Pharmacognosy Journal, 10*(1), 110–119. doi:10.5530/pj.2018.1.20

Rahimov, C., & Farzaliyev, I. (2011). Virtual bending of titanium reconstructive plates for mandibular defect bridging: Review of three clinical cases. *Craniomaxillofacial Trauma & Reconstruction, 4*(04), 223–234. doi:10.1055-0031-1293523 PMID:23205175

Raja, N. S. M., Fernandes, S. L., Dey, N., Satapathy, S. C., & Rajinikanth, V. (2018). Contrast enhanced medical MRI evaluation using Tsallis entropy and region growing segmentation. *Journal of Ambient Intelligence and Humanized Computing*, 1–12. doi:10.100712652-018-0854-8

Raja, N. S. M., Kavitha, G., & Ramakrishnan, S. (2012). Analysis of Vasculature in Human Retinal Images Using Particle Swarm Optimization Based Tsallis Multi-level Thresholding and Similarity Measures. *SEMCCO, 7677*, 380–387.

Raja, N. S. M., Sukanya, S. A., & Nikita, Y. (2015). Improved PSO based multi-level thresholding for cancer infected breast thermal images using Otsu. *Procedia Computer Science, 48*, 524–529. doi:10.1016/j.procs.2015.04.130

Raja, N., Rajinikanth, V., Fernandes, S. L., & Satapathy, S. C. (2017). Segmentation of Breast Thermal Images Using Kapur's Entropy and Hidden Markov Random Field. *Journal of Medical Imaging and Health Informatics*, *7*(8), 1825–1829. doi:10.1166/jmihi.2017.2267

Raja, N., Rajinikanth, V., & Latha, K. (2014). Otsu based optimal multilevel image thresholding using firefly algorithm. *Modelling and Simulation in Engineering*, *2014*, 37.

Rajendran, R. (2009). Shafer's textbook of oral pathology (6th ed.). Academic Press.

Rajinikanth, V., Aashiha, J. P., & Atchaya, A. (2014). Gray-level histogram based multilevel threshold selection with bat algorithm. *International Journal of Computers and Applications*, *93*(16).

Rajinikanth, V., & Couceiro, M. S. (2015). RGB histogram based color image segmentation using firefly algorithm. *Procedia Computer Science*, *46*, 1449–1457. doi:10.1016/j.procs.2015.02.064

Rajinikanth, V., & Couceiro, M. S. (2015b). Optimal multilevel image threshold selection using a novel objective function. In *Information Systems Design and Intelligent Applications* (pp. 177–186). New Delhi: Springer. doi:10.1007/978-81-322-2247-7_19

Rajinikanth, V., Dey, N., Satapathy, S. C., & Ashour, A. S. (2018). An approach to examine magnetic resonance angiography based on Tsallis entropy and deformable snake model. *Future Generation Computer Systems*, *85*, 160–172. doi:10.1016/j.future.2018.03.025

Rajinikanth, V., Fernandes, S. L., Bhushan, B., & Sunder, N. R. (2018). Segmentation and Analysis of Brain Tumor Using Tsallis Entropy and Regularised Level Set. In *Proceedings of 2nd International Conference on Micro-Electronics, Electromagnetics and Telecommunications* (pp. 313-321). Springer. 10.1007/978-981-10-4280-5_33

Rajinikanth, V., Raja, N. S. M., & Kamalanand, K. (2017). Firefly algorithm assisted segmentation of tumor from brain MRI using Tsallis function and Markov random field. *Journal of Control Engineering and Applied Informatics*, *19*(2), 97–106.

Rajinikanth, V., Raja, N. S. M., & Kamalanand, K. (2017). Firefly Algorithm Assisted Segmentation of Tumor from Brain MRI using Tsallis Function and Markov Random Field. *Journal of Control Engineering and Applied Informatics*, *19*(3), 97–106.

Rajinikanth, V., Raja, N. S. M., & Latha, K. (2014). Optimal multilevel image thresholding: An analysis with PSO and BFO algorithms. *Australian Journal of Basic and Applied Sciences*, *8*(9), 443–454.

Rajinikanth, V., Raja, N. S. M., & Satapathy, S. C. (2016). Robust color image multi-thresholding using between-class variance and cuckoo search algorithm. In *Information systems design and intelligent applications* (pp. 379–386). New Delhi: Springer. doi:10.1007/978-81-322-2755-7_40

Rajinikanth, V., & Satapathy, S. C. (2018). Segmentation of Ischemic Stroke Lesion in Brain MRI Based on Social Group Optimization and Fuzzy-Tsallis Entropy. *Arabian Journal for Science and Engineering*, 1–14.

Rajinikanth, V., & Satapathy, S. C. (2018). Segmentation of ischemic stroke lesion in brain MRI based on social group optimization and Fuzzy-Tsallis entropy. *Arabian Journal for Science and Engineering*, 1–14. doi:10.100713369-017-3053-6

Rajinikanth, V., Satapathy, S. C., Dey, N., & Vijayarajan, R. (2018). DWT-PCA Image Fusion Technique to Improve Segmentation Accuracy in Brain Tumor Analysis. In *Microelectronics, Electromagnetics and Telecommunications* (pp. 453–462). Singapore: Springer. doi:10.1007/978-981-10-7329-8_46

Rajinikanth, V., Satapathy, S. C., Fernandes, S. L., & Nachiappan, S. (2017). Entropy based segmentation of tumor from brain MR images–a study with teaching learning based optimization. *Pattern Recognition Letters*, *94*, 87–95. doi:10.1016/j.patrec.2017.05.028

Ramamurthy, R., Scheetz, J. P., Clark, S. J., & Farman, A. G. (2006). Effects of imaging system and exposure on accurate detection of the second mesio-buccal canal in maxillary molar teeth. *Oral Surgery, Oral Medicine, Oral Pathology, Oral Radiology, and Endodontics*, *102*(6), 796–802. doi:10.1016/j.tripleo.2006.02.009 PMID:17138184

Ramasubramanian, B., & Selvaperumal, S. (2016, April). A comprehensive review on various preprocessing methods in detecting diabetic retinopathy. In *Communication and Signal Processing (ICCSP), 2016 International Conference on* (pp. 0642-0646). IEEE. 10.1109/ICCSP.2016.7754220

Rao, K. (2000). *Materials for the database of medicinal plants*. Bangalore: Karnataka State Council for Science and Technology.

Rao, R. V. (2016). Jaya: A simple and new optimization algorithm for solving constrained and unconstrained optimization problems. *International Journal of Industrial Engineering Computations*, *7*(1), 19–34.

Rao, R. V., & More, K. C. (2017). Design optimization and analysis of selected thermal devices using self-adaptive Jaya algorithm. *Energy Conversion and Management*, *140*, 24–35. doi:10.1016/j.enconman.2017.02.068

Rao, R. V., & Saroj, A. (2017). A self-adaptive multi-population based Jaya algorithm for engineering optimization. *Swarm and Evolutionary Computation*, *37*, 1–26. doi:10.1016/j.swevo.2017.04.008

Rao, R. V., & Saroj, A. (2017a). Constrained economic optimization of shell-and-tube heat exchangers using elitist-Jaya algorithm. *Energy*, *128*, 785–800. doi:10.1016/j.energy.2017.04.059

Ratnaparkhe, V. R., Manthalkar, R. R., & Joshi, Y. V. (2009). Texture characterization of CT images based on ridgelet transform. *International Journal Graphics Vision Image Processing*, *8*, 43–50.

Reddy, V. V., & Sugandhan, S. (1994). A comparison of bitewing radiography and fibreoptic illumination as adjuncts to the clinical identification of approximal caries in primary and permanent molars. *Indian Journal of Dental Research*, *5*(2), 59-64.

Redmon, J., & Farhadi, A. (2017). YOLO9000: Better, Faster, Stronger. *IEEE Conference on Computer Vision and Pattern Recognition (CVPR)*, 6517-6525.

Rehman, F., Akram, M. U., Faraz, K., & Riaz, N. (2015, April). Human identification using dental biometric analysis. In *Digital Information and Communication Technology and its Applications (DICTAP), 2015 Fifth International Conference on* (pp. 96-100). IEEE. 10.1109/DICTAP.2015.7113178

Rehse, S. J. (2014). Biomedical applications of LIBS. In *Laser-Induced Breakdown Spectroscopy* (pp. 457–488). Berlin: Springer. doi:10.1007/978-3-642-45085-3_17

Ren, S., He, K., Girshick, R., & Sun, J. (2017). Faster R-CNN: Towards Real-Time Object Detection with Region Proposal Networks. *IEEE Transactions on Pattern Analysis and Machine Intelligence*, *39*(6), 1137–1149. doi:10.1109/TPAMI.2016.2577031 PMID:27295650

Rezaei, M., Yang, H., & Meinel, C. (2017). Deep Neural Network with l2-norm Unit for Brain Lesions Detection. In Neural Information Processing. ICONIP 2017. Lecture Notes in Computer Science (Vol. 10637, pp. 798-807). Berlin: Springer. doi:10.1007/978-3-319-70093-9_85

Ribeiro, M. R., Dias, M. A., de Best, R., da Silva, E. A., Neves, C. D. T. T., & Street, R. S. (2014). Enhancement and segmentation of dental structures in digitized panoramic radiography images. *International Journal of Applied Mathematics*, *27*(4), 387–406. doi:10.12732/ijam.v27i4.6

Ripa, L. W. (1985). The current status of pit and fissure sealants. *Journal - Canadian Dental Association*, *5*, 367–380. PMID:3160450

Roberts, H. W., Berzins, D. W., Moore, B. K., & Charlton, D. G. (2009). Metal-Ceramic Alloys in Dentistry: A Review. *Journal of Prosthodontics*, *18*(2), 188–194. doi:10.1111/j.1532-849X.2008.00377.x PMID:19178620

Robinson, C., Shore, R. C., Bonass, W. A., Brookes, S. J., Boteva, E., & Kirkham, J. (1998). Identification of human serum albumin in human caries lesions of enamel: The role of putative inhibitors of remineralisation. *Caries Research*, *32*(3), 193–199. doi:10.1159/000016452 PMID:9577984

Rohsenow, W. M., & Cho, Y. I. (1998). *Handbook of heat transfer* (J. P. Hartnett, Ed.; Vol. 3). New York: McGraw-Hill.

Ronneberger, O., Fischer, P., & Brox, T. (2015). U-Net: Convolutional Networks for Biomedical Image Segmentation. *Medical Image Computing and Computer Assisted Interventions Conference (MICCAI)*, *9351*, 234-241. 10.1007/978-3-319-24574-4_28

Roopini, I. T., Vasanthi, M., Rajinikanth, V., Rekha, M., & Sangeetha, M. (2018). Segmentation of tumor from brain MRI using fuzzy entropy and distance regularised level set. *Lecture Notes in Electrical Engineering*, *490*, 297–304. doi:10.1007/978-981-10-8354-9_27

Rose, E., & Svec, T. (2015). An evaluation of apical cracks in teeth undergoing orthograde root canal instrumentation. *Journal of Endodontics*, *41*(12), 2021–2024. doi:10.1016/j.joen.2015.08.023 PMID:26472677

Roser, S. M., Ramachandra, S., Blair, H., Grist, W., Carlson, G. W., Christensen, A. M., & Steed, M. B. (2010). The accuracy of virtual surgical planning in free fibula mandibular reconstruction: Comparison of planned and final results. *Journal of Oral and Maxillofacial Surgery*, *68*(11), 2824–2832. doi:10.1016/j.joms.2010.06.177 PMID:20828910

Rotimi, V. O., Laughon, B. E., Bartlet, J. G., & Mesodomi, H. A. (1988). Activities of Nigeria chewing stick against Bacteroides gingivalis and Bacteroides melaninogenicus. *Nig. J. Microbiol.*, *9*, 13–16.

Rozen, R., Bachrach, G., Bronshteyn, M., Gedalia, I., & Steinberg, D. (2001). The role of fructans on dental biofilm formation by *Streptococcus sobrinus, Streptococcus mutans, Streptococcus gordonii* and *Actinomyces viscosus. FEMS Microbiology Letters*, *195*(2), 205–210. doi:10.1111/j.1574-6968.2001.tb10522.x PMID:11179653

Rubin, C., Krishnamurthy, N., Capilouto, E., & Yi, H. (1983). Stress analysis of the human tooth using a three-dimensional finite element model. *Journal of Dental Research*, *62*(2), 82–86. doi:10.1177/00220345830620021701 PMID:6571871

Rubinstein, R., Zibulevsky, M., & Elad, M. (2010). Double sparsity: Learning sparse dictionaries for sparse signal approximation. *IEEE Transactions on Signal Processing*, *58*(3), 1553–1564. doi:10.1109/TSP.2009.2036477

Russell, M. W. (1979). Purification and properties of a protein surface antigen of *Streptococcus mutans. Microbios*, *25*, 7–18. PMID:393961

Russell, R. R. (1994). The application of molecular genetics to the microbiology of dental caries. *Caries Research*, *28*(2), 69–82. doi:10.1159/000261625 PMID:8156565

Saeki, Y., Ito, Y., Shibata, M., Sato, Y., Okuda, K., & Takazoe, I. (1989). Antimicrobial action of natural substances on oral bacteria. Bull.Tokyo Dent. Coll., 30(3), 129-135.

Saghlatoon, H., Soleimani, M., Moghimi, S., & Talebi, M. (2012, May). An experimental investigation about the heat transfer phenomenon in human teeth. In *Electrical Engineering (ICEE), 2012 20th Iranian Conference on* (pp. 1598-1601). IEEE. 10.1109/IranianCEE.2012.6292616

Sahana, A. S., Paramasivam, A., & Kamalanand, K. (2017, March). Experimental investigations on capacitive imaging of biological materials. In *Biosignals, Images and Instrumentation (ICBSII), 2017 Third International Conference on* (pp. 1-3). IEEE. 10.1109/ICBSII.2017.8082289

Sahoo, P. K., Soltani, S. A. K. C., & Wong, A. K. (1988). A survey of thresholding techniques. *Computer Vision Graphics and Image Processing*, *41*(2), 233–260. doi:10.1016/0734-189X(88)90022-9

Said, E., Fahmy, G. F., Nassar, D., & Ammar, H. (2004). Dental x-ray image segmentation. Proc. International society for optics and photonics defense and security, 409-417.

Said, Eldin, Nassar, Fahmy, & Ammar. (2006). Teeth Segmentation in Digitized Dental X-Ray Films Using Mathematical Morphology. *IEEE Transactions on Information Forensics and Security*, *1*(2).

Said, E. H., Nassar, D. E. M., Fahmy, G., & Ammar, H. H. (2006). Teeth segmentation in digitized dental X-ray films using mathematical morphology. *IEEE Transactions on Information Forensics and Security*, *1*(2), 178–189. doi:10.1109/TIFS.2006.873606

Said, E., Fahmy, G. F., Nassar, D., & Ammar, H. (2004, April). Dental x-ray image segmentation. *Proceedings of the Society for Photo-Instrumentation Engineers*, *5404*, 409–417. doi:10.1117/12.541658

Sakanaka, S., Kim, M., Taniguchi, M., & Yamamoto, T. (1989). Antibacterial substances in Japanese green tea extract against *Streptococcus mutans*, a cariogenic bacterium. *Agricultural and Biological Chemistry*, *53*, 2307–2311.

Sakanaka, S., Sato, T., Kim, M., & Yamamoto, T. (1990). Inhibitory effects of green tea polyphenols on glucan synthesis and cellular adherence of cariogenic *streptococci. Agricultural and Biological Chemistry*, *54*, 2925–2929.

Sakanaka, S., Shimura, N., Aizawa, M., Kim, M., & Yamamoto, T. (1992). Preventive effect of green tea polyphenols against dental caries in conventional rats. *Bioscience, Biotechnology, and Biochemistry*, *56*(4), 592–594. doi:10.1271/bbb.56.592 PMID:27280653

Salgueiro, M. I., & Stevens, M. R. (2010). Experience with the use of prebent plates for the reconstruction of mandibular defects. *Craniomaxillofacial Trauma & Reconstruction*, *3*(04), 201–208. doi:10.1055-0030-1268520 PMID:22132258

Santos, A. (2003). Evidence-based control of plaque and gingivitis. *Journal of Clinical Periodontology*, *30*(s5), 13–16. doi:10.1034/j.1600-051X.30.s5.5.x PMID:12787197

Sarkar, S., & Das, S. (2013). Multilevel image thresholding based on 2D histogram and maximum Tsallis entropy—a differential evolution approach. *IEEE Transactions on Image Processing*, *22*(12), 4788–4797. doi:10.1109/TIP.2013.2277832 PMID:23955760

Sasaki, H., Matsumoto, M., Tanaka, T., Maeda, M., Nakai, M., Hamada, S., & Ooshima, T. (2004). Antibacterial activity of polyphenol components in Oolong tea extract against *Streptococcus mutans. Caries Research*, *38*(1), 2–8. doi:10.1159/000073913 PMID:14684970

Satapathy, S. C., Raja, N. S. M., Rajinikanth, V., Ashour, A. S., & Dey, N. (2016). Multi-level image thresholding using Otsu and chaotic bat algorithm. *Neural Computing & Applications*, 1–23.

Sathya, P. D., & Kayalvizhi, R. (2010). Optimum multilevel image thresholding based on tsallis entropy method with bacterial foraging algorithm. *International Journal of Computational Science*, *7*(5), 336–343.

Sato, M., Fujiwara, S., Tsuchiya, H., Fujii, T., Tinuma, M., Tosa, H., & Ohkawa, Y. (1996). Flavones with antibacterial activity against cariogenic bacteria. *Journal of Ethnopharmacology*, *54*(2-3), 171–176. doi:10.1016/S0378-8741(96)01464-X PMID:8953432

Sato, M., Tanaka, H., Fujiwara, S., Hirata, M., Yamaguchi, R., Etoh, H., & Tokuda, C. (2003). Antibacterial property of isoflavonoids isolated from Erythrina variegata against cariogenic oral bacteria. *Phytomedicine*, *10*(5), 427–433. doi:10.1078/0944-7113-00225 PMID:12834009

Sauvola, J., & Pietikäinen, M. (2000). Adaptive document image binarization. *Pattern Recognition*, *33*(2), 225–236. doi:10.1016/S0031-3203(99)00055-2

Sawamura, S., Tonosaki, Y., & Hamada, S. (1992). Inhibitory effects of ellagic acid on glucosyltransferases from mutans streptococci. *Bioscience, Biotechnology, and Biochemistry*, *56*(5), 766–768. doi:10.1271/bbb.56.766 PMID:27286205

Scalbert, A. (1991). Antimicrobial properties of tannins. *Phytochemistry*, *30*(12), 3875–3883. doi:10.1016/0031-9422(91)83426-L

Scarfe, W. C., Farman, A. G., & Sukovic, P. (2006). Clinical applications of cone-beam computed tomography in dental practice. *Journal - Canadian Dental Association*, *72*(1), 75. PMID:16480609

Scarfe, W. C., Levin, M. D., Gane, D., & Farman, A. G. (2009). Use of cone beam computed tomography in endodontics. *International Journal of Dentistry*. PMID:20379362

Schilling, K. M., & Bowen, W. H. (1992). Glucans synthesized *in situ* in experimental salivary pellicle function as specific binding sites for *Streptococcus mutans*. *Infection and Immunity*, *60*, 284–295. PMID:1530843

Schilling, K., & Doyle, R. J. (1995). Bacterial adhesion to hydroxylapatite. *Methods in Enzymology*, *253*, 536–542. doi:10.1016/S0076-6879(95)53045-2 PMID:7476417

Schmidhuber, J. (2015). Deep learning in neural networks: An overview. *Neural Networks*, *61*, 85–117. doi:10.1016/j.neunet.2014.09.003 PMID:25462637

Schmidt, H. (1988). *Phenol oxidase (E.I.14.18.1), a marker enzyme for defense cells. Progress in histochemistry and cytochemistry* (Vol. 17). New York, N.Y.: Gustav Fischer.

Schneiderman, A., Elbaum, M., Shultz, T., Keem, S., Greenebaum, M., & Driller, J. (1997). Assessment of dental caries with digital imaging fiber-optic translllumination (DIFOTITM): In vitro Study. *Caries Research*, *31*(2), 103–110. doi:10.1159/000262384 PMID:9118181

Schour, I., & Massler, M. (1940). Studies in tooth development: The growth pattern of human teeth part II. *The Journal of the American Dental Association*, *27*(12), 1918–1931. doi:10.14219/jada.archive.1940.0367

Schubert, C., Van Langeveld, M. C., & Donoso, L. A. (2014). Innovations in 3D printing: A 3D overview from optics to organs. *The British Journal of Ophthalmology*, *98*(2), 159–161. doi:10.1136/bjophthalmol-2013-304446 PMID:24288392

Secilmis, A., Bulbul, M., Sari, T., & Usumez, A. (2013). Effects of different dentin thicknesses and air cooling on pulpal temperature rise during laser welding. *Lasers in Medical Science*, *28*(1), 167–170. doi:10.100710103-012-1108-1 PMID:22562450

Seetha, D. M., Devi, G. M., & Sunitha, D. K. (2012, February). Performance Assessment of Feature Based Image Retrieval using Neural Networks. In Issue ICTM 2011 (Vol. 2). Academic Press.

Selesnick, I. W. (1999). The slantlet transform. *IEEE Transactions on Signal Processing*, *47*(5), 1304–1313. doi:10.1109/78.757218

Sengun, A., Özbay, Y., Akdemir, B., Öztürk, B., Özer, F., & Baglar, S. (2013). Reliability of electronically detection of fissure caries (by using a prototype device): An alternative diagnostic electronic caries monitor device. *Journal of Restorative Dentistry*, *1*(1), 26. doi:10.4103/2321-4619.111230

Senthilkumaran, N., & Kirubakaran, C. (2014). Efficient implementation of Niblack thresholding for MRI brain image segmentation. *International Journal of Computer Science and Information Technologies*, *5*, 2173–2176.

Senthilkumaran, N., & Vaithegi, S. (2016). Image segmentation by using thresholding techniques for medical images. Computer Science & Engineering. *International Journal (Toronto, Ont.)*, *6*(1).

Sethian, J. A. (1999). *Level Set Methods and Fast Marching Methods*. Cambridge, UK: Cambridge University Press.

Sezgin, M. (2004). Survey over image thresholding techniques and quantitative performance evaluation. *Journal of Electronic Imaging*, *13*(1), 146–168. doi:10.1117/1.1631315

Shafarenko, L., Petrou, H., & Kittler, J. (1998). Histogram-based segmentation in a perceptually uniform color space. *IEEE Transactions on Image Processing*, *7*(9), 1354–1358. doi:10.1109/83.709666 PMID:18276345

Shafer's Tb. (2006). Textbook of Oral Pathology (6th ed.). Academic Press.

Shahin, K. A., Chatra, L., & Shenai, P. (2013). Dental and craniofacial imaging in forensics. *JOFRI*, *1*(2), 56–62.

Shah, N., Bansal, N., & Logani, A. (2014). Recent advances in imaging technologies in dentistry. *World Journal of Radiology*, *6*(10), 794–807. doi:10.4329/wjr.v6.i10.794 PMID:25349663

Shah, S., Abaza, A., Ross, A., & Ammar, H. (2006, September). Automatic tooth segmentation using active contour without edges. In *Biometric consortium conference, 2006 biometrics symposium: Special session on research at the* (pp. 1–6). IEEE. doi:10.1109/BCC.2006.4341636

Shapiro, S., Meier, A., & Guggenheim, B. (1994). The antimicrobiol activity of essential oils and essential oil components toward oral bacteria. *Oral Microbiology and Immunology*, *9*(4), 202–208. doi:10.1111/j.1399-302X.1994.tb00059.x PMID:7478759

Sharma, A., & Khunteta, A. (2016, September). Satellite Image Enhancement using Discrete Wavelet Transform, Singular Value Decomposition and its Noise Performance Analysis. In *Micro-Electronics and Telecommunication Engineering (ICMETE), 2016 International Conference on* (pp. 594-599). IEEE.

Shen, D., Wu, G., & Suk, H.-I. (2017). Deep Learning in Medical Image Analysis. *Annual Review of Biomedical Engineering*, *19*(1), 221–248. doi:10.1146/annurev-bioeng-071516-044442 PMID:28301734

Sherwood, L., Kell, R., & Ward, C. (2010). *Human physiology: From cells to systems*. Cengage Learning.

Sheu, C. W., & Freese, E. (1972). Effects of fatty acid on growth and envelope proteins of *Bacillus subtilis*. *Journal of Bacteriology*, *111*, 516–524. PMID:4626502

Shi, X., Sherry, L. X., Wang, X., Edward, G. X., & Niebur, G. L. (2010). Type and orientation of yielded trabeculae during overloading of trabecular bone along orthogonal directions. *Journal of Biomechanics*, *43*(13), 2460–2466. doi:10.1016/j.jbiomech.2010.05.032 PMID:20554282

Silva, I. M., Freitas, D. Q., Ambrosano, G. M., Bóscolo, F. N., & Almeida, S. M. (2012). Bone Density: Comparative evaluation of Hounsfield units in multislice and cone-beam computed tomography. *Brazilian Oral Research*, *26*(6), 550–556. doi:10.1590/S1806-83242012000600011 PMID:23184166

Simonyan, K., & Zisserman, A. (2015). Very Deep Convolutional Networks for Large-Scale Image Recognition. *International Conference on Learning Representations (ICLR)*.

Singh, M. M., & Mathur, R. (2013). *Thyroid Cancer Risk from Dental X-ray Diagnostics*. Academic Press.

Singh, T. R., Roy, S., Singh, O. I., Sinam, T., & Singh, K. (2012). *A new local adaptive thresholding technique in binarization.* arXiv preprint arXiv:1201.5227

Singh, V. (2014). Osteology of Head and Neck. In Anatomy: Head, Neck and Brain (vol. 3, pp. 24-40). Elsevier Publishers.

Singh, O. I., Sinam, T., James, O., & Singh, T. R. (2012). Local Contrast and Mean Thresholding in Image Binarization. *International Journal of Computers and Applications*, *51*(6).

Siu, A. S. C., Chu, F. C. S., Li, T. K. L., Chow, T. W., & Deng, F. (2010). *Imaging modalities for preoperative assessment in dental implant therapy: an overview. Hong Kong Dental Journal.*

Siyal, M. Y., & Yu, L. (2005). An intelligent modified fuzzy c-means based algorithm for bias estimation and segmentation of brainMRI. *Pattern Recognition Letters*, *26*(13), 2052–2062. doi:10.1016/j.patrec.2005.03.019

Smith, R., & Lehner, T. (1989). Characterisation of monoclonal antibodies binding to common protein epitopes on the cell surface of *Streptococcus mutans* and *Streptococcus sobrinus. Oral Microbiology and Immunology*, *4*(3), 153–158. doi:10.1111/j.1399-302X.1989.tb00243.x PMID:2639299

Smullen, J., Koutsou, G. A., Foster, H. A., Zumbe´, A., & Storey, D. M. (2007). The antibacterial activity of plant extracts containing polyphenols against *Streptococcus mutans. Caries Research*, *41*(5), 342–349. doi:10.1159/000104791 PMID:17713333

Sofowora, A. (1993). *Medicinal Plant and Traditional Medicine in Africa.* Ibadan, Nigeria: Spectrum Books.

Sofrata, A., Claesson, R., Lingstrom, P., & Gustafsson, A. (2008). Strong antibacterial effect of miswak against oral microorganisms associated with periodontitis and caries. *Journal of Periodontology*, *79*(8), 1474–1479. doi:10.1902/jop.2008.070506 PMID:18672998

Sofrata, A., Lingstrom, P., Baljoon, M., & Gustafsson, A. (2007). The effect of miswak extract on plaque pH. An *in vivo* study. *Caries Research*, *41*(6), 451–454. doi:10.1159/000107931 PMID:17823507

Song, J. H., Kim, S. K., Chang, K. W., Han, S. K., Yi, H. K., & Jeon, J. G. (2006). *In vitro* inhibitory effects of Polygonum cuspidatum on bacterial viability and virulence factors of *Streptococcus mutans* and *Streptococcus sobrinus. Archives of Oral Biology*, *51*(12), 1131–1140. doi:10.1016/j.archoralbio.2006.06.011 PMID:16914113

Song, Q., Zhao, L., Luo, X., & Dou, X. (2017). Using Deep Learning for Classification of Lung Nodules on Computed Tomography Images. *Journal of Healthcare Engineering*. PMID:29065651

Son, L. H., & Tuan, T. M. (2017). Dental segmentation from X-ray images using semi-supervised fuzzy clustering with spatial constraints. *Engineering Applications of Artificial Intelligence*, *59*, 186–195. doi:10.1016/j.engappai.2017.01.003

Son, L. H., Tuan, T. M., Fujita, H., Dey, N., Ashour, A. S., Ngoc, V. T. N., ... Chu, D. T. (2018). Dental diagnosis from X-Ray images: An expert system based on fuzzy computing. *Biomedical Signal Processing and Control*, *39*, 64–73. doi:10.1016/j.bspc.2017.07.005

Sparavigna, A. C. (2015). Tsallis entropy in bi-level and multi-level image thresholding. *International Journal of Sciences*, *4*(1), 40–49. doi:10.18483/ijSci.613

Srinivasan, S. S., & Swaminathan, R. (2014). Segmentation of Breast Tissues in Infrared Images Using Modified Phase Based Level Sets. In T. D. Pham, K. Ichikawa, M. Oyama-Higa, D. Coomans, & X. Jiang (Eds.), Biomedical Informatics and Technology. Communications in Computer and Information Science. 404. Academic Press. doi:10.1007/978-3-642-54121-6_14

Staat, R. H., Doyle, R. J., Langley, S. D., & Suddick, R. P. (1978). Modification of in vitro adherence of Streptococcus mutans by plant lectins. *Advances in Experimental Medicine and Biology*, *107*, 639–647. doi:10.1007/978-1-4684-3369-2_72 PMID:742505

Starck, J. L., Candès, E. J., & Donoho, D. L. (2002). The curvelet transform for image denoising. *IEEE Transactions on Image Processing*, *11*(6), 670–684. doi:10.1109/TIP.2002.1014998 PMID:18244665

Starck, J. L., Fadili, J., & Murtagh, F. (2007). The undecimated wavelet decomposition and its reconstruction. *IEEE Transactions on Signal Processing*, *16*(2), 297–309. PMID:17269625

Stark, J. A. (2000). Adaptive image contrast enhancement using generalizations of histogram equalization. *IEEE Transactions on Image Processing*, *9*(5), 889–896. doi:10.1109/83.841534 PMID:18255459

Stary, F. (1996). *The Natural Guide to Medicinal Herbs and Plants*. Barnes & Noble Inc.

Stathis, P., Kavallieratou, E., & Papamarkos, N. (2008). An Evaluation Technique for Binarization Algorithms. *J. UCS*, *14*(18), 3011–3030.

Stavropoulos, A., & Wenzel, A. (2007). Accuracy of cone beam dental CT, intraoral digital and conventional film radiography for the detection of periapical lesions. An ex vivo study in pig jaws. *Clinical Oral Investigations*, *11*(1), 101–106. doi:10.100700784-006-0078-8 PMID:17048029

Steinberg, D., Feldman, M., Ofek, I., & Weiss, E. I. (2004). Effect of a high-molecular-weight component of cranberry on constituents of dental biofilm. *The Journal of Antimicrobial Chemotherapy*, *54*(1), 86–89. doi:10.1093/jac/dkh254 PMID:15163648

Steinberg, D., Feldman, M., Ofek, I., & Weiss, E. I. (2005). Cranberry high molecular weight constituents promote *Streptococcus sobrinus* desorption from artificial biofilm. *International Journal of Antimicrobial Agents*, *25*(3), 247–251. doi:10.1016/j.ijantimicag.2004.10.014 PMID:15737520

Stephen, K. (1993). Caries in young populations–worldwide. In W. H. Bowen & L. A. Tabak (Eds.), *Cariology for the nineties* (pp. 71–84). Rochester, NY: University of Rochester Press.

Stern, J. L., Hagerman, A. E., Steinberg, P. D., & Mason, P. K. (1996). Phlorotannin-protein interactions. *Journal of Chemical Ecology*, *22*(10), 1887–1899. doi:10.1007/BF02028510 PMID:24227114

St-Onge, L., Kwong, E., Sabsabi, M., & Vadas, E. B. (2004). Rapid analysis of liquid formulations containing sodium chloride using laser-induced breakdown spectroscopy. *Journal of Pharmaceutical and Biomedical Analysis*, *36*(2), 277–284. doi:10.1016/j.jpba.2004.06.004 PMID:15496320

Stookey, G. K., Jackson, R. D., Zandona, A. G., & Analoui, M. (1999). Dental caries diagnosis. *Dental Clinics of North America*, *43*(4), 665–677. PMID:10553249

Strassler, H. E., & Sensi, L. G. (2008). Technology-enhanced caries detection and diagnosis. *Compendium of Continuing Education in Dentistry*, *29*(8), 464-5.

Subramanyam, R. B., Prasad, K. P., & Anuradha, B. (2014). Different Image Segmentation Techniques for Dental Image Extraction. *Int. Journal of Engineering Research and Applications*, *4*, 9622.

Suganthi, S. S., & Ramakrishnan, S. (2014). Anisotropic diffusion filter based edge enhancement for segmentation of breast thermogram using level sets. *Biomedical Signal Processing and Control*, *10*, 128–136. doi:10.1016/j.bspc.2014.01.008

Sun, X., Shu, F. L., Fei, Y. D., Shan, W., Christopher, P., James, H., ... Hong, W. D. (2006). Genetic and environmental correlations between bone geometric parameters and body compositions. *Calcified Tissue International*, *79*(1), 43–49. doi:10.100700223-006-0041-3 PMID:16868663

Syed, S. A., & Loesche, W. J. (1972). Survival of human dental plaque flora in various transport media. *Applied Microbiology*, *24*(4), 638–644. PMID:4628799

Szegedy, C., Vanhoucke, V., Ioffe, S., Shlens, J., & Wojna, Z. (2016). Rethinking the Inception Architecture for Computer Vision. *2016 IEEE Conference on Computer Vision and Pattern Recognition (CVPR)*, 2818-2826. 10.1109/CVPR.2016.308

Szilagyi, L., Szilagyi, S. M., & Benyo, Z. (2007). Efficient Feature Extraction for Fast Segmentation of MR Brain Images. *Lecture Notes in Computer Science*, *4522*, 611–620. doi:10.1007/978-3-540-73040-8_62

Tagashira, M., Uchiyama, K., Yoshimura, T., Shirota, M., & Uemitsu, N. (1997). Inhibition by hop bract polyphenols of cellular adherence and water-insoluble glucan synthesis *of mutans streptococci. Bioscience, Biotechnology, and Biochemistry*, *61*(2), 332–335. doi:10.1271/bbb.61.332 PMID:9058972

Taiwo, O., Xu, H., & Lee, S. (1999). Antibacterial activities of extracts from Nigerian chewing sticks. *Phytotherapy Research*, *13*(8), 675–679. doi:10.1002/(SICI)1099-1573(199912)13:8<675::AID-PTR513>3.0.CO;2-X PMID:10594937

Takaisikikuni, N. B., & Schilcher, H. (1994). Electron-microscopic and micro calorimetric investigations of the possible mechanism of the antibacterial action of a defined propolis provenance. *Planta Medica*, *60*(03), 222–227. doi:10.1055-2006-959463

Takarada, K., Kimizuka, R., Takahashi, N., Honma, K., Okuda, K., & Kato, T. A. (2004). Comparison of the antibacterial efficacies of essential oils against oral pathogens. *Oral Microbiology and Immunology*, *19*(1), 61–64. doi:10.1046/j.0902-0055.2003.00111.x PMID:14678476

Tao, W. B., Tian, J. W., & Liu, J. (2003). Image segmentation by three-level thresholding based on maximum fuzzy entropy and genetic algorithm. *Pattern Recognition Letters*, *24*(16), 3069–3078. doi:10.1016/S0167-8655(03)00166-1

Taweechaisupapong, S., Wongkham, S., Chareonsuk, S., Suparee, S., Srilalai, P. S., & Chaiyarak, S. (2000). Selective activity of Streblus asper on *Mutans streptococci. Journal of Ethnopharmacology*, *70*(1), 73–79. doi:10.1016/S0378-8741(99)00140-3 PMID:10720792

Tella, A. (1976). Analgesic and antimicrobial properties of *Vernonia amygdalina. British Journal of Clinical Pharmacology*, *7*, 295–297.

Thastrup, O., Knudsen, J. B., Lemmich, J., & Winther, K. (1985). Inhibitions of human platelet aggregation by dihydropyrano-and dihydrofuranocoumarins, a new class of cAMP phosphodiesterase inhibitors. *Biochemical Pharmacology*, *34*(12), 2137–2140. doi:10.1016/0006-2952(85)90407-1 PMID:2988567

Thomas, A. (1998). The Radiology History & Heritage Charitable Trust. *Newsletter of the Radiology History and Heritage Charitable Trust Winter, 10*, 5-24.

Thomas, G. P. (2013). *Amalgam-chemical composition, mechanical properties, and common applications*. Academic Press.

Tiwari, R. B., & Yardi, A. R. (2006). Dental radiograph image enhancement based on human visual system and local image statistics. *International Conference on Image Processing, Computer Vision and Pattern Recognition*, 100-108.

Tobias, R. S. (1988). Antibacterial properties of dental restorative materials: A review. *International Endodontic Journal*, *21*(2), 155–160. doi:10.1111/j.1365-2591.1988.tb00969.x PMID:3151895

Tognoni, E., Palleschi, V., Corsi, M., & Cristoforetti, G. (2002). Quantitative micro-analysis by laser-induced breakdown spectroscopy: A review of the experimental approaches. *Spectrochimica Acta. Part B, Atomic Spectroscopy*, *57*(7), 1115–1130. doi:10.1016/S0584-8547(02)00053-8

Tom, C. E., & Thomas, J. (2015). Segmentation of Tooth and Pulp from Dental Radiographs. *International Journal of Scientific & Engineering Research, 6.*

Tomasi, C., & Manduchi, R. (1998, January). Bilateral filtering for gray and color images. In *Computer Vision, 1998. Sixth International Conference on* (pp. 839-846). IEEE. 10.1109/ICCV.1998.710815

Tom, V. T., & Wolfe, G. J. (1982). Adaptive histogram equalization and its applications. *SPIE Applicat. Dig. Image Process. IV*, *359*, 204–209.

Torabinejad, M., & White, D. J. (1995). *U.S. Patent No. 5,415,547*. Washington, DC: U.S. Patent and Trademark Office.

Toro, C., Robiony, M., Costa, F., Zerman, N., & Politi, M. (2007). Feasibility of preoperative planning using anatomical facsimile models for mandibular reconstruction. *Head & Face Medicine*, *3*(1), 5. doi:10.1186/1746-160X-3-5 PMID:17224060

Toukairin, T., Uchino, K., Iwamoto, M., Murakami, S., Tatebayashi, T., Ogawara, H., & Tonosaki, Y. (1991). New polyphenolic 5′-nucleotidase inhibitors isolated from the wine grape "Koshu" and their biological effects. *Chemical & Pharmaceutical Bulletin*, *39*(6), 1480–1483. doi:10.1248/cpb.39.1480 PMID:1934168

Trongtokit, Y., Rongsriyam, Y., Komilamisra, N., Krisackphong, P., & Apiwathnasorn, C. (2004). Laboratory and field trial of developing medicinal local Thai plant products against four species of mosquito vectors. *The Southeast Asian Journal of Tropical Medicine and Public Health*, *35*, 325–333. PMID:15691131

Tsallis, C. (1988). Possible generalization of Boltzmann–Gibbs statistics. *Journal of Statistical Physics*, *52*(1), 479–487. doi:10.1007/BF01016429

Tsuchiya, H., Sato, M., Iinuma, M., Yokoyama, J., Ohyama, M., Tanaka, T., ... Namikawa, I. (1994). Inhibition of the growth of cariogenic bacteria *in vitro* by plant flavanones. *Experientia*, *50*(9), 846–849. doi:10.1007/BF01956469 PMID:7925853

Tuan, T. M. (2016). A cooperative semi-supervised fuzzy clustering framework for dental X-ray image segmentation. *Expert Systems with Applications*, *46*, 380–393. doi:10.1016/j.eswa.2015.11.001

Tuan, T. M., Son, L. H., & Dung, L. B. (2016). Dynamic semi-supervised fuzzy clustering for dental X-ray image segmentation: An analysis on the additional function. *Journal of Computer Science and Cybernetics*, *31*(4), 323.

Tuzoff, D., Tuzova, L., Bornstein, M. M., Krasnov, A., Kharchenko, M., Nikolenko, S., ... Bednenko, G. (2018). Teeth detection and numbering in panoramic radiographs using convolutional neural networks. *Dento Maxillo Facial Radiology*.

Tyndall, D. A., & Brooks, S. L. (2000). Selection criteria for dental implant site imaging: A position paper of the American Academy of Oral and Maxillofacial Radiology. *Oral Surgery, Oral Medicine, Oral Pathology, Oral Radiology, and Endodontics*, *89*(5), 630–637. doi:10.1067/moe.2000.106336 PMID:10807723

Tyndall, D. A., & Rathore, S. (2008). Cone-beam CT diagnostic applications: Caries, periodontal bone assessment, and endodontic applications. *Dental Clinics*, *52*(4), 825–841. doi:10.1016/j.cden.2008.05.002 PMID:18805231

Uijlings, J. R., Sande, K. E., Gevers, T., & Smeulders, A. W. (2013). Selective Search for Object Recognition. *International Journal of Computer Vision*, *104*(2), 154–171. doi:10.100711263-013-0620-5

Ursan, I. D., Chiu, L., & Pierce, A. (2013). Three-dimensional drug printing: A structured review. *Journal of the American Pharmacists Association*, *53*(2), 136–144. doi:10.1331/JAPhA.2013.12217 PMID:23571620

Usha, R., Sashidharan, S., & Palaniswamy, M. (2010). Antimicrobial Activity of a Rarely Known Species, Morinda citrifolia L. Ethnobotanical Leaflets, 2010(3).

Uzel, A., Sorkun, K., Oncag, O., Cogulu, D., Gencav, O., & Salih, B. (2005). Chemical compositions and antimicrobial activities of four different *Anatolian propolis* samples. *Microbial Res*, *160*(2), 189–195. doi:10.1016/j.micres.2005.01.002 PMID:15881836

Vaarkamp, J., Ten Bosch, J. J., & Verdonschot, E. H. (1995). *Propagation of light through human dental enamel and dentin*. Academic Press.

Vaarkamp, J., Ten Bosch, J. J., Verdonschot, E. H., & Tranaeus, S. (1997). Quantitative diagnosis of small approximal caries lesions utilizing wavelength-dependent fiber-optic transillumination. *Journal of Dental Research*, *76*(4), 875–882. doi:10.1177/00220345970760040901 PMID:9126184

Vaishnavi, G. K., Jeevananthan, K., Begum, S. R., & Kamalanand, K. (2014). Geometrical analysis of schistosome egg images using distance regularized level set method for automated species identification. *Journal of Bioinformatics and Intelligent Control*, *3*(2), 147–152. doi:10.1166/jbic.2014.1080

Van Meerbeek, B., Vanherle, G., Lesaffre, E., Braem, M., & Lambrechts, P. (1991). Trends in the selection of dental filling materials. *Journal of Dentistry*, *19*(4), 207–213. doi:10.1016/0300-5712(91)90118-I PMID:1787208

Vanka, A., Tandon, S., Rao, S. R., Udupa, N., & Ramkumar, P. (2001). The effect of indigenous Neem Azadirachta indica [correction of (Adirachta indica)] mouth wash on Streptococcus mutans and lactobacilli growth. *Indian Journal of Dental Research*, *12*(3), 133–144. PMID:11808064

Vasconcelos, L. C., Sampaio, F. C., Sampaio, M. C., Pereira, M. S. V., Higino, J. S., & Peixoto, M. H. P. (2006). Minimum inhibitory concentration of adherence of Punica granatum Linn (pomegranate) gel against *S. mutans, S. mitis* and *C. albicans. Brazilian Dental Journal*, *17*(3), 223–227. doi:10.1590/S0103-64402006000300009 PMID:17262129

Velasco, I., Vahdani, S., & Ramos, H. (2017). Low-cost method for obtaining medical rapid prototyping using desktop 3d printing: A novel technique for mandibular reconstruction planning. *Journal of Clinical and Experimental Dentistry*, *9*(9), e1103. PMID:29075412

Venditti, M., Baiocchi, P., Santini, C., Brandimarte, C., Serra, P., Gentile, G., ... Martino, P. (1989). Antimicrobial susceptibilities of Streptococcus species that cause septicemia in neutropenic patients. *Antimicrobial Agents and Chemotherapy*, *33*(4), 580–582. doi:10.1128/AAC.33.4.580 PMID:2729950

Ventola, C. L. (2014). Medical applications for 3D printing: Current and projected uses. *P&T*, *39*(10), 704. PMID:25336867

Vertucci, F. J. (1984). Root canal anatomy of the human permanent teeth. *Oral Surgery, Oral Medicine, and Oral Pathology*, *58*(5), 589–599. doi:10.1016/0030-4220(84)90085-9 PMID:6595621

Vijayakumari, B., Ulaganathan, G., & Banumathi, A. (2011). An effective shape extraction algorithm for dental radiographs using contour information. *International Journal of Computer Science and Telecommunications*, *2*(4), 311–316.

Vishnupriya, R., Raja, N. S. M., & Rajinikanth, V. (2017). An efficient clustering technique and analysis of infrared thermograms. In *Third International Conference on Biosignals, Images and Instrumentation (ICBSII)*. IEEE. doi:10.1109/ICBSII.2017.808227510.1109/ICBSII.2017.8082275

Wanat, R. (2011). A Problem of Automatic Segmentation of Digital Dental Panoramic X-Ray Images for Forensic Human Identification. *Central European Seminar on Computer Graphics*.

Wang, F. N., Liang, D., Cheng, Z. Y., & Tang, J. (2008). Nonsubsampled pyramid contourlet transform and its application, In *Proceedings of International Symposium on Information Science and Engineering* (*vol. 2*, pp. 567-570). Academic Press. 10.1109/ISISE.2008.220

Wang, Z. (2004). Image quality assessment from error measurement to structural similarity. *IEEE Transactions on Image Processing, 13*(4), 600-612.

Wang, C. W., Gosno, E., & Li, Y. (2015a). Fully automatic and robust 3D registration of serial-section microscopic images. *Nature-Scientific Reports*, *5*, 15051. doi:10.1038rep15051 PMID:26449756

Wang, C. W., Huang, C. T., Lee, J. H., Li, C. H., Chang, S. W., Siao, M. J., ... Fischer, P. (2016). A benchmark for comparison of dental radiography analysis algorithms. *Medical Image Analysis*, *31*, 63–76. doi:10.1016/j.media.2016.02.004 PMID:26974042

Wang, C. W., Huang, C.-T., Hsieh, M.-C., Li, C. H., Chang, S. W., Li, W. C., ... Ibragimov, B. (2015). Evaluation and comparison of anatomical landmark detection methods for cephalometric X-ray images: A Grand Challenge. *IEEE Transactions on Medical Imaging*, *34*(9), 1890–1900. doi:10.1109/TMI.2015.2412951 PMID:25794388

Wang, X., Shan, J., Niu, Y., Tan, L., & Zhang, S. X. (2014). Enhanced distance regularization for re-initialization free level set evolution with application to image segmentation. *Neurocomputing*, *141*, 223–235. doi:10.1016/j.neucom.2014.03.011

Wang, Z., Bovik, A. C., Sheikh, H. R., & Simoncelli, E. P. (2004). Image quality assessment: From error visibility to structural similarity. *IEEE Transactions on Image Processing*, *13*(4), 600–612. doi:10.1109/TIP.2003.819861 PMID:15376593

Wataha, J. C. (2000). Biocompatibility of dental casting alloys: A review. *The Journal of Prosthetic Dentistry*, *83*(2), 223–234. doi:10.1016/S0022-3913(00)80016-5 PMID:10668036

Weickert, J. (1998). *Anisotropic diffusion in image processing* (Vol. 1). Stuttgart, Germany: Teubner.

Weickert, J. (1999a). Coherence-enhancing diffusion of colour images. *Image and Vision Computing*, *17*(3), 201–212. doi:10.1016/S0262-8856(98)00102-4

Weickert, J. (1999b). Coherence-enhancing diffusion filtering. *International Journal of Computer Vision*, *31*(2-3), 111–127. doi:10.1023/A:1008009714131

Weintraub, J. A., Hilton, J. F., White, J. M., Hoover, C. I., Wycoff, K. L., Yu, L., ... Featherstone, J. D. B. (2005). Featherstone JD: Clinical trial of a plant-derived antibody on recolonization of *mutans streptococci. Caries Research*, *39*(3), 241–250. doi:10.1159/000084805 PMID:15914988

Weisleder, R., Yamauchi, S., Caplan, D. J., Trope, M., & Teixeira, F. B. (2009). The validity of pulp testing: A clinical study. *The Journal of the American Dental Association*, *140*(8), 1013–1017. doi:10.1109/ICACCI.2014.6968381 doi:10.14219/jada.archive.2009.0312 PMID:19654254

Weiss, E. L., Lev-Dor, R., Sharon, N., & Ofek, I. (2002). Inhibitory effect of a high-molecular-weight constituent of cranberry on adhesion of oral bacteria. *Critical Reviews in Food Science and Nutrition*, *42*(sup3), 285–292. doi:10.1080/10408390209351917 PMID:12058987

Wenzel, A. (1988). Effect of image enhancement for detectability of bone lesions in digitized intraoral radiographs. *European Journal of Oral Sciences*, *96*(2), 149–160. doi:10.1111/j.1600-0722.1988.tb01422.x PMID:3162600

Westergren, G., & Olsson, J. (1983). Hydrophobicity and adherence of oral *Streptococci* after repeated subculture *in vitro. Infection and Immunity*, *40*, 432–435. PMID:6832836

Whittaker, C. J., Klier, C. M., & Kolenbrander, P. E. (1996). Mechanisms of adhesion by oral bacteria. *Annual Review of Microbiology, 50*(1), 513–552. doi:10.1146/annurev.micro.50.1.513 PMID:8905090

Wielgus, A. R., Collier, R. J., Martin, E., Lih, F. B., Tomer, K. B., Chignell, C. F., & Roberts, J. E. (2010). Blue light induced A2E oxidation in rat eyes–experimental animal model of dry AMD. *Photochemical & Photobiological Sciences, 9*(11), 1505–1512. doi:10.1039/c0pp00133c PMID:20922251

Wielgus, A. R., & Roberts, J. E. (2012). Retinal photodamage by endogenous and xenobiotic agents. *Photochemistry and Photobiology, 88*(6), 1320–1345. doi:10.1111/j.1751-1097.2012.01174.x PMID:22582903

Wilkins, E. M., & McCullough, P. A. (1989). *Clinical practice of the dental hygienist* (Vol. 235). Lea &Febiger.

Wilson, A. D., & Kent, B. E. (1971). The glass-ionomer cement, a new translucent dental filling material. *Journal of Chemical Technology and Biotechnology (Oxford, Oxfordshire), 21*(11), 313–313.

Winder, J., & Bibb, R. (2005). Medical rapid prototyping technologies: State of the art and current limitations for application in oral and maxillofacial surgery. *Journal of Oral and Maxillofacial Surgery, 63*(7), 1006–1015. doi:10.1016/j.joms.2005.03.016 PMID:16003630

Wolinsky, L. E., Mania, S., Nachnani, S., & Ling, S. (1996). The inhibiting effect of aqueous azadirachta indica (neem) extract upon bacterial properties influencing *in vitro* plaque formation. *Journal of Dental Research, 75*(2), 816–822. doi:10.1177/00220345960750021301 PMID:8655780

Wolinsky, L. E., & Sote, E. O. (1984). Isolation of natural plaque-inhibiting substances from Nigerian chewing sticks. *Caries Research, 18*(3), 216–225. doi:10.1159/000260768 PMID:6584212

Wongkham, S., Laupattarakasaem, P., Pienthaweechai, K., Areejitranusorn, P., Wongkham, C., & Techanitiswad, T. (2001). Antimicrobial activity of *Streblus asper* leaf extract. *Phytotherapy Research, 15*(2), 119–121. doi:10.1002/ptr.705 PMID:11268109

Wriedt, S., Jaklin, J., Al-Nawas, B., & Wehrbein, H. (2012). Impacted upper canines: Examination and treatment proposal based on 3d versus 2d diagnosis. *Journal of Orofacial Orthopedics, 73*(1), 28–40. doi:10.100700056-011-0058-8 PMID:22246048

Wu, W., Wu, Y., & Huang, Q. (2012) An improved distance regularized level set evolution without re-initialization. *IEEE fifth International Conference on Advanced Computational Intelligence*, 631-636.

Wu, C. D., Darout, I. A., & Skaug, N. (2001). Chewing sticks: Timeless natural toothbrush for oral cleansing. *Journal of Periodontal Research, 36*(5), 275–28410. doi:10.1034/j.1600-0765.2001.360502.x PMID:11585114

Wu, M. a. (2015). Image recognition based on deep learning. *Chinese Automation Congress (CAC).*

Wunder, D., & Bowen, W. H. (1999). Action of agents on glucosyltransferases from *Streptococcus mutans* in solution and adsorbed to experimental pellicle. *Archives of Oral Biology, 44*(3), 203–214. doi:10.1016/S0003-9969(98)00129-0 PMID:10217511

Wu, Q., Merchant, F., & Castleman, K. (2010). *Microscope image processing*. Academic press.

Wu-Yuan, C. D., Green, L., & Birch, W. X. (1990). *In vitro* screening of Chinese medicinal toothpastes: Their effects on growth and plaque formation of *mutans streptococci. Caries Research, 24*(3), 198–202. doi:10.1159/000261265 PMID:2364405

Wu-Yuan, C. D., Tai, S., & Slade, H. D. (1978). Dextran/glucan binding by Streptococcus mutans: The role of molecular size and binding site in agglutination. *Advances in Experimental Medicine and Biology*, *107*, 737–748. doi:10.1007/978-1-4684-3369-2_83 PMID:742509

Xiao, J., Liu, Y., Zuo, Y. L., Li, J. Y., Ye, L., & Zhou, X. D. (2006). Effects of *Nidus Vespae* extract and chemical fractions on the growth and acidogenicity of oral microorganisms. *Archives of Oral Biology*, *51*(9), 804–813. doi:10.1016/j.archoralbio.2006.03.014 PMID:16723116

Xiao, J., Zhou, X. D., Feng, J., Hao, Y. Q., & Li, J. Y. (2007). Activity of *Nidus Vespae* extract and chemical fractions against *Streptococcus mutans* biofilms. *Letters in Applied Microbiology*, *45*(5), 547–552. doi:10.1111/j.1472-765X.2007.02230.x PMID:17916132

Xiao, Y., Liu, T. J., Huang, Z. W., Zhou, X. D., & Li, J. Y. (2004). The effect of natural medicine on adherence of *Streptococcus mutans* to salivary acquired pellicle. *Sichuan Da Xue Xue Baoxue Ban.*, *35*, 87–89.

Yadav, G., Maheshwari, S., & Agarwal, A. (2014, September). Contrast limited adaptive histogram equalization based enhancement for real time video system. In *Advances in Computing, Communications and Informatics (ICACCI), 2014 International Conference on* (pp. 2392-2397). IEEE. 10.1109/ICACCI.2014.6968381

Yamamoto, H., & Ogawa, T. (2002). Antimicrobial activity of perilla seed polyphenols against oral pathogenic bacteria. *Bioscience, Biotechnology, and Biochemistry*, *66*(4), 921–924. doi:10.1271/bbb.66.921 PMID:12036078

Yamanaka, A., Kimizuka, R., Kato, T., & Okuda, K. (2004). Inhibitory effects of cranberry juice on attachment of oral *streptococci* and biofilm formation. *Oral Microbiology and Immunology*, *19*(3), 150–154. doi:10.1111/j.0902-0055.2004.00130.x PMID:15107065

Yamashita, Y., Bowen, W. H., Burne, R. A., & Kuramitsu, H. K. (1993). Role of the Streptococcus mutans gtf genes in caries induction in the specific-pathogen-free rat model. *Infection and Immunity*, *61*, 3811–3817. PMID:8359902

Yamauchi, J., Masuhara, E., Nakabayashi, N., Shibatani, K., & Wada, T. (1981). *U.S. Patent No. 4,259,117*. Washington, DC: U.S. Patent and Trademark Office.

Yanagida, A., Kanda, T., Tanabe, M., Matsudaira, F., & Cordeiro, J. G. O. (2000). Inhibitory effects of apple polyphenols and related compounds on cariogenic factors of *Mutans Streptococci. Journal of Agricultural and Food Chemistry*, *48*(11), 5666–5671. doi:10.1021/jf000363i PMID:11087536

Yanagisawa, R., & Omachi, S. (2010, November). Extraction of 3D shape of a tooth from dental CT images with region growing method. In *International Workshop on Computational Forensics* (pp. 68-77). Springer.

Yatsuda, R., Rosalen, P. L., Cury, J. A., Murata, R. M., Rehder, V. L., Melo, L. V., & Koo, H. (2005). Effects of Mikania genus plants on growth and cell adherence of *mutans streptococci. Journal of Ethnopharmacology*, *97*(2), 183–189. doi:10.1016/j.jep.2004.09.042 PMID:15707750

Yau, H. T., Lin, Y. K., Tsou, L. S., & Lee, C. Y. (2008). An adaptive region growing method to segment inferior alveolar nerve canal from 3D medical images for dental implant surgery. *Computer-Aided Design and Applications*, *5*(5), 743–752. doi:10.3722/cadaps.2008.743-752

Yengopal, V., Harneker, S. Y., Patel, N., & Siegfried, N. (2009). Dental fillings for the treatment of caries in the primary dentition. *Cochrane Database of Systematic Reviews*, 2. PMID:19370602

Yeshwante, B., Patil, S., Baig, N., Gayakwad, S., Swami, A., & Doiphode, M. (2015). Dental implant- Classification, success and failure-An overview. *IOSR Journal of Dental and Medical Sciences*, *14*(5), 1–8.

Yilmaz, S. Y. Y., Misirlioglu, M., & Adisen, M. Z. (2014). A diagnosis of maxillary sinus fracture with Cone-Beam CT: Case report and literature review. *Craniomaxillofacial Trauma & Reconstruction*, *7*(2), 85–91. doi:10.1055-0034-1371550 PMID:25045417

You, Y. O., Shin, H. Y., Yu, H. H., Yoo, S. J., Kim, S. H., Kim, Y. K., ... Kim, H. M. (2004). Effect of Powerdental on caries-inducing properties of *Streptococcus mutans* and TNF-alpha secretion from HMC-1 cells. *Journal of Ethnopharmacology*, *92*(2-3), 331–335. doi:10.1016/j.jep.2004.03.023 PMID:15138020

Yovchev, D., Stanimirov, P., & Mihaylova, H. (2014). Lower jaw reconstruction using prototype from cone-beam computed tomography data. *Int Journal of Sciences and Research*, *3*, 57–59.

Yu, H. H., Lee, J. S., Lee, K. H., Kim, K. Y., & You, Y. O. (2007). Saussurea lappa inhibits the growth, acid production, adhesion, and water-insoluble glucan synthesis of *Streptococcus mutans*. *Journal of Ethnopharmacology*, *111*(2), 413–417. doi:10.1016/j.jep.2006.12.008 PMID:17234374

Zachrisson, B. U. (2015). Esthetics in tooth display and smile design. In Esthetics and Biomechanics in Orthodontics (2nd ed.; pp. 54-73). Academic Press.

Zahradnicek, O., Horacek, I., & Tucker, A. S. (2012). Tooth development in a model reptile: Functional and null generation teeth in the gecko Paroedura picta. *Journal of Anatomy*, *221*(3), 195–208. doi:10.1111/j.1469-7580.2012.01531.x PMID:22780101

Zak, J., Korzynska, A., Roszkowiak, L., Siemion, K., Walerzak, S., Walerzak, M., & Walerzak, K. (2018). The method of teeth region detection in panoramic dental radiographs. In *Advances in Intelligent Systems and Computing* (Vol. 578, pp. 298–307). Cham: Springer. doi:10.1007/978-3-319-59162-9_31

Zeng, P., Dong, H., Chi, J., & Xu, X. (2004, August). An approach for wavelet based image enhancement. In *Robotics and Biomimetics, 2004. ROBIO 2004. IEEE International Conference on* (pp. 574-577). IEEE.

Zeng, H., Liu, Y. Z., Fan, Y. M., & Tang, X. (2012). An improved algorithm for impulse noise by median filter. *AASRI Procedia*, *1*, 68–73. doi:10.1016/j.aasri.2012.06.014

Zero, D., Mol, A., Roriz, C. S., Spoon, M., Jacobs, A., Keem, S., & Elbaum, M. (2000). Caries detection using digital imaging fibre-optic transillumination (DIFOTITM): A preliminary evaluation. *Early Detection of Dental Caries*, *2*, 169–183.

Zhang, B., Fadili, J. M., & Starck, J. L. (2008). Wavelets, ridgelets, and curvelets for Poisson noise removal. *IEEE Transactions on Image Processing*, *17*(7), 1093–1108. doi:10.1109/TIP.2008.924386 PMID:18586618

Zhang, J., & Kashket, S. (1998). Inhibition of salivary amylase by black and green teas and their effects on the intraoral hydrolysis of starch. *Caries Research*, *32*(3), 233–238. doi:10.1159/000016458 PMID:9577990

Zhang, K., Zhang, L., Song, H., & Zhang, D. (2013). Reinitialization-free level set evolution via reaction diffusion. *IEEE Transactions on Image Processing*, *22*(1), 258–271. doi:10.1109/TIP.2012.2214046 PMID:22910114

Zhang, K., Zhang, L., & Zhang, S. (2010). A variational multiphase level set approach to simultaneous segmentation and bias correction. *Proc. 17th IEEE International Conference on Image Processing (ICIP)*, 4105-4108. 10.1109/ICIP.2010.5651554

Zhang, Y., & Wu, L. (2011). Optimal multi-level thresholding based on maximum Tsallis entropy via an artificial bee colony approach. *Entropy (Basel, Switzerland)*, *13*(4), 841–859. doi:10.3390/e13040841

Zhou, J. (2010). *Biometric recognition systems employing novel shape-based features* (Doctoral dissertation). University of Miami.

Zhou, J., & Abdel-Mottaleb, M. (2005). A content-based system for human identification based on bitewing dental X-ray images. *Pattern Recognition*, *38*(11), 2132–2142. doi:10.1016/j.patcog.2005.01.011

Zhou, Q., Li, Z., & Aggarwal, J. K. (2004). Boundary extraction in thermal images by edge map. *Proceedings of the ACM symposium on Applied computing*, 254-258. 10.1145/967900.967956

Zhou, S., Wang, J., Zhang, M., Cai, Q., & Gong, Y. (2017). Correntropy-based level set method for medical image segmentation and bias correction. *Neurocomputing*, *234*, 216–229. doi:10.1016/j.neucom.2017.01.013

Zhou, Y., & Bai, J. (2007). Atlas-based fuzzy connectedness segmentation and intensity nonuniformity correction applied to brain MRI. *IEEE Transactions on Biomedical Engineering*, *54*(1), 122–129. doi:10.1109/TBME.2006.884645 PMID:17260863

Zhuang, A. H., Valentino, D. J., & Toga, A. W. (2006). Skull-stripping magnetic resonance brain images using a model-based level set. *NeuroImage*, *32*(1), 79–92. doi:10.1016/j.neuroimage.2006.03.019 PMID:16697666

Zuiderveld, K. (1994). *Contrast limited adaptive histogram equalization. In Graphics gems IV* (pp. 474–485). San Diego, CA: Academic Press Professional, Inc.

About the Contributors

K. Kamalanand completed his Ph.D. at MIT Campus, Anna University in the field of Biomedical Engineering. At present he is an Assistant Professor at the Department of Instrumentation Engineering, Madras Institute of Technology Campus, Anna University, Chennai, India. He is well published with more than 70 publications to his credit. He has served as Guest Editor for several Journals. He is a member of the Council of Asian Science Editors. His research interests include Biomedical Engineering, Mathematical Modelling, Artificial Intelligence and Nanotechnology.

B. Thayumanavan received his M.D.S. from Tamil Nadu Government Dental College, Chennai and his B.D.S. from Sree Balaji Dental College and Hospital, Chennai. At present he is the Dean of Sathyabama University Dental College and Hospital, Chennai, India. He has teaching experience of more than 16 years. He has published several papers in international journals and conference proceedings. He has supervised several Dissertations. Also, he has organized several conferences and has chaired several scientific sessions.

P. Mannar Jawahar completed his Ph.D. at I.I.T. Delhi and subsequently did his Post Doctoral Research work at Warsaw University of Technology, Poland. He joined Anna University, Chennai, India, as a Faculty in 1978 and has put in more than three decades of service in the teaching profession in various capacities. He served as the Vice Chancellor of Anna University from June 2008 to June 2012. At present, he is the Vice Chancellor of Karunya Institute of Technology and Sciences, Coimbatore, India.His current research interest includes the application of computational methods in biomedical engineering.

* * *

Bennett T. Amaechi, BDS, MSc, PhD, MFDSRCPS (Glasg), FADI, the Professor and Director of Cariology in the Department of Comprehensive Dentistry at the University of Texas Health Science Center at San Antonio (UTHSCSA), specializes in three areas of dentistry, Prosthodontics, Preventive dentistry and Cariology (caries management). After receiving his dental degree at the University of Ife, Nigeria, Dr. Amaechi began his career as a dentist with specialist training and Master's degree in Prosthodontics and dental implantology at Guy's Hospital, London, UK, and he later gained a PhD degree in Cariology and Preventive Dentistry at the University of Liverpool, UK. His research interest, which is on dental caries and caries diagnostics, has been supported by NIH grants and Industry contracts. He has authored several journal articles, a book on dental erosion, 11 book chapters, and 12 conference proceedings. He is a highly accomplished dentist, scholar, educator and a dedicated mentor, who touches the careers of many faculty and students within and outside his institution. He was honored with Faculty Leadership award (2014) by the UTHSCSA Faculty Senate in recognition of his leadership qualities and accomplishments, teaching excellence award (2015) by UTHSCSA dental school, community service award (2015) by his local community, and Membership of Faculty of Dental Surgery by the Royal College of Physicians and Surgeons of Glasgow (2016). He is an active member and officer of various national and international professional and scientific organizations.

Christy Bobby is an Associate Professor at M.S. Ramaiah University of Applied Sciences, Bengaluru, Karnataka, India, with over 20 years of teaching experience. She has got over 15 years of research experience in the fields of Biomedical Image Processing, Signal Processing and Machine Learning. She is associate head of Centre for Biomedical Systems and 3D Printing Research, M.S. Ramaiah University of Applied Sciences Bengaluru. She received her Bachelor of Engineering in Electronics and Communication Engineering from Bharathidasan University, Tiruchirappalli, MS (By Research) and PhD from Anna University, Chennai, both specializing in Biomedical Image and Signal processing. She has published her research works in many International Journals and Conferences in the area of Medical Imaging. Her current research interests include Cognitive Neuroscience, Neuroimaging, Bio Signal and Image Processing and Bone Biomechanics.

Kavitha Ganesan is currently working as an Assistant Professor (Senior Grade) in the Department of Electronics Engineering, Madras Institute of Technology Campus, Anna University, Chennai, India. She completed her Ph.D. from Anna University, Chennai, India, in Retinal Image Processing and Master's degree in Applied Electronics from the same University. She has more than 13 years of teaching experience. She has published more than 70 papers in the journals and conferences. Her current interests are in biomedical image processing and soft computing techniques.

Feras N. Hasoon is currently working as senior lecturer, department of Electrical and Computer Engineering in Caledonian college of Engineering, Oman. He has published more than 30 research papers in various conferences and journals. His research interest includes the signal and image analysis.

Alexey Krasnov is a researcher in the field of radiology. More than 10 years of oral radiology experience. Lecturer, private oral radiology practitioner.

Latha M. is a Research Scholar at MIT, Anna University, Chennai. She has a teaching experience of more than 10 years. Her research interests focus on Bio medical signal and image processing.

Vijaya Madhavi received M.E degree from Anna University, Coimbatore, in 2005 and B.E. degree in Electronics and Communication Engineering from Bharathiar University, Coimbatore, in 2001. She is currently pursuing a PhD in biomedical image processing under VTU, Belgaum, Karnataka and has teaching experience of over 13 years. Research interests include biomedical image analysis and computer vision.

K. Suresh Manic completed his Ph.D from Anna University and currently working as senior lecturer, department of Electrical and Computer Engineering in Caledonian college of Engineering. His research interest includes, process modelling, image processing, controller design and heuristic algorithm based image processing. He has published more than 30 research articles in various journals and conferences.

M. Kayalvizhi, B.E (ECE), MS by (Research), Ph.D, Member IEEE, WIE, EMBS, Professor & Head of Agni college of technology, having expertise in Biomedical engineering, Biomedical signal processing, Biomedical image processing, Machine learning, Neuroscience, Nano science. She has published in scopus indexed journals and reviewer of Elsevier journals.

M. Muthulakshmi is a Research Scholar at MIT, Anna University, Chennai. She has a teaching experience of 4 years and industrial experience of 3 years. Her research interests are in Bio-medical image processing and embedded systems.

Dewan Najumnissa Jamal received her B.E. degree in Electrical and Electronics Engineering from Crescent Engineering College, Madras University and M.E. degree in Control and Instrumentation from College of Engineering, Guindy, Anna University, India in 1989 and 1992 respectively. She received her doctoral degree from Anna University during January 2012 in the field of Medical Electronics on the topic "Characterisation of Epileptic Seizure EEG signals using Wavelet Transform, Principal Component Analysis and Optimization Techniques She has been working at Crescent Engineering College in the Department of Electrical and Electronics Engineering, since January 1990 in various positions starting from teaching assistant, lecturer, senior lecturer, selection grade lecturer, assistant professor, associate professor and Professor. She had the opportunity to head the Instrumentation and control engineering department during 2006 – 2007 and Electronics and Instrumentation Engineering department during 2011 -2012. At present she is designated as Professor, Department of Electronics and Instrumentation Engineering, B.S.A. Crescent Institute of Science and technology, India. She has published 42 papers, nine in International Journals, one in national journal and 31 papers in International and National Conferences and two book chapters in reputed publications in the area of Biomedical Engineering, Intelligent Techniques, Control System, MEMs, Cloud computing and Bio Signal Processing. She has organised a number of workshops, conferences, and short term training programmes for both students and faculty members. She has visited countries like Singapore, Malaysia and Dubai. Her area of interest includes Bio-Medical Engineering, Wavelet Transforms, Principal Component Analysis and Signal processing and application of Artificial Intelligence Techniques in Medicine.

Chitimada Narendra Kumar is currently pursuing his ME in Aerospace Technology in MIT campus, Anna University, Chennai. His research interests includes CFD, High Speed Jet flows and Gas Dynamics. He has published 5 papers in international journals and conference proceedings.

Sergey Nikolenko is a researcher in the field of machine learning (deep learning, Bayesian methods, natural language processing and more) and analysis of algorithms (network algorithms, competitive analysis). Obtained his Ph.D. in 2009 from the Steklov Mathematical Institute at St. Petersburg. Authored more than 120 research papers, several books, courses "Machine learning", "Deep learning", and others. Has extensive experience with industrial projects in the field of machine learning, currently works as the Chief Research Officer of Neuromation.

E. Priya received her B.E. degree in Electronics and Communication Engineering from University of Madras in 1994 and M.E. degree in 2008 from Madras Institute of Technology, Anna University. She completed her Ph.D. in November 2013 under the faculty of Information and Communication from Anna University in the field of Biomedical Engineering. She is working as an Assistant Professor in the Department of ECE, Sri Sairam Engineering and has been in the Department since 2000. She has 16 years of teaching experience, 4 years of research experience and 3 years of industrial experience. She is a recipient of DST-PURSE fellow and a project participant of India-South African collaborative project titled "Development of computing tools for decision support in health assessment in rural areas". She has published papers in International Journals and Conferences in the area of medical imaging and infectious diseases. Her areas of interest include bio-medical imaging, image processing, signal processing, application of artificial intelligence and machine learning techniques.

Karthikeyan Ramalingam, M.Sc., M.Phil., Ph.D., Associate professor, BS, Abdur Rahman Crescent Institute of Science and Technology, Chennai, Tamil Nadu, India with over 15 years of laboratory knowledge, including academic, government, and industry projects handling experience with excellent supervisory skills and a strong record of scientific accomplishments and publications. He has strong scientific background in microbiology, including bacteria, fungi and Actinomyces, microbial pathogenesis, environmental microbiology, dental microbiology and cancer biology. He finished his doctoral degree from Centre of Advanced Study in Marine Biology, Annamalai University, Prangipettai, Tamil Nadu India. He has experienced and well trained at the University of Texas Health Science Center at San Antonio, Texas, USA on different aspects of dental microbiology, characterization and application for anticariogenic agents development and he has used nanoemulsion against battle field multi-resistant microorganism to prevent battle field related infections. In US Naval Medical Research Unit, San Antonio, Texas, USA, he had worked in different nanoemulsion and metal nano-particles against infectious disease causing microorganism. Then he worked in New York University, USA, their he worked in microbial derived modified Sophorolipids (biosurfactants) against eradication of different human and plant pathogenic microorganisms. His contribution in USA, research and supervision of undergraduate and pre-doctoral students has been appreciated and acknowledged by his mentors and he got 2012 Maria Yeung outstanding Postdoctoral Award with cash a prize and exceptional performance award at University of Texas Health Science Center at San Antonio, Texas, USA.

Prabha Sathees is an Associate Professor in the Department of Electronics and Communication Engineering at Hindustan Hindustan Institute of Technology and Science, India. She received her Ph.D. degree at College of Engineering, Guindy, Anna University Chennai, India in 2015 and M.E (Instrumentation Engineering) degree from Madras Institute of Technology and Science, Anna University Chennai, India in 2009. Her research interests include image processing, biomedical instrumentation, biometric security and cloud computing. She has published more than 15 research articles in International Journals and Conferences. She has published book chapters in computational image processing techniques. She is an active reviewer in the reputed International Journals and IEEE Journal of Biomedical and Health Informatics. She is a member of IEEE and IET.

Imad Saud Al Naimi is currently working as senior lecturer, department of Electrical and Computer Engineering in Caledonian college of Engineering, Oman. He is having 10 years of teaching experience. His area of research includes; controller design, image processing, heuristic algorithms and its engineering applications, etc.

Thanigaiarasu Subramanian M., after graduating from MIT Campus, Anna University, Chennai (B.Tech. Aeronautical Engineering, 1996), completed M.Tech with specialization in IC Engines and Gas Turbines at IIT-Madras in 1998. He finished his Doctorate (Aerospace Engineering-High Speed Jets) at MIT in 2009. He has been working as an Associate Professor in MIT since 2009. His interests are in High speed Jets and computational heat transfer. He also handled research projects in the field of composites. He is the recipient of Rajiv Gandhi Endorsement Award for two consecutive years from MIT, Chennai.

S. Arockia Sukanya received her M.E. degree from St. Joseph's college of engineering, Anna University, Chennai. At Present, she is working as a teaching fellow in the Department of Instrumentation Engineering at Madras Institute of Technology Campus, Anna University, Chennai. Her area of research is Image processing for Biomedical images.

Dmitry Tuzoff received his diploma cum laude from SSEU, Russia in 2000. Since then, he held senior positions in research and development at a number of international companies that provide software solutions to enterprises in healthcare, telecommunications, finance, and the government. In 2015, he started PhD research at Steklov Institute of Mathematics in St. Petersburg on the topic of machine learning and deep learning applications for the analysis of medical images.

Lyudmila Tuzova is a deep learning researcher in Denti.AI team. She received her MS in IT from the University of Communications and Informatics in 2007, Moscow, Russia. She has 10+ years software development experience including the position in leading Russian internet technologies company.

Rajinikanth V. is currently working as Professor, Department of EIE, St. Joseph's College of Engineering, Chennai, TN, India. He received his Ph.D titled Heuristic algorithm based system identification and controller design in 2013 from Anna University, TN, India. He has published over 60 papers in the field of heuristic algorithm applications. His research interest is heuristic algorithm assisted medical image and signal processing.

Index

J

L

M

N

O

P

Q

R

S

T

X

www.ingramcontent.com/pod-product-compliance
Lightning Source LLC
LaVergne TN
LVHW060416300926
849162LV00006B/1446